Lecture Notes in Computer Science

Lecture Notes in Bioinformatics 14248

The series Lecture Notes in Bioinformatics (LNBI) was established in 2003 as a topical subseries of LNCS devoted to bioinformatics and computational biology.

The series publishes state-of-the-art research results at a high level. As with the LNCS mother series, the mission of the series is to serve the international R & D community by providing an invaluable service, mainly focused on the publication of conference and workshop proceedings and postproceedings.

Xuan Guo · Serghei Mangul · Murray Patterson ·
Alexander Zelikovsky
Editors

Bioinformatics Research and Applications

19th International Symposium, ISBRA 2023
Wrocław, Poland, October 9–12, 2023
Proceedings

 Springer

Editors
Xuan Guo ⓘ
University of North Texas
Denton, TX, USA

Serghei Mangul ⓘ
University of Southern California
Los Angeles, CA, USA

Murray Patterson ⓘ
Georgia State University
Atlanta, GA, USA

Alexander Zelikovsky ⓘ
Georgia State University
Atlanta, GA, USA

ISSN 0302-9743   ISSN 1611-3349 (electronic)
Lecture Notes in Bioinformatics
ISBN 978-981-99-7073-5   ISBN 978-981-99-7074-2 (eBook)
https://doi.org/10.1007/978-981-99-7074-2

LNCS Sublibrary: SL8 – Bioinformatics

This Springer imprint is published by the registered company Springer Nature Singapore Pte Ltd.
The registered company address is: 152 Beach Road, #21-01/04 Gateway East, Singapore 189721, Singapore

Paper in this product is recyclable.

Preface

The 19th International Symposium on Bioinformatics Research and Applications (ISBRA 2023) was held on October 9–12, 2023, at Łukasiewicz Research Network - PORT Polish Center for Technology Development, Wrocław, Poland. The symposium provides a forum for the exchange of ideas and results among researchers, developers, and practitioners working on all aspects of bioinformatics and computational biology and their applications. The technical program of the symposium included 28 full papers and 16 short papers, selected by the Program Committee from 88 submissions received in response to the call for papers.

We would like to thank the Program Committee members and external reviewers for volunteering their time to review and discuss symposium papers. We would like to extend special thanks to the Steering and General Chairs of the symposium for their leadership, and to the Finance, Publication, Publicity, and Local Organization Chairs for their hard work in making ISBRA 2022 a successful event. Last but not least we would like to thank all authors for presenting their work at the symposium.

September 2023

Xuan Guo
Serghei Mangul
Murray Patterson
Alexander Zelikovsky

Organization

Program Committee

Derek Aguiar	University of Connecticut, USA
Tatsuya Akutsu	Kyoto University, Japan
Kamal Al Nasr	Tennessee State University, USA
Mukul S. Bansal	University of Connecticut, USA
Paola Bonizzoni	University of Milan-Bicocca, Italy
Hongmin Cai	South China University of Technology, China
Xing Chen	Jiangnan University, China
Alexandre G. de Brevern	INSERM UMR-S 1134, University of Paris, France
Pufeng Du	Tianjin University, China
Jorge Duitama	Universidad de los Andes, Colombia
Nadia El-Mabrouk	University of Montreal, Canada
Oliver Eulenstein	Iowa State University, USA
Pawel Gorecki	University of Warsaw, Poland
Xuan Guo (PC Chair)	University of North Texas, USA
Matthew Hayes	Xavier University of Louisiana, USA
Zengyou He	Dalian University of Technology, China
Steffen Heber	North Carolina State University, USA
Yury Khudyakov	Centers for Disease Control and Prevention, USA
Wooyoung Kim	University of Washington, USA
Danny Krizanc	Wesleyan University, USA
Soumya Kundu	Stanford University, USA
Min Li	Central South University, China
Yaohang Li	Old Dominion University, USA
Kevin Liu	Michigan State University, USA
Ion Mandoiu	University of Connecticut, USA
Serghei Mangul (PC Chair)	University of California, Los Angeles, USA
Marmar Moussa	University of Oklahoma, USA
Sheida Nabavi	University of Connecticut, USA
Murray Patterson (PC Chair)	Georgia State University, USA
Andrei Paun	University of Bucharest, Romania
Yuri Porozov	I.M. Sechenov First Moscow State Medical University, Russia
Russell Schwartz	Carnegie Mellon University, USA
Joao Setubal	University of São Paulo, Brazil

Jian-Yu Shi	Northwestern Polytechnical University, China
Xinghua Shi	Temple University, USA
Pavel Skums	Georgia State University, USA
Ileana Streinu	Smith College, USA
Emily Chia-Yu Su	Taipei Medical University, Taiwan
Sing-Hoi Sze	Texas A&M University, USA
Weitian Tong	Georgia Southern University, USA
Valentina Tozzini	Istituto Nanoscienze - CNR, Pisa, Italy
Jianrong Wang	Michigan State University, USA
Jianxin Wang	Central South University, China
Seth Weinberg	Ohio State University, USA
Fangxiang Wu	University of Saskatchewan, Canada
Alex Zelikovsky	Georgia State University, USA
Fa Zhang	Beijing Institute of Technology, China
Le Zhang	Sichuan University, China
Louxin Zhang	National University of Singapore, Singapore
Lu Zhang	Hong Kong Baptist University, China
Quan Zou	Tianjin University, China

Additional Reviewers

Adeniyi, Ezekiel
Ali, Sarwan
Arthur, Leonard
Ayyala, Ram
Campo, David S.
Chourasia, Prakash
Delabre, Mattéo
Ding, Yi
Doko, Rei
Gao, Meijun
Gascon, Mathieu
Kuzmin, Kiril
Lara, James
Lee, Byungho
Longmire, Atkinson
Markin, Alexey

Mohebbi, Fatemeh
Murad, Taslim
Mykowiecka, Agnieszka
Paszek, Jarosław
Sharma, Nitesh Kumar
Sims, Seth
Tayebi, Zahra
Vahed, Mohammad
Vijendran, Sriram
Wagle, Sanket
Wang, Zirui
Wuyun, Qiqige
Xu, Yu
Zeming, Li
Zheng, Julia

Contents

Unveiling the Robustness of Machine Learning Models in Classifying COVID-19 Spike Sequences

Sarwan Ali[1], Pin-Yu Chen[2], and Murray Patterson[1(✉)]

[1] Georgia State University, Atlanta, GA, USA
{sali85,mpatterson30}@gsu.edu

[2] IBM Research, IBM T. J. Watson Research Center, Yorktown Heights, NY, USA
pin-yu.chen@ibm.com

Abstract. In the midst of the global COVID-19 pandemic, a wealth of data has become available to researchers, presenting a unique opportunity to investigate the behavior of the virus. This research aims to facilitate the design of efficient vaccinations and proactive measures to prevent future pandemics through the utilization of machine learning (ML) models for decision-making processes. Consequently, ensuring the reliability of ML predictions in these critical and rapidly evolving scenarios is of utmost importance. Notably, studies focusing on the genomic sequences of individuals infected with the coronavirus have revealed that the majority of variations occur within a specific region known as the spike (or S) protein. Previous research has explored the analysis of spike proteins using various ML techniques, including classification and clustering of variants. However, it is imperative to acknowledge the possibility of errors in spike proteins, which could lead to misleading outcomes and misguide decision-making authorities. Hence, a comprehensive examination of the robustness of ML and deep learning models in classifying spike sequences is essential. In this paper, we propose a framework for evaluating and benchmarking the robustness of diverse ML methods in spike sequence classification. Through extensive evaluation of a wide range of ML algorithms, ranging from classical methods like naive Bayes and logistic regression to advanced approaches such as deep neural networks, our research demonstrates that utilizing k-mers for creating the feature vector representation of spike proteins is more effective than traditional one-hot encoding-based embedding methods. Additionally, our findings indicate that deep neural networks exhibit superior accuracy and robustness compared to non-deep-learning baselines. To the best of our knowledge, this study is the first to benchmark the accuracy and robustness of machine-learning classification models against various types of random corruptions in COVID-19 spike protein sequences. The benchmarking framework established in this research holds the potential to assist future researchers in gaining a deeper understanding of the behavior of the coronavirus, enabling the implementation of proactive measures and the prevention of similar pandemics in the future.

X. Guo et al. (Eds.): ISBRA 2023, LNBI 14248, pp. 1–15, 2023.
https://doi.org/10.1007/978-981-99-7074-2_1

Keywords: Adversarial Attack · Protein Sequences · k-mers ·
Classification

1 Introduction

In January 2020, a new RNA coronavirus was discovered, marking the onset of
the ongoing COVID-19 pandemic. Through the utilization of sequencing technology and phylogenetic analysis, the scientific community determined that this
novel coronavirus shares a 50% similarity with the Middle-Eastern Respiratory Syndrome Coronavirus (MERS-CoV), a 79% sequencing similarity with
the Severe Acute Respiratory Syndrome Coronavirus (SARS-CoV), commonly
known as "SARS," and over 85% similarity with coronaviruses found in bats.
Subsequent research confirmed bats as the probable reservoir of these coronaviruses; however, due to the ecological separation between bats and humans, it
is believed that other organisms might have served as intermediate hosts. Based
on comprehensive scientific evidence, the International Committee on Taxonomy
of Viruses officially designated the novel RNA virus as SARS-CoV-2 [26,31,32].

RNA viruses commonly introduce errors during the replication process,
resulting in mutations that become part of the viral genome after multiple repli-
cations within a single host. This leads to the formation of a diverse population
of viral quasispecies. However, SARS-CoV-2 possesses a highly effective proof-
reading mechanism facilitated by a nonstructural protein 14 (nsp14), resulting in
a mutation rate approximately ten times lower than that of typical RNA viruses.
On average, it is estimated that SARS-CoV-2 acquires 33 genomic mutations per
year. Some of these mutations confer advantages, leading to the emergence of
more infectious variants of SARS-CoV-2. These variants can be distinguished
by a small number of specific mutations, and the relatively slow accumulation
of mutations makes it unlikely for minor sequence perturbations or errors to
cause confusion between different variants. Additionally, most of these mutations occur in the S gene, which encodes the spike protein responsible for the
surface characteristics of the virus. Consequently, characterizing variants based
on the spike proteins transcribed from the genome is sufficient for the classification task [21,25,28].

The decreasing cost of next-generation sequencing (NGS) technology has
significantly facilitated SARS-CoV-2 whole-genome sequencing (WGS) by
researchers worldwide. The Centers for Disease Control and Prevention (CDC)
in the United States have provided comprehensive resources, tools, and protocols for SARS-CoV-2 WGS using various sequencing platforms such as Illumina, PacBio, and Ion Torrent. Additionally, the Global Initiative on Sharing
All Influenza Data (GISAID) hosts the largest SARS-CoV-2 genome sequencing
dataset to date, encompassing millions of sequences. This unparalleled volume
of genomic data generation and its easy accessibility have enabled researchers
to delve into the molecular mechanisms, genetic variability, evolutionary progression, and the potential for the emergence and spread of novel virus variants.
However, the sheer magnitude of this data surpasses the capabilities of existing

methods like Nextstrain [16] and even the more recent IQTREE2 [24] by several orders of magnitude, presenting a significant Big Data challenge. Consequently, alternative approaches focusing on clustering and classification of sequences have emerged in recent literature [2,3,5,23], demonstrating promising accuracy and scalability properties. These methods offer viable solutions for identifying major variants and addressing the challenges associated with the extensive volume of genomic data.

However, several challenges persist in studying the evolutionary and transmission patterns of SARS-CoV-2 [6,13] and other viruses. One of these challenges arises from sequencing errors, which can be mistakenly identified as mutations during analysis. These errors can result from various sources, including contamination during sample preparation, sequencing technology limitations, or genome assembly methodologies. To address this issue, computational biologists typically employ filtering techniques to remove sequences with errors or mask specific sequence fragments that exhibit errors. For instance, in the case of GISAID [14] sequences, each sequence represents a consensus sequence derived from the viral population within an infected individual. This consensus sequence averages out minor variations present in the population, providing a representative snapshot of the SARS-CoV-2 variant carried by the patient. Although this consensus sequence accurately captures the predominant variant, it comes at the cost of losing valuable information about these minor variations. However, given enough time and within an immunocompromised individual, these minor variations can undergo significant evolution and become dominant, as observed in the emergence of the Alpha variant [12].

What this means is that many machine learning approaches towards clustering and classification of sequences [2,3,5] have been operating under rather idealized conditions of virtually error-free consensus sequences. Moreover, these methods rely on a k-mer based feature vector representation — an approach that does not even rely on the alignment of the sequences, something which can also introduce bias [15]. Such a framework should easily cope with errors as well — something machine learning approaches can do very naturally [11]. There is hence a great need for some way to reliably benchmark such methods for robustness to errors, which is what we carry out in this paper.

We highlight the main contributions of this paper as follows:

- We propose several ways of introducing errors into spike sequences which reflect the error profiles of modern NGS technologies such as Illumina and PacBio;
- We demonstrate that the k-mer based approach is more robust to such errors when compared to the baseline (one-hot encoding); and
- We show that deep learning is generally more robust in handling these errors than machine learning models.

Moreover, we extend our error testing procedure as a framework for benchmarking the performance of different ML methods in terms of classification accuracy and robustness to different types of simulated random errors in the sequences. The two types of errors that we introduce are "consecutive" and "random" errors (see Sect. 3.4). Random errors are just point mutations, which happen uniformly at random along the protein sequence, simulating closely the behavior of Illumina sequencing technolgies [30]. Consecutive errors on the other hand, are small subsequences of consecutive errors, which can model insertion-deletion (indel) errors which are common in third generation long-read technologies such as Pacific Biosciences (PacBio) SMRT sequencing [10].

This paper is structured as follows. In Sect. 2, we discuss related work. In Sect. 3 we discuss some approaches we benchmark, and then how we benchmark: the type of adversarial attacks we use. Sect. 4 details the experiments, and Sect. 5 gives the results. Finally, we conclude this paper in Sect. 6.

2   Related Work

The evaluation and benchmarking of the robustness of machine learning (ML) and deep learning (DL) approaches through adversarial attacks have gained popularity in the field of image classification [17]. However, there are related works that focus more specifically on molecular data. For instance, in [29], the authors present a set of realistic adversarial attacks to assess methods that predict chemical properties based on atomistic simulations, such as molecular conformation, reactions, and phase transitions. Additionally, in the context of protein sequences, the authors of [18] demonstrate that deep neural network-based methods like AlphaFold [19] and RoseTTAFold [7], which predict protein conformation, lack robustness. These methods produce significantly different protein structures when subjected to small, biologically meaningful perturbations in the protein sequence. Although our approach shares similarities with these works, our goal is classification. Specifically, we investigate how a small number of point mutations, simulating errors introduced by certain types of next-generation sequencing (NGS) technologies, can impact the downstream classification performed by various ML and DL approaches.

When it comes to obtaining numerical representations, a common approach involves constructing a kernel matrix and using it as input for traditional machine learning classifiers like support vector machines (SVM) [20,22]. However, these methods can be computationally expensive in terms of space complexity. In related works [3,21], efficient embedding methods for spike sequence classification and clustering are proposed. Nevertheless, these approaches either lack scalability or exhibit poor performance when dealing with larger datasets. Although some effort has been made to benchmark the robustness of machine learning models using genome (nucleotide) sequences [4], no such study has been conducted on the (spike) protein sequences (to the best of our knowledge).

3   Proposed Approach

In this section, we start by explaining the baseline model for spike sequence classification. After that, we will explain our deep learning model in detail.

3.1 One-Hot Encoding (OHE) Based Embedding

Authors in [21] propose that classification of viral hosts of the coronavirus can be done by using spike sequences only. For this purpose, a fixed-length one-hot encoding-based feature vector is generated for the spike sequences. In the spike sequence, we have 21 unique characters (amino acids) that are *"ACDE-FGHIKLMNPQRSTVWXY"*. Also, note that the length of each spike sequence is 1273 with the stopping character '*' at the 1274th position. When we design the OHE-based numerical vector for the spike sequence, the length of the numerical vector will be $21 \times 1273 = 26733$. This high dimensionality could create the problem of "Curse of Dimensionality (CoD)". To solve the CoD problem, any dimensionality reduction method can be used, such as Principal Component Analysis [1]. After reducing the dimensions of the feature vectors, classical Machine Learning (ML) algorithms can be applied to classify the spike sequences. One major problem with such OHE-based representation is that it does not preserve the order of the amino acids very efficiently [2]. If we compute the pair-wise Euclidean distance between any two OHE-based vectors, the overall distance will not be affected if a random pair of amino acids are swapped for those two feature vectors. Since the order of amino acids is important in the case of sequential data, OHE fails to give us efficient results [2]. In this paper, we use OHE as a baseline embedding method.

3.2 *k*-mers Based Representation

A popular approach to preserve the ordering of the sequential information is to take the sliding window-based substrings (called mers) of length k. This k-mers-based representation is recently proven to be useful in classifying the spike sequences effectively [2]. In this approach, first, the k-mers of length k are computed for each spike sequence. Then a fixed length frequency vector is generated corresponding to each spike sequence, which contains the count of each k-mer in that sequence. One advantage of using k-mers based approach is that it is an *"alignment-free"* method, unlike OHE, which requires the sequences to be aligned to the reference genome.

Remark 1. Sequence alignment is an expensive process and requires reference sequence (genome) [8,9]. It may also introduce bias into the result [15].

The total number of k-mers in a given spike sequence are:

$$N - k + 1 \tag{1}$$

where N is the length of the sequence. The variable k is the user-defined parameter. In this paper, we take $k = 3$ (decided empirically). Since we have 1273 length spike sequences, the total number of k-mers that we can have for any spike sequence is $1273 - 3 + 1 = 1271$.

Frequency Vector Generation. After generating the k-mers, the next step is to generate the fixed-length numerical representation (frequency vector) for the set of k-mers in a spike sequence. Let the set of amino acids in the whole dataset is represented by alphabet Σ. Now, length of the frequency vector will be $|\Sigma|^k$ (all possible combinations of k-mers in Σ of length k). Recall that in our dataset, we have 21 unique amino acids in any spike sequence. Therefore, the length of frequency vector in our case would be $21^3 = 9261$ (when we take $k = 3$).

Note that CoD could be a problem in the case of k-mers based numerical representation of the spike sequence. To deal with this problem, authors in [3] use an approximate kernel method that map such vectors into low dimensional euclidean space using an approach, called Random Fourier Features (RFF) [27]. Unlike kernel trick, which compute the inner product between the lifted data points ϕ (i.e. $\langle\phi(a),\phi(b)\rangle = f(a,b)$, where $a,b \in \mathcal{R}^d$ and f(a,b) is any positive definite function), RFF maps the data into low dimensional euclidean inner product space. More formally:

$$z : \mathcal{R}^d \to \mathcal{R}^D \qquad (2)$$

RFF tries to approximate the inner product between any pair of transformed points.

$$f(a,b) = \langle\phi(a),\phi(b)\rangle \approx z(a)^T z(b) \qquad (3)$$

where z is low dimensional representation. Since z is the approximate representation, we can use it as an input for the classical ML models and analyse their behavior (as done in Spike2Vec [3]). However, we show that such approach performs poorly on larger size datasets (hence poor scalability).

3.3   Keras Classifier

We use a deep learning-based model called the Keras Classification model (also called Keras classifier) to further improve the performance that we got from Spike2Vec. For keras classifier, we use a sequential constructor. It contains a fully connected network with one hidden layer that contains neurons equals to the length of the feature vector (i.e. 9261). We use "rectifier" activation function for this classifier. Moreover, we use "softmax" activation function in the output layer. At last, we use the efficient Adam gradient descent optimization algorithm with "sparse categorical crossentropy" loss function (used for multi-class classification problem). It computes the crossentropy loss between the labels and predictions. The batch size and number of epochs are taken as 100 and 10, respectively for training the model. For the input to this keras classifier, we separately use OHE and k-mers based frequency vectors.

Remark 2. Note that we are using "sparse categorical crossentropy" rather than simple "categorical crossentropy" because we are using integer labels rather than one-hot representation of labels.

3.4   Adversarial Examples Creation

We use two types of approaches to generate adversarial examples so that we can test the robustness of our proposed model. These approaches are "Random Error generation" and "Consecutive Error generation".

In random error generation, we randomly select a fraction of spike sequences (we call them the set of errored sequences for reference) from the test set (i.e. 5%, 10%, 15%, and 20%). For each of the spike sequence in the set of errored sequences, we randomly select a fraction of amino acids (i.e. 5%, 10%, 15%, and 20%) and flip their value randomly. At the end, we replace these errored sequences set with the corresponding set of original spike sequences in the test set. The ideas is that this simulates the errors made by NGS technologies such as Illumina [30].

In consecutive error generation, the first step is the same as in random error generation (getting random set of spike sequences from the test set "set of errored sequences"). For this set of errored sequences, rather than randomly flipping a specific percentage of amino acid's values for each spike sequence (i.e. 5%, 10%, 15%, and 20%), we flip the values for the same fraction of amino acids but those amino acids are consecutive and at random position in the spike sequence. More formally, it is a consecutive set of amino acids (at random position) in the spike sequence for which we flip the values. At the end, we replace these errored sequences set with the corresponding set of original spike sequences in the test set. The idea is that this simulates indel errors, which are frequently found in third generation long-read technologies such as PacBio [10].

Using the two approaches to generate adversarial examples, we generate a new test set and evaluate the performance of the ML and deep learning models. To measure the performance, we also apply two different strategies. One strategy is called Accuracy and the other is called robustness. In the case of the Accuracy, we compute the average accuracy, precision, recall, F1 (weighted), F1 (Macro), and ROC-AUC for the whole test set including adversarial and non-adversarial examples. For our second strategy (robustness), we only consider the adversarial examples (set of errored spike sequences) rather than considering the whole test set and compute average accuracy, precision, recall, F1 (weighted), F1 (Macro), and ROC-AUC for them.

4   Experimental Setup

All experiments are conducted using an Intel(R) Xeon(R) CPU E7-4850 v4 @ 2.10 GHz having Ubuntu 64 bit OS (16.04.7 LTS Xenial Xerus) with 3023 GB memory. Our pre-processed data is also available online[1], which can be used after agreeing to terms and conditions of GISAID[2]. For the classification algorithms, we use 10% data for training and 90% for testing. Note that our data split and pre-processing follow those of [3].

---

[1] https://drive.google.com/drive/folders/1-YmIM8ipFpj-glr9hSF3t6VuofrpgWUa?usp=sharing.

[2] https://www.gisaid.org/.

4.1   Dataset Statistics

We used the (aligned) spike protein from a popular and publicly available database of SARS-CoV-2 sequences, GISAID. In our dataset, we have $2,519,386$ spike sequences along with the COVID-19 variant information. The total number of unique variants in our dataset are 1327. Since not all variants have significant representation in our data, we only select the variants having more than $10,000$ sequences. After this preprocessing, we are left with $1,995,195$ spike sequences [3].

5   Results and Discussion

In this section, we first show the comparison of our deep learning model with the baselines. We then show the results for the two approaches for adversarial examples generation and compare different ML and DL methods. Overall, we elucidate our key findings in the following subsections.

5.1   Effectiveness of Deep Learning

Table 1 contains the (accuracy) results for our keras classifier and its comparison with different ML models namely Naive Nayes (NB), Logistic Regression (LR), and Ridge Classifier (RC). For keras classifier, we use both OHE and k-mers-based embedding approaches separately. We can observe from the results that keras classifier with k-mers based frequency vectors is by far the best approach as compared to the other baselines.

Table 1. Variants Classification Results (10% training set and 90% testing set) for top 22 variants (1995195 spike sequences).

Approach	Embed. Method	ML Algo.	Acc.	Prec.	Recall	F1 weigh.	F1 Macro	ROC- AUC	Train. runtime (sec.)
ML Algo.	OHE	NB	0.30	0.58	0.30	0.38	0.18	0.59	1164.5
		LR	0.57	0.50	0.57	0.49	0.19	0.57	1907.5
		RC	0.56	0.48	0.56	0.48	0.17	0.56	709.2
	Spike2Vec	NB	0.42	0.79	0.42	0.52	0.39	0.68	1056.0
		LR	0.68	0.69	0.68	0.65	0.49	0.69	1429.1
		RC	0.67	0.68	0.67	0.63	0.44	0.67	694.2
Deep Learning	One-Hot	Keras Classifier	0.61	0.58	0.61	0.56	0.24	0.61	28971.5
	k-mers	Keras Classifier	**0.87**	**0.88**	**0.87**	**0.86**	**0.71**	**0.85**	13296.2

To test the robustness of these ML and DL methods, we use both "consecutive error generation" and "random error generation" separately. Table 2 shows the (accuracy) results (using keras classifier with k-mers because that was the best model from Table 1) for the consecutive error generation method (using different fraction of spike sequences from the test set and different fraction of amino acids

Table 2. Accuracy results for the whole test set (consecutive error seq.) for Keras Classifier with k-mers and different % of errors.

% of Seq.	% of Error in each Seq.	Acc.	Prec.	Recall	F1 (weighted)	F1 (Macro)	ROC-AUC	Train. runtime (sec.)
5%	5%	0.86	0.87	0.86	0.85	0.72	0.85	15075.72
	10%	0.83	0.88	0.83	0.83	0.68	0.83	15079.3
	15%	0.85	0.86	0.85	0.85	0.7	0.84	13747.5
	20%	0.86	0.87	0.86	0.85	0.71	0.84	11760.21
10%	5%	0.85	0.87	0.85	0.84	0.68	0.83	11842
	10%	0.8	0.86	0.8	0.82	0.69	0.83	14658.17
	15%	0.79	0.85	0.79	0.8	0.65	0.81	13783.92
	20%	0.84	0.84	0.84	0.82	0.67	0.82	13159.47
15 %	5%	0.85	0.86	0.85	0.84	0.68	0.83	15426.38
	10%	0.77	0.86	0.77	0.79	0.64	0.8	8156.5
	15%	0.76	0.86	0.76	0.79	0.65	0.81	16241.72
	20%	0.75	0.87	0.75	0.79	0.66	0.8	15321.63
20 %	5%	0.73	0.86	0.73	0.77	0.65	0.8	15930.54
	10%	0.76	0.87	0.76	0.79	0.64	0.79	14819.38
	15%	0.76	0.88	0.76	0.79	0.65	0.8	13764.76
	20%	0.8	0.81	0.8	0.77	0.63	0.78	10547.96

Table 3. Robustness (only error seq. in the test set) results (consecutive error seq.) for Keras Classifier with k-mers and different % of errors.

% of Seq.	% of Error in each Seq.	Acc.	Prec.	Recall	F1 (weighted)	F1 (Macro)	ROC-AUC	Train. runtime (sec.)
5%	5%	0.6	0.57	0.6	0.54	0.1	0.55	13882.03
	10%	0.15	0.62	0.15	0.1	0.04	0.52	8297.78
	15%	0.12	0.51	0.12	0.03	0.01	0.5	11943.5
	20%	0.6	0.44	0.6	0.48	0.07	0.53	13511.4
10%	5%	0.53	0.62	0.53	0.53	0.09	0.54	9475.31
	10%	0.31	0.62	0.31	0.32	0.1	0.54	7137.31
	15%	0.25	0.6	0.25	0.27	0.07	0.53	13399.18
	20%	0.55	0.35	0.55	0.42	0.06	0.52	13232.93
15%	5%	0.43	0.76	0.43	0.46	0.24	0.61	13588.11
	10%	0.54	0.58	0.54	0.54	0.11	0.56	14147.05
	15%	0.06	0.63	0.06	0.02	0.02	0.51	13729.87
	20%	0.06	0.64	0.06	0.03	0.02	0.5	13596.25
20%	5%	0.17	0.68	0.17	0.18	0.1	0.54	13503.11
	10%	0.48	0.63	0.48	0.48	0.09	0.55	10777.34
	15%	0.47	0.59	0.47	0.49	0.09	0.55	13550.35
	20%	0.49	0.39	0.49	0.33	0.03	0.5	11960.26

flips in each spike sequence). We can observe that keras classifier is able to perform efficiently even with higher proportion of error.

The robustness results for the consecutive error generation method are given in Table 3. Although we cannot see any clear pattern in this case, the keras classifier is giving us comparatively higher performance in some of the settings.

Table 4 contains the accuracy results for the keras classifier (with k-mers based frequency vectors as input) with random errored sequences approach. We can observe that our DL model is able to maintain higher accuracy even with

Table 4. Accuracy results for the whole test set (random error seq.) for Keras Classifier with k-mers and different % of errors.

% of Seq.	% of Error in each Seq.	Acc.	Prec.	Recall	F1 (weighted)	F1 (Macro)	ROC-AUC	Train. runtime (sec.)
5%	5%	0.85	0.88	0.85	0.84	0.69	0.84	9338.87
	10%	0.84	0.87	0.84	0.85	0.72	0.85	14365.47
	15%	0.84	0.87	0.84	0.83	0.70	0.84	14996.06
	20%	0.82	0.84	0.82	0.81	0.68	0.83	10958.00
10%	5%	0.84	0.86	0.84	0.84	0.69	0.83	15465.50
	10%	0.83	0.87	0.83	0.84	0.68	0.82	15135.49
	15%	0.79	0.87	0.79	0.82	0.67	0.82	14675.58
	20%	0.83	0.85	0.83	0.83	0.69	0.83	14758.57
15%	5%	0.77	0.83	0.77	0.77	0.64	0.80	16573.58
	10%	0.75	0.83	0.75	0.77	0.66	0.80	16472.99
	15%	0.76	0.86	0.76	0.79	0.65	0.80	16799.43
	20%	0.76	0.84	0.76	0.77	0.67	0.81	15495.56
20%	5%	0.77	0.87	0.77	0.81	0.67	0.81	15932.48
	10%	0.80	0.86	0.80	0.81	0.65	0.81	15823.57
	15%	0.82	0.83	0.82	0.80	0.64	0.79	14597.92
	20%	0.73	0.82	0.73	0.74	0.63	0.79	8885.70

Table 5. Robustness results for the whole test set (random error seq.) for Keras Classifier with k-mers and different % of errors.

% of Seq.	% of Error in each Seq.	Acc.	Prec.	Recall	F1 (weighted)	F1 (Macro)	ROC-AUC	Train. runtime (sec.)
5%	5%	0.18	0.64	0.18	0.16	0.10	0.55	11832.46
	10%	0.08	0.65	0.08	0.06	0.02	0.51	9405.35
	15%	0.28	0.58	0.28	0.28	0.07	0.53	9912.31
	20%	0.56	0.36	0.56	0.43	0.06	0.52	13029.71
10%	5%	0.62	0.59	0.62	0.54	0.10	0.55	13840.43
	10%	0.61	0.55	0.61	0.52	0.09	0.54	14016.53
	15%	0.49	0.54	0.49	0.49	0.07	0.54	14038.03
	20%	0.58	0.45	0.58	0.50	0.09	0.54	14202.82
15%	5%	0.27	0.65	0.27	0.26	0.06	0.54	14790.52
	10%	0.16	0.10	0.16	0.07	0.05	0.52	14539.64
	15%	0.19	0.56	0.19	0.18	0.04	0.52	13956.71
	20%	0.04	0.48	0.04	0.03	0.01	0.50	13321.69
20 %	5%	0.60	0.71	0.60	0.58	0.14	0.57	14172.11
	10%	0.22	0.58	0.22	0.19	0.08	0.53	12912.32
	15%	0.46	0.57	0.46	0.43	0.05	0.52	8594.60
	20%	0.03	0.59	0.03	0.01	0.01	0.50	13884.36

20% of the spike sequences having some fraction of error in the test set. Similarly, the robustness results are given in Table 5.

To visually compare the accuracy of the average accuracy for the consecutive errored sequences approach, we plot the average accuracies in Fig. 1a. Similarly, Fig. 2 contains the robustness results for the same two approaches.

The accuracy and the robustness results for the other ML models (using the Spike2Vec approach) are given in Tables 6 and Table 7, respectively. From the

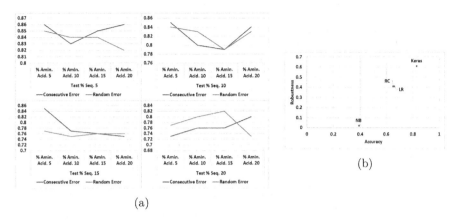

Fig. 1. (a) Accuracy Comparison of the Consecutive error generation and Random error generation approaches for different fractions of adversarial spike sequences. (b) Accuracy (x-axis) vs Robustness (y-axis) plot (for average accuracy values) for different ML and DL methods for 10% adversarial sequences from the test set with 10% amino acids flips.

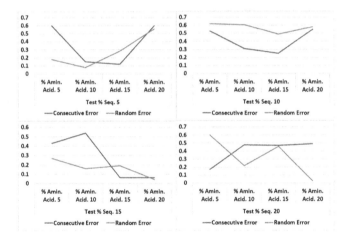

Fig. 2. Robustness Comparison of the Consecutive error generation and Random error generation approaches for different fractions of adversarial spike sequences.

results, we can conclude that our deep learning-based model is more accurate and robust than other compared machine learning models. Another interesting outcome from the results is that the k-mers-based feature vector is more robust than traditional one-hot embedding. This is a kind of "proof of concept" that since k-mers preserve the order of the amino acids in a spike sequence (as order matters in genomic sequence data), it outperforms the traditional OHE by a significant margin.

Table 6. Accuracy results for the whole test set (random error seq.) for ML models with Spike2Vec approach and different % of errors.

% of Seq.	% of Error in each Seq.	Acc.	Prec.	Recall	F1 (weighted)	F1 (Macro)	ROC-AUC	Train. runtime (sec.)
5%	5%	NB	0.40	0.80	0.40	0.51	0.40	0.68
		LR	0.68	0.68	0.68	0.64	0.68	0.69
		RC	0.67	0.67	0.67	0.62	0.67	0.67
	10%	NB	0.40	0.80	0.40	0.51	0.40	0.68
		LR	0.67	0.67	0.67	0.64	0.67	0.69
		RC	0.66	0.67	0.66	0.62	0.66	0.67
	15%	NB	0.40	0.75	0.40	0.50	0.40	0.68
		LR	0.68	0.68	0.68	0.64	0.68	0.68
		RC	0.67	0.67	0.67	0.62	0.67	0.67
	20%	NB	0.40	0.80	0.40	0.51	0.40	0.68
		LR	0.67	0.67	0.67	0.64	0.67	0.69
		RC	0.67	0.67	0.67	0.62	0.67	0.67
10%	5%	NB	0.38	0.79	0.38	0.49	0.38	0.67
		LR	0.67	0.67	0.67	0.63	0.67	0.68
		RC	0.66	0.66	0.66	0.61	0.66	0.66
	10%	NB	0.39	0.80	0.39	0.49	0.39	0.67
		LR	0.66	0.66	0.66	0.62	0.66	0.68
		RC	0.65	0.64	0.65	0.60	0.65	0.66
	15%	NB	0.38	0.80	0.38	0.49	0.38	0.67
		LR	0.67	0.67	0.67	0.63	0.67	0.68
		RC	0.65	0.66	0.65	0.61	0.65	0.66
	20%	NB	0.66	0.66	0.66	0.62	0.46	0.68
		LR	0.65	0.66	0.65	0.61	0.43	0.66
		RC	0.36	0.79	0.36	0.46	0.36	0.66
15%	5 %	NB	0.36	0.79	0.36	0.46	0.36	0.66
		LR	0.65	0.66	0.65	0.61	0.65	0.67
		RC	0.65	0.64	0.65	0.60	0.65	0.65
	10%	NB	0.36	0.75	0.36	0.46	0.36	0.66
		LR	0.66	0.66	0.66	0.61	0.66	0.67
		RC	0.65	0.65	0.65	0.60	0.65	0.65
	15%	NB	0.36	0.80	0.36	0.47	0.36	0.66
		LR	0.64	0.65	0.64	0.60	0.45	0.67
		RC	0.63	0.63	0.63	0.58	0.40	0.65
	20%	NB	0.36	0.75	0.36	0.46	0.35	0.66
		LR	0.65	0.65	0.65	0.61	0.44	0.67
		RC	0.64	0.64	0.64	0.59	0.41	0.65
20%	5%	NB	0.34	0.80	0.34	0.45	0.34	0.65
		LR	0.63	0.64	0.63	0.59	0.63	0.66
		RC	0.63	0.63	0.63	0.58	0.63	0.64
	10%	NB	0.34	0.75	0.34	0.45	0.34	0.65
		LR	0.64	0.65	0.64	0.60	0.64	0.66
		RC	0.63	0.63	0.63	0.58	0.63	0.64
	15%	NB	0.34	0.75	0.34	0.44	0.34	0.65
		LR	0.62	0.63	0.62	0.58	0.43	0.66
		RC	0.60	0.60	0.60	0.56	0.39	0.64
	20%	NB	0.34	0.8	0.34	0.45	0.34	0.65
		LR	0.64	0.64	0.64	0.6	0.43	0.66
		RC	0.61	0.61	0.61	0.56	0.39	0.64

Table 7. Robustness results for the whole test set (random error seq.) for ML models with Spike2Vec approach and different % of errors.

% of Seq.	% of Error in each Seq.	Acc.		Prec.	Recall	F1 (weighted)	F1 (Macro)	ROC-AUC	Train. runtime (sec.)
5%	5%	NB	0.02	0.03	0.02	0.01	0.02	0.50
		LR	0.46	0.29	0.46	0.35	0.46	0.50
		RC	0.46	0.28	0.46	0.34	0.46	0.50
	10%	NB	0.02	0.00	0.02	0.00	0.02	0.50
		LR	0.41	0.27	0.41	0.32	0.41	0.50
		RC	0.43	0.28	0.43	0.33	0.43	0.50
	15%	NB	0.02	0.05	0.02	0.01	0.02	0.50
		LR	0.46	0.28	0.46	0.34	0.46	0.50
		RC	0.46	0.28	0.46	0.34	0.46	0.50
	20%	NB	0.02	0.03	0.02	0.00	0.02	0.50
		LR	0.41	0.26	0.41	0.31	0.41	0.50
		RC	0.41	0.25	0.41	0.31	0.41	0.50
10%	5%	NB	0.02	0.01	0.02	0.01	0.02	0.50
		LR	0.46	0.29	0.46	0.35	0.46	0.50
		RC	0.47	0.30	0.47	0.36	0.47	0.50
	10%	NB	0.02	0.00	0.02	0.00	0.02	0.50
		LR	0.41	0.27	0.41	0.31	0.41	0.50
		RC	0.41	0.27	0.41	0.31	0.41	0.50
	15%	NB	0.02	0.01	0.02	0.01	0.02	0.50
		LR	0.41	0.26	0.41	0.31	0.41	0.50
		RC	0.42	0.26	0.42	0.31	0.42	0.50
	20%	NB	0.02	0.01	0.02	0.01	0.01	0.5
		LR	0.47	0.3	0.47	0.36	0.04	0.51
		RC	0.48	0.31	0.48	0.37	0.05	0.51
15%	5%	NB	0.02	0.03	0.02	0.01	0.02	0.50
		LR	0.46	0.29	0.46	0.35	0.46	0.50
		RC	0.41	0.26	0.41	0.31	0.41	0.50
	10%	NB	0.02	0.03	0.02	0.01	0.02	0.50
		LR	0.41	0.27	0.41	0.32	0.41	0.50
		RC	0.37	0.26	0.37	0.30	0.37	0.50
	15%	NB	0.02	0.01	0.02	0.01	0.01	0.50
		LR	0.48	0.31	0.48	0.36	0.05	0.51
		RC	0.48	0.31	0.48	0.36	0.05	0.51
	20%	NB	0.01	0.02	0.01	0.01	0.01	0.50
		LR	0.30	0.22	0.30	0.25	0.03	0.49
		RC	0.36	0.24	0.36	0.29	0.03	0.50
20 %	5%	NB	0.02	0.03	0.02	0.01	0.02	0.50
		LR	0.36	0.25	0.36	0.28	0.36	0.49
		RC	0.36	0.24	0.36	0.28	0.36	0.49
	10%	NB	0.02	0.02	0.02	0.01	0.02	0.50
		LR	0.36	0.25	0.36	0.29	0.36	0.50
		RC	0.36	0.26	0.36	0.28	0.36	0.49
	15%	NB	0.03	0.03	0.03	0.01	0.01	0.50
		LR	0.41	0.28	0.41	0.32	0.04	0.50
		RC	0.46	0.29	0.46	0.35	0.04	0.50
	20%	NB	0.02	0.06	0.02	0.01	0.01	0.50
		LR	0.47	0.31	0.47	0.36	0.04	0.51
		RC	0.42	0.29	0.42	0.33	0.04	0.50

The accuracy vs robustness comparison (for average accuracy values) of different ML and DL methods for 10% adversarial sequences from the test set with 10% amino acids flips is given in Fig. 1b. We can see that keras classifier performs best as compared to the other ML methods. This shows that not only our DL method show better predictive performance, but is also more robust as compared to the other ML models.

6   Conclusion

One interesting future extension is using other alignment-free methods such as Minimizers, which have been successful in representing metagenomics data. Since an intra-host viral population can be viewed as a metagenomics sample, this could be appropriate in this context. Another future direction is introducing more adversarial attacks resembling, in more detail, the error profiles of specific sequencing technologies. One could even fine-tune this to the particular experimental setting in which one obtained a sample, similar to sequencing reads simulators such as PBSIM (for PacBio reads).

References

1. Abdi, H., Williams, L.J.: Principal component analysis. Wiley Interdisc. Rev. Comput. Stat. **2**(4), 433–459 (2010)
2. Ali, S., Sahoo, B., Ullah, N., Zelikovskiy, A., Patterson, M.D., Khan, I.: A k-mer based approach for SARS-CoV-2 variant identification. In: International Symposium on Bioinformatics Research and Applications (ISBRA) (2021, accepted)
3. Ali, S., Patterson, M.: Spike2vec: an efficient and scalable embedding approach for COVID-19 spike sequences. CoRR arXiv:2109.05019 (2021)
4. Ali, S., Sahoo, B., Zelikovsky, A., Chen, P.Y., Patterson, M.: Benchmarking machine learning robustness in COVID-19 genome sequence classification. Sci. Rep. **13**(1), 4154 (2023)
5. Ali, S., Tamkanat-E-Ali, Khan, M.A., Khan, I., Patterson, M., et al.: Effective and scalable clustering of SARS-CoV-2 sequences. In: International Conference on Big Data Research (ICBDR) (2021, accepted)
6. Arons, M.M., et al.: Presymptomatic SARS-CoV-2 infections and transmission in a skilled nursing facility. N. Engl. J. Med. **382**(22), 2081–2090 (2020)
7. Baek, M., et al.: Accurate prediction of protein structures and interactions using a 3-track network. bioRxiv (2021)
8. Chowdhury, B., Garai, G.: A review on multiple sequence alignment from the perspective of genetic algorithm. Genomics **109**(5–6), 419–431 (2017)
9. Denti, L., et al.: Shark: fishing relevant reads in an RNA-Seq sample. Bioinformatics **37**(4), 464–472 (2021)
10. Dohm, J.C., Peters, P., Stralis-Pavese, N., Himmelbauer, H.: Benchmarking of long-read correction methods. NAR Genom. Bioinform. **2**(2) (2020). https://doi.org/10.1093/nargab/lqaa037
11. Du, N., Shang, J., Sun, Y.: Improving protein domain classification for third-generation sequencing reads using deep learning. BMC Genom. **22**(251) (2021)

12. Frampton, D., et al.: Genomic characteristics and clinical effect of the emergent SARS-CoV-2 b.1.1.7 lineage in London, UK: a whole-genome sequencing and hospital-based cohort study. Lancet Infect. Diseases **21**, 1246–1256 (2021). https://doi.org/10.1016/S1473-3099(21)00170-5

13. GISAID History (2021). https://www.gisaid.org/about-us/history/. Accessed 4 Oct 2021

14. GISAID Website (2021): https://www.gisaid.org/. Accessed 4 Sept 2021

15. Golubchik, T., Wise, M.J., Easteal, S., Jermiin, L.S.: Mind the gaps: evidence of bias in estimates of multiple sequence alignments. Mol. Biol. Evol. **24**(11), 2433–2442 (2007). https://doi.org/10.1093/molbev/msm176

16. Hadfield, J., et al.: NextStrain: real-time tracking of pathogen evolution. Bioinformatics **34**, 4121–4123 (2018)

17. Hendrycks, D., Dietterich, T.: Benchmarking neural network robustness to common corruptions and perturbations. arXiv (2019)

18. Jha, S.K., Ramanathan, A., Ewetz, R., Velasquez, A., Jha, S.: Protein folding neural networks are not robust. arXiv (2021)

19. Jumper, J., et al.: Highly accurate protein structure prediction with AlphaFold. Nature (2021)

20. Kuksa, P., Khan, I., Pavlovic, V.: Generalized similarity kernels for efficient sequence classification. In: SIAM International Conference on Data Mining (SDM), pp. 873–882 (2012)

21. Kuzmin, K., et al.: Machine learning methods accurately predict host specificity of coronaviruses based on spike sequences alone. Biochem. Biophys. Res. Commun. **533**, 553–558 (2020)

22. Leslie, C., Eskin, E., Weston, J., Noble, W.: Mismatch string kernels for SVM protein classification. In: Advances in Neural Information Processing Systems (NeurIPS), pp. 1441–1448 (2003)

23. Melnyk, A., et al.: Clustering based identification of SARS-CoV-2 subtypes. In: Jha, S.K., Măndoiu, I., Rajasekaran, S., Skums, P., Zelikovsky, A. (eds.) ICCABS 2020. LNCS, vol. 12686, pp. 127–141. Springer, Cham (2021). https://doi.org/10.1007/978-3-030-79290-9_11

24. Minh, B.Q., et al.: IQ-tree 2: New models and efficient methods for phylogenetic inference in the genomic era. Mol. Biol. Evol. **37**(5), 1530–1534 (2020)

25. Nelson, M.I.: Tracking the UK SARS-CoV-2 outbreak. Science **371**(6530), 680–681 (2021)

26. Park, S.E.: Epidemiology, virology, and clinical features of severe acute respiratory syndrome-coronavirus-2 (SARS-CoV-2; coronavirus disease-19). Clin. Exp. Pediatr. **63**(4), 119 (2020)

27. Rahimi, A., Recht, B., et al.: Random features for large-scale kernel machines. In: NIPS, vol. 3, p. 5 (2007)

28. SARS-CoV-2 Variant Classifications and Definitions (2021). https://www.cdc.gov/coronavirus/2019-ncov/variants/variant-info.html. Accessed 1 Sept 2021

29. Schwalbe-Koda, D., Tan, A., Gómez-Bombarelli, R.: Differentiable sampling of molecular geometries with uncertainty-based adversarial attacks. Nat. Commun. **12**(5104) (2021)

30. Stoler, N., Nekrutenko, A.: Sequencing error profiles of Illumina sequencing instruments. NAR Genom. Bioinform. **3**(1) (2021)

31. Wu, F., et al.: A new coronavirus associated with human respiratory disease in china. Nature **579**(7798), 265–269 (2020)

32. Zhang, Y.Z., Holmes, E.C.: A genomic perspective on the origin and emergence of SARS-CoV-2. Cell **181**(2), 223–227 (2020)

Efficient Sequence Embedding
for SARS-CoV-2 Variants Classification

Sarwan Ali[1]([✉]), Usama Sardar[2], Imdad Ullah Khan[2], and Murray Patterson[1]

[1] Georgia State University, Atlanta, GA, USA
{sali85,mpatterson30}@gsu.edu
[2] Lahore University of Management Sciences, Lahore, Pakistan
imdad.khan@lums.edu.pk

Abstract. Kernel-based methods, such as Support Vector Machines (SVM), have demonstrated their utility in various machine learning (ML) tasks, including sequence classification. However, these methods face two primary challenges:(i) the computational complexity associated with kernel computation, which involves an exponential time requirement for dot product calculation, and (ii) the scalability issue of storing the large $n \times n$ matrix in memory when the number of data points(n) becomes too large. Although approximate methods can address the computational complexity problem, scalability remains a concern for conventional kernel methods. This paper presents a novel and efficient embedding method that overcomes both the computational and scalability challenges inherent in kernel methods. To address the computational challenge, our approach involves extracting the k-mers/nGrams (consecutive character substrings) from a given biological sequence, computing a sketch of the sequence, and performing dot product calculations using the sketch. By avoiding the need to compute the entire spectrum (frequency count) and operating with low-dimensional vectors (sketches) for sequences instead of the memory-intensive $n \times n$ matrix or full-length spectrum, our method can be readily scaled to handle a large number of sequences, effectively resolving the scalability problem. Furthermore, conventional kernel methods often rely on limited algorithms (e.g., kernel SVM) for underlying ML tasks. In contrast, our proposed fast and alignment-free spectrum method can serve as input for various distance-based (e.g., k-nearest neighbors) and non-distance-based (e.g., decision tree) ML methods used in classification and clustering tasks. We achieve superior prediction for coronavirus spike/Peplomer using our method on real biological sequences excluding full genomes. Moreover, our proposed method outperforms several state-of-the-art embedding and kernel methods in terms of both predictive performance and computational runtime.

Keyword: Sequence Analysis; SARS-CoV-2; Spike Sequence Classification

X. Guo et al. (Eds.): ISBRA 2023, LNBI 14248, pp. 16–30, 2023.
https://doi.org/10.1007/978-981-99-7074-2_2

1 Introduction

The global impact of the coronavirus disease (COVID-19), caused by the severe acute respiratory syndrome coronavirus 2 (SARS-CoV-2), has been significant, affecting the lives of people worldwide. As of August 14th, the number of infections from this virus alone has reached approximately 595 million, with around 6 million deaths reported across 228 countries and territories[1]. In the United States, the number of confirmed cases stands at 92,560,911, with approximately 1,031,426 lives lost as of August 2022[2]. The rapid spread of the coronavirus has led to the collection of a massive amount of biological data, specifically SARS-CoV-2 genomic sequencing data. This data, freely available to researchers, presents an invaluable resource for studying the virus's behavior, developing effective vaccines, implementing preventive measures to curb transmission, and predicting the likelihood of future pandemics.

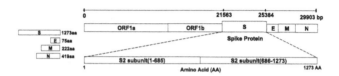

Fig. 1. An example of SARS-CoV-2 genome. Coronavirus exhibits a disproportionately high mutation rate in the S region.

In biology, it is widely recognized that a significant portion of the mutations associated with SARS-CoV-2 primarily occurs within the spike region, also known as the peplomer protein, of the complete genome [3,14]. The structure of the SARS-CoV-2 genome, including the spike region, is depicted in Fig. 1. With a genome length of approximately 30 kilobases (kb), the spike region spans the range of 21–25 kb and encodes a peplomer protein consisting of 1273 amino acids, which can be further divided into sub-units S1 and S2. Databases such as GISAID[3] provide freely accessible sequence data related to coronaviruses. Due to the disproportionate occurrence of mutations in the peplomer protein, analyzing protein data to gain insights into the virus's behavior poses considerable challenges. Focusing on the peplomer protein instead of the full-length genome can save computational time when analyzing coronavirus data due to the high occurrence of mutations in this region, the sheer volume of sequences, which amounts to millions, makes it challenging to apply traditional methods like phylogenetic trees for sequence analysis [10]. Consequently, ML techniques have emerged as an appealing alternative [14,18]. However, since most ML models operate on fixed-length numerical vectors (referred to as embeddings), utilizing

---

[1] https://www.worldometers.info/coronavirus/.
[2] https://www.cdc.gov/coronavirus/2019-ncov/index.html.
[3] https://gisaid.org/.

raw peplomer sequences as input is not feasible. Hence, the development of efficient embedding generation methods is a crucial step in ML-based classification pipelines [12]. Notable approaches, such as Spike2Vec [2] and PWM2Vec [1], have been proposed by researchers to address this challenge. Sequence alignment also plays a significant role in sequence analysis due to the importance of amino acid order. While alignment-based methods like one-hot encoding [14] have proven efficient in terms of predictive performance, there is growing interested among researchers in exploring alignment-free methods, such as Spike2Vec, to avoid the computationally expensive sequence alignment operations typically performed as a preprocessing step [5]. Alignment-free methods often employ the use of k-mers (or n-grams in the field of natural language processing) to generate spectra (vectors based on the frequency count of k-mers) [3]. Although existing alignment-free embedding methods show promising predictive results, they still pose computational challenges in terms of generation time and the high dimensionality of vectors, particularly for very long sequences and large values of k.

In contrast to traditional embedding generation methods, which are often referred to as "feature engineering" based methods, deep learning (DL) models offer an alternative approach for sequence classification [7]. However, DL methods have not achieved significant success when it comes to classifying tabular datasets. Tree-based models like random forests and XGBoost have consistently outperformed DL methods for tabular data [4].

The use of a kernel (gram) matrix in sequence classification, especially with kernel-based machine learning classifiers like SVM, has shown promising results, as reported in prior research [8]. Kernel-based techniques have displayed favorable outcomes when compared to traditional feature engineering-based approaches [3]. These methods operate by computing kernel (similarity) values between sequences, creating a matrix based on the matches and mismatches among k-mers [16]. This resulting kernel matrix can also be employed for data classification with non-kernel-based classifiers, such as decision trees (by using kernel PCA [11]). However, there are two challenges associated with the kernel-based approach.

- Computing pairwise sequence similarity is expensive
- Storing $n \times n$ dimensional kernel matrix in memory (where n represents number of sequences) is difficult when n is very large. Hence, the kernel-based method cannot be scaled on a large number of sequences.

The use of the "kernel trick" can address the challenge of computing pairwise sequence similarity. However, the storage problem of storing an $n \times n$ dimensional matrix in memory remains a significant concern. To tackle these issues, our paper proposes a novel hashing-based embedding technique. This method combines the advantages of kernel methods, particularly in computing pairwise similarity between sequences, while also addressing the storage limitation. When given a peplomer protein sequence as input, our method generates embeddings by utilizing k-mers/nGrams (substrings of consecutive characters) and computing sequence sketches, followed by a dot product operation to avoid full spectrum computation (frequency count). By operating on low-dimensional vectors (sketches) instead of an $n \times n$ matrix or full-length spectrum encompassing all possible k-mer combinations, our method enables scalability to handle a large number of sequences, effectively resolving the scalability problem. Our fast and alignment-free method seamlessly integrates with machine learning algorithms. It supports classification and clustering tasks, including k-nearest neighbors and decision trees. This paper presents the following contributions:

1. We introduce a fast, alignment-free, and efficient embedding method that rapidly computes low-dimensional numerical embeddings for protein sequences.
2. Our method combines kernel method benefits, addressing the scalability challenge for pairwise similarity.
3. Our Experiment achieves up to 99.6% faster embedding generation reduction in computation time compared to state-of-the-art (SOTA) methods.
4. Our results show that our method outperforms existing methods, achieving accuracy and ROC-AUC scores of up to 86% and 85%, respectively.
5. Visualization shows the similarity of our method's embeddings to SOTA.

In the upcoming sections, we provide literature review in Sect. 2, proposed model ins Sect. 3, experimental setup in Sect. 4, results in Sect. 5, and conclusion in Sect. 6.

2   Related Work

The feature engineering-based methods, such as Spike2Vec [2] and PWM2Vec [1], leverage the concept of using k-mers to achieve satisfactory predictive performance. However, these methods encounter the challenge known as the "curse of dimensionality." Specifically, as the value of k increases, the resulting spectrum (frequency count vector) becomes increasingly sparse, with smaller k-mers occurring less frequently. Consequently, the likelihood of encountering a specific k-mer again decreases. To address this issue of sparse vectors, the authors in [9] propose the utilization of gapped or spaced k-mers. Gapped k-mers involve extracting g-mers from larger k-mers, where g is a value smaller than k. In this approach, a g-mer consists of the first g characters (amino acids), while the remaining characters are disregarded.

The computation of pair-wise similarity between sequences using kernel matrices has been extensively studied in the field of ML [13], which can be computationally expensive. To address this issue, an approximate method was proposed by authors in [8], involving computing the dot product between the spectra of two sequences. The resulting kernel matrix can then be utilized as input for SVM or non-kernel classifiers using kernel PCA [11] for classification purposes. In a different approach, authors in [19] introduced a neural network-based model that employs the Wasserstein distance (WD) to extract features. It is worth noting that feature embeddings have applications in various domains, including product recommendations [12].

Feature engineering-based methods, while achieving improved predictive performance, often struggle to generalize effectively across different types of input sequences. Deep learning-based methods offer a potential solution to this problem [4]. In [22], authors use the ResNet for classification. However, deep learning methods generally do not yield promising results when applied to tabular data [20]. Converting protein sequences into images for input in existing image classification deep learning models is an alternative approach [17], yet preserving all pertinent information during the transformation remains a challenging task.

3   Proposed Approach

In this section, we delve into the specifics of our sequence representation and conduct an analysis of the computational complexities involved. As mentioned earlier, sequences typically exhibit varying lengths, and even when their lengths are identical, they might not be aligned. Consequently, treating them as straightforward vectors is not feasible. Even in the case of aligned sequences with equal lengths, treating them as vectors fails to account for the sequential order of elements and their continuity. To address these challenges comprehensively, one of the most successful approaches involves representing sequences with fixed-dimensional feature vectors. These feature vectors comprise the spectra or counts of all k-mers, which are strings of length k, found within the sequences [15].

Consider a scenario where we have at our disposal a collection of spike/peplomer protein sequences denoted as S, composed of amino acids. Now, for a fixed positive integer k, let's define Σ^k as the set comprising all strings with a length of k formed from characters in Σ (essentially, all conceivable k-mers). Consequently, there would be a total of $s = |\Sigma|^k$ possible k-mers in this set. The spectrum, denoted as $\Phi_k(X)$, associated with a sequence $X \in S$ can be envisioned as a vector spanning s dimensions. Each dimension corresponds to the count of a specific k-mer occurring within the sequence X. In more precise terms,

$$\Phi_k(X) = (\Phi_k(X)[\gamma])_{\gamma \in \Sigma^k} = \left(\sum_{\alpha \in X} I(\alpha, \gamma) \right)_{\gamma \in \Sigma^k}, \qquad (1)$$

where

$$I_k(\alpha, \gamma) = \begin{cases} 1, & \text{if } \alpha = \gamma \\ 0, & \text{otherwise} \end{cases} \qquad (2)$$

It's essential to note that $\Phi_k(X)$ represents a vector with $s = |\Sigma|^k$ dimensions. In this vector, each coordinate, denoted as γ and belonging to the set Σ^k, holds a value equal to the frequency of γ within the sequence X. Specifically, for peplomer sequences, where $|\Sigma| = 20$, the length of the feature vector grows exponentially with increasing k. However, in the case of other sequences like discretized music signals or text, Σ might be considerably larger, leading to a substantial increase in the space required to represent these sequences, which can become impractical.

In the realm of kernel-based machine learning, a vital component is the "kernel function." This function calculates a real-valued similarity score for a pair of feature vectors. Typically, this kernel function computes the inner product of the respective spectra.

$$K(i,j) = K(X_i, X_j) = \langle \Phi_k(X_i), \Phi_k(X_j) \rangle$$
$$= \Phi(X_i) \cdot \Phi(X_j) = \sum_{\gamma \in \Sigma^k} \Phi_k(X_i)[\gamma] \times \Phi_k(X_j)[\gamma] \qquad (3)$$

The utilization of a kernel matrix, also known as pairwise similarities, serves as input for a conventional support vector machine (SVM) classifier. This approach has proven to deliver outstanding classification performance across various applications [8]. However, the so-called *kernel trick* aims to bypass the explicit computation of feature vectors. While this technique is advantageous in terms of avoiding the quadratic space requirements for storing the kernel matrix, it encounters scalability issues when dealing with real-world sequence datasets.

Proposed Representation: In our proposed approach, denoted as $\Phi'_k(X)$ for a sequence X, we provide an approximation of the feature vector $\Phi_k(X)$. This approximation enables the application of machine learning methods based on vector space. Importantly, we establish a close relationship between the Euclidean distance of a pair of vectors and the aforementioned kernel-based proximity measure. For a sequence X within the set $\mathcal{S}$, we represent $\Phi'_k(X)$ as an approximate form of the spectrum $\Phi_k(X)$. To calculate $\Phi'(\cdot)$, we employ 2-universal hash functions.

Definition 1 (2-Universal Hash Function). *A family $\mathcal{H}$ of functions of the form $h : \Sigma^k \mapsto [w]$ is called a 2-universal family of hash functions, if for a randomly chosen $h \in \mathcal{H}$*

$$\forall \alpha \neq \beta \in \Sigma^k, \quad Pr[h(\alpha) = h(\beta)] = 1/w$$

Definition 2 (Linear Congruential Hash Functions). *For an integer w, let $p > w$ be a large prime number. For integers $0 < a < p$ and $0 \leq b < p$, and $\alpha \in \Sigma^k$ (represented as integer base $|\Sigma|$), the hash function $h_{a,b} : \Sigma^k \mapsto [w]$ is defined as*

$$h_{a,b}(\alpha) = \big((a(\alpha) + b) \mod p\big) \mod w$$

It is well-known that the family $\mathcal{H} = \{h_{a,b} : 0 < a < p, 0 \leq b < p\}$ is 2-universal. For an $0 < \epsilon < 1$, let $w > 2/\epsilon$ be an integer. Suppose $h_1 = h_{a_1,b_1} \in \mathcal{H}$ is randomly selected. $\Phi'_k(X)$ is a w-dimensional vector of integers. The ith coordinate of $\Phi'_k(X)$ is the cumulative frequency of all k-mers α that hash to the bucket i by h_1, i.e.

$$\Phi'_k(X)[i] = \sum_{\alpha:h(\alpha)=i} \Phi_k(X)[\alpha]. \qquad (4)$$

Next, we show that the dot-product between the approximate representation of a pair of sequences X and Y closely approximates the kernel similarity value given in (3). Then we extend this basic representation using multiple hash functions to amplify the goodness of the estimate.

We are going to show that for any pair of sequences X and Y, $\Phi'_k(X) \cdot \Phi'_k(Y) \simeq \Phi_k(X) \cdot \Phi_k(Y)$. For notational convenience let $\mathbf{u} = \Phi_k(X)$, $\mathbf{v} = \Phi_k(Y)$, $\mathbf{u}' = \Phi'_k(X)$, and $\mathbf{v}' = \Phi'_k(Y)$, we show that $\mathbf{u}' \cdot \mathbf{v}' \simeq \mathbf{u} \cdot \mathbf{v}$.

Theorem 1. $\mathbf{u} \cdot \mathbf{v} \leq \mathbf{u}' \cdot \mathbf{v}' \leq \mathbf{u} \cdot \mathbf{v} + \epsilon \|\mathbf{u}\|_1 \|\mathbf{v}\|_1$ *with probability* $\geq 1/2$

Theorem 1 Proof:

Proof.

$$\mathbf{u}' \cdot \mathbf{v}' = \sum_{i=1}^{w} u'_i v'_i = \sum_{i=1}^{s} u_i v_i + \sum_{i=1}^{s} \sum_{j>i}^{s} \mathbf{1}_{h(i)=h(j)} u_i v_j \qquad (5)$$

where $\mathbf{1}_{h(i)=h(j)}$ is the indicator function for the event $h(i) = h(j)$. Since entries in $\mathbf{u}$ and $\mathbf{v}$ are non-negative integers, we get that the first inequality holds certainly. For the second inequality, we estimate the error term $\sum_{i=1}^{s} \sum_{j>i}^{s} \mathbf{1}_{h(i)=h(j)} u_i v_j$

$$E\left[\sum_{i=1}^{s} \sum_{j>i}^{s} \mathbf{1}_{h(i)=h(j)} u_i v_j\right] = \sum_{i=1}^{s} \sum_{j>i}^{s} u_i v_j E\left[\mathbf{1}_{h(i)=h(j)}\right]$$

By the 2-university of h, we get $E[\mathbf{1}_{h(i)=h(j)}] = 1/w$ Using $w \geq 2/\epsilon$, we get that

$$E\left[\mathbf{u}' \cdot \mathbf{v}'\right] \leq \mathbf{u} \cdot \mathbf{v} + 2\epsilon \sum_{i=1}^{s} \sum_{j>i}^{s} u_i v_j \leq \mathbf{u} \cdot \mathbf{v} + \frac{\epsilon}{2} \|\mathbf{u}\|_1 \|\mathbf{v}\|_1,$$

where the last inequality uses the Cauchy-Shwarz inequality. By the Markov inequality, with probability at most $1/2$ the error is more than $\epsilon \|\mathbf{u}\|_1 \|\mathbf{v}\|_1$, hence the statement of the theorem follows. □

Mathematical Proofs: 1. asserts that $\hat{\mathbf{x}} \cdot \hat{\mathbf{y}}$ is an unbiased estimate for the kernel similarity. While 2. provides a bound on the deviation of the estimate.

$$E[\hat{\mathbf{x}} \cdot \hat{\mathbf{y}}] = E\left[\sum_{i=1}^{t} \hat{\mathbf{x}}_i \hat{\mathbf{y}}_i\right] = E\left[\sum_{i=1}^{t} \sum_{\alpha \in X} h^{(i)}(\alpha) \sum_{\beta \in Y} h^{(i)}(\beta)\right]$$

$$= E\left[\frac{1}{t} \sum_{i=1}^{t} \sum_{\alpha \in \Sigma^k} \mathbf{x}[\alpha] h^{(i)}(\alpha) \sum_{\beta \in \Sigma^k} \mathbf{y}[\beta] h^{(i)}(\beta)\right]$$

$$= E\left[\frac{1}{t} \sum_{i=1}^{t} \sum_{\alpha \in \Sigma^k} \mathbf{x}[\alpha] \mathbf{y}[\alpha] \left[h^{(i)}(\alpha)\right]^2 + \sum_{\alpha \neq \beta \in \Sigma^k} \mathbf{x}[\alpha] \mathbf{y}[\beta] h^{(i)}(\alpha) h^{(i)}(\beta)\right]$$

$$= E\Big[\frac{1}{t}\sum_{i=1}^{t}\big[\sum_{\alpha\in\Sigma^k}\mathbf{x}[\alpha]\mathbf{y}[\alpha]\times 1 + \sum_{\alpha\neq\beta\in\Sigma^k}\mathbf{x}[\alpha]\mathbf{y}[\beta]\times 0\big]\Big]$$

$$= E\Big[\frac{1}{t}\sum_{i=1}^{t}\mathbf{x}\cdot\mathbf{y}\Big] = \mathbf{x}\cdot\mathbf{y} = \Phi_k(X)\cdot\Phi_k(Y) = K(X,Y),$$

Note that the upper bound on the error is very loose, in practice we get far better estimates of the inner product. In order to enhance the result (so the error is concentrated around its mean), we use t hash functions $h_1,\ldots,h_t$ from the family $\mathcal{H}$. This amounts to randomly choosing t pairs of the integers a and b in the above definition. In this case, our representation for a sequence X is the scaled concatenation of t elementary representations. For $1 \le i \le t$, let $\Phi_k'^{(i)}(X)$ be the representation under hash function h_i (from Equation (4)). Then,

$$\Phi_k'(X) = \frac{1}{t}\ \big\|_{i=1}^{t}\ \Phi_k'^{(i)}(X) = \frac{1}{t}(\Phi_k'^{(1)}(X)\ \|\ \ldots\ \|\ \Phi_k'^{(t)}(X)), \tag{6}$$

where $\|$ is the concatenation operator. The quality bound on the approximation of (3) holds as in Theorem 1. Note that the definition of our representation of (4) is derived from the count-min sketch of [6], except for they take the minimum of the inner products over the hash functions and attain a better quality guarantee. Since we want a vector representation, we cannot compute the *"non-linear"* functions of min and median.

Remark 1. It's worth highlighting that our representation as expressed in (6) for a given sequence X can be computed efficiently in a single linear scan over the sequence, as detailed in Algorithm 1. Consequently, the runtime required to compute $\Phi_k'(X)$ is proportional to tn_x, where n_x denotes the number of characters in sequence X. Regarding the space complexity associated with storing $\Phi_k'(X)$, it can be described as $2t/\epsilon$, with both ϵ and t being parameters under the user's control. Additionally, it's important to note that within the error term, $|\mathbf{u}|_1 = n_x - k + 1$, with $\mathbf{u}$ representing the spectrum of the sequence X.

Subsequently, we demonstrate a close connection between the Euclidean or ℓ_2-distance commonly utilized in vector-space machine learning techniques such as k-NN classification or k-means clustering and the kernel similarity score defined in (3). This alignment allows our approach to harness the advantages of kernel-based learning without incurring the time complexity associated with kernel computation and the space complexity required for storing the kernel matrix. Through appropriate scaling of the vectors $\Phi_k'(\cdot)$, we derive detailed evaluations, which can be found in the supplementary material.

$$d^2(\mathbf{u}',\mathbf{v}') = 1 + 1 - 2\frac{\mathbf{u}\cdot\mathbf{v}}{\|u\|\|v\|} = 2 - 2\,Cos\,\theta_{uv}, \tag{7}$$

where $\mathbf{u},\mathbf{u}'$ are convenient notation for $\Phi_k(X)$ and $\Phi_k'(X)$, as defined above and $d(\cdot,\cdot)$ is the Euclidean distance. Thus, both the "Euclidean and cosine similarity" are proportional to the 'kernel similarity'.

The pseudocode of our method is given in Algorithm 1. For simplicity, we use a Python-like style for the pseudocode. Our algorithm takes a set of peplomer sequences S, integer k, m, p, alphabet Σ, and number of hash functions h. It outputs a sketch Φ, which is a low dimensional fixed-length numerical embedding corresponding to each peplomer sequence. The method starts by initializing Φ as an empty array (line 4), m with 2^{10} in line 5 (where we can take any integer power of 2), and p with 4999 (where p is any four-digit prime number and $p > m$). Now we iterate each sequence one by one (line 7) and generate a set of k-mers (line 8). Note that we take $k = 3$ here using the standard validation set approach. In the next step, for multiple hash functions h, where h is any integer value ≥ 1, we need to compute a numerical representation for each k-mer and store their frequencies in a local sketch list. The length of the local sketch list (for each sequence) equals to m (line 12). Since our idea to store the values in sketch is based on hashing, we initialize two variables $a1$ (random integer between 2 and $m - 1$) and $b1$ (random integer between 0 and $m - 1$) in line 13 and 14. To compute the integer number corresponding to each k-mer, we first compute each k-mers characters (amino acids) positions in the alphabet Σ (line 18), where Σ comprised of 21 characters $ACDEFGHIKLMNPQRSTVWXY$. We then sort the characters in k-mers (line 19) and note their position (line 20). Finally, we assign a numerical value to a character, which comprised of its position in Σ times $|\Sigma|^{\text{its position in the } k\text{-mer}}$ (line 21). This process is repeated for all characters within a k-mer (loop in line 17) and storing a running sum to get an index value for any given k-mer. Similarly, the same process is repeated for all k-mers within a sequence (loop in line 15). Now, we define the hash function (line 23):

$$(a1 * q + b1)(\mod p)(\mod m) \tag{8}$$

where q is the integer value assigned to the k-mer. After getting the hash value using Eq. 8, we increment that index (integer hash value) in the local sketch array by one (line 24). This process is repeated for all k-mers within a sequence. We normalize the local sketch list by first dividing it by the total sum of all values in the list and then dividing it by h, which is the number of hash functions (line 26). This process is repeated h times (loop in line 11) for all hash functions and all local sketch lists are concatenated to get the final sketch Φ_s for a single sequence.

4   Experimental Evaluation

In this section, we discuss the dataset used for experimentation and introduce state-of-the-art (SOTA) methods for comparing results. In the end, we show the visual representation of the proposed SOTA embeddings to get a better understanding of the data. All experiments are performed on a core i5 system (with a 2.40 GHz processor) having Windows OS and 32 GB memory. For experiments, we use the standard 70-30% split for training and testing sets, respectively. We repeat each experiment 5 times to avoid randomness and report average results.

Algorithm 1. Proposed method computation

1: **Input:** Set of Sequences S, integers k, m, p, Σ,h
2: **Output:** Φ
3: **function** COMPUTESKETCH(S, k, m, p, Σ,h)
4:     $\Phi = []$
5:     m = 2^{10} ▷ take integer power of 2
6:     p=4999   ▷ any 4 digit prime number, $p > m$
7:     **for** $s \in S$ **do**   ▷ for each sequence
8:         kmersSet = BUILDKMERS(s,k)
9:         LSketchArr = $[]$ ▷ Local Sketch Array
10:                       ▷ starting loop for multiple hash functions for each s
11:         **for** $hashLoop \leftarrow 1$ to h **do**   ▷ # of Hash Func.
12:             LocalSketch = $[0]$*m
13:             a1 = RANDOMINT(2, m-1)   ▷ range 2 to m-1
14:             b1 = RANDOMINT(0, m-1)   ▷ range 0 to m-1
15:             **for** $kmer \in kmersSet$ **do**   ▷ kmers in s
16:                 NumKmer = 0
17:                 **for** $kmersIndex \in kmer$ **do**
18:                     charPosition = Σ.index(kmersIndex)
19:                     sKmer = SORT(kmer)
20:                     position = sKmer.index(kmersIndex)
21:                     pos = charPosition $\times$ ($|\Sigma|^{position}$)
22:                     NumKmer = NumKmer + pos
23:                 hVal = ((a1 * NumKmer + b1) % p) % m
24:                 LocalSketch[hVal] ++
25:             denum = sum(LocalSketch) $\times$ h
26:             nLocalSketch = $\frac{LocalSketch}{denum}$ ▷ point-wise divide
27:             LSketchArr.Concat(nLocalSketch)
28:         Φ.append(LSketchArr)
29:     **return** Φ

Dataset Statistics: We extracted the spike sequences from a popular database called GISAID[4]. The total number of sequences we extracted is 7000, representing 22 coronavirus variants (class labels) [3].

Remark 2. Note that we use the lineages (i.e., B.1.1.7, B.1.617.2, etc., as the class labels for classification purposes). That is, given the embeddings as input, the goal is to classify the lineages using different ML classifiers.

For classification purposes, our study employs a variety of classifiers, including Support Vector Machine (SVM), Naive Bayes (NB), Multi-Layer Perceptron (MLP), K-Nearest Neighbors (KNN), Random Forest (RF), Logistic Regression (LR), and Decision Tree (DT) models. To assess the performance of these diverse models, we employ a range of evaluation metrics, including average accuracy, precision, recall, weighted F1 score, macro F1 score, Receiver Operator Characteristic Curve (ROC) Area Under the Curve (AUC), and training runtime.

---

[4] https://www.gisaid.org/.

In scenarios where metrics are designed for binary classification, we employ the one-vs-rest strategy for multi-class classification.

State-of-the-Art (SOTA) Models: We use five state-of-the-art methods, namely Spike2Vec [2], PWM2Vec [1], String Kernel [8], Wasserstein Distance Guided Representation Learning (WDGRL) [19], and Spaced k-mers [21], for comparison of results.

5   Results and Discussion

In this section, we report the classification results for our method and compare the results with SOTA methods.

Increasing Number of Hash Functions: Results for our method with an increasing number of hash functions is shown in Table 1. In this experimental setting, we use $k = 3$ for the k-mers. We can observe that although there is not any drastic change in results among different numbers of hash functions h, the random forest classifier with $h = 2$ outperforms all other classifiers and values of h for all but one evaluation metric. For training time, since the sketch length

Table 1. Classification results showing the effect of changing number of hash functions with k=3 for k-mers. Best values are shown in bold.

Parameter h	Algo.	Acc.	Prec.	Recall	F1 (Weig.)	F1 (Macro)	ROC AUC	Train Time (sec.)
Number of Hash Functions: 1	SVM	0.845	0.848	0.845	0.836	0.675	0.837	4.034
	NB	0.612	0.739	0.612	0.635	0.447	0.731	**0.708**
	MLP	0.819	0.820	0.819	0.813	0.604	0.800	12.754
	KNN	0.806	0.821	0.806	0.801	0.616	0.797	0.965
	RF	0.854	0.855	0.854	0.844	0.680	0.836	1.705
	LR	0.482	0.243	0.482	0.318	0.030	0.500	3.492
	DT	0.841	0.844	0.841	0.833	0.663	0.829	0.327
Number of Hash Functions: 2	SVM	0.848	0.858	0.848	0.841	0.681	0.848	9.801
	NB	0.732	0.776	0.732	0.741	0.555	0.771	1.440
	MLP	0.835	0.825	0.835	0.825	0.622	0.819	13.893
	KNN	0.821	0.818	0.821	0.811	0.616	0.803	1.472
	RF	**0.863**	**0.867**	**0.863**	**0.854**	**0.703**	**0.851**	2.627
	LR	0.500	0.264	0.500	0.333	0.031	0.500	11.907
	DT	0.845	0.856	0.845	0.841	0.683	0.839	0.956
Number of Hash Functions: 3	SVM	0.842	0.845	0.842	0.832	0.678	0.840	14.189
	NB	0.639	0.741	0.639	0.655	0.474	0.736	2.100
	MLP	0.817	0.816	0.817	0.809	0.608	0.802	18.490
	KNN	0.811	0.812	0.811	0.804	0.616	0.797	1.981
	RF	0.852	0.852	0.852	0.841	0.689	0.842	2.966
	LR	0.482	0.233	0.482	0.314	0.030	0.500	8.324
	DT	0.841	0.846	0.841	0.833	0.679	0.837	1.279

for $h = 1$ is the smallest among the others, it took the least amount of time to train classifiers.

Comparison With SOTA: A comparison of our method with SOTA algorithms is shown in Table 2. For these results, we report our method results for

Table 2. Classification performance of SOTA and Our methods.

Embeddings	Algo.	Acc.	Prec.	Recall	F1 (Weig.)	F1 (Macro)	ROC AUC	Train Time (sec.)
Spike2Vec [2]	SVM	0.855	0.853	0.855	0.843	0.689	0.843	61.112
	NB	0.476	0.716	0.476	0.535	0.459	0.726	13.292
	MLP	0.803	0.803	0.803	0.797	0.596	0.797	127.066
	KNN	0.812	0.815	0.812	0.805	0.608	0.794	15.970
	RF	0.856	0.854	0.856	0.844	0.683	0.839	21.141
	LR	0.859	0.852	0.859	0.844	0.690	0.842	64.027
	DT	0.849	0.849	0.849	0.839	0.677	0.837	4.286
PWM2Vec [1]	SVM	0.818	0.820	0.818	0.810	0.606	0.807	22.710
	NB	0.610	0.667	0.610	0.607	0.218	0.631	1.456
	MLP	0.812	0.792	0.812	0.794	0.530	0.770	35.197
	KNN	0.767	0.790	0.767	0.760	0.565	0.773	1.033
	RF	0.824	0.843	0.824	0.813	0.616	0.803	8.290
	LR	0.822	0.813	0.822	0.811	0.605	0.802	471.659
	DT	0.803	0.800	0.803	0.795	0.581	0.791	4.100
String Kernel [8]	SVM	0.845	0.833	0.846	0.821	0.631	0.812	7.350
	NB	0.753	0.821	0.755	0.774	0.602	0.825	0.178
	MLP	0.831	0.829	0.838	0.823	0.624	0.818	12.652
	KNN	0.829	0.822	0.827	0.827	0.623	0.791	0.326
	RF	0.847	0.844	0.841	0.835	0.666	0.824	1.464
	LR	0.845	0.843	0.843	0.826	0.628	0.812	1.869
	DT	0.822	0.829	0.824	0.829	0.631	0.826	0.243
WDGRL [19]	SVM	0.792	0.769	0.792	0.772	0.455	0.736	0.335
	NB	0.724	0.755	0.724	0.726	0.434	0.727	0.018
	MLP	0.799	0.779	0.799	0.784	0.505	0.755	7.348
	KNN	0.800	0.799	0.800	0.792	0.546	0.766	0.094
	RF	0.796	0.793	0.796	0.789	0.560	0.776	0.393
	LR	0.752	0.693	0.752	0.716	0.262	0.648	0.091
	DT	0.790	0.799	0.790	0.788	0.557	0.768	**0.009**
Spaced k-mers [21]	SVM	0.852	0.841	0.852	0.836	0.678	0.840	2218.347
	NB	0.655	0.742	0.655	0.658	0.481	0.749	267.243
	MLP	0.809	0.810	0.809	0.802	0.608	0.812	2072.029
	KNN	0.821	0.810	0.821	0.805	0.591	0.788	55.140
	RF	0.851	0.842	0.851	0.834	0.665	0.833	646.557
	LR	0.855	0.848	0.855	0.840	0.682	0.840	200.477
	DT	0.853	0.850	0.853	0.841	0.685	0.842	98.089
Ours (k = 3)	SVM	0.848	0.858	0.848	0.841	0.681	0.848	9.801
	NB	0.732	0.776	0.732	0.741	0.555	0.771	1.440
	MLP	0.835	0.825	0.835	0.825	0.622	0.819	13.893
	KNN	0.820	0.818	0.820	0.811	0.616	0.803	1.472
	RF	**0.863**	**0.867**	**0.863**	**0.854**	**0.703**	**0.851**	2.627
	LR	0.500	0.264	0.500	0.333	0.031	0.500	11.907
	DT	0.845	0.856	0.845	0.843	0.683	0.839	0.956

Table 3. Classification results showing the effect of k for k-mers with $h = 2$ for proposed method. The best values are shown in bold.

Parameter k	Algo.	Acc.	Prec.	Recall	F1 (Weig.)	F1 (Macro)	ROC AUC	Train Time (sec.)
k = 3	SVM	0.848	0.858	0.848	0.841	0.681	0.848	9.801
	NB	0.732	0.776	0.732	0.741	0.555	0.771	1.440
	MLP	0.835	0.825	0.835	0.825	0.622	0.819	13.893
	KNN	0.821	0.818	0.821	0.811	0.616	0.803	1.472
	RF	**0.863**	**0.867**	**0.863**	**0.854**	**0.703**	**0.851**	2.627
	LR	0.500	0.264	0.500	0.333	0.031	0.500	11.907
	DT	0.845	0.856	0.845	0.841	0.683	0.839	**0.956**
k = 5	SVM	0.850	0.847	0.850	0.836	0.680	0.839	8.827
	NB	0.640	0.715	0.640	0.640	0.463	0.721	1.432
	MLP	0.826	0.823	0.826	0.816	0.629	0.813	13.375
	KNN	0.818	0.824	0.818	0.812	0.621	0.801	1.319
	RF	0.857	0.853	0.857	0.843	0.690	0.842	2.322
	LR	0.483	0.237	0.483	0.315	0.030	0.500	7.219
	DT	0.844	0.840	0.844	0.833	0.667	0.834	0.987
k = 7	SVM	0.853	0.854	0.853	0.841	0.691	0.846	9.782
	NB	0.642	0.721	0.642	0.644	0.452	0.721	1.398
	MLP	0.831	0.826	0.831	0.821	0.634	0.818	13.363
	KNN	0.823	0.827	0.823	0.817	0.637	0.816	1.378
	RF	0.856	0.854	0.856	0.844	0.692	0.845	2.644
	LR	0.485	0.236	0.485	0.317	0.030	0.500	7.942
	DT	0.842	0.841	0.842	0.833	0.656	0.830	1.090
k = 9	SVM	0.849	0.847	0.849	0.838	0.676	0.836	10.099
	NB	0.644	0.714	0.644	0.651	0.437	0.707	1.540
	MLP	0.833	0.830	0.833	0.825	0.625	0.810	12.938
	KNN	0.820	0.826	0.820	0.815	0.622	0.802	2.842
	RF	0.853	0.852	0.853	0.842	0.679	0.835	3.127
	LR	0.485	0.236	0.485	0.317	0.030	0.500	8.140
	DT	0.836	0.836	0.836	0.828	0.647	0.821	1.127

$h = 2$ because it showed best performance in Table 1. We can observe that the proposed method (with $h = 2$ and $k = 3$) outperforms all the SOTA methods for all but one evaluation metric. In the case of training runtime, WDGRL performs the best.

Effect of k for k-mers: To evaluate the effect of k for k-mers on the results of our method, we report the results of our model with varying values of k in Table 3. We can observe that RF classifier with $k = 3$ outperforms other values of k for all but one evaluation metric. In terms of training runtime, decision tree classifiers take the least time.

Embeddings Generation Time: To illustrate the efficiency of our approach regarding embedding computation time, we conducted a runtime comparison with state-of-the-art (SOTA) methods, as summarized in Table 4. Our proposed method stands out by requiring the shortest time, completing the feature vector (sketch) generation in just 47.401 seconds, outperforming the SOTA alternatives. Among the alternatives, PWM2Vec, while not an alignment-free approach, emerges as the second-best in terms of runtime. In contrast, generating feature vectors with spaced k-mers consumes the most time. We also present the percentage improvement in runtime achieved by our method compared to PWM2Vec (the second-best in runtime) and Spaced k-mers (the slowest). To calculate this improvement, we use the formula: $\%improvement = \frac{R_{SOTA}-R_{Ours}}{R_{SOTA}} \times 100$. Here, R_{SOTA} represents the runtime of SOTA embedding methods (PWM2Vec and Spaced k-mers), while R_{Ours} corresponds to the runtime of our method's embedding computation. Table 4 clearly illustrates that our proposed method enhances runtime performance significantly, improving it by 70.9% and 99.6% compared to PWM2Vec and Spaced k-mers, respectively.

Table 4. Embedding generation runtime for different methods. The best value is shown in bold. The percentage improvement of the runtime is also given for our method.

Embeddings	Runtime (Sec.)
Spike2Vec [2]	354.061
PWM2Vec [1]	163.257
String Approx. [8]	2292.245
WDGRL [19]	438.188
Spaced k-mers [21]	12901.808
Ours	**47.401**
% Improv. of our method from PWM2Vec	70.9%
% Improv. of our method from Spaced k-mers	99.6%

6   Conclusion

This paper introduces a novel approach for rapidly generating protein sequence sketches that are both efficient and alignment-free, leveraging the concept of hashing. Our method not only exhibits swift sketch generation but also enhances classification outcomes when compared to existing methodologies. To validate our model, we conducted extensive experiments using real-world biological protein sequence datasets, employing a variety of evaluation metrics. Our method demonstrates an impressive 99.6% enhancement in embedding generation runtime compared to the state-of-the-art (SOTA) approach. Future endeavors will involve assessing our method's performance on more extensive sets of sequence data, potentially reaching multi-million sequences. Additionally, we aim to apply our approach to other virus data, such as Zika, to further explore its utility and effectiveness.

References

1. Ali, S., Bello, B., et al.: PWM2Vec: an efficient embedding approach for viral host specification from coronavirus spike sequences. Biology **11**(3), 418 (2022)

2. Ali, S., Patterson, M.: Spike2vec: an efficient and scalable embedding approach for COVID-19 spike sequences. In: IEEE Big Data, pp. 1533–1540 (2021)
3. Ali, S., Sahoo, B., et al.: A k-mer based approach for SARS-CoV-2 variant identification. In: ISBRA, pp. 153–164 (2021)
4. Borisov, V., Leemann, T., et al.: Deep neural networks and tabular data: a survey. IEEE Trans. Neural Netw. Learn. Syst. (2022)
5. Chowdhury, B., Garai, G.: A review on multiple sequence alignment from the perspective of genetic algorithm. Genomics **109**(5–6), 419–431 (2017)
6. Cormode, G., Muthukrishnan, S.: An improved data stream summary: the count-min sketch and its applications. J. Algorithms **55**(1), 58–75 (2005)
7. ElAbd, H., Bromberg, Y., Hoarfrost, A., Lenz, T., Franke, A., Wendorff, M.: Amino acid encoding for deep learning applications. Bioinformatics **21**(1), 1–14 (2020)
8. Farhan, M., et al.: Efficient approximation algorithms for strings kernel based sequence classification. In: NeurIPS, pp. 6935–6945 (2017)
9. Ghandi, M., Noori, M., Beer, M.: Robust k k-mer frequency estimation using gapped k-mers. J. Math. Biol. **69**(2), 469–500 (2014)
10. Hadfield, J., Megill, C., Bell, S., et al.: NextStrain: real-time tracking of pathogen evolution. Bioinformatics **34**, 4121–4123 (2018)
11. Hoffmann, H.: Kernel PCA for novelty detection. Pattern Recogn. **40**(3), 863–874 (2007)
12. Hu, W., Bansal, R., Cao, K., et al.: Learning backward compatible embeddings. In: Proceedings of the 28th ACM SIGKDD KDD, pp. 3018–3028 (2022)
13. Kuksa, P., Khan, I., et al.: Generalized similarity kernels for efficient sequence classification. In: SIAM International Conference on Data Mining (SDM) (2012)
14. Kuzmin, K., et al.: Machine learning methods accurately predicts host specificity of coronaviruses based on spike sequences alone. Biochem. Biophys. Res. Commun. **533**(3) (2020)
15. Leslie, C., Eskin, E., Noble, W.: The spectrum kernel: a string kernel for SVM protein classification. In: Symposium on Biocomputing, pp. 566–575 (2002)
16. Leslie, C., et al.: Mismatch string kernels for discriminative protein classification. Bioinformatics **20**(4), 467–476 (2004)
17. Löchel, H., et al.: Chaos game representation and its applications in bioinformatics. Comput. Struct. Biotechnol. J. **19**, 6263–6271 (2021)
18. Phylogenetic Assignment of Named Global Outbreak LINeages (Pangolin). https://cov-lineages.org/resources/pangolin.html
19. Shen, J., Qu, Y., Zhang, W., Yu, Y.: Wasserstein distance guided representation learning for domain adaptation. In: AAAI Conference on A.I (2018)
20. Shwartz-Ziv, R., Armon, A.: Tabular data: deep learning is not all you need. Inf. Fusion **81**, 84–90 (2022)
21. Singh, R., Sekhon, A., et al.: Gakco: a fast gapped k-mer string kernel using counting. In: Joint ECML and Knowledge Discovery in Databases, pp. 356–373 (2017)
22. Wang, Z., Yan, W., Oates, T.: Time series classification from scratch with deep neural networks: a strong baseline. In: IJCNN, pp. 1578–1585 (2017)

On Computing the Jaro Similarity Between Two Strings

Joyanta Basak[1]([✉]), Ahmed Soliman[1], Nachiket Deo[1], Kenneth Haase[2],
Anup Mathur[2], Krista Park[2], Rebecca Steorts[2,3], Daniel Weinberg[2],
Sartaj Sahni[4], and Sanguthevar Rajasekaran[1]

[1] Department of CSE, University of Connecticut, Storrs, CT 06268, USA
joyanta.basak@uconn.edu
[2] US Census Bureau, 4600 Silver Hill Rd, Hillcrest Heights, MD 20746, USA
[3] Departments of Statistical Science, Computer Science, Biostatistics and
Bioinformatics, Social Science, and Rhode Information Initiative,
Duke University, Durham, NC 27708, USA
[4] Department of CISE, University of Florida, Gainesville, USA

Abstract. Jaro similarity is widely used in computing the similarity (or
distance) between two strings of characters. For example, record linkage
is an application of great interest in many domains for which Jaro similarity is popularly employed. Existing algorithms for computing the Jaro
similarity between two given strings take quadratic time in the worst
case. In this paper, we present an algorithm for Jaro similarity computation that takes only linear time. We also present experimental results
that reveal that our algorithm outperforms existing algorithms.

Keywords: String similarity · Jaro similarity · Linear time
algorithm · Record linkage

1 Introduction

Several domains in science and engineering have to process string data. A general
problem in analyzing strings is that of computing the similarity (or distance)
between a pair of strings. For instance, in biology, quite frequently scientists
have to measure how similar two given genomic sequences are. Similarities can
be characterized as a function of the distance between the pair of strings.

Numerous distance metrics can be found in the literature for strings. Some
popular ones are: edit distance (also known as the Levenshtein distance), q-gram
distance, Hausdorff distance, etc. Jaro is one such popular distance metric that
is being widely used for applications such as record linkage [5,11,12].

Let R_1 and R_2 be any two strings of lengths ℓ_1 and ℓ_2, respectively. Algorithms known for computing the Jaro similarity between R_1 and R_2 take $\Omega(\ell_1 \ell_2)$
time in the worst case. In this paper, we offer an algorithm that takes only
$O(\ell_1 + \ell_2)$ time.

X. Guo et al. (Eds.): ISBRA 2023, LNBI 14248, pp. 31–44, 2023.
https://doi.org/10.1007/978-981-99-7074-2_3

We demonstrate the applicability and effectiveness of the proposed algorithm in two different applications, namely, record linkage and gene sequence similarity measurement.

The problem of record linkage is to take as input a collection of records and cluster them such that each cluster consists of records belonging to one and only one entity. An entity could refer to an individual, a family, a business, etc. A record can be thought of as a collection of attributes such as First Name, Last Name, Date of Birth, Gender, etc. If none of the records contains any errors in any of the (primary) attributes, the problem of record linkage is trivial to solve and database JOIN algorithms can be used. Unfortunately, records of the same person might look different owing to errors introduced by typing, phonetic similarity, differences in data collection, missing data, reversal of first and last names, etc. Thus record linkage is a challenging problem.

There are numerous applications for record linkage. For example, in the business domain, entrepreneurs' decisions about where to locate their businesses rely heavily on record linkage outputs. Record linkage can be employed to examine the variability of medical laboratory results by patient ethnicity and other variables [2]. Records from multiple data centers have been linked to identify disease origin and diversity [6]. Statistics about the U.S. population and economy obtained using entity resolution and record-linkage are routinely used for Congressional apportionment, redistricting, and distribution of public funds.

Computing the distance (or similarity) between two strings is a ubiquitous problem in bioinformatics. Pairwise distance computations are employed in solving many problems including biological data compression, phylogeny tree computation, metagenomic clustering, multiple sequence alignment, etc.

Note: In this paper we use the words distance and similarity interchangeably. Depending on the context, it will be clear which of these two is relevant.

2 Preliminaries

In this section, we define the Jaro similarity and describe the existing algorithms for computing this similarity.

Let s_1 and s_2 be any two strings from an alphabet Σ. Also let $\sigma = |\Sigma|$, $\ell_1 = |s_1|$, and $\ell_2 = |s_2|$.

The *Jaro* similarity between the strings s_1 and s_2 is defined as follows

$$\mathcal{S}_J(s_1, s_2) = \frac{1}{3}\left(\frac{m}{|s_1|} + \frac{m}{|s_2|} + \frac{(m-t)}{m}\right).$$

where m is the number of character matches and t is the number of transpositions divided by 2. A character from s_1 and a character from s_2 are matching if these two characters are the same and their positions do not differ by more than $r = \left\lfloor \frac{\max(|s_1|,|s_2|)}{2} \right\rfloor - 1$, which is referred as the range of search. A transposition refers to a pair of matching characters that are not in the right order.

When s_1 and s_2 do not have any matching characters, i.e., $m = 0$, then $\mathcal{S}_J(s_1, s_2) = 0$ and if these two strings are identical, then, $\mathcal{S}_J(s_1, s_2) = 1$.

As an example, let $s_1 = farming$ and $s_2 = misbegin$. In this case, the range of search is $r = 3$. The matching characters are $m, i, n,$, and g. Thus $m = 4$. The matching characters that are out of order are n and g and hence $t = 1$. As a result, $\mathcal{S}_J(s_1, s_2) = \frac{1}{3}\left[\frac{4}{7} + \frac{4}{8} + \frac{3}{4}\right] = 0.6071$.

Existing algorithms for computing Jaro similarity can be described in the following pseudocodes:

1)   Let $\ell_1 = |s_1|; \ell_2 = |s_2|; r = \left\lfloor \frac{\max(|s_1|, |s_2|)}{2} \right\rfloor - 1; m = 0;$

2)   Array $A[0 : \ell_1 - 1]$ has the characters of s_1 and the array

3) $B[0 : \ell_2 - 1]$ has the characters of s_2;

3)   Match1$[0 : \ell_1 - 1]$ and Match2$[0 : \ell_2 - 1]$ are bit arrays initialized to zeros;

4)   // Match1$[i]$ will be set to 1 if $A[i]$ has a match in s_2;

5)   // Match2$[j]$ will be set to 1 if $B[j]$ has a match in s_1;

6)   **for** $i = 0$ **to** $\ell_1 - 1$ **do**

7) $low = \max\{0, i - r\}; high = \min\{\ell_2 - 1, i + r\};$

8)   **for** $j = low$ **to** $high$ **do**

9)   **if** Match1$[i] \neq 1$ **and** Match2$[j] \neq 1$ **and** $A[i] = B[j]$ **then**

10)   Match1$[i] = 1$; Match2$[j] = 1$; $m = m + 1$; **break**;

The above pseudocode computes the value of m. Clearly, the run time of the above algorithm is $O(\ell r)$ where $\ell = \min\{\ell_1, \ell_2\}$. This run time is also $O(\ell_1 \ell_2)$, since $r = \Theta(\max\{\ell_1, \ell_2\})$. Please note that we can choose s_1 to be the shorter of the two input strings. If $\ell_1 = \ell_2 = \ell$, the run time is $O(\ell^2)$. We can compute the value of t in $O(\ell_1 + \ell_2)$ time as shown in the following pseudocode.

1) $t = 0; k = 0;$

2)   **for** $i = 0$ **to** $\ell_1 - 1$ **do**

3)   **if** (Match1$[i] = 1$) **then**

4)   **for** $j = k$ **to** $\ell_2 - 1$ **do**

5)   **if** Match2$[j] = 1$ **then** $\{k = j + 1;$ break;$\}$

6)   **if** $A[i] \neq B[j]$ **then** $t = t + 1$;

The above algorithmic segment takes $O(\ell_1 + \ell_2)$ time. As a result, we arrive at the following:

Lemma 1. *The Jaro similarity between two strings of length ℓ_1 and ℓ_2, respectively, can be computed in $O(\ell_1 \ell_2)$ time.*

3   A Linear Time Algorithm

In this section, we present the details of our linear time algorithm. Let the size of the alphabet under concern be σ. We will employ two arrays $P1$ and $P2$ of lists. Each of these arrays is of size σ. Each list in $P1$ ($P2$) corresponds to a character in the alphabet. For example, if the alphabet $\Sigma = \{a, b, \ldots, z\}$, the first lists of $P1$ and $P2$ will correspond to the character a, the second lists of $P1$ and $P2$ will correspond to the character b, and so on.

In one pass through the string s_1, for every character α present in s_1, we create a list of positions in s_1 in which the character α occurs. We store this list as $P1[\alpha]$. We do the same thing for the string s_2. For example, let $s_1 = mississippi$ and $s_2 = cincinnati$. There are 4 distinct characters in s_1 and hence there will be 4 lists. The list $P1[m]$ for the character m will be 1. The list $P1[i]$ for the character i will be $2, 5, 8, 11$. The list $P1[s]$ for s is $3, 4, 6, 7$. Finally, the list $P1[p]$ for p is $9, 10$.

We can create position lists for s_2 also in a similar manner. See Table 1. Followed by this, we scan through the characters in the shorter string. In our example, s_2 is shorter. For every character α in s_2 we merge the two lists $P1[\alpha]$ and $P2[\alpha]$ to get the number of matches corresponding to this character α. For example, when α is the character i, $P1[\alpha] = 2, 5, 8, 11$ and $P2[\alpha] = 2, 5, 10$. In our example, the range of search $r = 4$.

Table 1. Position lists for the strings $s_1 = mississippi$ and $s_2 = cincinnati$

Character	s_1	s_2
a		8
b		
c		1, 4
$\vdots$		
i	2, 5, 8, 11	2, 5, 10
$\vdots$		$\cdot$
m	1	
n		3, 6, 7
o		
p	9, 10	
$\vdots$		
s	3, 4, 6, 7	
t		9
$\vdots$		

In general let α be any character and $P1[\alpha] = i_0, i_1, i_2, \ldots, i_{u-1}$ and $P2[\alpha] = j_0, j_1, j_2, \ldots, j_{v-1}$. Without loss of generality, let $P1[\alpha]$ be shorter than $P2[\alpha]$. We start with i_0 in $P1[\alpha]$ and look for a match in $P2[\alpha]$. This is done by comparing i_0 with j_0. If $j_0 \in [i_0 - r, i_0 + r]$, then j_0 is a match for i_0. In this case, we move to i_1 and look for a match in $P2[\alpha]$ starting from position $j_0 + 1$. If j_0 is not a match for i_0, we see if j_1 is a match. If j_1 is a match, we move to i_1 and look for a match for i_1 in $P2[\alpha]$ starting from j_2, etc. If i_0 does not have a match in $P2[\alpha]$ we'll realize it when we compare j_k with i_0 (for some k). In this case, we move on to i_1 and look for a match in $P2[\alpha]$ starting from j_k.

For the example lists, we proceed as follows. We start with 2 in $P1[\alpha]$. It matches with 2 in $P2[\alpha]$. Thus we move to the next element in $P1[\alpha]$ which is 5. This 5 matches the 5 in $P2[\alpha]$. We move to the next element 8 in $P1[\alpha]$. This matches the 10 in $P2[\alpha]$. We move to 11 in $P1[\alpha]$ and there is no match for it. Thus we stop with the conclusion that there are three matches between s_1 and s_2 with respect to the character i. We do this for all the other characters. In our example, there are no matches with respect to any of the other characters, and hence $m = 3$. In this example, $t = 0$, and hence the Jaro similarity is 0.5242.

A pseudocode for this algorithm follows.

```
1)     m = 0; Let s₁ be the shorter of s₁ and s₂; Let ℓ₁ = |s₁| and ℓ₂ = |s₂|;
2)     The characters of s₁ and s₂ are stored in arrays A and B.
3)     For instance, A[i] is the character of s₁ in position i, 0 ≤ i ≤ ℓ₁ − 1;
4)     for i = 0 to (ℓ₁ − 1) do
5)          Insert i at the tail of the list P1[A[i]];
6)     for i = 0 to (ℓ₂ − 1) do
7)          Insert i at the tail of the list P2[B[i]];
8)     for every distinct character α in s₁ do
9)     // Merge the two lists P1[α] and P2[α];
10)    k = 0; Let P1[α] = i₀, i₁, i₂, …, iᵤ₋₁ and P2[α] = j₀, j₁, j₂, …, jᵥ₋₁;
11)         // Refer to the z th element in P1[α] (P2α) as
            // P1[α, z] (P2[α, z]), ∀z;
12)         for q = 0 to (u − 1) do
13)              while k < v do
14)                   if P2[α, k] ∈ [P1[α, q] ± r]
15)                   then {m = m + 1; k = k + 1; exit};
16)                   else if P2[α, k] > (P1[α, q] + r)
17)                        then exit;
18)                        else k = k + 1;
```

Lemma 2. *The above algorithm computes the number of matches correctly and in $O(\ell_1 + \ell_2)$ time.*

Proof. The **for** loops of lines 4 and 6 take a total of $O(\ell_1 + \ell_2)$ time. In the **for** loop of line 8, we process the lists for every distinct character in s_1. Consider any character α present in s_1. Let the corresponding lists be $P1[\alpha] = i_0, i_1, i_2, \ldots, i_{u-1}$ and $P2[\alpha] = j_0, j_1, j_2, \ldots, j_{v-1}$. In the **for** loop of line 12, whenever an element of $P1[\alpha]$ is compared with an element of $P2[\alpha]$, either the value of q is incremented by one, or the value of k is incremented by one, or the values of both q and k are incremented by one. This means that the entire **for** loop of line 12 runs in time $O(u + v)$. Adding this over all the characters in s_1, we realize that the **for** loop of line 8 has a run time of $O(\ell_1 + \ell_2)$. As a result, the run time of the entire algorithm is $O(\ell_1 + \ell_2)$.

Clearly, $P1[\alpha]$ and $P2[\alpha]$ are in sorted order. Also, with respect to the same character, the sequence of matches are monotonous. Specifically, if p_1 and p_1' from $P1[\alpha]$ are matched with p_2 and p_2', respectively, in $P2[\alpha]$ (for some character α and when $p_1 < p_1'$), then p_2 will be less than p_2'. In the above example, $P1[i] = 2, 5, 8, 11$ and $P2[i] = 2, 5, 10$. The matches are: $(2, 2), (5, 5)$ and $(8, 10)$. 5 and 8 from $P1[i]$ are matched with 5 and 10 from $P2[i]$. $5 < 8$ and $5 < 10$.

Consider a case in which $P2[\alpha, k] \in [P1[\alpha, q] \pm r]$ in line 14. In this case we increment both k and q by one. Due to monotonicity, the remaining matches between $P1[\alpha]$ and $P2[\alpha]$ will be to the right of $P1[\alpha, q]$ in $P1[\alpha]$ and to the right of $P2[\alpha, k]$ in $P2[\alpha]$ and hence this is a valid move.

In line 16, if $P2[\alpha, k] > (P1[\alpha, q] + r)$ it means that there is no possibility of finding a match for $P1[\alpha, q]$ in $P2[\alpha]$ and hence we exit from the **while** loop of line 13 and move onto the next value of q.

Let q be any integer in the range $[0, u - 1]$. If there is no match for $P1[\alpha, q]$ in $P2[\alpha]$, we'll get to know this if $P2[\alpha, k] > (P1[\alpha, q] + r)$ for some $k \leq (v - 1)$ or if $k \geq v$. In the former case, we move on to the next value $P1[\alpha, (q + 1)]$ and look for a match starting from $P2[\alpha, k]$. Can there be a match for $P1[\alpha, (q+1)]$ to the left of $P2[\alpha, k]$? This is not possible since there is no unmatched element in $P2[\alpha]$ that lies in the interval $[P1[\alpha, q] - r, \ P1[\alpha, q] + r]$ and $P2[\alpha, k]$ is the nearest element to $P1[\alpha, q] + r$ in $P2[\alpha]$. In other words, there is no unmatched element in $P2[\alpha]$ that lies in the interval $[P1[\alpha, q] - r, \ P2[\alpha, k])$. Note also that $P1[\alpha, q + 1] > P1[\alpha, q]$.

Line 18 applies when we are trying to find a match for $P1[\alpha, q]$ in $P2[\alpha]$ and for the current value of k there is no match and also $P2[\alpha, k] < (P1[\alpha, q] + r)$. This means that there could still be a match for $P1[\alpha, q]$ in $P2[\alpha]$ to the right of $P2[\alpha, k]$. Thus we increment k by 1.

4   Experimental Results

We have applied our algorithm in two different applications: record linkage of real people data and computing the similarity among genes of *Escherichia Coli* bacteria. We have compared the performance of our algorithm against an existing implementation [3] for the Jaro similarity computation. The applications of our interest involve the repeated computation of distances between pairs. We have implemented our algorithm in such a way that requires the initialization of the auxiliary data structures only once. Subsequently, values in these data structures are reset only when needed.

The existing implementation of the Jaro similarity [3] maintains two arrays of the size of the two comparing strings and initializes the value corresponding to each index to 0 for each comparison. For a fair evaluation, we have modified this implementation to set the size of those arrays to the maximum string length and initialized them only once for the whole experiment instead of once for each distance calculation. Also, we reset values at certain indices only if necessary similar to the implementation of our algorithm. Both implementations are available at https://github.com/joyantabasak13/LinearJaro.

All the experiments were carried out in a server machine with 6 Intel(R) Core(TM) i5-8400 2.80GHz CPU cores, 32GB DDR4 RAM, and 1TB of local storage running on Ubuntu 22.04.1 LTS. The programs were written in standard C++17.

4.1   Record Linkage

There are various algorithms for record linkage [7]. A key part of the record linkage algorithms is comparing the records to find their similarity. Linking records can be done in two phases. In the first phase, we define rules for deciding if two records belong to the same entity or not. For example, a rule could state that two records R_1 and R_2 belong to the same entity if and only if the distance between R_1 and R_2 is no more than a user-defined threshold τ. In the second phase we can use these rules to efficiently link the records.

Given a set of n records, a straightforward algorithm then could compute the similarity or distance between every pair of records and determine the linking pairs of records. The run time of this algorithm is $O(n^2)$. When n is large, this run time could be prohibitive. To speed up this algorithm several techniques have been proposed in the literature. One such technique is blocking (see e.g., [7]).

The idea of blocking is to group the records such that pairwise distance calculations are done only within the groups. The groups may or may not be disjoint. The grouping should be done in such a way that if two records belong to the same entity then they are very likely to fall together into at least one of the groups. Different blocking techniques are discussed in [7].

One of the blocking techniques we have frequently used employs k-mers.

1. Let R be a record. We think of each record as a string of characters. Each attribute in a record can be thought of as a string of characters and a record can be thought of as a concatenation of the attributes in it.
2. For some relevant k, we generate all the k-mers of R.
3. There will be a total of s^k blocks, where s is the size of the alphabet.
4. R will be placed in the blocks corresponding to all the k-mers in it.

The above blocking idea exploits the fact that if two records belong to the same individual, then, the records are very likely to share a k-mer (for a relevant value of k). Note that the blocks created above are not disjoint. Assume that we are able to group the records into 10 disjoint groups each of size $\frac{n}{10}$. In this case the number of distance calculations performed in each group is $\binom{n/10}{2}$. The total number of distance calculations across all the groups will be $10\binom{n/10}{2}$. If we did not do any blocking, the total number of distance calculations will be $\binom{n}{2}$, which is nearly 10 times more!

In our experiments we have employed the record linkage algorithm presented in [1]. The key steps in this algorithm can be summarized as follows:

1. Collect all the records from all the data sources and put them in a list L.
2. Concatenate all or some common attribute strings in the same order for each record in L. Sort L by the concatenated strings. Group records that are identical in all selected attributes. Take one representative record from each group and put it in a list L'.
3. Do blocking on L'. In our experiments, we used the superblocking technique [1] with the k-mer $k = 3$.
4. Do single linkage clustering of the records in each block obtained in step 3. Specifically, begin with each record in a separate cluster. If records r_1 and r_2 are assigned to a common block, compute the Jaro similarity between each common attribute of r_1 and r_2. If the similarity for each compared attribute is above a threshold τ, then merge the clusters of r_1 and r_2.

5. Do complete linkage clustering of the records in the single linkage clusters obtained in step 4. Given a single linkage cluster C containing records $r_1, r_2, ..., r_n$, compare all record pairs and obtain complete linkage clusters $C'_1, C'_2, ...C'_m$ where all the records in a complete linkage cluster C'_i have Jaro similarity over a threshold τ for attributes used in obtaining the single linkage cluster.

6. Add identical copies of a record found in step 2 to the corresponding complete linkage clusters.

7. Output the complete linkage clusters. Each record belongs to only one cluster as per the algorithm and the algorithm claims that all the records in each cluster represent one entity.

We applied both our linear time and existing Jaro similarity algorithms as the similarity computation procedure of the record linkage algorithm (steps 4 and 5). We applied the record linkage algorithm on two different datasets. The first dataset is collected from Soliman, et al. [9]. It contains 1 million records of deceased persons. In this dataset, each person has five records, one of which is corrupted. Each record in this dataset contains 5 attributes – id, first name, last name, date of birth, and date of death. The second dataset is collected from Saeedi, et al. [8]. It contains 5 million records including corrupted copies of records originally collected from North Carolina Voter information. In this dataset, each person may have up to 5 records where multiple of the records belonging to a person can be corrupted. Here, each record contains five attributes – id, first name, last name, suburb, and postcode. In both datasets, the attribute 'id' is only used for performance evaluation, not as an attribute in the record-linking process.

In both dataset 1 and dataset 2, any attribute string in a record can be comprised of characters from the English alphabet or digits or both. In our experiments, we set the alphabet size to 256, permitting any ASCII character to be present in the strings. In our experiments, we found the record linkage algorithm spends most of its time (>90%) in step 4. To keep the runtime feasible for the large datasets used in our experiments, we modified step 4 of the record linkage algorithm to distribute the workload over multiple cores in a single machine. This problem of load balancing (distributing blocks to cores) is known to be NP-hard. We applied an approximate load-balancing strategy. Let q be the number of blocks. Let n_i be the number of records in the i-th block (for $1 \leq i \leq q$) and m be the number of available cores. The number of pairwise similarity computations that have to be performed in the i-th block is $\binom{n_i}{2}$. This is the work to be done in the i-th block. The loads across the cores have to be balanced with respect to this metric. Hence, we sort the blocks in ascending order based on the corresponding possible number of similarity computations. Let $N = \sum_{i=1}^{q} \binom{n_i}{2}$. We assigned the largest p blocks from the end of the sorted list and the smallest r blocks from the start of the sorted list to core j such that $\sum_{i}^{p} n_i < \frac{N}{m}$ and $\sum_{i}^{r} n_i > \frac{N}{m} - \sum_{i}^{p} n_i$. Note that, the similarity comparisons done after this multicore modification are exactly the same for both the linear Jaro algorithm and the quadratic Jaro similarity computation algorithm.

We ran the modified multicore record linkage algorithm with 6 cores for both dataset 1 and dataset 2 with Jaro similarity thresholds $\tau = 0.95$ and $\tau = 0.8$, respectively. For both datasets, all the attributes except 'id' was used to compute the similarity between records. We report the comparative runtime performance of the linear Jaro similarity algorithm and the quadratic Jaro similarity algorithm in Table 2. Precision is the ratio of the true matches to the predicted matches, recall is the ratio of correctly predicted matches to the total predicted matches, and the reported F1-score is the harmonic mean between the precision and recall. As both Jaro similarity computation algorithms find similarities between the same records, the F1-score of the record linkage algorithm is the same for both algorithms on both datasets.

In our experiment with dataset 1, the 'last name' attribute is used as the blocking string. However, when comparing the records, the 'first name' and the 'last name' strings were concatenated to a single 'name' string as it results in a better f1-score. For dataset 2, the blocking string is formed by concatenating the first two characters of last name, suburb, postcode, and first name attributes respectively. These blocking strings as well as the value of similarity threshold τ were chosen as such that the resultant F1-score is comparable to the best reported in the literature. Soliman et al. [9] report an F1-score of 97% compared to 98.86% in our experiment on dataset 1. Saeedi et al. [8] provides a graph of the linking performance where the F1-score is around 87% on dataset 2. We achieved 87.34% F1-score.

Table 2. Comparative performance of linear and quadratic Jaro similarity computation for record linkage.

	Method	#Comparisions	Runtime (seconds)	F1-Score
Dataset 1	Linear Jaro	21.2 Billion	476.5	98.86%
	Quadratic Jaro		551.2	
Dataset 2	Linear Jaro	72.4 Billion	2579.7	87.34%
	Quadratic Jaro		2489.9	

The runtime performance of the Jaro similarity computation algorithms relates to the length of the strings to be compared. The linear algorithm is expected to run faster than the quadratic algorithm when strings are sufficiently large. However, in the record linkage problem, the comparing strings are generally small in length. Table 3 shows the lengths of the relevant attributes of dataset 1 and dataset 2. In the case of dataset 1, the average comparing strings is between 8 to 13 characters long. Record linkage with the linear algorithm runs 13.6% faster than the quadratic algorithm. The average length of comparing strings in dataset 2 is between 5 to 9 characters long and record linkage with the linear algorithm runs 3.6% slower than the quadratic algorithm. It shows that the linear Jaro similarity computation algorithm is a faster choice when string lengths are not very small (≥ 10) and its performance is comparable even when comparing strings that are small.

Table 3. Attribute string lengths of the records in Dataset 1 and Dataset 2.

	Attributes	Minimum Length	Maximum Length	Average Length
Dataset 1	Name	3	23	12.57
	Date of Birth	8	8	8
	Date of Death	8	8	8
Dataset 2	First Name	1	15	5.98
	Last Name	1	20	6.37
	Suburb	2	19	8.74
	Zip Code	4	9	4.99

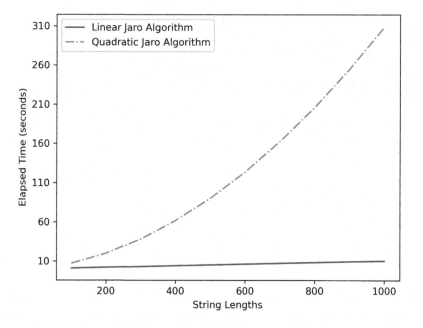

Fig. 1. Average elapsed time per million comparisons of certain lengths of *E. Coli* genome subsequences by linear time and quadratic time Jaro comparison algorithms.

4.2   Gene Sequence Similarity

Finding similarity or distance between genome or protein sequences is an important step for numerous biological problems such as phylogenetic tree reconstruction, motif search, genome assembly evaluation, etc. We collected the E.

Coli bacteria K-12 substrain genome from NCBI website (Genbank ascension no: U00096) and conducted four experiments on this genome. This genome has 4,639 annotated genes of varying sizes. In the first experiment, we considered each gene as a separate sequence and computed the Jaro similarity between all pairs of genes. As these genes can be several hundred to several thousand characters long, the range in the Jaro similarity algorithm can be large. In another set of experiments, we cut the genome sequence into non-overlapping subsequences of lengths 1000 to 100, decreasing the string length by 100 in each subsequent experiment, and computed Jaro similarity among all pairs of subsequences in each experiment. The results are shown in Table 4. Figure 1 shows the time taken per million comparisons by the two Jaro algorithms for experiments 2 to 11 in Table 4.

Table 4. Jaro similarity computation for *E.Coli* genome subsequences.

	Method	Average String Size	#Comparisions (millions)	Runtime (seconds)
Experiment 1	Linear Jaro	899.25	10.75	81.6
	Quadratic Jaro			2226.91
Experiment 2	Linear Jaro	1000	10.76	111.47
	Quadratic Jaro			3327.34
Experiment 3	Linear Jaro	900	13.29	126.03
	Quadratic Jaro			3391.06
Experiment 4	Linear Jaro	800	16.82	142.32
	Quadratic Jaro			3468.47
Experiment 5	Linear Jaro	700	21.97	163.44
	Quadratic Jaro			3573.70
Experiment 6	Linear Jaro	600	29.91	191.85
	Quadratic Jaro			3700.91
Experiment 7	Linear Jaro	500	43.08	228.26
	Quadratic Jaro			3891.68
Experiment 8	Linear Jaro	400	67.32	291.49
	Quadratic Jaro			4158.12
Experiment 9	Linear Jaro	300	119.68	376.86
	Quadratic Jaro			4534.68
Experiment 10	Linear Jaro	200	269.29	600.52
	Quadratic Jaro			5367.02
Experiment 11	Linear Jaro	100	1077.19	1231.23
	Quadratic Jaro			7763.37

Clearly, the linear Jaro similarity computation algorithm outperforms the quadratic one. The linear algorithm can compute millions of comparisons between large strings in several minutes, making experimenting with Jaro similarity a viable option for numerous biological problems.

Table 5. Runtime comparison between linear time and quadratic time Jaro similarity computation algorithms for datasets with varying string lengths and alphabet sizes.

String Length	Alphabet Size	Linear Algorithm runtime (seconds)	Quadratic Algorithm Runtime (seconds)
25	5	17.36	55.19
	10	21.60	62.49
	15	24.55	63.51
	20	25.54	62.03
	26	25.12	58.99
50	5	29.58	152.28
	10	33.82	178.66
	15	38.71	189.96
	20	43.63	195.43
	26	47.96	197.95
75	5	42.13	271.43
	10	44.93	326.37
	15	50.22	352.62
	20	55.93	367.65
	26	62.92	378.38
100	5	55.26	412.08
	10	56.37	500.10
	15	63.01	548.24
	20	68.48	577.49
	26	76.29	600.77
125	5	68.40	560.40
	10	68.43	678.48
	15	74.34	749.33
	20	80.46	797.54
	26	87.70	837.25

4.3   Effects of Alphabet Size

We conducted a set of experiments to assess the impact of alphabet size on the runtime of our proposed algorithm. In these experiments, we generated strings of certain lengths by randomly sampling from the first 5, 10, 15, 20, and 26 characters of the English alphabet. We fixed the string lengths to 25, 50, 100, and 125 characters. For each pair of alphabet size and string length, we generated 5 datasets each containing 10,000 randomly generated strings. All pairs of records in each dataset were compared among themselves to calculate Jaro similarity. This required 49.99 million comparisons for each dataset. The average runtime for the linear algorithm and the quadratic Jaro similarity computation algorithm is shown in Table 5.

Consider the quadratic algorithm for computing the Jaro similarity. Let R_1 and R_2 be any two strings. Let c be the character in position i of R_1 (for any i). The algorithm looks for a match for c in R_2 in positions in the range $[i-r, i+r]$, where r is the range defined in Sect. 2. If the characters are generated uniformly randomly from the alphabet, then we would expect to see a match for c within s characters in this range (where s is the size of the alphabet). As soon as a match is found for c, the algorithm moves to find a match for the next character in R_1. Thus we see that there is a linear dependence of the run time on s, in expectation.

5   Conclusions

In this paper we have presented a linear time algorithm for the computation of the Jaro similarity between two given strings. The previous best algorithm had a quadratic run time. Specifically, if the strings are of lengths ℓ_1 and ℓ_2, respectively, then the best previous run time was $\Omega(\ell_1\ell_2)$. In comparison, our algorithm takes only $O(\ell_1 + \ell_2)$ time. We have compared our algorithm with the previous algorithms in the context of two important applications namely record linkage and biological string similarities computation. Experiments demonstrate that our algorithm has significantly less run times, especially when the average length of the strings is large.

Pairwise distance calculation between strings is a common problem. Several distance measures have been introduced by scientists. Examples include edit distance, Jaro distance, Hausdorff distance, Jaccard distance, etc. Depending on the application, one distance measure might be more appropriate than the others. Traditionally, for each specific problem, scientists tend to favor a specific distance measure. It will be an interesting exercise to see the effect of other distance measures in solving the given problem. For example, in the bioinformatics domain, the edit distance happens to be popular. Not many works have been reported where other distance measures have been employed. We feel that experimentation with other distance measures is worthwhile.

The run times for computing these different distances vary as well. For example, the edit distance is typically computed using dynamic programming and takes quadratic time (see e.g., [4]). Until this paper, Jaro distance computation also took quadratic time. Even if one wants to employ edit distance, the Jaro distance can be used as a pruning technique since it takes much less time than the edit distance. For example, let R_1 and R_2 be two strings and assume that we are interested in computing the edit distance between them. There are many applications (including record linkage) where we are not interested in calculating the distances exactly. We are only interested in knowing whether the distance between R_1 and R_2 is less than a given threshold. In this case, we can first compute the Jaro distance between R_1 and R_2. If this distance is larger than τ', for some suitable τ', we can omit the edit distance calculation between R_1 and R_2. If the Jaro distance between R_1 and R_2 is $\leq \tau'$, we may not be sure if the edit distance between R_1 and R_2 will also be $\leq \tau$. In this case, we may compute the edit distance between R_1 and R_2. This idea of using another distance measure for pruning has been employed successfully before (see e.g., [10]).

Acknowledgements.. This work was partially supported by the United States Census Bureau under Award Number CB21RMD0160003. The content is solely the responsibility of the authors and does not necessarily represent the official views of the US Census Bureau.

References

1. Basak, J., Soliman, A., Deo, N., Rajasekaran, S.: SuperBlocking: an efficient blocking technique for record linkage, manuscript (2023)

2. Clark, D.E.: Practical introduction to record linkage for injury research. Injury Prevention BMJ J. **10**(3), 186–191 (2004)
3. GeeksforGeeks, "Jaro and Jaro-Winkler Similarity", 20 Jan. 2020. https://www.geeksforgeeks.org/jaro-and-jaro-winkler-similarity/
4. Horowitz, E., Sahni, S., Rajasekaran, S.: Computer Algorithms. Silicon Press (2008)
5. Jaro, M.A.: Advances in record linkage methodology as applied to the 1985 census of Tampa Florida. J. Am. Stat. Assoc. **84**(406), 414–20 (1989). https://doi.org/10.1080/01621459.1989.10478785
6. Maizlish, N., Herrera, L.: A record linkage protocol for a diabetes registry at ethnically diverse community health centers. J. Am. Med. Inform. Assoc. **12**, 331–337 (2005)
7. Papadakis, G., Ioannou, E., Thanos, E., Palpanas, T.: The four generations of entity resolution. Synthesis Lectures Data Manage. **16**, 1–170 (2021)
8. Saeedi, A., Peukert, E., Rahm, E.: Using link features for entity clustering in knowledge graphs. The Semantic Web: 15th International Conference, ESWC 2018, Heraklion, Crete, Greece, June 3–7, 2018, Proceedings 15, pp. 576–592 (2018)
9. Soliman, A., Rajasekaran, S.: FIRLA: a Fast Incremental Record Linkage Algorithm. J. Biomed. Inform. **130**, 104094 (2022)
10. Soliman, A., Rajasekaran, S.: A Novel String Map-Based Approach for Distance Calculations with Applications to Faster Record Linkage, manuscript (2023)
11. Winkler, W.E.: String comparator metrics and enhanced decision rules in the fellegi-sunter model of record linkage. In: Proceedings of the Section on Survey Research Methods, American Statistical Association: 354–359 (1990)
12. Winkler, W.E.: Overview of Record Linkage and Current Research Directions, Research Report Series, Statistical Research Division, U.S. Census Bureau, Washington, DC 20233 (2006)

Identifying miRNA-Disease Associations Based on Simple Graph Convolution with DropMessage and Jumping Knowledge

Xuehua Bi[1,2], Chunyang Jiang[3], Cheng Yan[4], Kai Zhao[3], Linlin Zhang[5(✉)], and Jianxin Wang[1(✉)]

[1] Hunan Provincial Key Lab on Bioinformatics,
School of Computer Science and Engineering, Central South University,
Changsha 410083, China
jxwang@mail.csu.edu.cn

[2] Medical Engineering and Technology College,
Xinjiang Medical University, Urumqi 830017, China

[3] School of Information Science and Engineering,
Xinjiang University, Urumqi 830046, China

[4] School of Informatics, Hunan University of Chinese Medicine,
Changsha 410208, China

[5] School of Software Engineering, Xinjiang University,
Urumqi 830091, China
zllnadasha@xju.edu.cn

Abstract. MiRNAs play an important role in the occurrence and development of human disease. Identifying potential miRNA-disease associations is valuable for disease diagnosis and treatment. Therefore, it is very urgent to develop efficient computational methods for predicting potential miRNA-disease associations in order to reduce the cost and time associated with biological wet experiments. In addition, although the good performance achieved by graph neural network methods for predicting miRNA-disease associations, they still face the risk of over-smoothing and have room for improvement. In this paper, we propose a novel model named nSGC-MDA, which employs a modified Simple Graph Convolution (SGC) to predict the miRNA-disease associations. Specifically, we first construct a bipartite attributed graph for miRNAs and diseases by computing multi-source similarity. Then we adapt SGC to extract the features of miRNAs and diseases on the graph. To prevent over-fitting, we randomly drop the message during message propagation and employ Jumping Knowledge (JK) during feature aggregation to enhance feature representation. Furthermore, we utilize a feature crossing strategy to get the feature of miRNA-disease pairs. Finally, we calculate the prediction scores of miRNA-disease pairs by using a fully connected neural network decoder. In the five-fold cross-validation, nSGC-MDA achieves a mean AUC of 0.9502 and a mean AUPR of 0.9496, outperforming six compared methods. The case study of cardiovascular disease also demonstrates the effectiveness of nSGC-MDA.

X. Guo et al. (Eds.): ISBRA 2023, LNBI 14248, pp. 45–57, 2023.
https://doi.org/10.1007/978-981-99-7074-2_4

Keywords: MiRNA-disease association · Simple Graph Convolution · DropMessage · Jumping Knowledge

1 Introduction

MicroRNA (miRNA), a single-stranded non-coding RNA, ranges in length from 18 to 26 nucleotides [1]. The abnormal expression of miRNAs plays a crucial role in the onset and progress of human diseases. Identifying miRNA-disease associations (MDAs) is essential for understanding the underlying pathogenic mechanism of diseases and developing new therapeutic approaches. Traditionally, biological wet experiments like microarray analysis [2] and northern blotting [3] have been used to identify MDAs. However, these methods generally are often associated with long experimental periods and high costs.

With the advancement of biotechnology, many databases related to miR-NAs and diseases have been constructed [4,5]. Based on these databases, some successful computational methods have been proposed for predicting MDAs. MiRNAs with similar functions are more likely to be associated with phenotypically similar diseases. Based on this hypothesis, researchers have proposed some similarity-based approaches [6,7]. Traditional machine learning and matrix factorization techniques are also effective models to predict MDAs. Li et al. [8] developed an effective computational model (called MCMDA) for MDAs predicition. This model aimed to predict miRNA-disease associations by updating a low-rank miRNA-disease association matrix. Chen et al. [9] utilized inductive matrix completion model to predict potential MDAs by integrating miRNA and disease similarity data. However, traditional machine learning methods are unable to capture the deep features of miRNA and disease nodes in predicting MDAs.

Recently, deep learning-based methods have been developed for MDAs prediction due to its powerful representational learning capabilities. For example, Xuan et al. [10] utilized convolutional neural networks to predict MDAs by fusing biological premises and similarity information of miRNAs and diseases. Ji et al. [11] proposed a novel approach in which two independent models were trained to extract the features of miRNA and disease nodes separately. In addition, graph neural network (GNN) has proven to be a successful approach in predicting MDAs due to its capacity to efficiently handle graph-structured data and extract intricate topological characteristics. Zhang et al. [12] proposed a node-level attention graph auto-encoder model to predict latent MDAs. Ning et al. [13] proposed a method based on attention aware multi-view similarity networks and hypergraph learning for MDA identification. Ding et al. [14] obtained the non-linear representations of miRNAs and disease by using variational graph auto-encoder with matrix factorization. Although the previous works have achieved good performances, they also have some limitations. On the one hand, most of models only focus on functional and Gaussian interaction profile kernel (GIP) similarity of miRNAs. The role of similarity information obtained from other similarity-based metric strategies is ignored. On the other hand, Graph Neural Networks (GNNs) have limitations in effectively handling noisy information in the graph and avoiding over-fitting. This is primarily due to their reliance on a fixed feature propagation and fusion method.

In this study, we propose a novel MDAs prediction model (named nSGC-MDA) based on Simple Graph Convolution (SGC) with message dropping strategy and Jumping Knowledge (JK). Firstly, we construct a bipartite attributed graph by integrating known miRNA-disease association information and multiple similarity matrices which includes miRNA sequence similarity, target similarity, family similarity and disease semantic similarity and Gaussian interaction profile kernel similarity. Secondly, after mapping miRNA and disease node features into a unified common space, we employ the Simple Graph Convolution to extract the features of miRNAs and diseases with a data augmentation strategy. To be specific, during node messages propagation, the messages are randomly dropped to prevent over-fitting. The final feature representations are obtained by using the Jumping Knowledge method, which incorporates the feature information from intermediate layers directly into the last layer to avoid information loss. Thirdly, we utilize dot product to obtain the interaction features between miRNAs and diseases, and then concatenate them. At last, the concatenated features of miRNA-disease pairs are fed into the decoder to calculate the association scores. Outstanding performance results of experiments demonstrate that our proposed model is beneficial for predicting potential MDAs.

2   Materials and Methods

2.1   Datasets

We obtain known miRNA-disease associations from the human miRNA disease database (HMDD v3.2), which provides experimentally validated miRNA-disease associations (MDAs) [4]. We also utilize other databases, including miRBase [15], mirTarBase [5] and MeSH descriptors [16], to calculate miRNA sequence and family similarity, miRNA functional similarity and disease semantic similarity. To ensure the data continuity and consistency, we preprocess the data by removing duplicates and isolated entities. At last, we obtain 14,550 miRNAs-disease associations between 917 miRNAs and 792 diseases.

2.2   The Overall Flow of the Model

In this paper, we propose a novel simple graph convolution method, which combines a message propagation mechanism and feature aggregation mechanism for miRNA-disease association prediction (nSGC-MDA). The flowchart of nSGC-MDA model is shown in Fig. 1, which consists of four steps: (1) constructing bipartite attributed graph by integrating known MDAs and multi-source similarity information; (2) applying feature propagation with random message dropping mechanism; (3) employing jumping knowledge network to aggregate features to enhance feature representations; (4) using feature-crossing to obtain interactions of miRNAs and disease and predict their associations using a MLP.

2.3 Problem Formulation

Given p miRNA nodes $M = \{m_1, \cdots m_p\}$ and q disease nodes $D = \{d_1, \cdots d_q\}$, the heterogeneous graph is expressed as $G = (V, E)$, where V is the set of miRNA and disease nodes defined as $V = \{M; D\}$. E is the set of all edges in G. The adjacency matrix of G is denoted as A. If miRNA i is associated with disease j, A_{ij} is set to 1, otherwise 0. Our task is to identify potential MDAs.

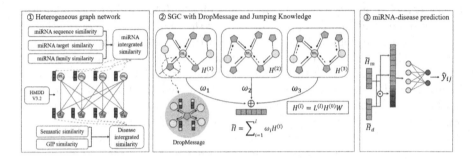

Fig. 1. The flowchart of nSGC-MDA.

2.4 Disease Semantic Similarity

We utilize Mesh descriptors to describe each disease as a directed acyclic graph (DAG), for a disease d, the DAG of it includes the ancestor nodes of d, the direct edges from parent to child nodes and d itself. Using this structure, we can calculate the contribution of any disease v in the DAG of d to d as follows:

$$\begin{cases} C_d(v) = 1, & \text{if } v = d \\ C_d(v) = \max\{\mu * C_d(v') \mid v' \in \text{children of } v\}, & \text{if } v \neq d \end{cases} \quad (1)$$

where μ is the semantic attenuation factor, which is set to 0.5 in this paper [17].

The larger the proportion of DAG shared by two diseases, the more semantically similar the two diseases are. Based on this hypothesis, the formula for calculating the similarity between disease d_i and d_j is shown as follows:

$$S_{sem}^d(d_i, d_j) = \frac{\sum_{v \in (N_{d_i} \cap N_{d_j})}(C_{d_i}(v) + C_{d_j}(v))}{\sum_{v \in N_{d_i}} C_{d_i}(v) + \sum_{v \in N_{d_j}} C_{d_j}(v)}. \quad (2)$$

2.5 Gaussian Interaction Profile Kernel (GIP) Similarity

The S_{sem}^d matrix is sparse, which impacts the quality of feature representation. Therefore, we fill the zero values in S_{sem}^d by calculating the GIP similarity. The

GIP similarity for diseases between disease d_i and disease d_j is calculated as follows:

$$S_{gip}^d(d_i, d_j) = exp(-\alpha_d ||IP(d_i) - IP(d_j)||^2), \tag{3}$$

where $IP(d_i)$ is the feature of d_i, which equals the i-th column in A. α_d controls the kernel bandwidth, which is calculated as:

$$\alpha_d = \alpha_d' / (\frac{1}{p}\sum_{i=1}^{p} ||IP(d_i)||^2), \tag{4}$$

where p indicates the number of diseases and α_d' is a bandwidth parameter.

Similarly, we can obtain the GIP similarity of miRNAs, denoted as S_{gip}^m.

2.6   MiRNA Sequence Similarity

To obtain sequence information of miRNAs, we utilize the data from miRBase [15] database. We calculate miRNA sequence similarity by using the Levenshtein distance [18]. For miRNAs m_i and m_j, the sequence similarity of them is calculated as follows:

$$S_{seq}(m_i, m_j) = 1 - \frac{distance(m_i, m_j)}{max(len(m_i), len(m_j))}, \tag{5}$$

where $distance(\cdot)$ denotes the Levenshtein distance between miRNA i and miRNA j, and $len(\cdot)$ represents the sequence length of miRNA.

2.7   MiRNA Functional Similarity

The functional similarity of miRNAs can be estimated by computing the similarity between the two gene sets corresponding to the two miRNAs. The target gene set of miRNAs is obtained from miRTarBase. We utilize Yu's method [19] to calculate miRNAs functional similarity, which infers gene sets similarity based on Best-Match Average (BMA). The calculation is defined as follows:

$$S_{fun}(m_i, m_j) = \frac{\sum_{1 \leq i \leq N_i} max(S_i') + \sum_{1 \leq j \leq N_j} max(S_j')}{N_i + N_j}, \tag{6}$$

where S_i' and S_j' are functional similarity of gene i and gene j in target gene sets, respectively; N_i and N_j are the target gene number of miRNA m_i and m_j, respectively.

2.8   MiRNA Family Similarity

The family information of miRNAs is also acquired from miRBase. When miRNA i and miRNA j belong to the same family, the corresponding position in the family similarity matrix S_{fam} is assigned as 1.

2.9   Integrated Similarity

To obtain disease integrated similarity matrix S_d, we combine the disease semantic similarity with the GIP similarity of diseases. The matrix S_d is defined as follows:

$$S_d(d_i, d_j) = \begin{cases} S_{gip}^d(d_i, d_j), & \text{if } S_{sem}^d(d_i, d_j) = 0 \\ S_{sem}^d(d_i, d_j), & \text{otherwise} \end{cases}, \tag{7}$$

where S_{sem}^d denotes the disease semantic similarity matrix and S_{gip}^d is the GIP similarity matrix of the diseases.

For miRNAs, we initially integrate miRNA sequence similarity, functional similarity and family similarity as S_{int} defined as below:

$$S_{int}(m_i, m_j) = \begin{cases} \lambda_1 S_{seq} + (1 - \lambda_1)S_{fun}, & \text{if } S_{fam}(m_i, m_j) = 0 \\ \lambda_2(\lambda_1 S_{seq} + (1 - \lambda_1)S_{fam}) + (1 - \lambda_2)S_{fun}, & \text{otherwise} \end{cases}, \tag{8}$$

where λ_1, λ_2 are parameters for balancing the similarity matrice, $S_{seq}, S_{fun}, S_{fam}$ are sequence similarity, functional similarity and family similarity, respectively. Subsequently, we integrate them and the GIP similarity of miRNAs using a disease-like approach. The final integrated similarity of miRNAs can be calculated as:

$$S_m(m_i, m_j) = \begin{cases} S_{gip}^m(m_i, m_j), & \text{if } S_{int}(m_i, m_j) = 0 \\ S_{int}(m_i, m_j), & \text{otherwise} \end{cases}, \tag{9}$$

where S_{gip}^m is GIP similarity of miRNAs.

2.10   Constructing Bipartite Attributed Graph

By utilizing the node similarities calculated in Subsect. 2.3 and the filtered miRNA-disease association information, we construct a miRNA-disease bipartite attributed graph that consists of 917 miRNA nodes and 792 disease nodes. The edges in the graph are defined with 14,550 validated miRNA-disease associations and the integrated miRNA similarity S_m and disease similarity S_d are regarded as their node features, respectively. To eliminate heterogeneity, we apply a node-specific transformation matrix W to project both miRNA nodes and disease nodes, which originally exist in different feature spaces, into a unified space. The specific process of miRNA nodes is shown as follows: $\tilde{X}_m = W_m \cdot S_m$. We obtain the transformation feature matrix of the disease in a similar way, which is defined as: $\tilde{X}_d = W_d \cdot S_d$.

2.11   Feature Propagation

In the message passing framework of GNN, nodes send messages to their neighbors while receiving messages from their neighbors. The propagated messages can be defined as a message matrix $M \in \mathbb{R}^{k \times c}$, where k is the edge number in the graph and c is the dimension number of the messages. The propagated message from node i to node j in the l-th layer can be expressed as:

$$M_i^{(l)} = AGG_{j \in N(i)}(h_i^{(l)}, h_j^{(l)}, e_{i,j}), \tag{10}$$

where $h_i^{(l)}$ denotes the hidden representation of node v_i in the l-th layer, and $N(i)$ is a set of nodes adjacent to node v_i; $e_{i,j}$ represents the edge between node i and j.

To avoid over-reliance on specific neighboring nodes during node feature propagation, we employ the DropMessage [20] strategy. This strategy involves randomly dropping some node messages on the propagated messages on the graph, which allows the same node to propagate different messages to its different neighbors. More specifically, DropMessage strategy conducts dropping on the message matrix with the dropping rate δ, which means that $\delta|M|$ elements of the message matrix will be masked in expectation. The process can be expressed as:

$$\tilde{M}_{i,j} = \frac{1}{1-\delta}\eta_{i,j}M_{i,j}, \tag{11}$$

where $\eta_{i,j} \sim Bernoulli(1\text{-}\delta)$ is an independent mask to determine whether element $M_{i,j}$ will be preserved or not. The node representations are updated based on node feature information and messages from neighbors, which can be formulated as: $h_i^{(l+1)} = h_i^{(l)} + \tilde{M}_i$.

We define the normalized adjacency matrix as in GCN, $L = \tilde{D}^{\frac{1}{2}}\tilde{A}\tilde{D}^{\frac{1}{2}}$, where $\tilde{A} = A + I$ and $\tilde{D}$ is the degree matrix of $\tilde{A}$. The l-th layer information of the nodes can be obtained as $H^k = LH^{k-1}W^k$, where $H^0 = \tilde{X}$.

Inspired by Simple Graph Convolution (SGC) [21], in order to emphasize the benefit arises from the local averaging, we remove the nonlinear transition functions between each layer. The final softmax is kept to obtain probabilistic outputs, and the l-th simple graph convolution can be defined as:

$$H^{(l)} = softmax(L\ldots LL\tilde{X}W^{(1)}\ldots W^{(l)}). \tag{12}$$

The repeated multiplication with the normalized adjacency matrix L can be collapsed in to a single matrix L^l. The $H^{(l)}$ can be simply expressed as $H^{(l)} = softmax(L^{(l)}\tilde{X}W)$, where $W = W^{(1)}W^{(2)}\ldots W^{(l)}$ is the cumulative multiplication of parameters of each layer.

2.12   Feature Aggregation

In neighbor-hood aggregation networks, each layer increases the size of the influence distribution by aggregating neighborhoods from the previous layer. We employ the Jumping Knowledge (JK) network [22] to aggregate the representations of intermediate layer to the last layer (named *jumping*), which is defined as:

$$\tilde{H} = \sum_{i=1}^{l}\omega_i H^{(i)}, \tag{13}$$

where ω_i is the weight coefficient of layer i when aggregating features. This selective combination of different aggregations can adaptively adjust the influence range of each node.

2.13   Feature Crossing and Association Prediction

To extract cross-feature information of each miRNA-disease pair, we employ the element-wise dot product operation. The formula is $H_{cross} = \tilde{H}_m \odot \tilde{H}_d$, where $\tilde{H}_m$ and $\tilde{H}_d$ are the final feature representations of miRNAs and diseases, respectively.

The final prediction is obtained by a fully connected neural network decoder:

$$\hat{Y} = sigmoid(f(\tilde{H}_m \oplus \tilde{H}_d \oplus H_{cross})), \qquad (14)$$

where $f(\cdot)$ indicates the fully connected layer. Subsequently, we utilize the cross-entropy loss function to compute the loss between predicted value and the true label:

$$\mathcal{L}(A, \hat{Y}) = - \sum_{i,j \in (Y^+ \cup Y^-)} (A_{i,j} log(\hat{Y}_{i,j}) + (1 - A_{i,j}) log(1 - \hat{Y}_{i,j})), \qquad (15)$$

where $A_{i,j}$ denotes the true association between miRNA i and disease j, while $\hat{Y}_{i,j}$ denotes the result predicted by our model.

3   Experiments and Results

3.1   Experiment Settings

Our model is implemented in the DGL framework of PyTorch and Adam algorithm is applied to optimize the model. In addition, we determine the optimal combination of parameters through a grid search. After the search, we set the learning rate to 0.001, dropout rate to 0.5, and layer number l to 4 to obtain the best results.

We take 14,550 known associations as positive samples. In order to ensure the balance of samples, we randomly select the same number miRNA-disease pairs from unknown associations as negative samples. The 5-fold cross-validation method is used to evaluate the model's performance. In each fold, we use 20% of the samples as the test set and the rest as the training set. To quantify the average performance of the model, we employ AUC and AUPR as the primary evaluation metrics. In addition, we also use four evaluation metrics accuracy (Acc), precision (Pre), recall and F1-score (F1) to comprehensively evaluate the model performance.

3.2   Comparison with Baseline Methods

To demonstrate the superior performance of nSGC-MDA, we compare it with six compared models that also utilize the HMDD v3.2 dataset. TDRC [23] is a matrix decomposition-based method. MDA-CF [24] is a method which based on traditional machine learning. AGAEMD [12], AMHMDA [13], GRPAMDA [25] and VGAMF [14] are all GNN-based methods. In addition, VGAMF uses the linear features extracted by matrix decomposition to enhance the feature representation of miRNA-disease pairs. Both TDRC and AGAEMD use only two types of miRNA similarities and other models use multiple similarity information.

To ensure the fairness of comparison, the experimental results of these models are conducted on condition of the optimal parameters provided in their original articles. The MDA-CF does not provide a code implementation, so we directly utilize the results from its paper. The comparison results are shown in Table 1, where the best results are in bold and the second are underline. From Table 1 we can see the followings: (1) All the other models are better than TDRC and AGAEMD, which indicates that fusing more similarity information can improve the performance of the model. (2) VGAMF achieved the second highest AUC and AUPR, suggesting that additional features can improve the quality of miRNA-disease pair feature representations. (3) The performance of nSGC-MDA is superior to other GNN-based methods (AGAEMD, AMHMDA, GRPAMDA and VGAMF). The reason is that nSGC-MDA uses the modified SGC, which effectively mitigates the effect of noise in the graphs. Moreover, the utilization of jumping knowledge mechanism enhances the representation of nodes.

Table 1. Comparison results with baseline methods.

	AUC	AUPR	Acc	Pre	Recall	F1
TDRC	0.9109	0.9246	0.8544	0.8522	0.8578	0.8549
AGAEMD	0.9205	0.9283	0.8617	0.8695	0.8513	0.8603
AMHMDA	0.9401	0.9364	0.8626	0.8549	0.8733	0.8651
GRPAMDA	0.9443	0.9404	0.8719	0.8644	**0.8749**	0.8706
VGAMF	<u>0.9470</u>	<u>0.9409</u>	0.8616	0.8729	0.8669	0.8714
MDA-CF	0.9464	0.9401	<u>0.8766</u>	<u>0.8818</u>	0.8698	<u>0.8757</u>
nSGC-MDA	**0.9502**	**0.9496**	**0.8812**	**0.8863**	<u>0.8747</u>	**0.8804**

3.3   Performance Comparison of Different Components

To verify the effectiveness of each componets in nSGC-MDA, we also conduct ablation experiments. Firstly, we test the contribution of the Jumping Knowledge (JK). Furthermore, we compare the impact of different message dropping methods, including Dropout, DropEdge, DropNode, and without dropout (WOD). Dropout performs random dropping operation on the node feature matrix. DropEdge and DropNode discard edges and nodes, respectively. WOD means that there are no discard operations. The comparison graphs of AUPR and AUC are shown in Fig. 2.

As can be seen from the results, the performance of the model with JK is significantly higher than without JK, with an AUC of 95.02%. Random dropping

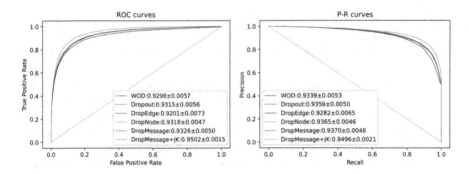

Fig. 2. Comparison results of different dropping methods

method consistently outperforms SGC without random dropping. Besides, the model prediction performance vary over different dropping methods. DropMessage achieves overall better results on all metrics and Dropout achieves model performance second only to DropMessage. These indicate that the perturbation strategy can enhance the feature representation when aggregating node feature information, while DropMessage retains the integrity of the features when discarding information. DropEdge and DropNode causes a lot of damage to the structure information of the original graph due to the high discarding rate.

3.4   Case Study

To validate the utility of nSGC-MDA in practical application, we conduct case study and choose cardiovascular disease as the subject of our case study. We divide the edges between miRNAs and diseases into two sets: the test set and the training set. The test set comprises all the edges related to cardiovascular disease, while the training set consists of the remaining edges. The top 10 predicted miRNAs with the highest association scores are considered as the model's predictions. We conduct verification by published literatures in PubMed. The experimental results are presented in Table 2.

Table 2. The results of the case study.

num	miRNA	evidence (PMID)	num	miRNA	evidence (PMID)
1	hsa-mir-21	19043405, 28944900	6	hsa-mir-126	28065883
2	hsa-mir-155	24475727, 34506226	7	hsa-mir-145	28379027
3	hsa-mir-146a	25865299, 33297927	8	hsa-mir-17	24212931, 30338905
4	hsa-mir-223	24573468, 29845432	9	hsa-mir-150	30260095, 35071356
5	hsa-mir-34a	20627091	10	hsa-mir-221	28379027, 35668131

MicroRNA 21 (hsa-mir-2) promotes the development of cardiac fibrosis, hypertrophy and heart failure. Zhang et al. [26] showed that in human failing heart tissues, microRNA 21 levels were elevated. They also demonstrated that circulating microRNA 21 has potential to be a biomarker of heart failure by statistically analyzing relevant patients. This shows that nSGC-MDA has potential and application value in predicting the miRNA-disease association.

4 Conclusion and Future Work

Identifying the associations between miRANs and diseases is crucial for understanding the underlying pathogenic mechanism of diseases. In this study, we propose an novel MDAs prediction method based on Simple Graph Convolution with feature perturbation strategy. Random dropping methods can mitigate over-fitting in GNNs by randomly masking part of the input. During the message passing, the propagated messages are directly discarded to preserve the graph structure and provide comprehensive feature representations for the prediction task. Feature aggregation is achieved by adopting the Jumping Knowledge Network, which aggregates features from each layer based on adaptive weights to obtain a final, more informative feature representation compared to higher-order features. The case study shows the reliability of our model.

However, bipartite graphs have limitations in terms of the information they can provide, and leveraging more comprehensive networks can enhance the accuracy and effectiveness of predictions. In future MDA prediction tasks, it would be beneficial to utilize more complex heterogeneous networks that provide a better representation of the biological associations between miRNAs and diseases.

Acknowledgements. This work was supported by the National Natural Science Foundation of China (Nos. 12061071, 61962050), the Key R&D Program of Xinjiang Uygur Autonomous Region (No. 2022B03023), the Natural Science Foundation of Xinjiang Uygur Autonomous Region (Nos. 2022D01C427, 2022D01C429 and 2020D01C028), the Natural Science Foundation of Hunan Province of China (No. 2022JJ30428). This work was carried out in part using computing resources at the High Performance Computing Center of Central South University.

References

1. Wronska, A.: The role of microRNA in the development, diagnosis, and treatment of cardiovascular disease: recent developments. J. Pharmacol. Exp. Ther. **384**(1), 123–132 (2023)
2. Baskerville, S., Bartel, D.P.: Microarray profiling of microRNAs reveals frequent coexpression with neighboring miRNAs and host genes. RNA **11**(3), 241–247 (2005)
3. Pall, G.S., Hamilton, A.J.: Improved northern blot method for enhanced detection of small RNA. Nat. Protocols **3**(6), 1077–1084 (2008)
4. Huang, Z., Shi, J., Gao, Y., et al.: HMDD v3. 0: a database for experimentally supported human microRNA-disease associations. Nucleic Acids Res. **47**(D1), D1013–D1017 (2019)

5. Huang, H.-Y., Lin, Y.-C.-D., Cui, S., et al.: mirtarbase update 2022: an informative resource for experimentally validated miRNA-target interactions. Nucleic Acids Res. **50**(D1), D222–D230 (2022)
6. Jiang, Q., Hao, Y., Wang, G., et al.: Prioritization of disease microRNAs through a human phenome-microRNAome network. BMC Syst. Biol. **4**(1), 1–9 (2010)
7. Chen, X., Yan, C.C., Zhang, X., et al.: WBSMDA: within and between score for miRNA-disease association prediction. Sci. Rep. **6**(1), 1–9 (2016)
8. Li, J.-Q., Rong, Z.-H., Chen, X., Yan, G.-Y., You, Z.-H.: MCMDA: matrix completion for miRNA-disease association prediction. Oncotarget **8**(13), 21187 (2017)
9. Chen, X., Wang, L., Jia, Q., et al.: Predicting miRNA-disease association based on inductive matrix completion. Bioinformatics **34**(24), 4256–4265 (2018)
10. Xuan, P., Sun, H., Wang, X., et al.: Inferring the disease-associated miRNAs based on network representation learning and convolutional neural networks. Int. J. Mol. Sci. **20**(15), 3648 (2019)
11. Ji, C., Gao, Z., Ma, X., et al.: AEMDA: inferring miRNA-disease associations based on deep autoencoder. Bioinformatics **37**(1), 66–72 (2021)
12. Zhang, H., Fang, J., Sun, Y., et al.: Predicting miRNA-disease associations via node-level attention graph auto-encoder. IEEE ACM Trans. Comput. Biol. Bioinform. **20**(2), 1308–1318 (2023)
13. Ning, Q., Zhao, Y., Gao, J., et al.: AMHMDA: attention aware multi-view similarity networks and hypergraph learning for miRNA-disease associations identification. Brief. Bioinform. **24**(2) (2023)
14. Ding, Y., Lei, X., Liao, B., et al.: Predicting miRNA-disease associations based on multi-view variational graph auto-encoder with matrix factorization. IEEE J. Biomed. Health Inform. **26**(1), 446–457 (2022)
15. Kozomara, A., Birgaoanu, M., Griffiths-Jones, S.: miRBASE: from microRNA sequences to function. Nucleic Acids Res. **47**(D1), D155–D162 (2019)
16. Lipscomb, C.E.: Medical subject headings (mesh). Bull. Med. Libr. Assoc. **88**(3), 265 (2000)
17. Wang, D., Wang, J., Ming, L., et al.: Inferring the human microRNA functional similarity and functional network based on microRNA-associated diseases. Bioinformatics **26**(13), 1644–1650 (2010)
18. Levenshtein, V.I., et al.: Binary codes capable of correcting deletions, insertions, and reversals. In: Soviet Physics Doklady, vol. 10, pp. 707–710. Soviet Union (1966)
19. Guangchuang, Yu., et al.: A new method for measuring functional similarity of microRNAs. J. Integr. Omics **1**(1), 49–54 (2011)
20. Fang, T., Xiao, Z., Wang, C., Jiarong, X., Yang, X., Yang, Y.: DropmesSage: unifying random dropping for graph neural networks. In: Proceedings of the AAAI Conference on Artificial Intelligence, vol. 37, pp. 4267–4275 (2023)
21. Wu, F., Souza, A., Zhang, T., Fifty, C., Yu, T., Weinberger, K.: Simplifying graph convolutional networks. In: International Conference on Machine Learning, pp. 6861–6871. PMLR (2019)
22. Xu, K., Li, C., Tian, Y., Sonobe, T., Kawarabayashi, K., Jegelka, S.: Representation learning on graphs with jumping knowledge networks. In: International Conference on Machine Learning, pp. 5453–5462. PMLR (2018)
23. Huang, F., Yue, X., Xiong, Z., et al.: Tensor decomposition with relational constraints for predicting multiple types of microRNA-disease associations. Brief. Bioinform. **22**(3) (2021)
24. Dai, Q., Chu, Y., Li, Z., et al.: MDA-CF: predicting miRNA-disease associations based on a cascade forest model by fusing multi-source information. Comput. Biol. Med. **136**, 104706 (2021)

25. Zhong, T., Li, Z., You, Z.-H., et al.: Predicting miRNA-disease associations based on graph random propagation network and attention network. Brief. Bioinform. **23**(2), bbab589 (2022)
26. Zhang, J., et al.: Circulating miRNA-21 is a promising biomarker for heart failure. Mol. Med. Rep. **16**(5), 7766–7774 (2017)

Reconciling Inconsistent Molecular Structures from Biochemical Databases

Casper Asbjørn Eriksen$^{(\boxtimes)}$, Jakob Lykke Andersen, Rolf Fagerberg,
and Daniel Merkle

Department of Mathematics and Computer Science, University of Southern Denmark,
Odense, Denmark
{casbjorn,jlandersen,rolf,daniel}@imada.sdu.dk

Abstract. Information on the structure of molecules, retrieved via biochemical databases, plays a pivotal role in various disciplines, such
as metabolomics, systems biology, and drug discovery. However, no
such database can be complete, and the chemical structure for a given
compound is not necessarily consistent between databases. This paper
presents STRUCTRECON, a novel tool for resolving unique and correct
molecular structures from database identifiers. STRUCTRECON traverses
the cross-links between database entries in different databases to construct what we call an identifier graph, which offers a more complete
view of the total information available on a particular compound across
all the databases. In order to reconcile discrepancies between databases,
we first present an extensible model for chemical structure which supports multiple independent levels of detail, allowing standardisation of
the structure to be applied iteratively. In some cases, our standardisation
approach results in multiple structures for a given compound, in which
case a random walk-based algorithm is used to select the most likely
structure among incompatible alternates. We applied STRUCTRECON to
the *EColiCore2* model, resolving a unique chemical structure for 85.11%
of identifiers. STRUCTRECON is open-source and modular, which enables
the potential support for more databases in the future.

Keywords: Standardisation · Chemical structure identifiers ·
Small-molecule databases · Cheminformatics

1 Introduction

As the volume of available biochemical information grows, databases have
become indispensable resources for researchers, enabling advances in various
fields, including metabolomics, systems biology, and drug discovery. These
databases are curated and maintained by different organisations and research
groups, each employing their own data collection methods, annotation standards,
and quality control procedures. However, as the collective amount of information
stored in the databases expands, so does the amount of errors within databases,
and in particular, inconsistencies between them [1,24]. Discrepancies between
biochemical databases pose a significant challenge to researchers performing

X. Guo et al. (Eds.): ISBRA 2023, LNBI 14248, pp. 58–71, 2023.
https://doi.org/10.1007/978-981-99-7074-2_5

large-scale analyses, in particular when integrating data from multiple databases [25]. In this work, we focus on incompleteness and inconsistencies in the chemical structures within and between database entries, which can pose a significant problem in applications such as drug discovery, quantitative structure-activity relationship, and atom tracing [6,26].

The entries in each database may contain quantitative information about the compounds, structural information on these compounds, as well as references to related entries in other databases. A starting observation behind our contribution is that the cross-database references can be traversed in order to get a fuller view of the properties of a compound. In the best case, the databases complement each other, making up for the incompleteness of each and allowing the identification a chemical structure of each compound of interest, even if not all of these are contained in any single database.

However, integrating entries from several databases will invariably introduce discrepancies in the chemical structure. In many cases, these discrepancies are simply caused by a difference in the representation of what is intended to be identical chemical structures [15]. Other times, different structural isomers are present under the same name, for example 5-deoxy-D-ribose, which appears in cyclic and linear forms, depending on the database, as depicted in Fig. 1. Over the years, considerable efforts have been directed towards the development of standards and guidelines for chemical structure representation. Structural identifiers such as Standard InChI [10] and Standard SMILES [21–23] aim to provide an unambiguous and standardised structural identifier. However, ambiguity is not completely prevented, as sources may wish to denote chemical structures in varying levels of detail, e.g., whether to denote stereochemistry, tautomerism, charge, and more. SMILES makes no distinction whether such features are explicitly represented, while the layered structure of InChI [20] makes it clear which features are to be explicitly represented in some cases.

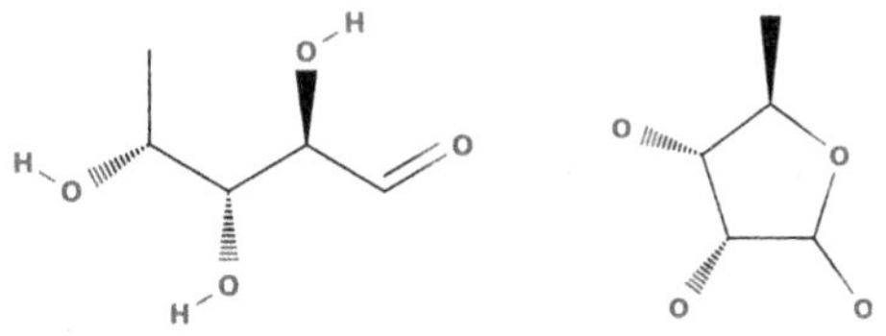

Fig. 1. An example of structural discrepancy between databases: the linear (PubChem, ECMDB) and cyclic (ChEBI, MetaNetX, MetaCyc, KEGG) form of 5-deoxy-D-ribose.

The problem of comparing chemical structures from entries with different notation and level of standardisation is one of the main challenges of this work. It is evident that for biochemical problems, an automatic method for retrieving correct chemical structures, up to some degree of standardisation, is needed. Our goal is to use the combined resources from several databases in order to create a more complete mapping between database identifiers and chemical structures than any single database provides, while automatically handling discrepancies between these identifiers.

To our knowledge, there are no other tools which give such a consolidated view of the structural information on compounds in databases. While some databases, such as ChEBI and PubChem, present data collected from other databases, this is still susceptible to issues such as incompleteness and incorrectness of individual entries. Furthermore, given a type of identifier, e.g. BiGG, this may not be present in a given database. For this reason, it is desirable to develop a flexible and extensible system which can in principle be made compatible with any database, while taking into account potential discrepancies.

In Sect. 2, we establish a *model* for representing the chemical structure of compounds, with the goal of being able to describe, and compare across, the various levels of detail to which chemical structures are given by databases. Next, in Sect. 3, we present STRUCTRECON, a *tool* for programmatically retrieving chemical structures from database identifiers, by traversing database cross-references and using cheminformatics methods for analysing, comparing, and standardising structural representations based on the model developed in Sect. 2. Finally, in Sect. 4, we apply the tool to a set of compounds established by genome sequencing of *E. coli* and analyse the resulting network of identifiers and structural representations.

2   Multi-level Modelling of Chemical Structures

In this section, we introduce a model for chemical structures which allows representation at multiple levels of detail and formalises the standardisation functions which transform structures between these levels. Compared to established models, such as SMILES and InChI, this model places a particular focus on extensibility, formal specification, and standardisation of structures.

We call the levels of detail *features*. The seven features used throughout this work will be introduced one-by one as the necessary theory is established. We categorise identifiers into two classes: *structural identifiers*, which directly encode a chemical structure, and from which the structure can be recovered algorithmically (e.g. InChI, SMILES), and *symbolic identifiers*, which are generated more or less arbitrarily, and do not carry direct meaning, but reference an entry in the corresponding database (e.g. PubChem CID, BiGG ID).

Depending on the application, the exact definition of a chemical structure may vary. We wish to model the connectivity of atoms in a molecule, and optionally stereo-chemical information, while the full spatial information is not taken into account. For this application, the classical method in cheminformatics is the graph-based approach: a structure is, in its most basic form, an unlabelled graph $G = (V, E)$, in which atoms are represented by vertices, and (covalent) bonds are represented by edges. A graph-based structure is linearly encoded by established structural identifiers, such as SMILES and InChI; even the systematic IUPAC name may be described as a graph-based identifier [11]. We may wish to describe additional features, e.g., charge, isotopic labelling, stereochemistry, or tautomerism. In broad terms, a chemical feature is a class of information about a structure which we may or may not wish to account for when modelling, depending on the input and application.

We will establish a unified model for graph-based chemical structures. This provides a consistent view of chemical structure, regardless of the features represented, to which structural identifiers can be mapped. The underlying simple graph structure is, for our purposes, always assumed to be present, but even basic information such as the chemical element of each atom and the order of covalent bonds, are considered optional features. First, we formally define the notion of a feature.

Definition 1 (Feature). *A feature $\mathbf{\Phi}$ is a pair $\mathbf{\Phi} = (\mathbf{\Phi}_V, \mathbf{\Phi}_E)$, where $\mathbf{\Phi}_V$ and $\mathbf{\Phi}_E$ are sets of possible values for the attribute on atoms and bonds respectively, each of which must contain a special 'nil'-element, ϵ, indicating that the value is unspecified or not applicable.*

We will define and apply seven such features in this work. Starting with the most essential, the element of each atom can be expressed as a feature $\mathbf{E}$, with

$$\mathbf{E} := (\{\epsilon, H, He, Li, \dots\}, \{\epsilon\}).$$

For a given structure, each vertex will either be assigned ϵ, indicating that no indication is given as to the element of that atom, or it will be assigned a specific element. Edges can only be assigned ϵ, indicating that this feature assigns no attribute to edges. Similarly, bond types are expressed as the feature

$$\mathbf{B} := (\{\epsilon\}, \{\epsilon, -, =, \equiv\}).$$

This feature can be expanded to also indicate other bond types, such as aromatic or ionic, if needed. The isotope of each atom can be stated as

$$\mathbf{I} := (\{\epsilon\} \cup \mathbb{N}, \{\epsilon\}),$$

where the vertex attribute indicates the atomic weight of each atom. Before describing the remaining features, we need a precise definition of chemical structure. Combining the sets of values for all features, we can define the overall feature space, which is needed for a formal definition of a chemical structure.

Definition 2 (Feature space). *Given a set of features $\{\mathbf{\Phi}_1, \dots, \mathbf{\Phi}_n\}$, the feature space for vertices and edges, $\mathcal{F}_V$ and $\mathcal{F}_e$, respectively, is the combined attribute space of the features, where $\mathcal{F}_V = \mathbf{\Phi}_{1V} \times \cdots \times \mathbf{\Phi}_{nV}$ and $\mathcal{F}_E = \mathbf{\Phi}_{1E} \times \cdots \times \mathbf{\Phi}_{nE}$.*

Definition 3 (Chemical structure). *A chemical structure is an undirected graph $G = (V, E, A_V, A_E)$, where V is the set of atoms, E is the set of covalent bonds, and $A_V : V \to \mathcal{F}_V$ (resp. $A_E : E \to \mathcal{F}_E$) is an attribute function, assigning to each vertex (resp. edge) a value for each feature in the feature space.*

Let $\mathcal{G}$ be the set of all such chemical structures. Simultaneously working in several levels of detail, i.e., features, naturally raises the problem of how to compare equivalence of structures between different sets of features. For this, we introduce, for each feature, a *standardisation function*.

Definition 4 (Standardisation function). *Given a feature* Φ*, the standardisation function w.r.t.* Φ *is* $S_\Phi : \mathcal{G} \to \mathcal{G}$*. The function is required to be idempotent, and all vertices and edges of the resulting structure should have* ϵ *as value for feature* Φ *in the attribute function.*

We extend the definition to sequences of features, resulting in a composition of standardisations, i.e., $S_{\Phi_1...\Phi_n} = S_{\Phi_n} \circ \cdots \circ S_{\Phi_1}$.

For a structure G, the image $S_\Phi(G)$ is the corresponding standardised structure, which does not contain any information about feature Φ. For a sequence of features $\Phi_1, \ldots, \Phi_n$, we define $\mathcal{G}_{\Phi_1...\Phi_n} \subseteq \mathcal{G}$ as the set of structures which are standardised according to the of features. That is, all applicable atom and bond attributes are ϵ these features, and they are their own image in the standardisation function:

$$G \in \mathcal{G}_{\Phi_1...\Phi_n} \iff S_{\Phi_1...\Phi_n}(G) = G.$$

For the features described so far, **E**, **B**, and **I**, the *trivial standardisation function* is sufficient. This function simply erases the attributes by setting them to ϵ. In some cases, the trivial standardisation function is not sufficient. For example, we define the charge feature as

$$\mathbf{C} := (\{\epsilon\} \cup \mathbb{Z}, \{\epsilon\}),$$

where the vertex attribute indicates the charge of the atom. The process of standardising the charge of a given molecule may be limited by chemical constraints, such as is the case in RDKit [17] and InChI [20]. The function may, among other modifications, add and/or remove hydrogen atoms, trying to remove charges from each atom in a chemically valid way. We will not further discuss the intricacies of this operation, as it is implementation-dependant.

For some features, the definition even depends entirely on the standardisation function. E.g., the fragment feature [15]:

$$\mathbf{F} := (\{\epsilon\}, \{\epsilon\}).$$

In this case, no information is explicitly encoded in the graph, but we still want a way to standardise w.r.t. fragments. We define $S_\mathbf{F}(G)$ to be the largest connected component of G (measured by the number of non-hydrogen atoms), with an implementation-specific method for breaking ties based on other attributes of each connected component.

Tautomerism is similarly difficult to define due to the complexities involved in determining the tautomers of any given structure. Let

$$\mathbf{T} := (\{\epsilon\}, \{\epsilon\}).$$

Again, we have no features, instead relying on the standardisation function: assume a chemical oracle, which given a structure G, returns the set of all tautomeric structures, according to some definition. Then, let a deterministic method choose a canonical representative from among these structures. The standardised structure $S_\mathbf{T}(G)$ is this canonical tautomer.

For stereochemistry, there are multiple methods for encoding information about local geometry at the vertex and edge level [2, 3, 16] Any method for encoding such information can be used to generate the feature S. The trivial standardisation function will in most cases be sufficient for standardising structures with respect to stereochemistry.

Finally, we define equivalence of chemical structures: With the notation $G_1 =_{\Phi_1 \ldots \Phi_n} G_2$, we denote that the structures G_1 and G_2 are equivalent up to standardisation of $\Phi_1, \ldots, \Phi_n$, i.e., they are equal when the features $\Phi_1, \ldots, \Phi_n$ are not considered. Formally

$$G_1 =_{\Phi_1 \ldots \Phi_n} G_2 \iff S_{\Phi_1, \ldots, \Phi_n}(G_1) = S_{\Phi_1, \ldots, \Phi_n}(G_2).$$

As an example, consider the structures for methanol and methoxide, where $CH_3OH \neq_{\mathbf{FICTS}} CH_3O^-$, but $CH_3OH =_{\mathbf{FITS}} CH_3O^-$.

We have now described the features which are considered in this contribution, based on the FICTS features [19]: ($\mathbf{E}$) elements, ($\mathbf{B}$) bonds, ($\mathbf{F}$) fragments, ($\mathbf{I}$) isotopes, ($\mathbf{C}$) charge, ($\mathbf{T}$) tautomerism, and ($\mathbf{S}$) stereoisomerism. It should be noted that some of these features depend upon each other, e.g., it would not make sense to specify the isotope of an atom without also specifying the element. The standardisation functions can not be expected commute with each other in general. For this reason, STRUCTRECON needs a defined order in which the standardisation functions will be applied.

3 Algorithms and Implementation

In this section, we describe the ideas and algorithms of STRUCTRECON, based on the model developed in Sect. 2.

Data Sources. We used six sources of data: BiGG [14], ChEBI [5], the E. Coli Metabolome Database (ECMDB) [8, 18], KEGG [12], MetaNetX [7], and PubChem [13]. These were selected based on their programmatic accessibility and relevance to metabolic modelling. STRUCTRECON is modular, making it easy to add more data sources in the future.

MetaNetX uses a versioning system in which entries present in one version are not necessarily present in another. Inter-database references do not typically specify which version of MetaNetX is referenced, but MetaNetX keeps a record of deprecated entries. Following the chain of deprecations, it is possible to obtain the newest entries corresponding to any given entry. The deprecation relationship can both split and merge entries in between versions.

Construction of the Identifier Graph. The interconnected nature of identifiers within and between databases is represented as a directed graph, called the *identifier graph*. In the identifier graph, vertices correspond to identifiers, while arcs represent relationships between these. Each vertex contains as attributes, the type of ID (e.g. PubChem CID, BiGG ID, SMILES) called the *identifier class*, as well as the actual ID. Each edge is annotated with the source database.

A number of *procedures* are specified, each being a subroutine which takes identifiers as input, and finds associated identifiers in chemical databases. By executing the procedures in an order which seeks to minimise overhead, the identifier graph is built iteratively, starting with the *input vertices* which are directly obtained as input. The resulting graph will contain symbolic identifiers as well as the structural identifiers as they appear in the respective databases. For an example of a complete identifier graph, refer to Fig. 2.

Structure Standardisation. When a structural identifier (SMILES, InChI) is added to the identifier graph, it is first converted to an internal graph-based representation, in accordance with Definition 3. From this point, we assume that the atom (**A**) and bond (**B**) features are always implicitly represented, and will therefore refer to the remaining features as **FICTS**. As the standardisation functions are not expected to commute, in general, we enforce a particular order on the features defined by the user, by default $\mathbf{F}, \mathbf{I}, \mathbf{C}, \mathbf{T}, \mathbf{S}$. This was chosen as the default ordering, as it produces the greatest number of uniquely resolved structures which a lesser degree of standardisation.

We assign to each structural identifier G, an attribute specifying for which features a structure is standardised. For features $\mathbf{F}$, $\mathbf{I}$, and $\mathbf{S}$, it is simple to guarantee that it is standardised by inspecting the structure. Checking $\mathbf{F}$ is simply examining the connectivity of the graph, and $\mathbf{I}, \mathbf{S}$, inspecting for all vertices and edges whether they have the equivalent of ϵ as attribute. For $\mathbf{C}$ and $\mathbf{T}$, we check whether a structure is standardised by applying the respective standardisation functions.

When the links to new databases is exhausted, in many cases, there will be multiple different structural identifiers associated with each compound. We aim to achieve a unified representation of the compounds by iteratively applying the standardisation functions: For any structure G, which is not fully standardised, that is, $G \notin \mathcal{G}_{\mathbf{FICTS}}$, let $\mathbf{\Phi_k}$ be the first feature in which G is not standardised according to the feature ordering. That is, $S_{\mathbf{\Phi_1}...\mathbf{\Phi_{k-1}}}(G) = G$, but $S_{\mathbf{\Phi_1}...\mathbf{\Phi_k}}(G) \neq G$. Then produce $G' = S_{\mathbf{\Phi_k}}(G)$, adding this new structure to the identifier graph, with an arc (G, G'). If G' is equivalent to an existing structure H, then no new vertices are created, but the arc (G, H) is added instead. The standardisation process is visualised in the output of STRUCTRECON (Fig. 2) in which the blue nodes represent structures, and purple nodes represent maximally standardised structures.

Structure Selection. While the identifier graph is a general digraph, considering only the structural identifiers yields a forest of in-trees, as standardisation functions are many-to-one, and may therefore merge structures, but never split. For each input, the transitive and reflexive closure of the forest of in-trees imposes a partial order on the reachable structures. In this partial order, structure G_1 precedes G_2 if G_2 is a more standardised identifier, reachable from G_1

by applying standardisation functions. Maximal elements are fully standardised structures, $G \in \mathcal{G}_{\textbf{FICTS}}$.

For each input identifier i, STRUCTRECON should resolve the input to a single structure. The vertices reachable from i represent symbolic and structural identifiers which are related to i through database links, as well as structural identifiers which can be derived from these by standardisation. Denote by $\mathcal{S}(i)$ the partial order of structures reachable from i.

If $\mathcal{S}(i)$ has one maximal (greatest) element, then we say that $\mathcal{S}(i)$ is resolved and *consistent*—all sources for the compound can agree on a structure, at least up to the highest degree of standardisation. That is, $G =_{\textbf{FICTS}} G'$ for any pair of structures G, G' in the reachable database entries. If there are multiple maximal elements, then the sources cannot agree, and we call $\mathcal{S}(i)$ *inconsistent*. If $\mathcal{S}(i)$ contains no structures at all, then the compound is *unresolved*. In the case where $\mathcal{S}(i)$ is consistent, we want to select the most specific element on which all sources agree. In the partial order, this is the supremum of the set of structures which were found directly in the databases.

If $\mathcal{S}(i)$ is inconsistent, resolving i requires a choice between the maximal structures. This choice can be made automatically by a scoring algorithm which, for each input, assigns a confidence score to all vertices reachable from the input vertex. The scoring algorithm essentially computes the probability that a random walk in the identifier graph, starting at the input vertex, will arrive at each vertex. The algorithm is based on PageRank [4], with the key differences that the initial probability distribution is 1 for the input vertex, and 0 for all other vertices, and that sink vertices only loop back to the input vertex, rather than all vertices. After assigning a confidence score to each vertex, the confidence scores of each maximally standardised structure is evaluated. The *confidence ratio* is computed, the confidence of the second-most likely structure over the confidence of the most likely structure. If this value is below a given threshold (0.5 by default), then the vertex with the highest confidence score is automatically selected. Otherwise, no structure can be chosen with high enough confidence, and the input is marked for manual disambiguation.

Implementation. STRUCTRECON was implemented in Python, and is available at https://github.com/casbjorn/structrecon. The accompanying web interface can be accessed at https://cheminf.imada.sdu.dk/structrecon. Our model for chemical structure is implemented using RDKit [17], which furthermore provides functions for uncharging structures and computing the canonical tautomer.

4   Results

The tool was tested on the metabolic network model *EColiCore2* [9]. The model contains 2138 identified compounds, with associated BiGG IDs We chose this dataset for evaluation, as the selected databases have a particular focus on biochemistry, and because of the well-established nature of the *E. coli* genome.

Of the 2138 inputs, 136 (6.36%) were identified as macromolecules, based on string-matching BiGG IDs and names found in databases to an incomplete list of substrings associated with macromolecules, such as "tRNA" and the names of various proteins and enzymes. We consider the handling of macromolecules to be out of the scope of this work. In the identifier graph, an average of 31.70 vertices are reachable from each input vertex. Of the non-macromolecule inputs, 1459 (72.88%) resolved to exactly one structure up to maximal **FICTS** standardisation, while only 492 (24.58%) had only one structure up to **FICT** standardisation.

Of the non-macromolecule inputs, 57 (2.85%) yielded no structure at all. Examples of this category includes bis-molybdopteringuaninedinucleotide (BiGG: M_bmocogdp), Hexadecanoyl-phosphate(n-C16:1) (BiGG: M_hdceap), and 2-tetradec-7-enoyl-sn-glycerol-3-phosphate (BiGG: M_2tdec7eg3p).

A total of 486 inputs (24.28%) yielded multiple maximally standardised structures, and needed to be disambiguated based on the *confidence ratio*. In our experimentation, we found 0.5 to be a reasonable threshold, meaning that we select the structure with the highest confidence if it has at least twice the confidence of any other structure. With a threshold of 0.5, an additional 245 inputs were uniquely resolved, for a total of 1704 consistent inputs (85.11%), leaving 241 compounds for manual disambiguation. The effect of different choices of confidence ratio threshold is displayed in Fig. 3.

We will proceed to describe some concrete examples of identifier graphs which serve to demonstrate both the problem of database inconsistency and the solution provided by STRUCTRECON. One example is 5-methylthio-D-ribose. The associated identifier graph is displayed in Fig. 2. Database interconnections do not necessarily make distinctions between this compound and S-methyl-5-thio-D-ribose. The confidence ratio between these two maximised structures is 0.59, indicating a relatively high degree of interconnection between the associated database entries. While the correct structure has the highest confidence score, the default threshold of 0.5 would mark this discrepancy for manual disambiguation.

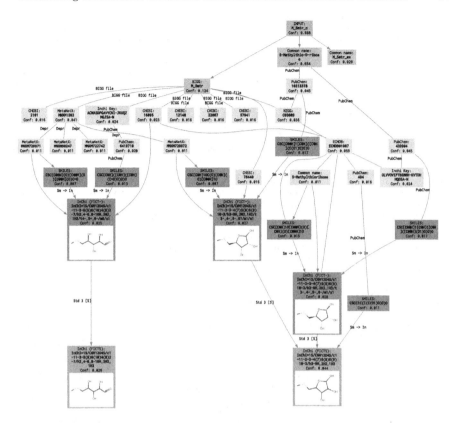

Fig. 2. The identifier graph generated by the BiGG ID M_5mtr_c. Each vertex displays the type of identifier, the identifier itself, and the confidence assigned by the scoring algorithm. For structures, the set of features in which the structure is standardised is also displayed, along with a graphical representation. The green vertex is the input vertex. The turquoise vertices are symbolic identifiers found directly within the input file, in this case the *EColiCore2* model. The light blue vertices are other symbolic identifiers. The dark blue vertices represent structural identifiers, either found in databases, or obtained by standardisation. The violet vertices represent maximally standardised structures. Arcs with no direction are shorthand for one arc in each direction. (Color figure online)

Unexpectedly, the simplest and most prevalent molecules turns out to be inconsistent, but easy to reconcile based on the confidence ratio. A good example is water, as displayed in Appendix A. The conventional structure H20 is found in a multitude of databases, however, the ChEBI identifier 29356 (oxide(2-)) is associated with the generic BiGG identifier for water through the BiGG database. However, as this is the only connection, that structure is assigned a smaller confidence score than the conventional structure by the scoring algorithm. This graph therefore has a low confidence ratio of 0.07, representing a high degree of support for the conventional structure, which is chosen by STRUCTRECON.

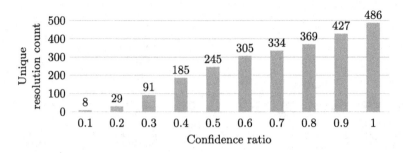

Fig. 3. For several choices of confidence ratio threshold (the confidence of the second-most likely structure over the confidence of the most likely structure), shows the number of inputs, out of 486, which resolve to a unique structure in the ECC2 model. STRUC-TRECON uses a default threshold of 0.5. Setting the threshold to 0.0 would mean only choosing a structure if no alternatives exist, while a threshold of 1.0 results in picking one of the structures with the highest confidence arbitrarily.

5   Conclusion

In this work, we propose a model for chemical structure, which supports multiple levels of standardisation. Based on this model, we present STRUCTRECON, a novel tool which identifies and reconciles the chemical structure of compounds based on the traversal of interconnections between biochemical databases. We applied the tool to *EColiCore2*, a metabolic model of *E. coli*. In 85.11% of cases, a chemical structure could be uniquely identified with reasonable confidence, demonstrating that STRUCTRECON can be a valuable tool for structure-based approaches in bioinformatics and related fields. STRUCTRECON is open-source and developed with modularity in mind, making integration of additional databases and procedures possible.

Acknowledgements. This work is supported by the Novo Nordisk Foundation grants NNF19OC0057834 and NNF21OC0066551.

Appendix A

See Fig. 4.

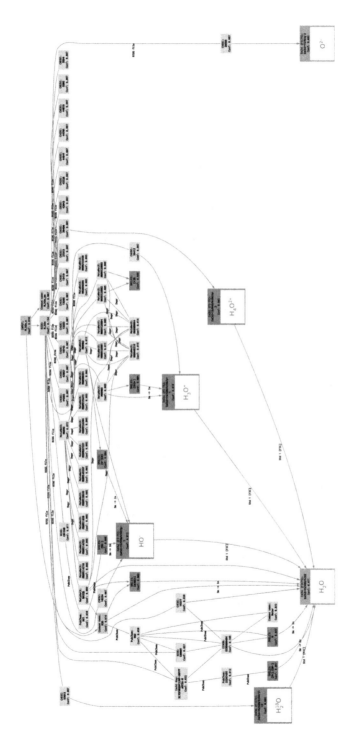

Fig. 4. The identifier graph generated by the BiGG ID M_h20_c. For a description of how to read this graph, see the caption of Fig. 2. This figure clearly demonstrates the impact of the scoring algorithm, as it chooses the conventional structure, H_2O, at a confidence ratio of 0.07.

References

1. Akhondi, S.A., Kors, J.A., Muresan, S.: Consistency of systematic chemical identifiers within and between small-molecule databases. J. Cheminform. **4**, 35 (2012). https://doi.org/10.1186/1758-2946-4-35
2. Akutsu, T.: A new method of computer representation of stereochemistry. Transforming a stereochemical structure into a graph. J. Chem. Inf. Comput. Sci. **31**(3) (1991). https://doi.org/10.1021/ci00003a008
3. Andersen, J.L., Flamm, C., Merkle, D., Stadler, P.F.: Chemical graph transformation with stereo-information. In: de Lara, J., Plump, D. (eds.) ICGT 2017. LNCS, vol. 10373, pp. 54–69. Springer, Cham (2017). https://doi.org/10.1007/978-3-319-61470-0_4
4. Brin, S., Page, L.: The anatomy of a large-scale hypertextual web search engine. Comput. Netw. ISDN Syst. **30**(1) (1998). https://doi.org/10.1016/S0169-7552(98)00110-X
5. Degtyarenko, K., et al.: ChEBI: a database and ontology for chemical entities of biological interest. Nucleic Acids Res. **36**(Database issue), D344–D350 (2008). https://doi.org/10.1093/nar/gkm791
6. Fourches, D., Muratov, E., Tropsha, A.: Trust, but verify: on the importance of chemical structure curation in cheminformatics and QSAR modeling research. J. Chem. Inf. Model. **50**(7), 1189–1204 (2010). https://doi.org/10.1021/ci100176x
7. Ganter, M., Bernard, T., Moretti, S., Stelling, J., Pagni, M.: MetaNetX.org: a website and repository for accessing, analysing and manipulating metabolic networks. Bioinformatics **29**(6), 815–816 (2013). https://doi.org/10.1093/bioinformatics/btt036
8. Guo, A.C., et al.: ECMDB: the e.coli metabolome database. Nucleic Acids Res. **41**(Database issue), D625–630 (2013). https://doi.org/10.1093/nar/gks992
9. Hädicke, O., Klamt, S.: Ecolicore2: a reference network model of the central metabolism of escherichia coli and relationships to its genome-scale parent model. Sci. Rep. **7**(11) (2017). https://doi.org/10.1038/srep39647
10. Heller, S.R., McNaught, A., Pletnev, I., Stein, S., Tchekhovskoi, D.: InChI, the IUPAC international chemical identifier. J. Cheminform. **7**(1), 1–34 (2015). https://doi.org/10.1186/s13321-015-0068-4
11. International Union of Pure and Applied Chemistry Commission on the Nomenclature of Organic Chemistry, Klesney, S.P.: Nomenclature of Organic Chemistry (1979)
12. Kanehisa, M., Furumichi, M., Tanabe, M., Sato, Y., Morishima, K.: KEGG: new perspectives on genomes, pathways, diseases and drugs. Nucleic Acids Res. **45**(D1), D353–D361 (2017). https://doi.org/10.1093/nar/gkw1092
13. Kim, S., et al.: PubChem 2023 update. Nucleic Acids Res. **51**(D1), D1373–D1380 (2023). https://doi.org/10.1093/nar/gkac956
14. King, Z.A., et al.: BiGG models: a platform for integrating, standardizing and sharing genome-scale models. Nucleic Acids Res. **44**(D1), D515–D522 (2016). https://doi.org/10.1093/nar/gkv1049
15. Muresan, S., Sitzmann, M., Southan, C.: Mapping between databases of compounds and protein targets. Methods Mol. Biol. (Clifton, N.J.) **910**, 145–164 (2012). https://doi.org/10.1007/978-1-61779-965-5_8
16. Petrarca, A.E., Lynch, M.F., Rush, J.E.: A method for generating unique computer structural representations of stereoisomers. J. Chem. Doc. **7**(3) (1967). https://doi.org/10.1021/c160026a008

17. RDKit: Open-source cheminformatics software. https://www.rdkit.org/
18. Sajed, T., et al.: ECMDB 2.0: a richer resource for understanding the biochemistry of e.coli. Nucleic Acids Res. **44**(D1), D495–501 (2016). https://doi.org/10.1093/nar/gkv1060
19. Sitzmann, M., Filippov, I., Nicklaus, M.: Internet resources integrating many small-molecule databases1. SAR QSAR Environ. Res. **19**(1–2), 1–9 (2008). https://doi.org/10.1080/10629360701843540
20. Stein, S.E., Heller, S.R., Tchekhovskoi, D.V.: The IUPAC Chemical Identifier - Technical Manual (2011)
21. Weininger, D.: SMILES, a chemical language and information system. 1. Introduction to methodology and encoding rules. J. Chem. Inf. Comput. Sci. **28**(1), 31–36 (1988). https://doi.org/10.1021/ci00057a005
22. Weininger, D.: SMILES. 3. DEPICT. Graphical depiction of chemical structures. J. Chem. Inf. Comput. Sci. **30**(3), 237–243 (1990). https://doi.org/10.1021/ci00067a005
23. Weininger, D., Weininger, A., Weininger, J.L.: SMILES. 2. Algorithm for generation of unique SMILES notation. J. Chem. Inf. Comput. Sci. **29**(2), 97–101 (1989). https://doi.org/10.1021/ci00062a008
24. Williams, A.J., Ekins, S.: A quality alert and call for improved curation of public chemistry databases. Drug Discov. Today **16**(17), 747–750 (2011). https://doi.org/10.1016/j.drudis.2011.07.007
25. Williams, A.J., Ekins, S., Tkachenko, V.: Towards a gold standard: regarding quality in public domain chemistry databases and approaches to improving the situation. Drug Discov. Today **17**(13), 685–701 (2012). https://doi.org/10.1016/j.drudis.2012.02.013
26. Young, D., Martin, T., Venkatapathy, R., Harten, P.: Are the chemical structures in your QSAR correct? QSAR Comb. Sci. **27**(11–12), 1337–1345 (2008). https://doi.org/10.1002/qsar.200810084

Deep Learning Architectures for the Prediction of YY1-Mediated Chromatin Loops

Ahtisham Fazeel Abbasi[1,2]([✉]), Muhammad Nabeel Asim[1], Johan Trygg[3,4], Andreas Dengel[1,2], and Sheraz Ahmed[1]

[1] German Research Center for Artificial Intelligence (DFKI), Kaiserslautern, Germany
`ahtisham.abbasi@dfki.de`
[2] University of Kaiserslautern-Landau, Kaiserslautern (RPTU), Germany
[3] Computational Life Science Cluster (CLiC), Umeå University, Umeå, Sweden
[4] Sartorius Corporate Research, Sartorius Stedim Data Analytics, Umeå, Sweden

Abstract. YY1-mediated chromatin loops play substantial roles in basic biological processes like gene regulation, cell differentiation, and DNA replication. YY1-mediated chromatin loop prediction is important to understand diverse types of biological processes which may lead to the development of new therapeutics for neurological disorders and cancers. Existing deep learning predictors are capable to predict YY1-mediated chromatin loops in two different cell lines however, they showed limited performance for the prediction of YY1-mediated loops in the same cell lines and suffer significant performance deterioration in cross cell line setting. To provide computational predictors capable of performing large-scale analyses of YY1-mediated loop prediction across multiple cell lines, this paper presents two novel deep learning predictors. The two proposed predictors make use of Word2vec, one hot encoding for sequence representation and long short-term memory, and a convolution neural network along with a gradient flow strategy similar to DenseNet architectures. Both of the predictors are evaluated on two different benchmark datasets of two cell lines HCT116 and K562. Overall the proposed predictors outperform existing DEEPYY1 predictor with an average maximum margin of 4.65%, 7.45% in terms of AUROC, and accuracy, across both of the datases over the independent test sets and 5.1%, 3.2% over 5-fold validation. In terms of cross-cell evaluation, the proposed predictors boast maximum performance enhancements of up to 9.5% and 27.1% in terms of AUROC over HCT116 and K562 datasets.

Keywords: YY1 · Chromatin loops · Gene regulation · Convolutional Networks · Word2vec · One hot encoding · LSTM

This work was supported by Sartorius Artificial Intelligence Lab.

1 Introduction

In living organisms proteins are essential to perform diverse types of cellular activities and dysregulation of proteins lead towards development of multifarious diseases such as cancer, neurological, and immunological disorders [1]. Primarily, the production of proteins depends upon the regulation of genes. The process of gene regulation is mediated by different regulatory elements that are distributed in the DNA i.e., enhancers, and promoters. Mainly, extra-cellular signals or transcription factors bind with the enhancer regions to regulate gene expression of nearby or distant genes by forming physically connected chromatin (DNA) loops among enhancers and promoters [2]. These interactions between proximal promoters and distal enhancers often lead to higher order chromatin structure known as topologically associated domains which may contain several chromatin loops [3]. These chromatin loops play a substantial role in performing insulation function to stop the process of transcription.

Different transcription factors are involved in the formation of chromatin loops, such as 11-zinc finger protein (CTCF), and Ying Yang-1 protein (YY1). "Figure 1" illustrates the formation of chromatin loops with the involvement of two different transcription factors CTCF and YY1. YY1-mediated chromatin loops are usually shorter in length, and may bind to smaller motifs of size 12, on the other hand CTCF mediated chromatin loops are larger in length and they form in the regions containing CTCF motifs of 19 nucleotide bases.

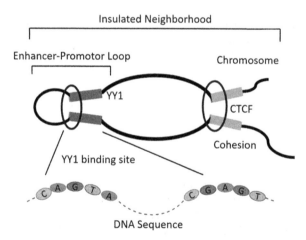

Fig. 1. The generic phenomenon of chromatin loop formation due to the interactions between enhancers, promoters, DNA binding proteins (CTCF or YY1), and Cohesion protein.

Several studies reveal the importance of YY1-mediated chromatin loops in certain disorders such as, neurodevelopmental disorders [4], Gabriele–de Vries

Syndrome [5], ischemic damage, Parkinson and Alzheimer disease [6]. In addition, YY1 can act as a tumor suppressor or stimulator in the case of pancreatic cancer, melanoma, and glioma [6]. These disorders or diseases arise due to the dysregulation at the genetic level caused by the YY1 protein. Despite the importance of gene regulation, cell identity and cell development, the roles of YY1 transcription factor and YY1 mediated chromatin loops are not properly characterized and understood yet.

The identification of generic enhancer promoter interactions provide an abstract level information related to the gene regulation, but such identifications lack information regarding the DNA binding proteins that are involved in initiating such interactions. On the other hand, TF specific enhancer-promoter (EP) interaction and chromatin loops identification can assist in understanding underlying phenomena such as cell-cell communication, and extracellular signalling, which may help out in dealing with complex disorders and cancers or tumors.

Different in-silico and wet lab methods are utilized to identify chromatin interactions and DNA proteins binding sites such as, chromatin interaction analysis by pair-end tagging (ChIA-PET) [7], Chromatin immunoprecipitation (ChIP) [8], High-through chromosome conformation capture (Hi-C), protein-centric chromatin conformation method (HiChIP), chromatin-interacting protein mass spectrometry (ChIP-MS) [6], and proteomics of isolated chromatin segments (PICh) [9]. However, it is laborious, expensive, and time-consuming to identify chromatin interactions at large number of cell types purely based on such experiments. Particularly, with the avalanche of the sequence data produced at the DNA sequence level, it is highly compelling to develop computational methods for fast, and effective analyses of chromatin loops.

Artificial intelligence (AI) has been a key area of research in genomics to analyze DNA sequence data. Several AI-based methods have been developed with an aim to analyze different chromatin interactions. Majority of these approaches are based on the identification of CTCF mediated loops along with their genomic signatures [10,11] or generic enhancer promoter interactions. In comparison, there is a scarcity of AI-based predictors for YY1 mediated chromatin loops, for instance only one AI-based method has been developed for the prediction of YY1 mediated chromatin loops namely, DEEPYY1 [12]. DEEPYY1 made use of Word2vec embeddings for encoding DNA sequences and a convolutional neural network for the prediction of YY1 mediated chromatin loops. DEEPYY1 predictor was evaluated on DNA sequence data obtained from HiChip experiments related to two different cell types i.e., HCT116 (colon cell line), and K562 (lymphoblasts from bone marrow). DEEPYY1 predictor failed to produce better performance over the same and cross-cell line data. Therefore, there is a need to develop a more generic and high-performance YY1-chromatin loop predictor to perform analysis over different cell lines data.

The paper in hand proposed two deep learning based approaches named densely connected neural network (DCNN) and hybrid neural network (hybrid). Following the observations of different researchers that deep learning based predictors perform better when they are trained on large datasets and considering the unavailability of DNA sequences annotated against YY1-mediated chromatin Loops, DCNN-YY1 predictor utilize the idea of transfer learning to generate pretrained k-mer embeddings. Further, in order to extract diverse types of discriminative features, DCNN-YY1 makes use of convolutional layers in two different settings shallow and deep. With an aim to reap the benefits of both types of layers and to avoid gradient vanishing and exploding problems in the process of training, we provide alternative paths for gradient flow among different layers through identity functions which are commonly used in DenseNet architectures [13]. On the other hand, in order to capture positional information of nucleotides the hybrid model makes use of one hot encoding approach for sequence representation. Whereas, the hybrid model itself is comprised of convolution neural network (CNN) and long short terms memory unit (LSTM), to capture discriminative higher spatial and nucleotide level information.

Proposed predictors are evaluated over two different cell lines benchmark datasets. Jointly, over both datasets, experimental results reveals the superiority of the proposed predictors over state-of-the-art DEEPYY1 predictor with average maximum margin of 4.65%, 7.45% in terms of AUROC, and accuracy, across both of the datasets over the independent test sets and 5.1%, 3.2% over 5-fold validation. To explore whether proposed predictors are capable to predict YY1-mediated chromatin loops in different cell lines, we also performed experimentation in cross domain setting in which predictors are trained over sequences of one cell line and evaluation is performed over sequences of other cell line. In cross domain setting proposed predictors outperformed state-of-the-art predictor by a maximum margin of 28% and 10.7% in terms of AUROC, and 22.4%, and 7.01% across accuracy, over HCT116 and K562 datasets.

2   Material and Methods

This section briefly demonstrates the working paradigm of the proposed YY1 chromatin loop predictors, benchmark datasets, and the evaluation metrics.

2.1   CNN and LSTM Based YY1-Mediated Chromatin Loop Predictor

The working paradigm of the proposed hybrid (CNN+LSTM) model can be divided in two main stages. At the first stage, one hot encoded vectors are generated from the nucleotides of DNA sequences. At the second stage, the hybrid model utilizes these vectors to predict YY1-mediated chromatin loops. The working paradigm of one hot encoding and the hybrid model are discussed in the following sections.

One Hot Encoding. One hot encoding is a simplified yet effective way of representing genomic sequences for classification, which may encode the nucleotide composition information. In this method, out of four different nucleotides, each

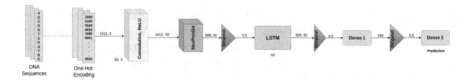

Fig. 2. The graphical illustration of the proposed (hybrid) architecture.

nucleotide is represented by a vector of size 4 with 1 representing the presence and 0 referring to the absence of a particular nucleotide.

LSTM and CNN. Convolutional neural networks (CNNs) are used widely in the domain of computer vision and natural language processing [14]. A CNN is comprised of three different layers: convolution, pooling, and fully connected layers, which allow it to capture spatial features from the input. The convolution operation leads to the formation of feature maps, whereas the pooling operation reduces the dimension of the feature maps by taking either average or maximum value. CNNs are commonly used for DNA analysis problems to learn local features such as, motifs in the case of DNA sequences and are often referred as local feature learning layers. On the other hand, CNN ignores the dependence present within the inputs which deteriorates their power to model NLP-based problems accurately.

Therefore, recurrent neural networks and their variants i.e., long short-term memory (LSTM), and gated recurring units (GRUs) are used to learn such long-term dependencies. LSTM contains a series of memory cells, which are dependent on three different gates to compute the output i.e., input, forget, and output gate. The input gate adds the input to the current cell state by the use of two non-linear activation functions i.e., sigmoid and tanh. Sigmoid assigns a probability to the inputs where tanh transforms them in the range of -1 and 1. The forget gate is responsible for the removal of undesired information from the inputs, it achieves this by taking two different inputs i.e., h_{t-1}, and X_t. These inputs are multiplied with the weights and a bias is added in them. The output of the multiplication operation is followed by a sigmoid operation that transforms these values in the range of 0 (forget) and 1 (remember). The output gate gives the output of the LSTM cell based on the sigmoid and tanh activation functions.

In the current setting, we make use of one hot encoding to represent DNA sequences in the form of vectors. One hot encoded DNA sequences are passed through the convolution, and max-pooling layers to extract the local features, which is followed by the LSTM layer to learn long-term sequence dependencies and a fully connected layer for classification.

2.2 Densely Connected Neural Network Based YY1-Mediated Chromatin Loop Predictor

The working paradigm of the proposed DCNN-YY1 predictor can be divided in two different stages. At the first stage, pretrained k-mer embeddings are generated in an unsupervised manner using well known Word2vec model [15]. In second stage, DCNN-YY1 utilizes pretrained k-mer embeddings and raw DNA sequence to predict YY1-mediated chromatin loops.

The working paradigm of Word2vec algorithm is summarized in the Subsect. 2.2. Furthermore, CNNs and dense connectivity are comprehensively discussed in Subsect. 2.2.

Word2vec. Word2vec is a two-layered neural network model that is capable to learn associations of k-mers from the raw DNA sequence data [16]. Word2vec takes DNA sequences as an input and transforms the sequence data into a numerical vector space, where each k-mer is represented by a N-dimensional vector. Such vectors include important characteristics related to k-mers with respect to four unique nucleotides i.e., semantic relationships and contexts. Moreover, the k-mers that are similar or semantically close to each other they lie closer in the continuous vector space.

Two common methods are used in a Word2vec model for the generation of embeddings namely, common bag of words (CBOW), and skipgram [16]. CBOW works by predicting the target k-mer when provided with the distributed representations of the sequence k-mers. Whereas, the skipgram model tries to predict the context of a k-mer which is opposite to the working paradigm of CBOW model.

We generate 7 different overlapping k-mers from range 1 to 7. Iteratively, for each k-mer we generate 100 dimensional k-mer embeddings using CBOW model. Based on the size of k-mer we obtain 100 dimensional vectors associated to each unique k-mer. For instance, for 1-mer there exist 4 unique 1-mers A, C, G, T, so we obtain 4, 100 dimensional vectors. For 2-mers we obtain 16 and for 3-mers we obtain 64 vectors and so on. K-mer embeddings are generated separately for enhancer and promoter sequences.

Convolutional Neural Networks and Dense Connectivity. We utilize CNN based architecture for YY1-mediated chromatin loop prediction, "Figure 3" shows the complete architecture of the proposed predictor. The network consists of three 1-dimensional (1-D) convolutions, 2 dropout and 4 fully connected layers.

We generate k-mers of enhancer and promoter sequences and transform enhancer k-mer sequences to 100 dimensional statistical vectors by taking average of k-mers pretrained vectors that are generated over enhancer sequences. Similarly, promoter k-mer sequences are transformed to 100 dimensional vectors by utilizing precomputed k-mer vectors over promoter sequences. Statistical vectors of both sequences are concatenated to generate single 200 dimensional vector for each sample. This statistical vector is passed through 1-D convolutional

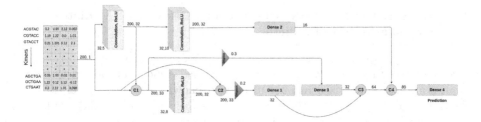

Fig. 3. The graphical illustration of the proposed (DCNN) architecture.

layers to extract robust and meaningful features for further processing i.e., the output size (200 × 32) remains same across each convolution due to the usage of paddings. Each 1-D convolution layer is followed by an activation layer which uses rectified linear unit (ReLU) as an activation function throughout the network except for the last fully connected layer which utilizes Sigmoid function for binary classification.

We further amplify the representational power of the proposed YY1-mediated chromatin loop predictor by the use of dense connectivity that is inspired by the concept of identity mapping or skip connections [17]. Skips connections allow to train the network in a more efficient way. In skip connections the input of a layer is added to the output of a layer, but in terms of dense connections [13] the input of a layer is concatenated to the output of the layer thus offering multiple advantages such as, less vanishing-gradient problem, better feature propagation, and substantial reduction in the number of parameters. "Figure 3" illustrates 4 dense connections namely, C1, C2, C3, and C4 for better feature propagation throughout the network.

2.3   Experimental Setup

The proposed predictors are implemented in Keras. Adam is used as an optimizer with a learning rate of 0.01, and binary cross-entropy is used as the loss function. For the experiments of this study, the DCNN model is trained only for 6 epochs with a batch size of 32. Whereas, the hybrid model is trained for 20 epochs in cross-validation and independent test settings. In addition, the parameters of CNN and LSTM layers are provided in the Figs. 3, and 2.

2.4   Dataset

We utilized the datasets of DEEPYY1 [12] related to two different cell lines i.e., K562, and HCT116. The datasets were collected from the HiChip experiments of Weintrub et al., [18]. As the details related to the preprocessing of the DNA sequences are given in the study of DEEPYY1 [12], therefore we summarize here the number of positive and negative samples in the train and test sets of HCT116 and K562 cell lines. The datasets of both cells are well balanced in terms of positive and negative samples, and "Table 1" demonstrates the number of positive and negative samples in train and test sets of HCT116 and K562.

Table 1. Statistics of benchmark datasets.

Cell Type	Set	$+_{EPIs}$	$-_{EPIs}$
K562	Train	3863	3866
	Test	1657	1657
HCT116	Train	2095	2097
	Test	898	898

3   Evaluation

To evaluate the integrity and predictive performance of the proposed predictor, following the evaluation criteria of the state-of-the-art [12], we utilized two different measures, i.e., accuracy (ACC), and area under the receiver operating characteristic curve (AUROC).

Accuracy (ACC) measures the proportion of correct predictions in relation to all predictions. Area under receiver operating characteristics (AUROC) calculates performance score by using true positive rate (TPR) and false positive rate (FPR) at different thresholds, where true positive rate (TPR) gives the proportion of correct predictions in predictions of positive class and false positive rate (FPR) is the proportion of false positives among all positive predictions (the sum of false positives and true negatives).

$$f(x) = \begin{cases} \text{ACC} = (T_P + T_N)/(T_P + T_N + F_P + F_N) \\ \text{TPR} = T_P/(T_P + F_N) \\ \text{FPR} = F_P/(T_N + F_P) \end{cases} \tag{1}$$

In the above cases, T_P denotes the true positive predictions, T_N shows true negative predictions. Whereas, F_P and F_N refer to the false predictions related to the positive and negative classes.

4   Results

This section demonstrates the performance of the proposed and state-of-the-art DEEPYY1 [13] predictor using two benchmark datasets sets in 5-fold cross-validation setting and independent test sets.

4.1 Proposed DCNN Predictor Performance

Figure 4 shows the effect of different k-mers on the performance values of the proposed predictor. Lower-sized k-mers yield frequent and less unique patterns in the DNA sequences which lead to low performance of the proposed predictor i.e., $K = 1, \cdots, 3$. As, K-mer size is increased the performance scores also increase, where the performance scores are maximum at K=6, as higher k-mers lead to unique patterns in the DNA sequences which are crucial for the generation of discriminative sequence representations. For K=7 the performance deteriorates due to the formation of rare patterns. Hence, the value of K=6 is selected for further experiments and performance comparisons.

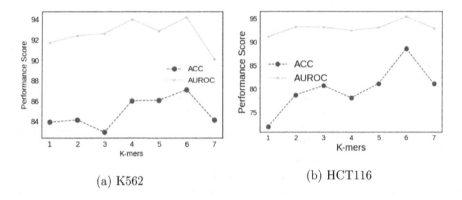

(a) K562   (b) HCT116

Fig. 4. Proposed predictor performance analyses at different k-mers.

4.2 Predictive Performance Analyses over K562 and HCT116

Figures 5 and 6 compare accuracy and AUROC values produced by proposed (hybrid, and DCNN) and state-of-the-art DEEPYY1 predictors at 5-fold cross-validation and independent test sets

Figure 5 shows that the proposed hybrid (CNN+LSTM) predictor achieves AUROC values of 98.2% and 95.7% across K562 and HCT116 datasets. Whereas, in terms of accuracy the proposed predictor achieves 92.9% and 88.5% across K562 and HCT116 datasets. Overall, in comparison to the state-of-the-art DEEPYY1 [12], the proposed predictor shows performance improvements in terms of AUROC and ACC with the average margins of 3.02% and 5.05% across both K5652 and HCT116 datasets, via 5-fold cross-validation. Figure 5 shows that the proposed DCNN predictor achieves AUROC values of 94.1% and 95.4% across K562 and HCT116. Whereas, in terms of accuracy the proposed predictor achieves 86.5% and 87.0% across K562 and HCT116 datasets. Overall, in comparison to the state-of-the-art DEEPYY1 [12], the proposed predictor shows performance improvements across AUROC and ACC with the average margin of 1% and 1.1% across both of the datasets, via 5-fold cross-validation.

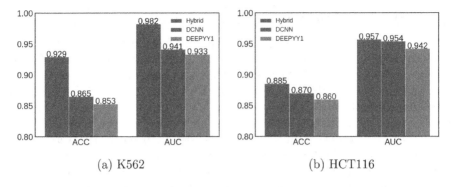

(a) K562   (b) HCT116

Fig. 5. Performance comparison of the proposed and state-of-the-art predictors on two benchmark datasets using 5-fold cross-validation.

Figure 6 shows proposed hybrid (CNN+LSTM) predictor produces 98.5%, and 98.3% AUROC values over K562 and HCT116 independent test sets. Whereas, in terms of accuracy the proposed hybrid predictor shows ACC values of 94.0% and 93.0%. Overall, the proposed predictor provides performance improvements in terms of ACC and AUROC with significant margins i.e., 7.45% and 4.65% in comparison to the state-of-the-art DEEPYY1 predictor. Figure 6 shows proposed DCNN predictor produced 94.1%, and 95.7% AUROC values over K562 and HCT116 independent test sets. Whereas, in terms of accuracy the proposed DCNN predictor shows ACC values of 86.1% and 88.3%. Overall, the proposed predictor provides average performance improvements over ACC and AUROC i.e., 2.3% and 1.15% in comparison to the state-of-the-art DEEPYY1 predictor.

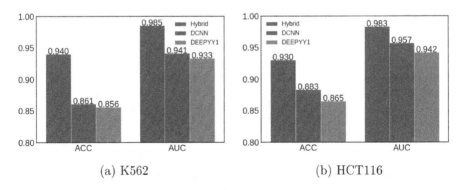

(a) K562   (b) HCT116

Fig. 6. Performance comparison of the proposed and state-of-the-art predictors on two independent test sets.

4.3   Performance Analyses over Cross Cell Data

To assess the generalizability of the models over different cell lines data, we train the models for chromatin loops in a cross-cell manner such that the models are trained on K562 cell data and predictions are performed on second cell HCT116 data and vice versa. Table 2 demonstrates the AUROC and ACC performance values of proposed and DEEPYY1 predictors.

The proposed hybrid predictor outperforms DEEPYY1 by a margin of 9.5% in terms of AUROC over train-test set of K562-HCT116 and 27.1% over a train-test set of HCT116-K562. Overall, the proposed hybrid predictor shows consistent and better performance in cross-cell evaluation in terms of AUROC. Similarly, in terms of accuracy, the proposed hybrid predictor outperforms DEEPYY1 by 11.0% over HCT116 test set and 25.2% over K562 test set, which shows the generalizability power of the proposed approach.

The proposed DCNN predictor outperforms DEEPYY1 by a margin of 7.01% in terms of AUROC over train-test set of K562-HCT116 and 22.4% over train-test set of HCT116-K562. Overall, the proposed DCNN predictor shows consistent and better performance in cross cell evaluation in terms of AUROC. Similarly, in terms of accuracy, the proposed predictor outperforms DEEPYY1 by 6% over HCT116 test set and 23.2% over K562 test set, which shows the generalizability power of the proposed approach.

The better performance offered by the proposed DCNN predictor is subjected to the use of dense connectivity in the CNN, as it allows the model to learn in a more suitable manner due to the better feature propagation in the deeper layers. In comparison, the lower performance of the DEEPYY1 is because of the 1 layer CNN which does not possess enough learning power for complex genomic sequences, and the use of max-pool layer which often ignores very crucial information in terms of textual data. Similarly, the proposed hybrid approach takes in to account nucleotide composition information and learns the dependencies of nucleotides with an incorporated LSTM which makes it superior to the other approaches.

Table 2. Performance values of the proposed and existing predictors on the basis cross cell line testing.

Method	Training Data	Testing Data	Accuracy	AUC
DEEP-YY1	K562	HCT116	80.0	87.9
	HCT116	K562	66.0	70.1
DCNN	K562	HCT116	86.0	94.1
	HCT116	K562	84.2	92.5
Hybrid	K562	HCT116	**91.0**	**97.4**
	HCT116	K562	**91.2**	**97.2**

5 Conclusion

This study presents two new YY1-mediated chromatin loop predictors based on CNNs, and RNNs and dense connectivity. It illustrates the impact of different k-mers on the predictive performance of the proposed DCNN predictor, where 6-mer lead to the best performance. The analyses shows that both the proposed predictors are able to generalize well on similar and cross-cell datasets. It also demonstrates that the proposed predictors offer performance superiority over the state-of-the-art DEEPYY1. Overall the proposed predictors outperform existing DEEPYY1 predictor with an average maximum margin of 4.65%, 7.45% in terms of AUROC, and accuracy, across both of the datases over the independent test sets and 5.1%, 3.2% over 5-fold validation. In terms of cross-cell evaluation, the proposed predictors boast maximum performance enhancements of up to 9.5% and 27.1% in terms of AUROC over HCT116 and K562 datasets. The proposed predictors can assist in understanding the process of transcriptional regulation, and multiple disorders which are related the YY1-mediated chromatin loops. Furthermore, in the future, this approach can be leveraged for large-scale cellular chromatin loops analyses and also for other chromatin loops predictions.

References

1. Hepler, J.R., Gilman, A.G.: G proteins. Trends Biochem. Sci. **17**(10), 383–387 (1992)
2. Bonev, B., Cavalli, G.: Organization and function of the 3d genome. Nat. Rev. Genet. **17**(11), 661–678 (2016)
3. Dixon, J.R.: Topological domains in mammalian genomes identified by analysis of chromatin interactions. Nature **485**(7398), 376–380 (2012)
4. He, Y., Casaccia-Bonnefil, P.: The yin and yang of yy1 in the nervous system. J. Neurochem. **106**(4), 1493–1502 (2008)
5. Carminho-Rodrigues, M.T., et al.: Complex movement disorder in a patient with heterozygous yy1 mutation (gabriele-de vries syndrome). Am. J. Med. Genet. Part A **182**(9), 2129–2132 (2020)
6. Verheul, T.C.J., van Hijfte, L., Perenthaler, E., Barakat, T.S.: The why of yy1: mechanisms of transcriptional regulation by yin yang 1. Frontiers in cell and developmental biology, p. 1034 (2020)
7. Wang, R., Wang, Y., Zhang, X., Zhang, Y., Xiaoyong, D., Fang, Y., Li, G.: Hierarchical cooperation of transcription factors from integration analysis of dna sequences, chip-seq and chia-pet data. BMC Genomics **20**(3), 1–13 (2019)
8. Leina, L., Sun, K., Xiaona Chen, Yu., Zhao, L.W., Zhou, L., Sun, H., Wang, H.: Genome-wide survey by chip-seq reveals yy1 regulation of lincrnas in skeletal myogenesis. EMBO J. **32**(19), 2575–2588 (2013)
9. Kan, S.L., Saksouk, N., Déjardin, J.: Proteome characterization of a chromatin locus using the proteomics of isolated chromatin segments approach. In Proteomics, pp. 19–33. Springer (2017)
10. Lv, H., Dao, F.-Y., Zulfiqar, H., Su, W., Ding, H., Liu, L., Lin, H.: A sequence-based deep learning approach to predict ctcf-mediated chromatin loop. Briefings Bioinform. **22**(5), bbab031 (2021)

11. Cao, F., et al.: Chromatin interaction neural network (chinn): a machine learning-based method for predicting chromatin interactions from dna sequences. Genome Biol. **22**(1), 1–25 (2021)
12. Dao, F.-Y., Lv, H., Zhang, D., Zhang, Z.-M., Liu, L., Lin, H.: Deepyy1: a deep learning approach to identify yy1-mediated chromatin loops. Briefings in bioinformatics, 22(4):bbaa356, 2021
13. Huang, G., Liu, Z., Van Der Maaten, L., Weinberger, K.Q.: Densely connected convolutional networks. In Proceedings of the IEEE Conference on Computer Vision and Pattern Recognition, pp. 4700–4708 (2017)
14. Albawi, S., Mohammed, T.A., Al-Zawi, S.: Understanding of a convolutional neural network. In: 2017 International Conference on Engineering and Technology (ICET), pp. 1–6. IEEE (2017)
15. Kenneth Ward Church: Word2vec. Nat. Lang. Eng. **23**(1), 155–162 (2017)
16. Zhang, R., Wang, Y., Yang, Y., Zhang, Y., Ma, J.: Predicting ctcf-mediated chromatin loops using ctcf-mp. Bioinformatics **34**(13), i133–i141 (2018)
17. He, K., Zhang, X., Ren, S., Sun, J.: Deep residual learning for image recognition. In: Proceedings of the IEEE Conference on Computer Vision and Pattern Recognition, pp. 770–778 (2016)
18. Weintraub, A.S., et al.: Yy1 is a structural regulator of enhancer-promoter loops. Cell **171**(7), 1573–1588 (2017)

MPFNet: ECG Arrhythmias Classification Based on Multi-perspective Feature Fusion

Yuxia Guan[1], Ying An[2], Fengyi Guo[1], and Jianxin Wang[1($\boxtimes$)]

[1] School of Computer Science and Engineering,
Central South University, Changsha 410083, China
jxwang@mail.csu.edu.cn

[2] Big Data Institute, Central South University, Changsha 410083, China

Abstract. Arrhythmia is a common cardiovascular disease that can cause sudden cardiac death. The electrocardiogram (ECG) signal is often used to diagnose the state of the heart. However, most existing ECG diagnostic methods only use information from a single perspective, ignoring the extraction of fusion information. In this paper, we propose a novel Multi-Perspective feature Fusion Network (MPFNet) for ECG arrhythmia classification. In this model, two independent feature extraction modules are first deployed to learn one-dimensional and two-dimensional ECG features from the original one-dimensional ECG signals and its corresponding recurrence plots. At the same time, an interactive feature extraction module based on bidirectional encoder-decoder is designed to further capture the interrelationships between one-dimensional and two-dimensional perspectives, and combine them with independent features from two different perspectives to enhance the completeness and accuracy of the final representation by utilizing the correlation and complementarity between perspectives. We evaluate our method on a large public ECG dataset and the experimental results demonstrate that MPFNet outperforms the state-of-the-art approaches.

Keywords: Arrhythmia classification · Attention mechanism · Bidirectional encoder-decoder · Multi-perspective

1 Introduction

Arrhythmia is a common type of cardiovascular disease, and electrocardiogram (ECG) is usually used as one of the most important diagnostic criteria in clinical practice [1]. Traditional ECG analysis is mainly subjectively judged by doctors based on their own experience and knowledge. However, many subtle features may be hidden in ECG signals that do not conform to traditional medical knowledge and are challenging to detect with the naked eye [2]. It makes how to quickly and accurately detect arrhythmias based on ECG a difficult task with important clinical significance. In recent years, the advancement of computer and artificial intelligence technology has opened a new era of healthcare.

X. Guo et al. (Eds.): ISBRA 2023, LNBI 14248, pp. 85–96, 2023.
https://doi.org/10.1007/978-981-99-7074-2_7

Many computer-assisted diagnostic systems based on machine learning have been widely applied to automatic ECG analysis and related clinical decision-making, and have achieved promising results.

Early classification of arrhythmia mainly relies on manual feature extraction by signal processing and statistical techniques, which requires the high medical expertise of researchers. For example, many ECG intervals are important time-domain features that are most commonly extracted and applied [3,4]. In addition to the waveform characteristics of the ECG structure itself, the time domain signal characteristics can also be represented by the statistical characteristics of the maximum value, minimum value, standard deviation, and average value over time to realize the ECG classification [5,6]. Due to the inherent non-stationary characteristics of ECG signals, there are also some studies using frequency-based technology to extract time-frequency features for ECG analysis [7–9]. However, these methods typically require sophisticated feature design and selection, which is extremely cumbersome and highly dependent on knowledge and experience. Therefore, it greatly affects their effectiveness and generalization.

In recent years, the development of deep learning technology provides a new idea for the automatic classification of arrhythmia. In most existing studies, the original one-dimensional ECG signal is usually directly used as the input of the deep model to achieve end-to-end classification. For example, Sellami et al. [10] adopted a CNN with a dynamic batch weighting function, which achieve high-performance classification in both inter-patient and intra-patient paradigm. Niu et al. [11] proposed a multi-view convolutional neural network that used symbolic representations of heartbeats to automatically learn features and classify arrhythmias. Yildirim et al. [12] used a new wavelet sequence model based on deep bidirectional LSTM network to classify five types of heart rhythm in the MIT-BIH arrhythmia database. Jin et al. [13] used a dual attention convolutional long short-term memory network to detect intra-patient and inter-patient atrial fibrillation on the MIT-BIH atrial fibrillation database. Hannun et al. [14] used a 34-layer DNN model similar to ResNet to classify 12 types of heart rhythm, and achieved a recognition level similar to that of cardiologists.

In addition, some studies have found that the two-dimensional processing of ECG signals can provide more abundant potential features, and is conducive to exerting the feature learning ability of deep models such as two-dimensional convolutional neural networks. For example, Xia et al. [15] converted one-dimensional ECG signals into two-dimensional signals through short-time Fourier transforms and stationary wavelet transforms and then input them into a two-dimensional CNN model for feature extraction and classification. Naz et al. [16] converted ECG signals into two-dimensional images and then adopted transfer learning to give full play to the advantages of classical models such as VGG to achieve arrhythmia classification. Huang et al. [17] converted the original ECG signals into time-frequency spectra through a short-time Fourier transform and then fed them into a two-dimensional CNN to classify five arrhythmias. In fact, the original one-dimensional ECG signals and their corresponding two-dimensional images can be seen as different perspectives on the heart health status of patients during the same period. Most of the above methods only consider extracting relevant features hidden in ECG data from a single perspective without fully utilizing the complementarity between the information contained in the two perspectives, thus limiting the comprehensiveness and accuracy of ECG feature representation to some extent.

In order to make full use of the characteristics of one-dimensional perspective and two-dimensional perspective, we propose a novel Multi-Perspective feature Fusion Network for ECG arrhythmias classification, which fully integrates multi-perspective features from the raw one-dimensional ECG signal and its corresponding two-dimensional image to effectively improve the accuracy of ECG feature representation. The main contributions of this paper are summarized as follows:

(1) We design two independent feature extraction modules to learn ECG features from different perspectives in the raw one-dimensional ECG signal and its corresponding two-dimensional image, respectively.

(2) We propose an interactive feature extraction module based on bidirectional encoder-decoder to capture the interrelationships between one-dimensional and two-dimensional ECG data, and combine them with independent features from two different perspectives to obtain a more accurate and comprehensive ECG representation.

The rest of this paper is organized as follows: Section 2 introduces the methodology, while Sect. 3 presents experimental details and results. In Sect. 4, we discuss data imbalance and interpretability issues. Finally, Sect. 5 concludes the paper and outlines future research directions.

2 Method

As shown in Fig. 1, the overall architecture of MPFNet consists of four components. The one-dimensional perspective feature extraction module and the two-dimensional perspective feature extraction module extract ECG features from the original one-dimensional ECG signals and their corresponding two-dimensional spectral images. Moreover, the interaction feature extraction module captures the interaction relationship between one-dimensional and two-dimensional perspective features through an improved Transformer based on a bidirectional encoder-decoder structure. Finally, the independent features from the two perspectives are combined with their interactive features and fed into the classification module to obtain the final classification results. The detailed structure of each module is described below.

2.1   One-Dimensional Perspective Feature Extraction Module

We first use a 1D-CNN to convert the original one-dimensional ECG signal into the initial one-dimensional input embedding X_s. Then, a Conv_BiLSTM network is applied to capture a one-dimensional perspective ECG feature. Specifically, the initial one-dimensional input embedding X_s will be sequentially fed into five

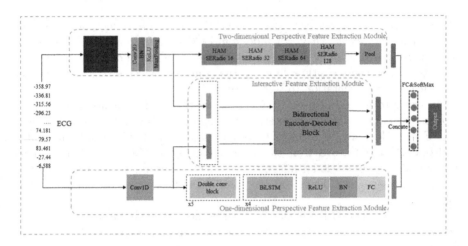

Fig. 1. The overall architecture of MPFNet model for ECG arrhythmia classification

Double Conv blocks to extract corresponding spatial features. The Double Conv block consists of a two-tier 1D-CNN module with a batch normalization layer and a LeakyReLU activation layer in between. The LeakyReLU and MaxPooing layers follow the second convolution layer. Then, four stacked BiLSTM layers are employed to learn the relevant temporal features. Finally, a ReLU and a Layer-Norm followed by a full connection layer are used to obtain the one-dimensional perspective feature vector G_s.

2.2   Two-Dimensional Perspective Feature Extraction Module

To facilitate the extraction of two-dimensional perspective features, we first convert the original one-dimensional ECG signal into the corresponding recurrence plot (RP) [18]. Let $q(t) \in R_d$ be a time series, the recurrence plot can be defined as equation (1).

$$RP = \theta(\varepsilon - ||q(i) - q(j)||) \tag{1}$$

where ε is a threshold, θ is the Heaviside function, i and j are different points in time series. Then, we utilize the HA-ResNet model, proposed in our previous work [19], to learn the hidden ECG features in two-dimensional RP images. Specifically, in the HA-ResNet model, an embedding block consisting of a 2D convolutional layer followed by a BatchNorm layer, a ReLU layer, and a Max-Pooling layer is first employed to generate the initial two-dimensional input embedding X_d of the RP image. After that, X_d will be fed into four hidden attention modules (HAMs) with different parameters to extract the deep spatiotemporal features. Each HAM contains a residual block and a bi-directional ConvLSTM(BConvLSTM) block. We adopt the Squeeze-and-Excitation(SE) block to give different weights to different channels of the feature maps. After the SE block feature extraction, we begin to focus on timing-dependent features of

ECG signals. So, we use BConvLSTM to consider the data dependency of the two directions. Finally, the deep feature representation is sequentially input to an average pooling layer to obtain the two-dimensional perspective feature vector G_d.

2.3 Interactive Feature Extraction Module

To better integrate the ECG features learned above, inspired by Bert [20], we design an interactive feature extraction module based on an improved Transformer, which utilizes a bidirectional encoder-decoder network to capture the bidirectional interaction relationships between one-dimensional and two-dimensional perspective features. Specifically, the bidirectional encoder-decoder consists of two structurally identical parts (a forward association extractor and a backward association extractor). Take the forward association extractor as an example, the encoder part consists of three identical encoder blocks, each consisting of a 1D CNN layer and a GLU layer (as shown in Fig. 2(c)). This structure can ensure feature extraction and avoid the overfitting phenomenon caused by the high complexity of the model. The decoder part composes of three identical decoder blocks, each of which mainly includes a multi-head self-attention module, a dropout layer, a residual connection layer, and a feedforward neural network (as shown in Fig. 2(b)).

We first take the one-dimensional embedding vector X_s and the two-dimensional embedding vector X_d as the encoder and decoder inputs of the forward association extractor, respectively, and learn the forward association feature G_{fwd} from one-dimensional to two-dimensional perspectives. Similarly, we take the two-dimensional embedding vector X_d and one-dimensional embedding vector X_s as the encoder and decoder inputs of the backward association extractor, respectively, to obtain the backward association feature G_{bwd} from the two-dimensional perspective to the one-dimensional perspective. Finally, we splice forward and backward correlation features to get the final interaction feature G_{inter}.

2.4 Arrhythmia Classification Module

Finally, the one-dimensional and two-dimensional perspective features and their interactive features are concatenated and input into a fully connected layer with a Softmax activation function to generate the corresponding classification result $\hat{Y}$, as shown in (2).

$$\hat{Y} = \delta(W_{out}[G_s; G_{inter}; G_d] + b_{out}) \tag{2}$$

where δ is the Softmax function, W_{out} and b_{out} is the learnable parameter, $\hat{Y}$ is the predicted label.

3   Experiments and Results

3.1   Data Description

We collect experimental data from a large ECG dataset published by Zheng et al. [21], which includes 12-lead ECG signals with a high sampling rate of 500 Hz for 10,646 individual subjects. We first filter out the invalid (only containing zero values) or incomplete (missing lead values) recordings and exclude four categories of arrhythmia signals (AT, AVNRT, AVRT, and SAAWR) with too small a sample size. The final dataset included 438 atrial flaps(AF), 1,780 atrial fibrillation(AFIB), 397 sinusoidal irregular rhythms(SI), 3,888 sinus bradycardia(SB), 1,825 sinus rhythm(SR), 1,564 sinus tachycardia(ST), and 544 supracentricular tachycardia(SVT), for a total of 10,436 ECG records.

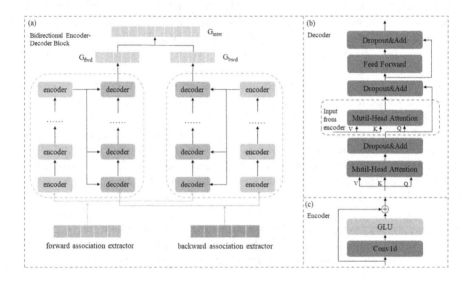

Fig. 2. Overview of the two-dimensional perspective feature extraction module

3.2   Experimental Setting

In our study, all the methods are implemented in Python 3.6.0 with Keras 2.2.2 and trained by Adam optimizer with a learning rate 0.0001. We conduct the experiments on a server with Intel(R) Xeon(R) Gold 6,230 CPU @2.10 GHz, 126 GB memory, and six GeForce RTX cards. We train each model by 10-fold cross-validation to enhance the generalization performance of models. The microaverage Accuracy, Precision, Recall, and F1 Score are used to evaluate the performance of our methods. Adasyn is only applied to the training set during each fold to address the data imbalance issue without compromising the test set. For all the methods, we repeatedly conduct experiments ten times and finally report the mean evaluation metrics.

3.3   Baseline Method

CNN+LSTM [22]: It mainly comprises six convolution layers and a 128-unit LSTM block is included in the model for sequential learning. This model efficiently captures the ECG signal's local and global features.

HIT [7]: It utilizes a directed acyclic graph based on Homeomorphically Irreducible Tree (HIT) for feature generation and maximum absolute pooling for signal decomposition. Finally, they use SVM to classify arrhythmia.

SVM [5]: It proposes the fusion algorithm of fractional-Fourier-transform and two-event related moving-averages to detect R, P, and T peaks and then extract different features in ECG data, such as PR, and RT interval, and finally use SVM for classification.

ResNet50+Logistic Regression [23]: It transforms ECG signals into wavelet transform images and ECG grayscale images for fine-tuning the pre-trained ResNet-50 model. Logistic regression is employed as meta-learners for training using the stacking integration approach.

3.4   Performance Comparison with Baselines

Table 1 presents the performance comparison results of our proposed MPFNet with other baselines. It can be observed that our method achieves the best performance compared to the baselines, with an F1 Score of 94.2%. The main reason is that our method fully combines the original one-dimensional signal from the electrocardiogram and the relevant features of the two-dimensional image to effectively improve the integrity of the patient's electrocardiogram representation, thereby enhancing the classification ability of the model. In contrast, several other comparison methods based on single perspective ECG features are significantly inferior to our method. Among them, the F1 Score of the three methods (SVM, HIT, and CNN+LSTM) that only consider the one-dimensional perspective features in the original ECG signals do not exceed 86%, which is more than 8.9% lower than that of our method. HA-ResNet+RP and ResNet50+Logistic Regression achieve relatively better classification performance by using two-dimensional perspective features in ECG images. Especially for the latter, it combines the two-dimensional perspective features from multiple images (wavelet transform images and ECG grayscale images) converted from the original ECG signals and obtains an F1 Score of 93.6%, which is second only to our method. This to some extent indicates that the two-dimensional transformation of ECG signals can indeed improve the richness of feature information in ECG data.

3.5   Ablation Experiments

To further investigate the impact of various modules in our method on the final classification performance, we compare MPFNet with its several variants.

MPFNet (only_1d): It is a model obtained by removing the interactive feature extraction module and the two-dimensional perspective feature extraction module from our MPFNet.

Table 1. Performance comparison between mpfnet and existing advanced methods

Methods	Acc(std)	Precision(std)	Recall(std)	F1 Score(std)
CNN+LSTM [22]	92.2(0.006)%	80.3(0.007)%	80.2(0.006)%	80.2(0.003)%
HIT [7]	93.0(0.02)%	90.2(0.01)%	81.0(0.02)%	85.3(0.01)%
SVM [5]	82.7(0.05)%	78.0(0.02)%	80.2(0.05)%	80.1(0.03)%
HA-ResNet+RP [19]	88.2(0.01)%	87.6(0.04)%	88.2(0.02)%	88.0(0.007)%
ResNet50+Logistic Regression [23]	93.9(0.006)%	93.7(0.007)%	94.0(0.006)%	93.6(0.003)%
MPFNet	94.3(0.002)%	94.0(0.003)%	94.3(0.01)%	94.2(0.02)%

MPFNet (only_2d): It is a model obtained by removing the interactive feature extraction module and the one-dimensional perspective feature extraction module from our MPFNet.

MPFNet (only_inter): It is a model obtained by removing the one-dimensional perspective feature extraction module and the two-dimensional perspective feature extraction module while only retaining the interactive feature extraction module.

MPFNet (no_inter): It is a model obtained by removing the interactive feature extraction module from MPFNet.

From the results in Fig. 3, it can be seen that the two variant models MPFNet(only_1d) and MPFNet(only_2d) that only use a single perspective feature perform relatively poorly, and their performance are significantly lower than that of other models which combine two perspectives of ECG features. It fully proves that integrating features from one-dimensional and two-dimensional perspectives can achieve effective complementarity, thereby improving the accuracy of patient ECG representation. MPFNet (only_inter) directly utilizes the interactive feature extraction module to learn the interrelationships between one-dimensional and two-dimensional perspective features and achieves multi-perspective feature fusion representation. Therefore, its performance is significantly superior to the above two variant models that only use single perspective features. However, compared to MPFNet, it lacks independent extraction modules for one-dimensional and two-dimensional perspective features, which to some extent reduces its ability to capture the specificity of corresponding perspective features, thus affecting the final classification performance. In addition, when we retain the feature extraction modules for one-dimensional and two-dimensional perspectives and remove the interactive feature extraction module, the performance of MPFNet (no_inter) also shows a significant decrease compared to MPFNet. This also strongly demonstrates the important role of extracting interactive features between different perspectives in improving the effectiveness of feature fusion representation.

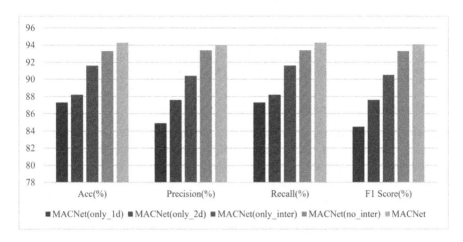

Fig. 3. Experimental results of MPFNet and its variants

4 Discussion

4.1 Comparative Analysis of Imbalance Processing Methods

In this section, we select four representative imbalance processing methods (class weight adjustment [24], Adasyn [25], Smote [26], Facolloss [27]) and analyze the impact of different imbalance processing methods on the performance of our model. The experimental results are shown in Fig. 4.

We can see from the figure that not all imbalance processing methods can have a positive impact on the final performance of the model. For instance, self-defined weight and Facolloss do not exhibit promising results for this study and even reduce the overall performance. On the contrary, after data synthesis using

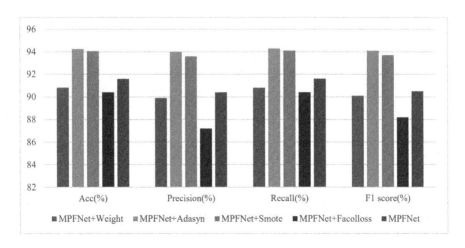

Fig. 4. Results of different imbalance processing methods

Smote and Adasyn, the performance of the model in terms of all metrics shows a significant improvement. Among them, Adasyn brings the greatest performance increase. Therefore, in all performance evaluation experiments in this paper, we ultimately choose it to alleviate the problem of data imbalance.

4.2 Visualization Analysis

To further demonstrate the effectiveness of our method in feature representation, we used the t-distributed random neighborhood embedding (t-SNE) [28] to visualize the feature representations of data samples obtained by HA-ResNet and MPFNet, respectively. As shown in Fig. 5, the points with different colors represent different categories of data samples. From the results of the t-SNE diagrams, it can be seen that our method can more clearly distinguish samples of different categories and make samples of the same class gather relatively closely. This fully proves that MPFNet can indeed enhance the effectiveness of ECG feature representation by utilizing the complementarity and fusion of multi-perspective features.

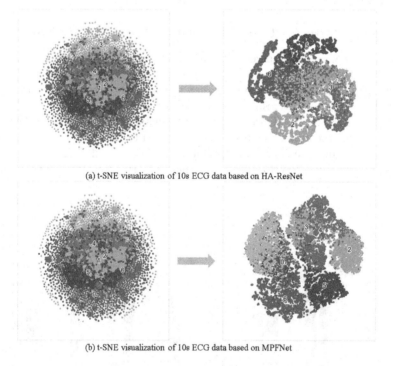

(a) t-SNE visualization of 10s ECG data based on HA-ResNet

(b) t-SNE visualization of 10s ECG data based on MPFNet

Fig. 5. t-SNE visualization; 0–6 represents atrial flutter (orange), atrial fibrillation (yellow), sinus bradycardia (green), sinusoidal irregular rhythm (dark green), sinus rhythm (blue), sinus tachycardia (purple), and supracentral tachycardia (pink)

5   Conclusion

In this model, while extracting independent perspective features from the original one-dimensional ECG signal and two-dimensional ECG image, we also design an interactive feature extraction module to further capture the interrelationships between different perspectives and achieve effective fusion of multi-perspective features. In this way, it fully utilizes the complementarity between different perspectives to enhance the integrity and accuracy of ECG feature representation, thereby effectively improving the model's ability to classify arrhythmia. The experimental results demonstrate the effectiveness of our method. In the future, we will research further to improve the classification performance by fusing more perspectives of information or knowledge into the embedding representation of heartbeat.

Acknowledgement. This work was supported in part by the National Key Research and Development Program of China (No. 2021YFF1201200), the Science and Technology Major Project of Changsha (No. kh2202004), and Changsha Municipal Natural Science Foundation (kq2202106). This work was also carried out in part using computing resources at the High Performance Computing Center of Central South University.

References

1. Versaci, M., Angiulli, G., La Foresta, F.: A modified heart dipole model for the generation of pathological ECG signals. Computation **8**(4), 92 (2020)
2. Wang, J.: Automated detection of atrial fibrillation and atrial flutter in ECG signals based on convolutional and improved elman neural network. Knowl.-Based Syst. **193**, 105446 (2020)
3. Devi, R.L., Kalaivani, V.: Machine learning and IoT-based cardiac arrhythmia diagnosis using statistical and dynamic features of ECG. J. Supercomput. **76**(9), 6533–6544 (2020)
4. Kuila, S., Dhanda, N., Joardar, S.: ECG signal classification and arrhythmia detection using elm-RNN. Multimed. Tools Appl. **81**(18), 25233–25249 (2022)
5. Aziz, S., Ahmed, S., Alouini, M.S.: ECG-based machine-learning algorithms for heartbeat classification. Sci. Rep. **11**(1), 18738 (2021)
6. Pham, T.H., et al.: A novel machine learning framework for automated detection of arrhythmias in ECG segments. J. Ambient Intell. Humanized Comput., 1–18 (2021)
7. Baygin, M., Tuncer, T., Dogan, S., Tan, R.S., Acharya, U.R.: Automated arrhythmia detection with homeomorphically irreducible tree technique using more than 10,000 individual subject ecg records. Inf. Sci. **575**, 323–337 (2021)
8. Yang, J., Yan, R.: A multidimensional feature extraction and selection method for ECG arrhythmias classification. IEEE Sens. J. **21**(13), 14180–14190 (2020)
9. Al-Yarimi, F.A.: Arrhythmia prediction on optimal features obtained from the ECG as images. Comput. Syst. Sci. Eng. **44**(1), 129–142 (2023)
10. Sellami, A., Hwang, H.: A robust deep convolutional neural network with batch-weighted loss for heartbeat classification. Expert Syst. Appl. **122**, 75–84 (2019)
11. Niu, J., Tang, Y., Sun, Z., Zhang, W.: Inter-patient ECG classification with symbolic representations and multi-perspective convolutional neural networks. IEEE J. Biomed. Health Inform. **24**(5), 1321–1332 (2019)

12. Yildirim, Ö.: A novel wavelet sequence based on deep bidirectional LSTM network model for ECG signal classification. Comput. Biol. Med. **96**, 189–202 (2018)
13. Jin, Y., Qin, C., Huang, Y., Zhao, W., Liu, C.: Multi-domain modeling of atrial fibrillation detection with twin attentional convolutional long short-term memory neural networks. Knowl.-Based Syst. **193**, 105460 (2020)
14. Hannun, A.Y., et al.: Cardiologist-level arrhythmia detection and classification in ambulatory electrocardiograms using a deep neural network. Nat. Med. **25**(1), 65–69 (2019)
15. Xia, Y., Wulan, N., Wang, K., Zhang, H.: Detecting atrial fibrillation by deep convolutional neural networks. Comput. Biol. Med. **93**, 84–92 (2018)
16. Naz, M., Shah, J.H., Khan, M.A., Sharif, M., Raza, M., Damaševičius, R.: From ECG signals to images: a transformation based approach for deep learning. PeerJ Comput. Sci. **7**, e386 (2021)
17. Huang, J., Chen, B., Yao, B., He, W.: ECG arrhythmia classification using STFT-based spectrogram and convolutional neural network. IEEE access **7**, 92871–92880 (2019)
18. Eckmann, J.P., Kamphorst, S.O., Ruelle, D., et al.: Recurrence plots of dynamical systems. World Sci. Ser. Nonlinear Sci. Ser. A **16**, 441–446 (1995)
19. Guan, Y., An, Y., Xu, J., Liu, N., Wang, J.: Ha-resnet: residual neural network with hidden attention for ECG arrhythmia detection using two-dimensional signal. IEEE/ACM Trans. Comput. Biol. Bioinform. (2022)
20. Devlin, J., Chang, M.W., Lee, K., Toutanova, K.: Bert: Pre-training of deep bidirectional transformers for language understanding. arXiv preprint arXiv:1810.04805 (2018)
21. Zheng, J., Zhang, J., Danioko, S., Yao, H., Guo, H., Rakovski, C.: A 12-lead electrocardiogram database for arrhythmia research covering more than 10,000 patients. Sci. Data **7**(1), 48 (2020)
22. Yildirim, O., Talo, M., Ciaccio, E.J., San Tan, R., Acharya, U.R.: Accurate deep neural network model to detect cardiac arrhythmia on more than 10,000 individual subject ECG records. Comput. Methods Programs Biomed. **197**, 105740 (2020)
23. Yoon, T., Kang, D.: Multi-modal stacking ensemble for the diagnosis of cardiovascular diseases. J. Personalized Med. **13**(2), 373 (2023)
24. Mehari, T., Strodthoff, N.: Self-supervised representation learning from 12-lead ECG data. Comput. Biol. Med. **141**, 105114 (2022)
25. He, H., Bai, Y., Garcia, E.A., Li, S.: Adasyn: adaptive synthetic sampling approach for imbalanced learning. In: 2008 IEEE International Joint Conference on Neural Networks (IEEE World Congress on Computational Intelligence), pp. 1322–1328. IEEE (2008)
26. Mousavi, S., Afghah, F.: Inter-and intra-patient ECG heartbeat classification for arrhythmia detection: a sequence to sequence deep learning approach. In: ICASSP 2019–2019 IEEE International Conference on Acoustics, Speech and Signal Processing (ICASSP), pp. 1308–1312. IEEE (2019)
27. Lin, T.Y., Goyal, P., Girshick, R., He, K., Dollár, P.: Focal loss for dense object detection. In: Proceedings of the IEEE International Conference on Computer Vision, pp. 2980–2988 (2017)
28. Arora, S., Hu, W., Kothari, P.K.: An analysis of the T-SNE algorithm for data visualization. In: Conference on Learning Theory, pp. 1455–1462. PMLR (2018)

PCPI: Prediction of circRNA and Protein Interaction Using Machine Learning Method

Md. Tofazzal Hossain[1], Md. Selim Reza[2], Xuelei Li[4], Yin Peng[3], Shengzhong Feng[4], and Yanjie Wei[4(✉)]

[1] Department of Statistics, Bangabandhu Sheikh Mujibur Rahman Science and Technology University, Gopalganj 8100, Bangladesh
[2] Bioinformatics Lab, Department of Statistics, Rajshahi University, Rajshahi, Bangladesh
[3] Department of Pathology, The Shenzhen University School of Medicine, Shenzhen 518060, Guangdong, People's Republic of China
[4] Shenzhen Key Laboratory of Intelligent Bioinformatics and Center for High Performance Computing, Shenzhen Institute of Advanced Technology, Chinese Academy of Sciences, Shenzhen 518055, China
{sz.feng,yj.wei}@siat.ac.cn

Abstract. Circular RNA (circRNA) is an RNA molecule different from linear RNA with covalently closed loop structure. CircRNAs can act as sponging miR-NAs and can interact with RNA binding protein. Previous studies have revealed that circRNAs play important role in the development of different diseases. The biological functions of circRNAs can be investigated with the help of circRNA-protein interaction. Due to scarce circRNA data, long circRNA sequences and the sparsely distributed binding sites on circRNAs, much fewer endeavors are found in studying the circRNA-protein interaction compared to interaction between linear RNA and protein. With the increase in experimental data on circRNA, machine learning methods are widely used in recent times for predicting the circRNA-protein interaction. The existing methods either use RNA sequence or protein sequence for predicting the binding sites. In this paper, we present a new method PCPI (Predicting CircRNA and Protein Interaction) to predict the interaction between circRNA and protein using support vector machine (SVM) classifier. We have used both the RNA and protein sequences to predict their interaction. The circRNA sequences were converted in pseudo peptide sequences based on codon translation. The pseudo peptide and the protein sequences were classified based on dipole moments and the volume of the side chains. The 3-mers of the classified sequences were used as features for training the model. Several machine learning model were used for classification. Comparing the performances, we selected SVM classifier for predicting circRNA-protein interaction. Our method achieved 93% prediction accuracy.

Keywords: circular RNA · circRNA-protein interaction · machine learning methods

X. Guo et al. (Eds.): ISBRA 2023, LNBI 14248, pp. 97–106, 2023.
https://doi.org/10.1007/978-981-99-7074-2_8

1 Introduction

Circular RNA (circRNA) is an RNA molecule different from linear RNA with covalently closed loop structure produced by joining a downstream 3′ splice donor site and an upstream 5′ splice acceptor site [1–3]. CircRNAs are more stable than linear RNAs and conserved across species [4]. They can act as sponging miRNAs [5] and can interact with RNA binding protein [6]. Several studies have identified that circRNAs play important role in the development of different diseases [7–13]. The functions of majority of the identified circRNAs are unknown. The study of interaction between circRNA and protein is helpful in exploring the biological functions of the circRNAs [14, 15].

Several computational methods have been developed for predicting the RNA-protein interaction based on high throughput sequencing data. For example, RBPgroup is a soft-clustering method to group RNA binding proteins that bind to same RNAs based on CLIP-seq data [16]. Another approach is used to identify regulatory interactions between circRNAs and protein through cross-linking and immunoprecipitation followed by CLIP-seq data [17]. Machine learning methods are also used for predicting the RNA-protein interactions by large scale experimental data [18]. So far developed methods for the prediction of RNA-protein interactions can be classified into two categories: identification of binding sites in RNA chain and protein chain respectively. A large number of studies perform prediction in protein domain utilizing the abundant domain knowledge of protein databases [19–22]. The prediction in the RNA domain is relatively difficult due to limited information resources of RNA sequences and a few efforts have been made in RNA domain for predicting RNA-protein interactions [23, 24]. Instead of linear RNA our focus is on circRNA-protein interaction prediction. Compared to linear RNA-protein interaction, much fewer studies are conducted to predict the binding sites on circRNAs [25, 26]. The reasons behind this are scarce circRNA data, long circRNA sequences and the sparsely distributed binding sites.

RNA binding proteins are involved in generation [27], post transcriptional regulation [28], and functional execution [29] of circRNAs. Several previous studies have revealed that the circRNA-protein interaction has an important impact on disease development and may be potential biomarkers for different diseases [30–33]. Therefore, predicting circRNA-protein interaction is of great interest. Gradually, the circRNA research is drawing more attention and the experimental data on circRNA are increasing. There is a circRNA-protein interaction database CircInteractome [29] which gathered RBP/miRNA-binding sites on human circRNAs. Due to the availability of data, machine/deep learning methods are being used in predicting circRNA-protein interactions. CRIP is a deep learning method which predicts the circRNA-RBP binding sites using the RNA sequences by a hybrid convolution neural networks and recurrent neural networks [34]. PASSION [25] is another method for identifying the binding sites of RBPs on circRNAs which combines several statistical sequence features and selects important features by feature selection methods to improve the prediction accuracy. GraphProt [35] learns from the secondary structure of proteins and uses SVM for predicting binding sites and affinity of RBPs. RNAcommender [36] applies the recommender system to predict RBP-RNA interactions utilizing the protein domain composition and the RNA predicted secondary structure. Deep-Bind is a deep learning method which uses convolutional neural network (CNN) to predict DNA and RNA binding proteins and achieves better performance [37]. iDeepE is another deep learning method and it uses the global CNN to predict the binding sites utilizing the RNA sequences [38]. DeCban predicts the circRNA-RBP interaction sites by a hybrid double embeddings sequence representation scheme and a cross-branch attention neural network [39]. CRBPDL predicts the circRNA-RBP interaction sites by an ensemble neural network approach [40]. The above mentioned methods either use RNA sequences or protein sequences. But none of these methods use both RNA and protein sequences for predicting the interactions.

In this paper, we developed a new method PCPI (predicting circRNA and protein interaction) using SVM algorithm to predict the interaction between circRNA and protein. The existing methods for predicting the binding sites on circRNAs use only circRNA sequences and predicted whether there is binding sites on the circRNA or not. In our method, we utilized both circRNA and protein sequences and predicted whether the circRNA and protein sequences are interacted with a specified probability of interaction. The alternative representation of the RNA and protein sequences were adopted for feature generation. The amino acid sequences of protein were converted into 7 letters sequences according to the 7 classes based on dipole moments and the volume of the side chains. The circRNA sequences were represented as pseudo peptide sequences. Then, the pseudo peptide sequences were further converted as 8 letters (7 amino acid classes plus 1 stop codon) sequences. The 3-mers of the 7 letters and 8 letters sequences were used as features for training the model. Several machine learning algorithms were used and compared their prediction performances. Finally, we selected SVM algorithm for predicting the circRNA-protein interaction utilizing both circRNA and protein sequences.

2 Methodology

2.1 Data Description

To assess the effectiveness of our method PCPI, we collected circRNA-protein interaction data from the circinteractome (https://circinteractome.nia.nih.gov/) database [29]. As our model is based on sequence information and the sequence similarity may produce incorrect interaction prediction, we removed the redundant sequences by CD-Hit [41] with the threshold similarity parameter 0.9. After removing the redundant sequences, we got a total of 29623 circRNAs interacted with 35 proteins. We considered 29623 interactions with 29623 circRNAs and 35 proteins as positive samples. Then, with the 29623 circRNAs and 35 proteins, we randomly selected 29622 interactions as negative samples in such a way that the same interaction in positive sample was not selected as negative sample. Finally, we got a total $29623 + 29622 = 59245$ interactions with the positive-to-negative interaction ratio 1:1. We used 41471 (70%) interactions for training and 17774 (30%) interactions for testing the model. The circRNA and protein sequences were downloaded from circbase (http://circbase.org/) and UniPort (https://www.uniprot.org/) databases respectively.

2.2 Feature Generation

The amino acids of the protein sequences were classified according to dipole moments and the volume of the side chains. According to these criteria, the amino acids were classified into 7 categories as shown in Table 1. The 20 letter amino acids sequences were then converted into 7 letter sequences. Then, 3-mers of the 7 letter sequences were generated. It produced the $7 \times 7 \times 7 = 343$ all possible 3-mers. A total of 343 features were used to encode a protein. The frequencies of these 343 features were used as value of the features.

Table 1. Classification of amino acid sequences.

Category	Symbol	Short Name	Full Name
1	A, G, V	ala, gly, val	Alanine, Glycine, Valine
2	I, L, F, P	ile, leu, phe, pro	Isoleucine, Leucine, Phenylalanine, Proline
3	Y, M, T, S	tyr, met, thr, ser	Tyrosine, Methionine, Threonine, Serine
4	H, N, Q, W	his, asn, gln, trp	Histidine, Asparagine, Glutamine, Tryptophan
5	R, K	arg, lys	Arginine, Lysine
6	D, E	asp, glu	Aspartic Acid, Glutamic Acid
7	C	cys	Cysteine

For each circRNA, the 4 letter nucleotide sequences were converted into 21 letter (20 amino acids and 1 stop codon) pseudo peptides. Next, the 21 letters pseudo peptide sequences were converted into 8 (7 amino acid classes plus 1 stop codon) letters sequences. Then, 3-mers of the 8 letter sequences produced a total of $8 \times 8 \times 8 = 512$ all possible 3-mers. In total, 512 features were used to encode a circRNA. The frequencies of these 512 features were used as value of the features.

2.3 Machine Learning Models

After feature generation, we used several machine learning models including Support vector machine (SVM), the least absolute shrinkage and selection operator (LASSO), Decision tree (DT) and Naïve Bayes (NB) for predicting circRNA-protein interaction. Based on the prediction performance, we selected the best classification model.

2.4 Evaluation Metrics

We used the training data to fit different machine learning models and used the test data to evaluate the model performances. We used four evaluation metrics precision, sensitivity, accuracy and F-measure for the evaluation of different models defined as follows.

$$Precision = \frac{TP}{TP + FP}$$

$$Sensitivity = \frac{TP}{TP + FN}$$

$$Accuracy = \frac{TP + TN}{TP + TN + FP + FN}$$

$$F - measure = \frac{2 * precision * sensitivity}{precision + sensitivity}$$

Here *TP, TN, FP* and *FN* were the number of true positives, true negatives, false positives and false negatives respectively.

3   Results

3.1   Prediction of Interaction Using Different Models

We got a total of 855 (343 + 512) features from the alternative representation of the circRNA and protein sequences. We counted the frequencies of the features and constructed the data matrix. We observed that there were zero column sums for several columns of the data matrix which implied that for several features all the values were zero. We deleted such type of features which resulted a total of 500 features. We used these 500 features for training different models.

At first, we used support vector machine to predict the interaction between circRNA and protein. In SVM, there are two types of parameter to be optimized: the accuracy parameter C and the kernel type K. Radial basis function (RBF) kernel was used in this study. The parameter optimization was done using grid search. A total of 5763 support vectors were used for classification. The training and test accuracies were 93.76% and 93.52% respectively. The misclassification errors for the training and test data were 6.24% and 6.48% respectively.

Then, we used LASSO model for predicting the circRNA-protein interaction. In LASSO model, the regression parameters are estimated by minimizing the sum of squared residuals with the constraint that $\sum_{i=1}^{k} |\beta_i| \leq \lambda$, where $\lambda > 0$ is a parameter indicating the amount of shrinkage. The best lambda was obtained using cross validation. The training and test accuracies were 93.22% and 93.06% respectively. The misclassification errors were 6.78% and 6.94% for training and test data respectively.

Next, we used decision tree utilizing the *tree* r package for fitting decision tree model and making prediction. The accuracy and misclassification error for training data were 92.79% and 7.21% whereas the accuracy and misclassification error for test data were 92.49% and 7.51%.

Finally, we used Naïve Bayes classifier using *e1071* r package. The accuracy for the training and test data were 87.83% and 88.02% respectively whereas the misclassification error for training and test data were 12.17% and 11.98% respectively.

3.2   Comparison of Different Models' Performance

We provided the circRNA and protein sequences as inputs in our model. Using the alternative representation of the sequences, a total of 855 features were obtained. With the frequencies of the features, we constructed the data matrix. We modified the data matrix by deleting the columns (features) having zero column sum. To compare the performance of the models, we used the evaluation metrics precision, sensitivity, accuracy and F-measure. The values of the different evaluation metrics were given in Table 2.

Table 2. Calculation of different evaluation metrics.

Model	Training/Test	TP	FP	TN	FN	Precision (%)	Sensitivity (%)	Accuracy (%)	F-measure
SVM	training	19444	1295	19440	1292	93.76	93.77	93.76	0.94
	test	8309	574	8313	578	93.54	93.50	93.52	0.94
LASSO	training	19198	1276	19459	1538	93.77	92.58	93.21	0.93
	test	8223	570	8317	664	93.52	92.53	93.06	0.93
DT	training	19752	2007	18728	984	90.78	95.25	92.79	0.93
	test	8470	917	7970	417	90.23	95.31	92.49	0.93
NB	training	18166	2476	18259	2570	88.01	87.61	87.83	0.88
	test	7812	1055	7832	1075	88.10	87.90	88.02	0.88

From Table 2, we observed that the training and test precisions of SVM (93.76% and 93.54%) were higher than that of other models (Table 2). The sensitivities for training and test data of DT (95.25% and 95.31%) were higher than that of other models. The accuracies of the training and test data of SVM (93.76% and 93.52%) were greater than that of other models. Again, SVM achieved higher F-measure (0.94) than other models.

The precision of SVM was greater than that of other models (see Fig. 1A). The sensitivity of DT was higher than that of other models (see Fig. 1B). Again, SVM achieved

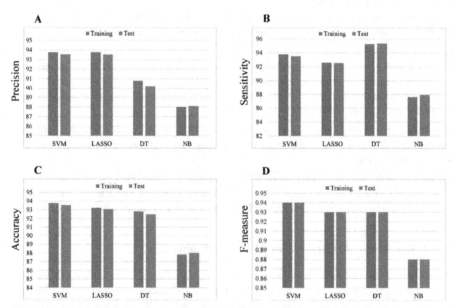

Fig. 1. Performance comparison of different models. (A) Precision, (B) Sensitivity, (C) Accuracy and (D) F-measure for different models in training and test datasets.

highest accuracy (see Fig. 1C). And SVM gained largest F-measure among all the models (see Fig. 1D). Overall, we found that SVM classifier showed better performance than other models.

4 Discussion

CircRNAs play important role in gene regulation and development of different diseases. The study of circRNA-protein interaction can reveal the biological functions of the circRNAs. Several computation methods have been developed for the study of circRNA-protein interactions. These methods mainly fall into two categories: identification of binding sites in RNA chain and protein chain. No existing methods use both RNA and protein sequences for predicting the interaction. We utilized both the circRNA and protein sequences for predicting their interactions.

Different schemes are developed for the representation of biological sequences. The sequence representation methods fall into two categories: one is feature engineering and the other is sequence encoding. In some studies [42, 43], the k-mer frequencies of the sequences were used as feature and the SVM and random forest models were used as classifiers. Some other studies [37, 44, 45] used one-hot encoding for feature extraction and deep learning methods for classification. One-hot encoding has several limitations including low dimensional feature representation and sequence context information is not encoded properly. With the increase in machine learning and deep learning methods, the traditional feature extraction methods have been replaced by the sequence encoding schemes. We used alternative representation of the circRNA and protein sequences.

The existing methods for predicting the binding sites on circRNAs use only circRNA sequences and predicted whether there is binding sites on the circRNA or not. In our method, we utilized both circRNA and protein sequences and predicted whether the circRNA and protein sequences are interacted with a specified probability of interaction. As we used a new approach, our method is not comparable to the existing methods. We computed the performance of our method and found that our method achieved 93% prediction accuracy. We believe that our method will add an additional value in the study of circRNA and protein interactions.

The limitation of our method is that there is a limited number of proteins in the training dataset. So far, there is only one circRNA-protein interaction database named circinteractome. After deleting redundancy, there are 29623 interactions with 29623 circRNAs and 35 proteins. The number of proteins are very limited compared to the number of circRNAs. This is the limitation of the current method.

5 Conclusion

We developed a method PCPI to predict the interaction between circRNA and protein. The existing methods use either RNA sequences or protein sequences to predict the interaction. But our method PCPI used both the circRNA and protein sequences to predict the interaction between them. Several machine learning algorithms were used for classifying the circRNA-protein interaction. Among them, SVM performed better than the other algorithms and we selected SVM as the classifier for our method. Our method gained 93% prediction accuracy. We developed an R package for our method which is freely available at https://github.com/tofazzalh/PCPI.

Funding. This work was partly supported by the Key Research and Development Project of Guangdong Province under grant No. 2021B0101310002, National Key Research and Development Program of China Grant No. 2021YFF1200100, Strategic Priority CAS Project XDB38050100, National Science Foundation of China under grant No. 62272449, the Shenzhen Basic Research Fund under grant, No. RCYX20200714114734194, KQTD20200820113106007 and ZDSYS20220422103800001. We would also like to thank the funding support by the Youth Innovation Promotion Association (Y2021101), CAS to Yanjie Wei.

References

1. Li, L., et al.: Comprehensive analysis of CircRNA expression Bprofiles in humans by RAISE. Int. J. Oncol. **51**, 1625–1638 (2017). https://doi.org/10.3892/ijo.2017.4162
2. Kristensen, L.S., Andersen, M.S., Stagsted, L.V.W., Ebbesen, K.K., Hansen, T.B., Kjems, J.: The biogenesis, biology and characterization of circular RNAs. Nat. Rev. Genet. **20**, 675–691 (2019)
3. Li, X., Yang, L., Chen, L.L.: The biogenesis, functions, and challenges of circular RNAs. Mol. Cell **71**, 428–442 (2018)
4. Jeck, W.R., et al.: Circular RNAs are abundant, conserved, and associated with ALU repeats. RNA **19**, 141–157 (2013). https://doi.org/10.1261/rna.035667.112
5. Hansen, T.B., et al.: Natural RNA circles function as efficient microRNA sponges. Nat. **495**, 384–388 (2013). https://doi.org/10.1038/nature11993
6. Hentze, M.W., Preiss, T.: Circular RNAs: splicing's enigma variations. EMBO J. **32**, 923–925 (2013). https://doi.org/10.1038/emboj.2013.53
7. Li, G.F., Li, L., Yao, Z.Q., Zhuang, S.J.: Hsa_circ_0007534/MiR-761/ZIC5 Regulatory loop modulates the proliferation and migration of glioma cells. Biochem. Biophys. Res. Commun. **499**, 765–771 (2018). https://doi.org/10.1016/j.bbrc.2018.03.219
8. Han, D., et al.: Circular RNA CircMTO1 acts as the sponge of MicroRNA-9 to suppress hepatocellular carcinoma progression. Hepatol. **66**, 1151–1164 (2017). https://doi.org/10.1002/hep.29270
9. Huang, W.J., et al.: Silencing Circular RNA Hsa_circ_0000977 suppresses pancreatic ductal adenocarcinoma progression by stimulating MiR-874-3p and inhibiting PLK1 expression. Cancer Lett. **422,** 70–80 (2018)
10. Chen, J., et al.: Circular RNA profile identifies CircPVT1 as a proliferative factor and prognostic marker in gastric cancer. Cancer Lett. **388**, 208–219 (2017). https://doi.org/10.1016/j.canlet.2016.12.006
11. Xu, T., Wu, J., Han, P., Zhao, Z., Song, X.: Circular RNA expression profiles and features in human tissues: a study using RNA-Seq data. BMC Genomics **18** (2017). https://doi.org/10.1186/s12864-017-4029-3
12. Tucker, D., Zheng, W., Zhang, D.-H., Dong, X.: Circular RNA and its potential as prostate cancer biomarkers. World J. Clin. Oncol. **11**, 563–572 (2020). https://doi.org/10.5306/wjco.v11.i8.563
13. Li, Z., Chen, Z., Hu, G.H., Jiang, Y.: Roles of circular RNA in breast cancer: present and future. Am. J. Transl. Res. **11**, 3945–3954 (2019)
14. Du, W.W., Zhang, C., Yang, W., Yong, T., Awan, F.M., Yang, B.B.: Identifying and characterizing CircRNA-Protein interaction. Theranostics **7**, 4183–4191 (2017)

15. Zang, J., Lu, D., Xu, A.: The interaction of CircRNAs and RNA binding proteins: an important part of CircRNA maintenance and function. J. Neurosci. Res. **98**, 87–97 (2020)
16. Li, Y.E., et al.: Identification of high-confidence RNA regulatory elements by combinatorial classification of RNA-protein binding sites. Genome Biol. **18** (2017). https://doi.org/10.1186/s13059-017-1298-8
17. Yang, Y.C.T., et al.: CLIPdb: a CLIP-Seq database for protein-RNA interactions. BMC Genomics **16** (2015). https://doi.org/10.1186/s12864-015-1273-2
18. Li, J.H., Liu, S., Zhou, H., Qu, L.H., Yang, J.H.: StarBase v2.0: decoding MiRNA-CeRNA, MiRNA-NcRNA and protein-RNA interaction networks from large-scale CLIP-Seq data. Nucleic Acids Res. **42** (2014). https://doi.org/10.1093/nar/gkt1248
19. Zhao, H., Yang, Y., Zhou, Y.: Prediction of RNA binding proteins comes of age from low resolution to high resolution. Mol. Biosyst. **9**, 2417–2425 (2013)
20. Fornes, O., Garcia-Garcia, J., Bonet, J., Oliva, B.: On the use of knowledge-based potentials for the evaluation of models of Protein-Protein, Protein-DNA, and Protein-RNA interactions. Adv. Protein Chem. Struct. Biol. **94**, 77–120 (2014). ISBN 9780128001684
21. Kauffman, C., Karypis, G.: Computational tools for Protein-DNA interactions. Wiley Interdiscip. Rev. Data Min. Knowl. Discov. **2**, 14–28 (2012)
22. Liu, L.A., Bradley, P.: Atomistic modeling of Protein-DNA interaction specificity: progress and applications. Curr. Opin. Struct. Biol. **22**, 397–405 (2012)
23. Choi, S., Han, K.: Predicting Protein-binding RNA nucleotides using the feature-based removal of data redundancy and the interaction propensity of nucleotide triplets. Comput. Biol. Med. **43**, 1687–1697 (2013). https://doi.org/10.1016/j.compbiomed.2013.08.011
24. Panwar, B., Raghava, G.P.S.: Identification of Protein-Interacting nucleotides in a RNA sequence using composition profile of Tri-Nucleotides. Genomics **105**, 197–203 (2015). https://doi.org/10.1016/j.ygeno.2015.01.005
25. Jia, C., Bi, Y., Chen, J., Leier, A., Li, F., Song, J.: PASSION: an ensemble neural network approach for identifying the binding sites of RBPs on CircRNAs. Bioinformatics **36**, 4276–4282 (2020). https://doi.org/10.1093/bioinformatics/btaa522
26. Wang, Z., Lei, X.: Matrix factorization with neural network for predicting CircRNA-RBP interactions. BMC Bioinform. **21** (2020). https://doi.org/10.1186/s12859-020-3514-x
27. Conn, S.J., et al.: The RNA binding protein quaking regulates formation of CircRNAs. Cell **160**, 1125–1134 (2015). https://doi.org/10.1016/j.cell.2015.02.014
28. Abdelmohsen, K., et al.: Identification of HuR target circular RNAs uncovers suppression of PABPN1 translation by CircPABPN1. RNA Biol. **14**, 361–369 (2017). https://doi.org/10.1080/15476286.2017.1279788
29. Dudekula, D.B., Panda, A.C., Grammatikakis, I., De, S., Abdelmohsen, K., Gorospe, M.: Circinteractome: a web tool for exploring circular RNAs and their interacting proteins and MicroRNAs. RNA Biol. **13**, 34–42 (2016). https://doi.org/10.1080/15476286.2015.1128065
30. Okholm, T.L.H., et al.: Transcriptome-wide profiles of circular RNA and RNA-binding protein interactions reveal effects on circular RNA biogenesis and cancer pathway expression. Genome Med. **12** (2020). https://doi.org/10.1186/s13073-020-00812-8
31. Zhou, H.L., Mangelsdorf, M., Liu, J.H., Zhu, L., Wu, J.Y.: RNA-binding proteins in neurological diseases. Sci. China Life Sci. **57**, 432–444 (2014)
32. Pereira, B., Billaud, M., Almeida, R.: RNA-binding proteins in cancer: old players and new actors. Trends in Cancer **3**, 506–528 (2017)
33. Prashad, S., Gopal, P.P.: RNA-binding proteins in neurological development and disease. RNA Biol. **18**, 972–987 (2021)
34. Zhang, K., Pan, X., Yang, Y., Shen, H.: Bin CRIP: predicting CircRNA-RBP-binding sites using a codon-based encoding and hybrid deep neural networks. RNA **25**, 1604–1615 (2019). https://doi.org/10.1261/rna.070565.119

35. Maticzka, D., Lange, S.J., Costa, F., Backofen, R.: GraphProt: modeling binding preferences of RNA-binding proteins. Genome Biol. **15** (2014). https://doi.org/10.1186/gb-2014-15-1-r17

36. Corrado, G., Tebaldi, T., Costa, F., Frasconi, P., Passerini, A.: RNAcommender: genome-wide recommendation of RNA-Protein interactions. Bioinformatics **32**, 3627–3634 (2016). https://doi.org/10.1093/bioinformatics/btw517

37. Alipanahi, B., Delong, A., Weirauch, M.T., Frey, B.J.: Predicting the sequence specificities of DNA- and RNA-binding proteins by deep learning. Nat. Biotechnol. **33**, 831–838 (2015). https://doi.org/10.1038/nbt.3300

38. Pan, X., Shen, H.: Bin predicting RNA-Protein binding sites and motifs through combining local and global deep convolutional neural networks. Bioinformatics **34**, 3427–3436 (2018). https://doi.org/10.1093/bioinformatics/bty364

39. Yuan, L., Yang, Y.: DeCban: prediction of CircRNA-RBP interaction sites by using double embeddings and cross-branch attention networks. Front. Genet. **11** (2021). https://doi.org/10.3389/fgene.2020.632861

40. Niu, M., Zou, Q., Lin, C.: CRBPDL: identification of CircRNA-RBP interaction sites using an ensemble neural network approach. PLoS Comput. Biol. **18** (2022). https://doi.org/10.1371/journal.pcbi.1009798

41. Fu, L., Niu, B., Zhu, Z., Wu, S., Li, W.: CD-HIT: accelerated for clustering the next-generation sequencing data. Bioinformatics **28**, 3150–3152 (2012). https://doi.org/10.1093/bioinformatics/bts565

42. Shen, J., et al.: Predicting protein-protein interactions based only on sequences information. Proc. Natl. Acad. Sci. U.S.A. **104**, 4337–4341 (2007). https://doi.org/10.1073/pnas.0607879104

43. Muppirala, U.K., Honavar, V.G., Dobbs, D.: Predicting RNA-protein interactions using only sequence information. BMC Bioinform. **12** (2011). https://doi.org/10.1186/1471-2105-12-489

44. Pan, X., Shen, H.: Bin RNA-protein binding motifs mining with a new hybrid deep learning based cross-domain knowledge integration approach. BMC Bioinform. **18** (2017). https://doi.org/10.1186/s12859-017-1561-8

45. Pan, X., Rijnbeek, P., Yan, J., Shen, H.: Bin prediction of RNA-protein sequence and structure binding preferences using deep convolutional and recurrent neural networks. BMC Genomics **19** (2018). https://doi.org/10.1186/s12864-018-4889-1

Radiology Report Generation via Visual Recalibration and Context Gating-Aware

Xiaodi Hou, Guoming Sang, Zhi Liu, Xiaobo Li, and Yijia Zhang[✉]

School of Information Science and Technology, Dalian Maritime University, Dalian 116026, Liaoning, China
zhangyijia@dlmu.edu.cn

Abstract. The task of radiology report generation aims to analyze medical images, extract key information, and then assist medical personnel in generating detailed and accurate reports. Therefore, automatic radiology report generation plays an important role in medical diagnosis and healthcare. However, radiology medical data face the problems of visual and text data bias: medical images are similar to each other, and the normal feature distribution is larger than the abnormal feature distribution; second, the accurate location of the lesion and the generation of accurate and coherent long text reports are important challenges. In this paper, we propose Visual Recalibration and Context Gating-aware model (VRCG) to alleviate visual and textual data bias for enhancing report generation. We employ a medical visual recalibration module to enhance the key lesion feature extraction. We use the context gating-aware module to combine lesion location and report context information to solve the problem of long-distance dependence in diagnostic reports. Meanwhile, the context gating-aware module can identify text fragments related to lesion descriptions, improve the model's perception of lesion text information, and then generate coherent, consistent medical reporting. Extensive experiments demonstrate that our proposed model outperforms existing baseline models on a publicly available IU X-Ray dataset. The source code is available at: https://github.com/Eleanorhxd/VRCG.

Keywords: Medical Visual Recalibration · Context Gating-aware · Report Generation

1 Introduction

Radiology report generation has emerged as a systematic research area in medical imaging technology and clinical decision-making tasks. In clinical practice, the conventional approach of radiology report generation involves trained and experienced doctors manually reviewing images and then composing detailed reports to describe their findings. This process typically takes an average of 10 min or more for a radiologist to complete a full report [1]. Moreover, with the exponential growth of medical imaging data, the inconsistent proficiency of clinicians in editing medical diagnoses poses significant challenges to the quality of radiology report generation [3].

X. Guo et al. (Eds.): ISBRA 2023, LNBI 14248, pp. 107–119, 2023.
https://doi.org/10.1007/978-981-99-7074-2_9

However, recent advancements in deep learning have profoundly impacted the field of medicine and healthcare, offering innovative solutions for automated radiology report generation. Leveraging generative models based on deep neural architectures can automatically extract key information from images, generating accurate and consistent reports. This technological breakthrough significantly enhances the efficiency and precision of radiology report generation, thereby reducing the likelihood of human misdiagnosis and missed diagnoses.

Most existing models [3, 7, 8] employ image captioning methods and adopt encoder and decoder frameworks [9, 10]. Oriol et al. [4] designed a model, Neural Image Caption (NIC), which utilizes a Convolutional Neural Network (CNN) as an encoder to represent image features compactly and leverages a Recurrent Neural Network (RNN) for decoding to generate corresponding text descriptions of the image. In contrast to NIC, Chen et al. [5] incorporated patch features learned from CNN into the transformer encoder layer to capture rich visual information in medical images. The authors utilized the transformer decoder to align visual features with corresponding text features. Notably, the authors designed a relational memory module to record key information during training and ultimately integrated it into the transformer decoder. Subsequently, Chen et al. [6] enhanced their previous work by introducing memory query and memory response mechanisms to facilitate sufficient interaction between images and texts.

However, automatic radiology report generation still faces numerous challenges, one of which is the issue of text and visual data bias [11]. In clinical practice, radiologists often tend to describe the overall coarse-grained features of the image, with a predominant focus on describing normal areas. The problem of text data bias arises from the imbalanced distribution of normal and abnormal descriptions, which may lead the model to excessively emphasize normal descriptions during training, causing it to struggle to generate descriptions accurately specific to anomaly discovery. Additionally, pathological or abnormal areas in radiologic images typically occupy a small portion of the entire image, and these abnormal regions may sometimes appear somewhat blurry. Severe visual bias issues significantly hamper the model's ability to identify abnormal areas accurately.

To address the above challenges, we propose a Visual Recalibration and Context Gating-aware (VRCG) model to mitigate text and visual data bias issues. This paper introduces a Medical Vision Recalibration module (MVR) to extract advanced features from radiological images, thereby enhancing the representation of lesion areas. Moreover, considering the significant text data bias in radiology reports, we design a Context Gating-aware Module (CGM) to recognize disease-specific text descriptions while capturing the contextual dependencies of input sequences. Our specific contributions can be categorized into the following three aspects:

This paper designs a Medical Vision Recalibration module (MVR) to capture rich visual information in radiological images, which can enhance the model's ability to represent visual features and lesion areas.

The Context Gating-aware Module (CGM) is designed to recognize disease-specific text descriptions, capture the contextual dependencies of input sequences, and generate coherent and consistent medical reports.

The experimental results show that our model has been compared with several strong baseline models on different evaluation indicators on the widely used IU X-Ray dataset, achieving state-of-the-art (SOTA) performance.

2   Method

Automatic report generation is a multi-modal process that combines image and text information. This task utilizes deep learning methods to analyze and understand input radiology medical images $Img = \{i_1, i_2, \ldots, i_{N_d}\}$ to extract key information and generate corresponding radiology medical reports $Y = \{y_1, y_2, \ldots, y_{N_r}\}$. The N_d and N_r represent the total of the radiology medical images and reports.

This paper proposes a Visual Recalibration and Context Gating-aware model (VRCG) for radiology report generation, which composes a Vision Features Extractor (VFE), a Medical Vision Recalibration module (MVR), a Context Gating-aware Module (CGM) and a Report Generator (RG). The overall architecture is shown in Fig. 1.

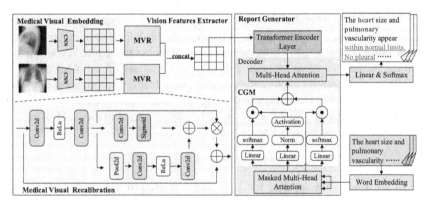

Fig. 1. The overview architecture of VRCG

2.1   Visual Features Extractor

The visual features extractor employs the **M**edical **V**isual **E**mbedding module (MVE) and the **M**edical **V**isual **R**ecalibration module (MVR) to focus on the lesion position in the medical images and enhance the fine-grained extraction of abnormal visual features, which could alleviate the data bias phenomenon.

Medical Visual Embedding
The MVE takes a patient's anteroposterior and lateral radiological medical images $Img = \{i_m^1, i_m^2\} \in \mathbb{R}^D$ as input, and the size of each image is $3 \times 224 \times 224$. Following previous work [5], we use the pre-trained model ResNet-101 [12] as the encoder of the CNN to extract the visual features of radiological medical images. Typically, the MVE extracts anteroposterior and lateral visual features v_1 from the last convolutional

layer. Consequently, we acquire a series of visual feature representations, denoted as $V = \{v_1, v_2, \ldots, v_{N_d}\} \in \mathbb{R}^d$, where v_{N_d} represents extracted visual features and d is the size of the feature vector. The visual embedding module's process is as follows:

$$V = \{v_1, v_2, \ldots, v_{N_d}\} = f_{vem}(Img) \tag{1}$$

where $f_{vem}(\cdot)$ denotes the visual embedding module.

Medical Visual Recalibration

After MVE, we extract a series of visual features of radiological medical images. Then, we introduce a Medical Visual Recalibration module (MVR) to strengthen the model's ability to perceive key structures and lesion locations in medical images and improve the accuracy and reliability of the visual features on lesion locations.

Due to data bias and the unbalanced distribution of features in medical images, normal visual features occupy most areas of medical images, and abnormal parts only account for a small number. As the network depth becomes deeper, low-level image features are easily erased by too-deep layers. However, residual blocks introduce skip connections, allowing the network to pass input features to subsequent layers when needed directly. This mechanism extends the representation capability of the network, enabling the network to capture low-level features and high-level semantic information simultaneously. Therefore, MVR introduces residual blocks to capture key information in medical images for more accurate diagnosis and analysis of radiological medical images. We can effectively capture the positive visual features information in medical images with the residual block, thus improving the model's performance.

We feed the obtained visual features $V = \{v_1, v_2, \ldots, v_{N_d}\}$ to the residual module. The calculation process is as follows:

$$U = \{u_1, u_2, \ldots, u_{N_d}\} = ReLu(f_{res}(V))\# \tag{2}$$

where $f_{res}(\cdot)$ denotes the residual function, which is used to learn radiological visual feature information.

In addition, MVR also adopts channel attention and spatial attention to enhance the model's ability to extract abnormal features, which can adaptively increase the attention to key features in the input visual features. Channel attention can adaptively learn the lesion visual features of visual features to better capture the key information related to diseases in medical images and improve the expressiveness and discrimination of visual features. For the channel attention module, we send the visual features obtained through the residual module to the channel attention module. The process is as follows:

$$\widehat{U} = AvgPool(U)\# \tag{3}$$

$$\widehat{X} = F_{c2}(F_{c1}(\widehat{U}))\# \tag{4}$$

$$X = \sigma\left(\widehat{X}\right)\# \tag{5}$$

where $AvgPool(\cdot)$ represents the adaptive average pooling operation, $F_c(\cdot)$ represents the full connection, and $\sigma(\cdot)$ represents the *sigmoid* activation.

Spatial attention can preserve the spatial structure information of the original image, can better utilize the local spatial information in the image, and preserve the structural details and context information. For the spatial attention module, we send the visual features obtained through the residual module to the spatial attention module. The process is as follows:

$$Y = Conv_{3\times3}(U)\#$$ (6)

where $Conv_{3\times3}(\cdot)$ denotes the 3×3 convolution.

The joint action of the attention mechanism helps to capture the abnormal visual features in medical images accurately.

$$H = V \oplus ((X \oplus Y) \otimes U)\#$$ (7)

where $\otimes$ is element-wise product, $\oplus$ is element-wise addition. The $H = \{h_1, h_2, \ldots, h_{N_d}\} \in \mathbb{R}^d$.

2.2 Report Generator

The report generator is composed of a standard transformer encoder and transformer decoder [13]. In the transformer decoder, we use the Context Gating-aware Module (CGM) to combine lesion characteristics and key information of the report context to solve the problems of long-distance dependence and data bias in generating reports.

Transformer Encoder

The input medical image data are further encoded and feature extracted through the transformer encoder to provide strong support for radiological analysis and diagnosis. Each encoding layer consists of a multi-head self-attention mechanism and a fully connected feed-forward network. The multi-head self-attention mechanism captures key disease information of visual features by calculating the internal structure of input features.

We feed the enhanced visual features $H = \{h_1, h_2, \ldots, h_{N_d}\}$ into the encoder. First, the input visual feature sequence is linearly transformed to obtain query$Q \in \mathbb{R}^{G_q \times d_k}$, key $K \in \mathbb{R}^{G_k \times d_k}$ and value $V \in \mathbb{R}^{G_k \times d_k}$.

$$Q = HW_q K = HW_k V = HW_v\#$$ (8)

where W_q, W_k, W_v are learnable weight matrices.

Next, we calculate the attention score for each attention head and then use the attention score to perform a weighted summation to obtain the output of each attention head. The calculation process is as follows:

$$Att_i(Q, K) = softmax\left(\frac{QK^T}{\sqrt{d_k}}\right)\#$$ (9)

$$head_i = Att_i V\#$$ (10)

where *softmax* is used to normalize the attention score to a probability distribution, and $\sqrt{d_k}$ is the scaling factor.

Finally, the output of all attention heads is concatenated to obtain the final output of multi-head self-attention.

$$M = MHA(head_1, head_2, \ldots, head_h)W_O \#$$ (11)

where W_O is the learnable weight matrix.

After the multi-head self-attention mechanism, the final output is obtained through residual connection and layer normalization operations:

$$V' = LN(Res(M + V)) \#$$ (12)

where $Res(\cdot)$ and $LN(\cdot)$ denotes residual connection and layer normalization operations, respectively.

Transformer Decoder

We employ the transformer decoder to generate medical reports. At the decoder level, we introduce CGM to capture positive disease contextual information in medical report sequences to improve the model's ability to extract lesion information. Therefore, the decoding process is expressed as:

$$y_t = f_d\left(v'_1, \ldots, v'_{N_d}, CGM\left(y_1, \ldots, y_{N_r-1}\right)\right) \#$$ (13)

where $f_d(\cdot)$ denotes the transformer decoder. The detailed process of CGM is introduced in the following subsections.

After the above process, the entire report generation process can be written as follows:

$$p(Y|Img) = \prod_{t=1} p\left(y_{N_r}|y_1, y_2, \ldots, y_{N_r-1}, Img\right) \#$$ (14)

where $Y = \{y_1, y_2, \ldots, y_{N_r}\}$ is the target sequence.

Context Gating-Aware Module

We employ a Context Gating-aware Module (CGM) in the decoder layer to enhance the transformer encoder's ability to perceive contextual information. CGM could alleviate the text data bias and the long-distance dependency problem in text reports. The CGM introduces information gate and state gate to solve the long text dependency problem and capture the context information of text reports. Due to the unbalanced distribution of lesion information, CGM can identify and correlate this information, improve the model's ability to extract key information about the disease (including lesion location, shape, size, etc.,), and then generate accurate and coherent radiology medical imaging reports.

The decoder first feeds the report word embedding to Masked Multi-Head Attention to obtain the contextual representation of the position of each word $E = \{e_1, \ldots, e_{N_r}\}$ in the medical text report. After obtaining the contextual representation of each location, it enters the CGM.

At time step t, an information gate is entered to control information flow and enhance the importance of lesion information. The calculation process is as follows:

$$E = MHA(y_1, \ldots, y_{N_r-1}) \#$$ (15)

$$E^l = \sigma(W_i \cdot [y_i, e_i] + b_i)\# \tag{16}$$

$$Y^c = Act(N(L(Y)))\# \tag{17}$$

$$C = Y^c \odot E^l \# \tag{18}$$

where $\sigma(\cdot)$, $L(\cdot)$, $N(\cdot)$ and $\odot$ denotes *sigmoid* function, linear layer, normalization operation, element-wise multiplication, respectively. Y^c represents the sequence obtained by context attention, C is the context representation adjusted by information gating.

Then, the C is taken into the state gate. It is introduced to adjust the encoder state of the current position and determine how much information about the current position is retained. It produces an adjusted encoder state by element-wise multiplying the context representation with the previous encoder state. The calculation process is as follows:

$$S = \sigma\left(W_s \cdot [y_i, c_i] + b_s\right)\# \tag{19}$$

$$S' = Y \odot S\# \tag{20}$$

where W_i, b_i, W_s, b_s are learnable parameters.

Finally, the output of the context gating-aware module is:

$$G = C \oplus S' \tag{21}$$

where $\oplus$ is element-wise addition. G is the output of the entire process at the step t.

3 Experiment Settings

This section introduces the datasets required for our experiments, evaluation metrics and implementation details.

3.1 Datasets

IU X-Ray: We conducted experiments using the publicly available IU X-Ray dataset [14]. IU X-Ray is a widely used medical imaging dataset collected by Indiana University, comprising 7,470 medical radiological images and 3,955 related medical reports. Within this dataset, most patients have provided both a medical report and lateral and anteroposterior medical images. Each medical image contains a diverse range of disease categories and information, including *pneumonia, tuberculosis, pneumothorax, abnormal lung texture,* and more. The medical reports consist of findings, impressions and other relevant details.

To adhere to the settings of our previous work, the proposed model VRCG randomly divides the IU X-Ray dataset into training, testing, and validation sets, with proportions of 70%, 20%, and 10%, respectively.

3.2 Evaluation Metrics

To assess the quality of the reports generated by the proposed VRCG model, we employed several natural language generation (NLG) indicators for evaluation, including BLEU [15], METEOR [16], and ROUGE-L [17]. BLEU encompasses multiple variants, including BLEU-1, BLEU-2, BLEU-3, and BLEU-4, which are utilized to gauge the similarity between the generated medical reports and the ground truth. METEOR evaluates the resemblance between the generated text and the reference text by calculating matches at both the word and phrase levels. ROUGE-L measures the quality of automatic summarization and text generation tasks.

3.3 Implementation Details

To ensure consistency between our proposed VRCG model and previous methods, we employed two radiological medical images as inputs to generate reports on the IU X-Ray dataset. For the radiology visual feature extraction, we utilized the pre-trained ResNet-101 model as the backbone CNN model. MVR extracted each patch feature with a dimensionality of $d = 2048$. As for the report generator module, we adopted a transformer architecture with three layers and eight attention heads as the encoder-decoder. During the training process, we employed the Adam [18] optimizer and utilized the cross-entropy loss function to optimize our proposed model VRCG. The visual extractor parameters were set to $5e - 5$, while the other parameters were set to $1e - 4$.

4   Results and Analysis

4.1 Performance Comparison with Other Models

We compared our experimental results with the previous radiology report generation models on the IU X-Ray dataset. As shown in Table 1, our proposed model VRCG achieved state-of-the-art (SOTA) performance on the IU X-Ray dataset. According to Table 1, when compared to the PPKED model, our model outperformed it in several key metrics. Specifically, BLEU-1, BLEU-2, BLEU-3, BLEU-4 are increased by 1.6%, 2.8%, 1.9%, and 1.2%, respectively. These results indicate that our model effectively filters out irrelevant information and places more emphasis on generating accurate and comprehensive disease descriptions.

Furthermore, our model shows improvement in other evaluation indicators as well. The METEOR score is 2.8% higher than that of PPKED, suggesting that our model excels in generating radiology medical reports that are both accurate and fluent. The ROUGE-L evaluation metric also reveal a 3.5% improvement for VRCG over PPKED, demonstrating that our model has made significant progress in generating coherent disease descriptions. In summary, our proposed VRCG model exhibits advancements in capturing longer phrases, and enhancing accuracy, fluency, and coherence in disease descriptions.

Table 1. Comparison of the proposed model with previous studies on the test sets of IU X-Ray with respect to NLG metrics.

MODEL	BL-1	BL-2	BL-3	BL-4	MET	RG-L
NIC* [4]	0.216	0.124	0.087	0.066	–	0.306–
SA&T* [20]	0.339	0.251	0.168	0.118	–	0.323
HRGR [2]	0.438	0.298	0.208	0.151	–	0.322
COATT [3]	0.455	0.288	0.205	0.154	–	0.369
R2Gen [5]	0.470	0.304	0.219	0.165	0.187	0.371
CMN [6]	0.475	0.309	0.222	0.170	0.191	0.375
PPKED [9]	0.483	0.315	0.224	0.168	0.190	0.376
VRCG	**0.499**	**0.343**	**0.243**	**0.180**	**0.218**	**0.411**

The best values are highlighted in bold.

4.2 Ablation Experiments

To verify the effectiveness of the medical visual recalibration and context gating-aware module, we conducted experiments on each module on the dataset IU X-Ray.

Effect of the Medical Visual Recalibration Module

To verify the effect of the medical visual recalibration module, we only add the visual recalibration module to the base model and use BLEU, METEOR and ROUGE-L to evaluate the module. It can be seen from Table 2 that after adding the medical visual recalibration module, BLEU-1, BLEU-2, BLEU-3, and BLEU-4 are 9.6%, 7.7%, 6.4%, and 4.5% higher than BASE, respectively. In addition, METEOR and ROUGE-L are 3.4% and 2.6% higher than BASE, respectively. Among them, BLEU-4 reaches the highest, which shows that the medical visual recalibration module can capture abnormal findings in radiology medical images, strengthen the visual features of abnormal findings, and alleviate data bias.

Effect of the Context Gating-Aware Module

To prove the effect of the context gating-aware module, we removed the medical visual recalibration module from the model and only included CGM experiments on the IU X-Ray dataset. The results in Table 2 clearly demonstrate the effectiveness of incorporating the context gating-aware module. With the inclusion of this module, we observe notable improvements across various metrics. Specifically, BLEU-1, BLEU-2, BLEU-3, and BLEU-4 show increases of 7.9%, 6.9%, 5.4%, and 3.9%, respectively. Furthermore, METEOR is higher 5.5% and ROUGE-L is higher by 5% than BASE. The experimental results show that the CGM module can effectively utilize the contextual information between reports to generate coherent and consistent radiology reports.

Table 2. Result of the ablation experiment.

MODEL	BL-1	BL-2	BL-3	BL-4	MET	RG-L
BASE	0.396	0.254	0.179	0.135	0.164	0.342
BASE + MVR	0.492	0.331	0.243	**0.180**	0.198	0.368
BASE + CGM	0.475	0.323	0.233	0.174	0.219	0.392
VRCG	**0.499**	**0.343**	**0.243**	**0.180**	**0.223**	**0.411**

The best values are highlighted in bold. The Base denotes transformer model.

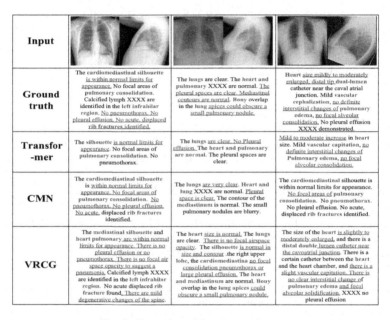

Fig. 2. Examples of the generated reports

5 Case Study and Visualization

To verify the effectiveness of our proposed model, we randomly selected three examples of generating reports on the test set, as shown in Fig. 2. We compare the proposed VRCG model with ground truth, transformer and CMN models. In the report, the red font indicates the organs found in the report, and the orange font indicates that it is the same as the ground truth.

The first example shows that the report generated by VRCG is closer to the ground truth, while the transformer model does not accurately identify "*cardio medical*". It is worth noting that VRCG can identify the key abnormal findings in medical images: "*There are mild degenerative changes of the spine*", which are not found in the other two models. This shows that the proposed model VRCG pays more attention to anomaly detection in medical images.

In the second example, we can observe that the report generated by VRCG gives a more comprehensive and detailed description of the findings in medical images than the other two models. In addition, our model can also identify *"bony overlap in the lung apices could observe a small monetary nodule"*, which is not recognized and generated by the transformer model.

In the third example, the generated report is still more specific and accurate than the reports of the other two models and can identify and generate exceptions not found in the transformer and CMN models: *"fatal tip dual lumen catheter near the caval atmospheric junction"*.

Through the above analysis, the proposed model VRCG in this paper can capture lesion information and describe abnormal findings in radiology medical images in detail. The generated radiology medical report can consider the report context information and accurately locate abnormal findings.

6 Future Work

In recent years, deep learning has performed well in radiology report generation. However, further research is needed in many areas, such as the interpretability of models and data privacy protection. The radiology report generation system should provide interpretable results and reasoning processes. The interpretability generation process will help doctors and clinical teams understand and trust the reports generated by the system, promoting their application in clinical decision-making. In addition, radiology reports contain sensitive medical information. Researchers need to develop appropriate privacy protection measures to ensure the safe storage, transmission and processing of medical images and related data.

7 Conclusion

This paper proposes a novel VRCG model, which utilizes a visual recalibration module and a context-gated perception module to address visual and text data bias issues in radiology report generation. We designed a medical visual recalibration module to enhance the model's ability and capture lesion features, thereby alleviating visual bias and improving model performance. We use a context gating-aware module to aggregate relevant contextual parts and extract text descriptions related to lesions, which makes the model suppress irrelevant information and alleviate the problem of text data bias. The experimental results on the IU X-Ray dataset indicate that our proposed model significantly exceeds other advanced baselines.

Acknowledgment. This work is supported by grant from the Natural Science Foundation of China (No. 62072070).

References

1. Alfarghaly, O., Khaled, R., Elkorany, A., et al.: Automated radiology report generation using conditioned transformers. Inform. Med. Unlocked **24**, 100557 (2021)
2. Li, Y., Liang, X., Hu, Z., et al.: Hybrid retrieval-generation reinforced agent for medical image report generation. In: Advances in Neural Information Processing Systems, vol. 31 (2018)
3. Jing, B., Xie, P., Xing, E.: On the automatic generation of medical imaging reports. In: Proceedings of the 56th Annual Meeting of the Association for Computational Linguistics, Melbourne, Australia, pp. 2577–2586. Association for Computational Linguistics (2018)
4. Vinyals, O., Toshev, A., Bengio, S., et al.: Show and tell: a neural image caption generator. In: Proceedings of the IEEE Conference on Computer Vision and Pattern Recognition, pp. 3156–3164 (2015)
5. Chen, Z., Song, Y., Chang, T.-H., Wan, X.: Generating radiology reports via memory-driven transformer. In: Proceedings of the 2020 Conference on Empirical Methods in Natural Language Processing (EMNLP), pp. 1439–1449. Association for Computational Linguistics (2020)
6. Chen, Z., Shen, Y., Song, Y., Wan, X.: Cross-modal memory networks for radiology report generation. In: Proceedings of the 59th Annual Meeting of the Association for Computational Linguistics and the 11th International Joint Conference on Natural Language Processing, pp. 5904–5914. Association for Computational Linguistics (2021)
7. Liang, X., Hu, Z., Zhang, H., et al.: Recurrent topic-transition GAN for visual paragraph generation. In: Proceedings of the IEEE International Conference on Computer Vision, pp. 3362–3371 (2017)
8. Yu, H., Wang, J., Huang, Z., et al.: Video paragraph captioning using hierarchical recurrent neural networks. In: Proceedings of the IEEE Conference on Computer Vision and Pattern Recognition, pp. 4584–4593 (2016)
9. Liu, F., Wu, X., Ge, S., et al.: Exploring and distilling posterior and prior knowledge for radiology report generation. In: Proceedings of the IEEE/CVF Conference on Computer Vision and Pattern Recognition, pp. 13753–13762 (2021)
10. Lu, J., Xiong, C., Parikh, D., et al.: Knowing when to look: adaptive attention via a visual sentinel for image captioning. In: Proceedings of the IEEE Conference on Computer Vision and Pattern Recognition, pp. 375–383 (2017)
11. Li, J., Li, S., Hu, Y., et al.: A self-guided framework for radiology report generation. In: Wang, L., Dou, Q., Fletcher, P.T., Speidel, S., Li, S. (eds.) MICCAI 2022. LNCS, vol. 13438, pp. 588–598. Springer, Cham (2022). https://doi.org/10.1007/978-3-031-16452-1_56
12. He, K., Zhang, X., Ren, S., Sun, J.: Deep residual learning for image recognition. In: Proceedings of the IEEE Conference on Computer Vision and Pattern Recognition, pp. 770–778 (2016)
13. Vaswani, A., et al.: Attention is all you need. In: Advances in Neural Information Processing Systems, vol. 30 (2017)
14. Demner-Fushman, D., Kohli, M.D., Rosenman, M.B., et al.: Preparing a collection of radiology examinations for distribution and retrieval. J. Am. Med. Inform. Assoc. **23**(2), 304–310 (2016)
15. Papineni, K., Roukos, S., Ward, T., et al.: BleU: a method for automatic evaluation of machine translation. In: Proceedings of the 40th Annual Meeting of the Association for Computational Linguistics, pp. 311–318 (2002)
16. Banerjee, S., Lavie, A.: METEOR: an automatic metric for MT evaluation with improved correlation with human judgments. In: Proceedings of the ACL Workshop on Intrinsic and Extrinsic Evaluation Measures for Machine Translation and/or Summarization, pp. 65–72 (2005)

17. Wu, Y., Zeng, M., Fei, Z., Yu, Y., Wu, F.-X., Li, M.J.N.: Kaicd: a knowledge attention-based deep learning framework for automatic ICD coding, **469**, 376–83 (2022)
18. Lin, C.Y.: Rouge: a package for automatic evaluation of summaries. In: Text Summarization Branches Out, pp. 74–81 (2004)
19. Reddi, S.J., Kale, S., Kumar, S.: On the convergence of Adam and beyond. arXiv preprint arXiv:1904.09237 (2019)
20. Xu, K., Ba, J., Kiros, R., et al.: Show, attend and tell: neural image caption generation with visual attention. In: International Conference on Machine Learning, pp. 2048–2057. PMLR (2015)

Using Generating Functions to Prove Additivity of Gene-Neighborhood Based Phylogenetics - Extended Abstract

Guy Katriel[1], Udi Mahanaymi[2], Christoph Koutschan[3], Doron Zeilberger[4], Mike Steel[5], and Sagi Snir[2(✉)]

[1] Department of Mathematics, Ort Braude, Israel
[2] Department of Evolutionary and Environmental Biology, University of Haifa, Haifa, Israel
ssagi@research.haifa.ac.il
[3] RICAM, Austrian Academy of Sciences, Linz, Austria
[4] Department of Mathematics, Rutgers University, Piscataway, USA
[5] School of Mathematics and Statistics, University of Canterbury, Christchurch, New Zealand

Abstract. Prokaryotic evolution is often described as the *Spaghetti of Life* due to massive genome dynamics (GD) events of gene gain and loss, resulting in different evolutionary histories for the set of genes comprising the organism. These different histories, dubbed as *gene trees* provide confounding signals, hampering the attempt to reconstruct the *species tree* describing the main trend of evolution of the species under study. The *synteny index* (SI) between a pair of genomes combines gene order and gene content information, allowing comparison of unequal gene content genomes, together with order considerations of their common genes. Recently, GD has been modelled as a continuous-time Markov process. Under this formulation, the distance between genes along the chromosome was shown to follow a birth-death-immigration process. Using classical results from birth-death theory, we recently showed that the SI measure is consistent under that formulation. In this work, we provide an alternative, stand alone combinatorial proof of the same result. By using generating function techniques we derive explicit expressions of the system's probabilistic dynamics in the form of rational functions of the model parameters. This, in turn, allows us to infer analytically the expected distances between organisms based on a transformation of their SI. Although the expressions obtained are rather complex, we establish additivity of this estimated evolutionary distance (a desirable property yielding phylogenetic consistency). This approach relies on holonomic functions and the Zeilberger Algorithm in order to establish additivity of the transformation of SI.

Keywords: Genome Dynamics · Markovian Processes · Generating Functions · Phylogenetics · Holonomic Functions

SS was supported by the Israel Science Foundation (grant No. ISF 1927/21) and the by the American/Israeli Binational Science Foundation (grant no. BSF 2021139).

© The Author(s), under exclusive license to Springer Nature Singapore Pte Ltd. 2023
X. Guo et al. (Eds.): ISBRA 2023, LNBI 14248, pp. 120–135, 2023.
https://doi.org/10.1007/978-981-99-7074-2_10

1 Introduction

The dramatic advancements in sequencing technologies have made realistic biological tasks seemed imaginary only a decade ago. Inferring the evolutionary history of thousands of species, is among the most fundamental tasks in biology with implications to medicine, agriculture, and more. Such a history is nd is called a *phylogeny*. Leaves of that tree correspond to contemporary (i.e. extant) species and the tree edges (or branches) represent evolutionary relationships. Despite the impressive advancement in the extraction of such molecular data, and of ever increasing quality, finding the underlying phylogenetic tree is still a major challenge requiring reliable approaches for inferring the true evolutionary distances between the species at the tips (leaves) of the tree. The tree sought should preserve the property that the length of the path between any two organisms at its leaves equals the inferred pairwise distance between these two organisms. When such a tree exists, these distances are called *additive*, as does the distance matrix storing them.

Modern approaches in systematics rely on statistical modelling in which a model fitting optimally the data is sought. The challenges under this framework, are both statistical, i.e. accurately modelling the data, and computational for efficient model inference and selection from given data. In phylogenetics, *maximum likelihood* seeks for a tree under which the probability of observing the given leaf sequences is maximised [8,9,12–14]. Normally, the data for this task is taken from few ubiquitous genes, such as ribosomal genes, that reside in every species and are immune for GD events. Such genes are typically highly conserved by definition and hence cannot provide a strong enough signal to distinguish the shallow branches of the prokaryotic tree. Nevertheless, GD events, gene gain in the form of horizontal gene transfer (HGT), a mechanism by which organisms transfer genetic material not through vertical inheritance, and gene loss, seem to provide valuable evolutionary information that can be harnessed for classification [7,20,23]. Approaches relying on GD are mainly divided into gene-order-based and gene-content-based techniques. Under the gene-order-based approach [11,24,33], two genomes are considered as permutations over the gene set, and distance is defined as the minimal number of operations needed to transform one genome to the other. The gene-content-based approach [10,29,30] ignores entirely gene order, and similarity is defined as the size of the set of shared genes. Although a statistical framework was devised for part of these models [4,25,26,31] to the best of our knowledge no such framework accounted for HGT.

The *synteny index* (SI) [1,27,28] captures both existence and locality, i.e. gene content and order respectively, by summarising gene neighbourhoods across the entire genome. An attractive property of the SI measure is the relaxation of the equal gene content requirement, in which genomes are permutations of the gene set. Under the attempt to model SI in a statistical framework, the *Jump model* was defined to account for gene order variation between evolving genomes. The *Jump operation* moves a gene to a random location in the genome. In the *Jump model*, every gene jumps, in a Poisson process. Under that framework, a genome is defined as a continuous-time Markov process (CTMP) [2]. Consequently, gene distance along the genome can be described as a (critical) birth-death-immigration process. The setting poses intrinsic hurdles such as overlapping neighbourhoods, non-stationarity, confounding factors, and more. Therefore, trees were constructed from evolutionary distances inferred heuristically based on exponential decay modelling.

In a recent paper [19] we have used classical tools for the birth-death field such as spectral theory and orthogonal polynomials, to derive analytical expressions for deriving the model transition probabilities and hence expected evolutionary distances. These analytical expressions yielded *model consistency* - an attractive property in systematics, implying that a measure infers accurate distances under a given model of evolution.

In this work, we provide an alternative, standalone combinatorial derivation for the model parameter and the proof of consistency. We first define the system in terms of a generating function, and extract transition probabilities as a function of time since divergence. However, the complexity of the expressions obtained to infer distances, could not readily imply consistency for the SI. By showing that these expressions satisfy the conditions for holonomic functions [32,35] and applying the Zeilberger Algorithm [34] we prove consistency of the SI measure under the jump model. We believe that this alternative proof, besides its independent interest, confers better understandings of the system and might prove useful for future extensions of the model, handling richer models such as unequal gene content or jumps of several genes.

Due to space considerations, several proofs were omitted and will appear at the journal version.

2   Preliminaries

We provide preliminary definitions and concepts to be used throughout the paper. We start with the Jump Model that comprises of a Jump operation operating on a single gene, and a stochastic process acting on the genomic ensemble of genes.

The *Jump* Model. In this work we consider the genome as a gene list, that is, the basic unit of resolution is an entire gene. Let $\mathcal{G}^{(n)} = (g_1, g_2, \ldots, g_n)$ be a sequence of 'genes' (see Fig. 1). For the sake of ignoring the tips of the sequence $\mathcal{G}^{(n)}$, we assume n is large enough compared to other sizes defined below.

Let $\mathcal{G}^{(n)}(0)$ be a genome at time zero and WLOG let $\mathcal{G}^{(n)}(0) = (g_1, g_2, \ldots, g_n)$. Now consider the following continuous-time Markovian process $\mathcal{G}^{(n)}(t), t \geq 0$ on the state space of all $n!$ permutations of $g_1, g_2, \ldots, g_n$. Each gene g_i is independently subject to a Poisson process transfer event (at constant rate λ) in which g_i is moved (or simply *Jumps*) to a different location in the sequence, with each of these possible $n - 1$ locations selected uniformly at random (see Fig 2).

For example, if $\mathcal{G}^{(n)}(t) = (g_1, g_2, g_3, g_4, g_5)$, and then g_1 jumps and lands between g_3 and g_4 then the sequence yielded is $\mathcal{G}^{(n)}(t + \delta) = (g_2, g_3, g_1, g_4, g_5)$. Note, that g_i can also move to one of the tips of the genome.

Fig. 1. A Genome as Gene List: The basic unit of resolution is a gene and a genome is defined as a sequence of genes.

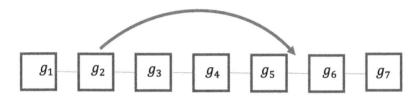

Fig. 2. The Jump operation: Gene g_2 jumps into the space between genes g_5 and g_6.

Since the model assumes a Poisson process, the probability that g_i is transferred to a different position between times t and $t + \delta$ is $\lambda\delta + o(\delta)$, where the $o(\delta)$ term accounts for the possibilities of more than one transfer occurring in the δ time period (these are of order δ^2 and so are asymptotically negligible compared to terms of order δ as $\delta \to 0$). Moreover, a single transfer event always results in a different sequence.

The Synteny Index. Let k be any constant positive integer (note it may be possible to allow k to grow slowly with n but we will not explore such an extension here). Then, for $j \in k + 1, \ldots, n - k$ the $2k$-neighbourhood of gene g_j in a genome $\mathcal{G}^{(n)}$, $N_{2k}(g_j, \mathcal{G}^{(n)})$ is the set of $2k$ genes (different from g_j) that have distance, in terms of separating genes along the chromosome, at most k from g_j in $\mathcal{G}^{(n)}$. Consider genomes $\mathcal{G}_1^{(n)}$ and $\mathcal{G}_2^{(n)}$, with the restriction that $\mathcal{G}_1^{(n)}$ and $\mathcal{G}_2^{(n)}$ share the same gene set. Let $SI_j(\mathcal{G}_1^{(n)}, \mathcal{G}_2^{(n)})$ be the relative intersection size between $N_{2k}(g_j, \mathcal{G}_1^{(n)})$ and $N_{2k}(g_j, \mathcal{G}_2^{(n)})$, or formally

$$SI_j(\mathcal{G}_1^{(n)}, \mathcal{G}_2^{(n)}) = \frac{1}{2k} |N_{2k}(g_j, \mathcal{G}_1^{(n)}) \cap N_{2k}(g_j, \mathcal{G}_2^{(n)})|$$

(this is also called *the Jaccard index* between the two neighbourhoods [15]). See Fig. 3 for example of a gene neighbourhood and the synteny index of a particular gene.

For the special case of our stochastic process, we define $SI_j(t)$ to be the SI for gene g_j between $\mathcal{G}^{(n)}(0)$ and $\mathcal{G}^{(n)}(t)$, $\frac{1}{2k}|N_{2k}(g_j, \mathcal{G}^{(n)}(0)) \cap N_{2k}(g_j, \mathcal{G}^{(n)}(t))$.

Let $\overline{SI}(\mathcal{G}_1^{(n)}, \mathcal{G}_2^{(n)})$ be the average of these $SI_j(\mathcal{G}_1^{(n)}, \mathcal{G}_2^{(n)})$ values over all genes g_j for j between $k+1$ and $n-k$. That is,

$$\overline{SI}(\mathcal{G}_1^{(n)}, \mathcal{G}_2^{(n)}) = \frac{1}{n-2k} \sum_{j=k+1}^{n-k} SI_j(\mathcal{G}_1^{(n)}, \mathcal{G}_2^{(n)}).$$

Finally, we equivalently define $\overline{SI}(\mathcal{G}^{(n)}(0), \mathcal{G}^{(n)}(t)$ be the average of these $SI_j(t)$ values between $\mathcal{G}^{(n)}(0)$ and $\mathcal{G}^{(n)}(t)$, over all j from $k+1$ to $n-k$.

$$\overline{SI}(\mathcal{G}_1^{(n)}, \mathcal{G}_2^{(n)}) = \frac{1}{n-2k} \sum_{j=k+1}^{n-k} SI_j(t). \tag{1}$$

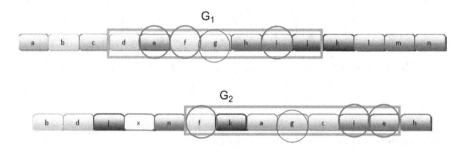

Fig. 3. The synteny Index The two gene neighbourhoods induced by gene g in genomes G_1 and G_2 and the synteny Index between G_1 and G_2 for gene g, $SI_g(\mathcal{G}_1^{(n)}, \mathcal{G}_2^{(n)}) = \frac{1}{2k}|N_{2k}(g, \mathcal{G}_1^{(n)}) \cap N_{2k}(g, \mathcal{G}_2^{(n)})|$. As genes e, f and i are shared between the two neighbourhoods induced by gene g in G_1 and G_2, we obtain $SI_g(\mathcal{G}_1^{(n)}, \mathcal{G}_2^{(n)}) = \frac{1}{2}$.

In the sequel, when time t does not matter, we simply use $\overline{SI}$ or simply SI where it is clear from the context.

2.1 Genome Permutations as a State Space

We now introduce a random process, that will play a key role in the analysis of the random variable $\overline{SI}(\mathcal{G}^{(n)}(0), \mathcal{G}^{(n)}(t))$. Consider the location of a gene g_i, not being transferred during time period t, with respect to another gene $g_{i'}$. WLG assume $i > i'$ and let $j = i - i'$. Now, there are j 'slots' between $g_{i'}$ and g_i in which a transferred gene can be inserted, but only $j - 1$ genes in that interval, that can be transferred. Obviously, a transfer into that interval moves $g_{i'}$ one position away from g_i, and transfer from that interval, moves $g_{i'}$ closer to g_i. The above can be modelled as a continuous-time random walk on state space $1, 2, 3, \ldots$ with transitions from j to $j + 1$ at rate $j\lambda$ (for all $j \geq 1$) and from j to $j - 1$ at rate $(j - 1)\lambda$ (for all $j \geq 2$), with all other transition rates 0. This is thus a (generalised linear) birth-death process, and the process is illustrated in Fig 4. As the process is not affected by the specific values of i and i' (rather by their difference), we can ignore them and let X_t denote the random variable that describes the state of this random walk (a number $1, 2, 3$ etc.) at time t.

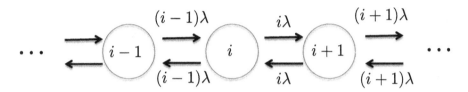

Fig. 4. The Markov Chain as a Birth-Death process Transitions between the states in the linear birth-death process with linear rate's growth/decrease.

The process X_t is slightly different from the much-studied critical linear birth-death process, for which the rate of birth and death from state j are both equal to j (here the rate of birth is j but the rate of death is $j - 1$), and for which 0 is an absorbing state (here there are no absorbing states). However, this stochastic process is essentially a translation of a critical linear birth-death process with immigration rate equal to the birth-death rate λ. This connection is key to the analysis of divergence times that we establish below.

Phylogenetic Trees and Distances. For a set of species (denoted *taxa*) $\mathcal{X}$, a phylogenetic $\mathcal{X}$-tree T is a tree $T = (V, E)$ for which there is a one-to-one correspondence between $\mathcal{X}$ and the set $\mathcal{L}(T)$ of leaves of T. A tree T is *weighted* if there is a weight (or length) function associating non-negative weights (lengths) to the edges of T. Along this work we will use the term length as it corresponds to number of events or time span. Edge lengths are naturally extended to *paths* where path length is the sum of edge lengths along the path. For a tree T over n leaves, let $D(T)$ (or simply D) be a symmetric $n \times n$ matrix where $[D]_{i,j}$ holds the path length (distance) between leaves i and j in T. A matrix D' is called *additive* if there is a tree T' such that $D(T') = D'$. A distance measure is considered *additive on a model M* if it can be transformed (or *corrected*) to the expected number of events generated under M.

3   Asymptotic Estimation of the Model Parameters

In order to reconstruct maximum likelihood trees, we need to estimate the model parameters, in a way that maximises their likelihood. We here establish the main theoretical result of this work, by defining the problem parameters as a generating function and use rules from this area. That in turn yields an analytical expression of divergence times. Recall that we wish to link SI to our model parameter X_t which is the expected value (state) of the model. Such a linkage was established in [27] that we restate explicitly below. While this expression is essential for the analysis, it is not stated in terms of the parameters of the model, specifically the time since divergence, and therefore has limited power.

We start with some essential definitions that are central in the analysis. Let $p_{i,j}(t)$ be the transition probability for X_t to be at state j given that at time 0 it was at state i. Formally,

Definition 1. *For each ordered pair* $i, j \in \{1, 2, 3 \dots, \}$ *let* $p_{i,j}(t) = \mathbb{P}(X_t = j \mid X_0 = i)$.

$p_{i,j}(t)$ is the most basic variable and on which more special variables are defined.
Now denote

$$q_{i,k}(t) = \sum_{j=1}^{k} p_{i,j}(t) \tag{2}$$

the conditional probability that $X_t \leq k$ given that $X_0 = i$, as $q_{i,k}(t)$.

Also let $q_k(t)$ denote the probability that for a gene at an initial state i (i.e., distance from a reference gene) chosen uniformly at random between 1 and k, the process X_* is still between 1 and k after time t, or formally:

$$q_k(t) := \frac{1}{k} \sum_{i=1}^{k} q_{i,k}(t) = \frac{1}{k} \sum_{i=1}^{k} \sum_{j=1}^{k} p_{i,j}(t). \tag{3}$$

Having defined these variables, we can restate the fundamental theorem we proved in [27]:

Theorem 1. *For any given value of* t, *and as* n *grows:*

$$\overline{SI}(\mathcal{G}^{(n)}(0), \mathcal{G}^{(n)}(t)) \xrightarrow{p} \exp(-2\lambda t) q_k(t),$$

where $\xrightarrow{p}$ *denotes convergence in probability.*

Theorem 1 is important as it links between SI, event rate, and probabilities of genes staying at their original neighbourhoods. Nevertheless, these factors are confounded in the sense that $q_k(t)$ depends on t, and therefore it would be desirable to arrive at an expression stated in the parameters of the model, i.e. time and rate solely, so divergence times, or alternatively number of events, can be estimated and trees can be reconstructed. The rest of the section is devoted to this.

3.1 Finding the Model Transition Probabilities

The transition probabilities of the Markov Model defined above are fundamental for our goal - analytical expressions of the expected synteny index between two genomes in terms of their divergence time. Hence, finding an explicit expression, in terms of i, j, and t, is our first task.

Theorem 2.

$$p_{i,j}(t) = \frac{1}{(t+1)^{i+j-1}} \cdot \sum_{\ell=1}^{\min(i,j)} \frac{(i+j-\ell-1)!}{(i-\ell)!(j-\ell)!(\ell-1)!} \left(1-t^2\right)^{\ell-1} t^{i+j-2\ell}. \quad (4)$$

The next step uses a lemma from [27] that adapts the Forward Kolmogorov Equation [2] to our special setting.

Lemma 1. *[27]*

(a) The transition probabilities $p_{i,j}(t)$ satisfy the following tri-diagonal differential system

$$\frac{1}{\lambda}\frac{dp_{i,j}(t)}{dt} = -(2j-1)p_{i,j}(t) + jp_{i,j+1}(t) + (j-1)p_{i,j-1}(t), \quad (5)$$

subject to the initial condition:

$$p_{i,j}(0) = \begin{cases} 1, & \text{if } i = j; \\ 0, & \text{if } i \neq j. \end{cases}$$

(b) The expected value of X_t grows as a linear function of t. Specifically,

$$\mathbb{E}[X_t \mid X_0 = i] = i + t\lambda. \quad (6)$$

Our first aim is to solve the above infinite system of differential Eq. (5).

Without loss of generality, we assume $\lambda = 1$ and introduce the following definition making use of a generating function.

Definition 2. *For each $i \geq 1$, we define a generating function $f_i(t,z) = \sum_{j=1}^{\infty} p_{i,j}(t)z^j$.*

Lemma 2.

$$f_i(t,z) = \frac{t^{i-1}}{(t+1)^i} \sum_{j=1}^{\infty} (-1)^{j-1} z^j \left(\frac{t}{t+1}\right)^{j-1} \sum_{\ell=1}^{\min(i,j)} \binom{i-1}{\ell-1}\binom{-i}{j-\ell}\left(\frac{t^2-1}{t^2}\right)^{\ell-1} \quad (7)$$

The full proof of Lemma 2 is deferred to the journal version.

We are now in a position to prove Theorem 2.

Proof of Theorem 2: From Definition 2 and Lemma 2 above, we have two equal power series. Therefore, by the uniqueness of generating functions [22], we conclude that the coefficients are pairwise equal, hence:

$$p_{i,j}(t) = (-1)^{j-1} \frac{t^{i-1}}{(t+1)^i} \left(\frac{t}{t+1}\right)^{j-1} \sum_{\ell=1}^{\min(i,j)} \binom{i-1}{\ell-1}\binom{-i}{j-\ell}\left(\frac{t^2-1}{t^2}\right)^{\ell-1}$$

$$= (-1)^{j-1} \left(\frac{1}{t+1}\right)^{i+j-1} \sum_{\ell=1}^{\min(i,j)} \binom{i-1}{\ell-1}\binom{-i}{j-\ell}\left(t^2-1\right)^{\ell-1} t^{i+j-2\ell}$$

$$= \left(\frac{1}{t+1}\right)^{i+j-1} \sum_{\ell=1}^{\min(i,j)} \binom{-i}{j-\ell}\frac{(i-1)!}{(i-\ell)!(\ell-1)!}(-1)^{j-1}\left(t^2-1\right)^{\ell-1} t^{i+j-2\ell} \quad (8)$$

Recalling that the generalisation of the binomial coefficient to negative integers $-n$ is:

$$\binom{-n}{k} = (-1)^k \binom{n+k-1}{k}$$

we obtain from (8):

$$p_{i,j}(t) = \left(\frac{1}{t+1}\right)^{i+j-1} \sum_{\ell=1}^{\min(i,j)} (-1)^{j-\ell}\frac{(i+j-\ell-1)!}{(i-1)!(j-\ell)!}\frac{(i-1)!}{(i-\ell)!(\ell-1)!}(-1)^{j-1}\left(t^2-1\right)^{\ell-1} t^{i+j-2\ell}$$

$$= \frac{1}{(t+1)^{i+j-1}} \cdot \sum_{\ell=1}^{\min(i,j)} \frac{(i+j-\ell-1)!}{(i-\ell)!(j-\ell)!(\ell-1)!}(-1)^{2j-\ell-1}\left(t^2-1\right)^{\ell-1} t^{i+j-2\ell}$$

$$= \frac{1}{(t+1)^{i+j-1}} \cdot \sum_{\ell=1}^{\min(i,j)} \frac{(i+j-\ell-1)!}{(i-\ell)!(j-\ell)!(\ell-1)!}\left(1-t^2\right)^{\ell-1} t^{i+j-2\ell}. \quad (9)$$

□

From Theorem 2 we can see the following:

Corollary 1. *For any i, j, and t it holds that $p_{i,j}(t) = p_{j,i}(t)$.*

3.2   Expectation and Variance of X_t

Having explicit expression for $p_{i,j}(t)$ allows us to confirm other derived values. Therefore we here note by passing the expected value and variance of X_t. By the definition of X_t we have

$$E(X_t \mid X_0 = i) = \sum_{j=1}^{\infty} j p_{i,j}(t)$$

.

Also, from Definition 2 we have

$$\frac{d}{dz} f_i(t,z) = \frac{d}{dz}\sum_{j=1}^{\infty} p_{i,j}(t)z^j = \sum_{j=1}^{\infty} j p_{i,j}(t)z^{j-1}$$

Using the generating functions we have

$$E(X_t \mid X_0 = i) = \frac{d}{dz}f_i(t,z)\Big|_{z=1} = i+t \quad (10)$$

in agreement with Lemma 1b.

We also have

$$E(X_t(X_t - 1) \mid X_0 = i) = \left. \frac{d^2}{dz^2} f_i(t, z) \right|_{z=1} = 2t^2 - 2t + (4t - 1)i + i^2.$$

Hence

$$E(X_t^2 \mid X_0 = i) = E(X_t(X_t - 1) \mid X_0 = i) + E(X_t \mid X_0 = i) = 2t^2 - t + 4ti + i^2.$$

Hence

$$\begin{aligned}
Var(X_t \mid X_0 = i) &= E(X_t^2 \mid X_0 = i) - E(X_t \mid X_0 = i)^2 \\
&= 2t^2 - t + 4ti + i^2 - (i + t)^2 \\
&= t^2 + (2i - 1)t.
\end{aligned} \tag{11}$$

3.3   Explicit Expression for $q_k(t)$

As stated above, Theorem 1 (originally from [27]) gives an explicit expression for SI between two genomes, $\mathcal{G}_0$ and $\mathcal{G}_t$. Nevertheless we could not derive an expression only in terms of the number of events occurred during time t, or alternatively a path along the tree of length λt "separating" genomes $\mathcal{G}_i$ and $\mathcal{G}_j$, as we could not arrive at an explicit expression for q_k (also in terms of (λt). As here we obtained explicit expression for $p_{i,j}(t)$ we can aim now at expressing q_k.

Lemma 3.

$$q_k(t) = \frac{1}{k} \sum_{\ell=0}^{k-1} \sum_{i=0}^{k-\ell-1} \sum_{j=0}^{k-\ell-1} \frac{(i + j + \ell)!}{i! j! \ell!} t^{i+j} (t + 1)^{-i-j-2\ell-1} \left(1 - t^2\right)^{\ell}. \tag{12}$$

The full proof of Lemma 3 is deferred to the journal version.

4   Additivity of the SI Measure

Our goal now is to prove the monotonicity of the SI measure for any t and, by Theorem 1, of the expression $h_k(t) = e^{-2t} q_k(t)$ in $t \in [0, \infty)$. In fact we will prove that $q_k(t)$ itself is monotone decreasing, which obviously implies that $h_k(t)$ is monotone decreasing. To do so we first obtain expressions for $q_k'(t)$. As $q_{i,k}(t) = \sum_{j=1}^{k} p_{i,j}(t)$, we get $\frac{dq_{i,k}(t)}{dt} = \sum_{j=1}^{k} \frac{dp_{i,j}(t)}{dt}$.

Lemma 4.

$$\frac{dq_{i,k}(t)}{dt} = k\lambda[p_{i,k+1}(t) - p_{i,k}(t)]. \tag{13}$$

The full proof of Lemma 4 is deferred to the journal version.

Now, as by Lemma 4 we have $q'_{i,k}(t) = k[p_{i,k+1}(t) - p_{i,k}(t)]$, hence

$$q'_k(t) = \frac{1}{k}\sum_{i=1}^{k} q'_{i,k}(t) = \sum_{i=1}^{k}[p_{i,k+1}(t) - p_{i,k}(t)] \tag{14}$$

Using the explicit expressions for $p_{i,j}(t)$, we have, for $i \le k$:

$$p_{i,k}(t) = \frac{1}{(t+1)^{i+k-1}} \cdot \sum_{\ell=1}^{i} \frac{(i+k-\ell-1)!}{(i-\ell)!(k-\ell)!(\ell-1)!}\left(1-t^2\right)^{\ell-1} t^{i+k-2\ell}, \tag{15}$$

and

$$p_{i,k+1}(t) = \frac{1}{(t+1)^{i+k}} \cdot \sum_{\ell=1}^{i} \frac{(i+k-\ell)!}{(i-\ell)!(k+1-\ell)!(\ell-1)!}\left(1-t^2\right)^{\ell-1} t^{i+k+1-2\ell}$$

$$= \frac{1}{(t+1)^{i+k-1}} \cdot \sum_{\ell=1}^{i} \frac{(i+k-\ell-1)!}{(i-\ell)!(k-\ell)!(\ell-1)!} \cdot \frac{i+k-\ell}{k+1-\ell} \cdot \frac{t}{t+1}\left(1-t^2\right)^{\ell-1} t^{i+k-2\ell} \tag{16}$$

hence

$$p_{i,k+1}(t) - p_{i,k}(t) = \frac{1}{(t+1)^{i+k-1}} \cdot \sum_{\ell=1}^{i} \frac{(i+k-\ell-1)!}{(i-\ell)!(k-\ell)!(\ell-1)!} \cdot \left(\frac{k+i-\ell}{k+1-\ell}\cdot\frac{t}{t+1} - 1\right)\left(1-t^2\right)^{\ell-1} t^{i+k-2\ell}$$

so that

$$q'_k(t) = \sum_{i=1}^{k}[p_{i,k+1}(t) - p_{i,k}(t)] \tag{17}$$

$$= \sum_{i=1}^{k}\frac{1}{(t+1)^{i+k-1}} \cdot \sum_{\ell=1}^{i} \frac{(k+i-\ell-1)!}{(i-\ell)!(k-\ell)!(\ell-1)!} \cdot \left(\frac{k+i-\ell}{k+1-\ell}\cdot\frac{t}{t+1} - 1\right)\left(1-t^2\right)^{\ell-1} t^{i+k-2\ell}.$$

We would like to prove that $q'_k(t) < 0$ for all $k \ge 1$, $t > 0$. This is not clear from the above expression. In the next section we prove this by advanced computer algebra tools.

4.1   Computer Proof of a Double-Sum Identity

This section is dedicated to the proof of the following identity:

$$q'_k(t) = -\frac{1}{(t+1)^{2k}} \sum_{m=0}^{k-1} \binom{k-1}{m}\binom{k}{m} t^{2m}. \tag{18}$$

By (17) we need to prove that

$$\sum_{i=1}^{k}(t+1)^{k-i} \sum_{\ell=1}^{i} \frac{(i+k-\ell-1)!}{(i-\ell)!(k-\ell)!(\ell-1)!} \cdot \left(1 - \frac{i-1}{k+1-\ell}\cdot t\right)\left(1-t^2\right)^{\ell-1} t^{i+k-2\ell} = \sum_{m=0}^{k-1}\binom{k-1}{m}\binom{k}{m}t^{2m} \tag{19}$$

The strategy is as follows: we first prove that the right-hand side of (19) satisfies a second-order recurrence in k (Lemma 5), then we derive a recurrence equation for the left-hand side (Lemmas 7 and 8). Since it turns out that these two recurrences are the same, the equality is established by comparing a few initial values. A key component in the proof is Zeilberger's algorithm [34]. It takes as input a parametric sum of the form $F(n) := \sum_k f(n, k)$ where n is a (discrete) parameter and k runs from $-\infty$ to $+\infty$, or between summation bounds that are linear expressions in n (the most common situation is $k = 0, \ldots, n$). Moreover, the summand $f(n, k)$ needs to be hypergeometric in both variables, that means, the quotients $f(n+1, k)/f(n, k)$ and $f(n, k+1)/f(n, k)$ are bivariate rational functions in n and k. As output, Zeilberger's algorithm produces a linear recurrence equation with polynomial coefficients for $F(n)$, i.e., a linear relation of the form $c_d(n)F(n + d) + \ldots + c_1(n)F(n + 1) + c_0(n)F(n) = 0$ where the c_i are polynomials in n, that is satisfied for all $n \in \mathbb{N}$.

For our calculations below, we have employed the Mathematica package HolonomicFunctions [21].

Theorem 3. *For all $k \in \mathbb{N}$ and t a parameter, identity (19) holds.*

Before we prove Theorem 3 we state few auxiliary lemmas.

Lemma 5. *The right-hand side of (19), i.e., the expression*

$$R_k(t) := \sum_{m=0}^{k-1} \binom{k-1}{m} \binom{k}{m} t^{2m} \tag{20}$$

satisfies the recurrence

$$(k + 2)(2k + 1)R_{k+2} - 2\left(2k^2 t^2 + 2k^2 + 4kt^2 + 4k + 2t^2 + 1\right) R_{k+1} + k(2k + 3)(t - 1)^2(t + 1)^2 R_k = 0$$

for all $k \in \mathbb{N}$.

The full proof of Lemma 5 is deferred to the journal version.

The proof of the following lemma uses the same strategy as the one of Lemma 5.

Lemma 6. *The inner sum of the left-hand side of (19), i.e., the expression*

$$M_{k,i}(t) := \sum_{\ell=1}^{i} \frac{(t+1)^{k-i}(i+k-\ell-1)!\,(k+1-\ell-(i-1)t)(1-t^2)^{\ell-1}t^{i+k-2\ell}}{(i-\ell)!\,(k-\ell+1)!\,(\ell-1)!}$$

satisfies the following bivariate recurrences:

$$(k + 1)t(i - k)M_{k+1,i} - it(t + 1)^2(i - k - t - 1)M_{k,i+1}$$
$$+ (t - 1)(t + 1)^2 \left(i^2 - 2ik - it - i + k^2 + k\right) M_{k,i} = 0,$$
$$(i + 1)t(t + 1)(i - k)M_{k,i+2}$$
$$+ \left(-2i^2 t^2 + i^2 + 2ikt^2 - 2ik - 2it^2 + i + k^2 + kt^2 - kt - k\right) M_{k,i+1}$$
$$+ i(t - 1)t(i - k + 1)M_{k,i} = 0.$$

Lemma 7. *The left-hand side of* (19) *can be simplified to a single sum, i.e., the following identity holds for all* $k \in \mathbb{N}$:

$$\sum_{i=1}^{k} M_{k,i}(t) = L_k(t) := \sum_{\ell=1}^{1+k} \frac{k(1-t)^{\ell-1}t^{1+2k-2\ell}(1+t)^{\ell-2}(2k-\ell)!}{(1+k-\ell)!(2+k-\ell)!(\ell-1)!}$$
$$\times \left((1+k-\ell)(2+k-\ell)+(2+k-\ell)t+\left(1-k^2\right)t^2\right)$$

with $M_{k,i}(t)$ *as introduced in Lemma* 6.

The full proof of Lemma 7 is deferred to the journal version.

The proof of the following lemma uses the same strategy as the one of Lemma 5.

Lemma 8. *The sum* $L_k(t)$ *defined in Lemma* 7 *satisfies the recurrence*

$$(k+2)(2k+1)L_{k+2} - 2\left(2k^2t^2+2k^2+4kt^2+4k+2t^2+1\right)L_{k+1} + k(2k+3)(t-1)^2(t+1)^2L_k = 0$$

for all $k \in \mathbb{N}$.

Proof (Proof of Theorem 3). We have shown that both sides of (19) satisfy the same second-order linear recurrence equation (Lemma 5 and Lemma 8). Since the leading coefficient $(k+2)(2k+1)$ is nonzero for all $k \in \mathbb{N}$, it suffices to verify that (19) holds for $k=0$ (indeed: both sides evaluate to 0) and for $k=1$ (indeed: both sides evaluate to 1).

Theorem 3 justifies Eq. (18), which in turn implies that the function $h_k(t) = \exp(-\lambda t)q_k(t)$ is monotone decreasing with t and thus has an inverse (h_k^{-1}). Moreover, $h_k(t)$ can be exactly calculated (using the explicit expression for $q_k(t)$ given by Eq. (12)), and so, by Theorem 1, the time separating two sequences of genes involving n genes (where n is large) can be estimated by applying h_k^{-1} to the $\overline{SI}$ for the two gene sequences. Since the expected number of transfer events is additive on the tree (and proportional to t), we conclude the following:

Corollary 2. *The topology of the underlying unrooted tree* T *can be reconstructed in a statistically consistent way from the* $\overline{SI}$ *values by applying the transformation* h_k^{-1}, *followed by a consistent distance-based tree reconstruction method such as Neighbour-Joining (NJ).*

5   Conclusions

In this paper, we have provided an alternative derivation for the system variables of the birth-death formulation of the synteny index (SI) distance measure. The classical approach for this task uses the so-called Karlin-McGregor spectral representation, that is based on a sequence of orthogonal polynomials and a spectral measure [3, 16–18]. The approach presented here is a self-contained derivation, based on generating functions representation and a subsequent combinatorial treatment, leading to an application of tools from symbolic algebra.

Although the biological contribution of this work is seemingly less pronounced, as it merely arrives at the same expressions for the transition probabilities as the traditional approach, we believe that the derivation presented here not only has independent interest for mathematical biology, but may also be key to future rigorous extensions of the Jump model.

One such immediate follow-up extension we see as important is to augment the pure Jump process, with more realistic genome dynamic events such as external gene gain, in which a novel gene is acquired from a different genome, leading to an extension of the gene repertoire of the organism, and events of gene loss. Both these evens potentially cause a divergence in genome content between the analysed genomes, and require special treatment that, based on initial attempts, is non-trivial.

Regarding the mathematical aspect, the symbolic algebra tools as we apply here, have been proved useful in other applications of mathematical biology [5,6] leading to accurate expressions of quantities known to be derived heuristically before. We are hopeful that this new derivation is a basis for the extensions we consider in the future.

References

1. Adato, O., Ninyo, N., Gophna, U., Snir, S.: Detecting horizontal gene transfer between closely related taxa. PLoS Comput. Biol. **11**(10), e1004408 (2015)
2. Allen, L.J.: An Introduction to Stochastic Processes with Applications to Biology. Chapman and Hall/CRC, Boca Raton (2010)
3. Anderson, W.J.: Continuous-Time Markov Chains: An Applications-Oriented Approach. Springer, Cham (2012). https://doi.org/10.1007/978-1-4612-3038-0
4. Biller, P., Guéguen, L., Tannier, E.: Moments of genome evolution by double cut-and-join. BMC Bioinform. **16**(14), S7 (2015)
5. Chor, B., Hendy, M.D., Snir, S.: Maximum likelihood jukes-cantor triplets: analytic solutions. Mol. Biol. Evol. **23**(3), 626–632 (2006)
6. Chor, B., Khetan, A., Snir, S.: Maximum likelihood on four taxa phylogenetic trees: analytic solutions. In: Proceedings of the Seventh annual International Conference on Computational Molecular Biology (RECOMB), Berlin, Germany, April 2003, pp. 76–83 (2003)
7. Doolittle, W.F.: Phylogenetic classification and the universal tree. Science **284**(5423), 2124–2128 (1999)
8. Felsenstein, J.: Cases in which parsimony or compatibility methods will be positively misleading. Syst. Zool. **27**(4), 401–410 (1978)
9. Felsenstein, J.: Evolutionary trees from DNA sequences: a maximum likelihood approach. J. Mol. Evol. **17**(6), 368–376 (1981)
10. Fitz Gibbon, S.T., House, C.H.: Whole genome-based phylogenetic analysis of free-living microorganisms. Nucleic Acids Res. **27**(21), 4218–4222 (1999)
11. Hannenhalli, S., Pevzner, P.A.: Transforming cabbage into turnip: polynomial algorithm for sorting signed permutations by reversals, **46**, 1–27 (1999). ACM

12. Hendy, M.D., Penny, D.: A framework for the quantitative study of evolutionary trees. Syst. Zool. **38**(4), 297–309 (1989)
13. Hendy, M.D., Penny, D.: Spectral analysis of phylogenetic data. J. Classif. **10**(1), 5–24 (1993)
14. Hendy, M.D., Penny, D., Steel, M.: A discrete Fourier analysis for evolutionary trees. Proc. Natl. Acad. Sci. **91**(8), 3339–3343 (1994)
15. Jaccard, P.: Étude comparative de la distribution florale dans une portion des alpes et des jura. Bull. Soc. Vaudoise Sci. Nat. **37**, 547–579 (1901)
16. Karlin, S., McGregor, J.: The classification of birth and death processes. Trans. Am. Math. Soc. **86**(2), 366–400 (1957)
17. Karlin, S., McGregor, J.: A characterization of birth and death processes. Proc. Natl. Acad. Sci. **45**(3), 375–379 (1959)
18. Karlin, S., McGregor, J.L.: The differential equations of birth-and-death processes, and the stieltjes moment problem. Trans. Am. Math. Soc. **85**(2), 489–546 (1957)
19. Katriel, G., et al.: Gene transfer-based phylogenetics: analytical expressions and additivity via birth–death theory. System. Biol. (2023, accepted)
20. Koonin, E.V., Makarova, K.S., Aravind, L.: Horizontal gene transfer in prokaryotes: quantification and classification. Annu. Rev. Microbiol. **55**(1), 709–742 (2001)
21. Koutschan, C.: HolonomicFunctions (user's guide). Technical report 10-01, RISC Report Series, Johannes Kepler University, Linz, Austria (2010). https://www.risc.jku.at/research/combinat/software/HolonomicFunctions/
22. Miller, S.: The Probability Lifesaver: All the Tools You Need to Understand Chance. Princeton Lifesaver Study Guides, Princeton University Press (2017). https://books.google.co.il/books?id=VwtHvgAACAAJ
23. Ochman, H., Lawrence, J.G., Groisman, E.A.: Lateral gene transfer and the nature of bacterial innovation. Nature **405**(6784), 299 (2000)
24. Sankoff, D.: Edit distance for genome comparison based on non-local operations. In: Apostolico, A., Crochemore, M., Galil, Z., Manber, U. (eds.) CPM 1992. LNCS, vol. 644, pp. 121–135. Springer, Heidelberg (1992). https://doi.org/10.1007/3-540-56024-6_10
25. Sankoff, D., Nadeau, J.H.: Conserved synteny as a measure of genomic distance. Discret. Appl. Math. **71**(1–3), 247–257 (1996)
26. Serdoz, S., et al.: Maximum likelihood estimates of pairwise rearrangement distances. J. Theor. Biol. **423**, 31–40 (2017)
27. Sevillya, G., Doerr, D., Lerner, Y., Stoye, J., Steel, M., Snir, S.: Horizontal gene transfer phylogenetics: a random walk approach. Mol. Biol. Evol. **37**(5), 1470–1479 (2019). https://doi.org/10.1093/molbev/msz302
28. Shifman, A., Ninyo, N., Gophna, U., Snir, S.: Phylo SI: a new genome-wide approach for prokaryotic phylogeny. Nucleic Acids Res. **42**(4), 2391–2404 (2013)
29. Snel, B., Bork, P., Huynen, M.A.: Genome phylogeny based on gene content. Nat. Genet. **21**(1), 108 (1999)
30. Tekaia, F., Dujon, B.: Pervasiveness of gene conservation and persistence of duplicates in cellular genomes. J. Mol. Evol. **49**(5), 591–600 (1999)
31. Wang, L.S., Warnow, T.: Estimating true evolutionary distances between genomes. In: Proceedings of the Thirty-Third Annual ACM Symposium on Theory of Computing, pp. 637–646. ACM (2001)
32. Wilf, H.S., Zeilberger, D.: An algorithmic proof theory for hypergeometric (ordinary and "q") multisum/integral identities. Invent. Math. **108**(1), 575–633 (1992)
33. Yancopoulos, S., Attie, O., Friedberg, R.: Efficient sorting of genomic permutations by translocation, inversion and block interchange. Bioinformatics **21**(16), 3340–3346 (2005)

34. Zeilberger, D.: A fast algorithm for proving terminating hypergeometric identities. Discret. Math. **80**(2), 207–211 (1990). https://doi.org/10.1016/0012-365X(90)90120-7
35. Zeilberger, D.: A holonomic systems approach to special functions identities. J. Comput. Appl. Math. **32**(3), 321–368 (1990)

TCSA: A Text-Guided Cross-View Medical Semantic Alignment Framework for Adaptive Multi-view Visual Representation Learning

Hongyang Lei, Huazhen Huang, Bokai Yang, Guosheng Cui, Ruxin Wang, Dan Wu[✉], and Ye Li

Shenzhen Institute of Advanced Technology, Chinese Academy of Sciences, Shenzhen 518055, China
dan.wu@siat.ac.cn

Abstract. Recently, in the medical domain, visual-language (VL) representation learning has demonstrated potential effectiveness in diverse medical downstream tasks. However, existing works typically pre-trained on the one-to-one corresponding medical image-text pairs, disregarding fluctuation in the quantity of views corresponding to reports (e.g., chest X-rays typically involve 1 to 3 projection views). This limitation results in sub-optimal performance in scenarios with varying quantities of views (e.g., arbitrary multi-view classification). To address this issue, we propose a novel Text-guided Cross-view Semantic Alignment (TCSA) framework for adaptive multi-view visual representation learning. For arbitrary number of multiple views, TCSA learns view-specific private latent sub-spaces and then maps them to a scale-invariant common latent subspace, enabling individual treatment of arbitrary view type and normalization of arbitrary quantity of views to a consistent scale in the common sub-space. In the private sub-spaces, TCSA leverages word context as guidance to match semantic corresponding sub-regions across multiple views via cross-modal attention, facilitating alignment of different types of views in the private sub-space. This promotes the combination of information from arbitrary multiple views in the common sub-space. To the best of our knowledge, TCSA is the first VL framework for arbitrary multi-view visual representation learning. We report the results of TCSA on multiple external datasets and tasks. Compared with the state of the art frameworks, TCSA achieves competitive results and generalize well to unseen data.

Keywords: representation learning · medical images · multi-modal

This work was supported by the National Key Research and Development Program (2021YFF0703704); Guangdong Basic and Applied Basic Research Foundation (2022A1515011217 and 2022A1515011557)); National Natural Science Foundation of China (U1913210 and 62206269); Shenzhen Science and Technology Program (JSGG20211029095546003 and JCYJ20190807161805817).

X. Guo et al. (Eds.): ISBRA 2023, LNBI 14248, pp. 136–149, 2023.
https://doi.org/10.1007/978-981-99-7074-2_11

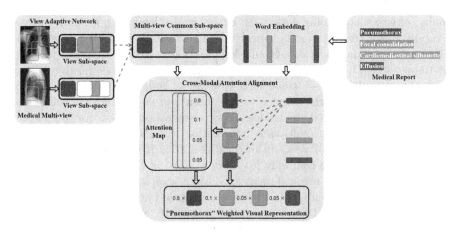

Fig. 1. Illustration of text-guided cross-view medical semantic alignment framework TCSA. Same color represent consistent semantics between words and sub-regions, while white indicate inactive features.

1   Introduction

Large-scale annotated medical image datasets facilitate the advancement of medical image understanding based on deep learning. However, the inherent challenges of establishing large-scale annotated data in the medical field pose significant obstacles to the progress of current research [1–4]. Recently, deep learning has witnessed the rise of visual-language (VL) representation learning, which is pre-trained on large-scale naturally occurring paired image-text and demonstrates general effectiveness across various VL downstream tasks with limited labeling (e.g., image-text retrieval) [5]. Inspired by the remarkable success of VL representation learning in natural domain, researchers attempt to transfer it to the medical domain to alleviate the inherent challenges of establishing large-scale annotated data [6–8].

However, migrating VL representation learning frameworks from the natural domain to the medical domain is challenging due to the following reasons: (1) Unlike natural domain where text and image mostly exhibit one-to-one correspondence, medical imaging examinations involve the acquisition of diverse numbers of views based on each patient's distinct clinical attributes and individualized clinical needs. (2) Pathology in medical images typically occupies a minuscule fraction. Capturing subtle yet crucial visual cues in medical images is essential. In summary, for medical VL representation learning, it is crucial to handle the fluctuation in the number of medical views, capture subtle yet crucial visual cues on medical images, and achieve fine-grained semantic alignment among multiple views.

Existing works have attempted to address the aforementioned issues through various approaches. To address the fluctuation in the number of views, one effective approach is to selectively employ a subset of studies that maintain a definite number of views as training data [6–8]. However, this leads to poor performance in various downstream tasks that involve arbitrary numbers of views (e.g., arbitrary multi-view classification). GLoRIA [7] and ConVIRT [9] are pre-trained on one-to-one image-text pairs, and in tasks involving multiple views, they either select one of multi-view or treat each view as an independent sample, which results in its poor performance due to the absence of visual information. While the works based on fixed multi-view report generation allows for generating reports by incorporating multi-view information, it fails to cope with the fluctuation of the number of views in medical scenarios [10,11]. Furthermore, the fixed-view approach also results in low data utilization rates. To capture subtle yet crucial visual cues in medical images, GLoRIA leverage CNNs [12] and attention to align word and sub-region to capture subtle but critical visual cues on medical images. MGCA [13] achieves multi-granularity alignment by incorporating instance-wise alignment, token-wise alignment, and cross-modal disease-level alignment. However, these works only consider the correspondence between individual view and text, whereas most downstream tasks usually require aligning fine-grained semantics among multiple views to effectively integrate information and make accurate predictions.

To address these issues, we propose a Text-guided Cross-view medical Semantic Alignment (TCSA) framework for adaptive multi-view visual representation learning. As shown in Fig. 1, to address the challenge of arbitrary multiple views, we introduce the View Adaptive Network, which aims to learn view-specific private latent sub-spaces tailored to different types of views and efficiently filter out irrelevant information through attention mechanisms, followed by aggregation in a multi-view common sub-space. To align fine-grained semantics across multiple views, TCSA maps view-specific sub-spaces to a scale-invariant common latent sub-space based on guidance at both the report-level and word-level. TCSA leverages Cross-Modal Attention Alignment to match word context with semantically consistent sub-regions across multiple views, enabling arbitrary multi-view fine-grained semantic alignment via word-level guidance improving sensitivity of the representations. As shown in the Cross-Modal Attention Alignment of Fig. 1, we utilize the word embedding of "Pneumothorax" to weight multi-view features via attention map, obtaining "Pneumothorax" weighted visual representation. Report-level guidance capture more comprehensive information. The main contributions of this paper can be summarized as follows:

1. We propose a text-guided cross-view medical semantic alignment framework TCSA for adaptive multi-view visual representation learning, which leverages text guidance to align the fine-grained semantics across multiple views, promoting the extraction of both consistency and complementary features among the views. To the best of our knowledge, this is the first VL representation framework for medical arbitrary multi-view visual representation learning.
2. We offer an effective solution for handling arbitrary numbers of multi-view problems in real-world medical scenarios. TCSA provides flexibility in processing any number of multi-view based on diagnostic requirements, enhancing the applicability of VL representation learning in real medical settings.

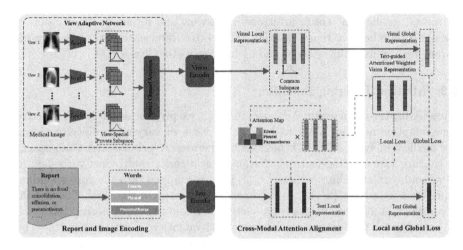

Fig. 2. Overview of TCSA framework.

We pre-trained TCSA on a large-scale dataset of medical image-report pairs i.e., MIMIC-CXR [14] and validated it on multiple external datasets and downstream tasks. Experimental results consistently demonstrate that TCSA outperforms state of the art models across various tasks with stability.

2 Method

2.1 Overview

The main objective of this work is to learn generalized visual representations for arbitrary multi-view by aligning fine-grained semantics of arbitrary multi-view based on text guidance, benefiting various VL downstream tasks with limited labeled data. As shown in Fig. 2, this framework is primarily divided into three components. The first component is report and image encoding, the second component is Cross-Modal Attention Alignment (CMAA), and the third component is the overall training objective. In the image encoding, we propose a View Adaptive Network (VAN) to address the issue of varying numbers of views. Furthermore, we extract image and sub-regions representations to facilitate global and local alignment. In the text encoding, we also extract report representation and word representations as global and local representations, respectively. After obtaining global and local representations of images and text, we input the local representations into the CMAA for fine-grained semantic alignment across arbitrary multi-view. CMAA contrasts the words in the report with attention-weighted sub-regions from multiple views, allowing for the matching of sub-regions of interest across multiple views based on the given word. This enables word-guided fine-grained semantic alignment. The global representation is utilized to align the entire arbitrary multi-view and report, aiming to obtain a more comprehensive representation. The entire framework optimizes network parameters by minimizing local and global contrastive losses.

2.2 Medical Imaging Encoding

View Adaptive Network: As shown in Fig. 2, we propose a simple but effective view adaptive network VAN to handle the issue of varying number of views. For medical imaging with K types, VAN generates K private latent sub-spaces to capture view-specific features. Given an arbitrarily multi-view with $k \leq K$ views and corresponding reports $\{x_v, x_t\}$. x_v is mapped into the corresponding k private sub-spaces: $\{z^1, z^2, ..., z^k\}$ based on the type of the view. The critical information in medical images is typically subtle and not easily perceptible, such as pathological features, while a substantial quantity of information may be irrelevant to the task. After extracting view-specific features, we utilize spatial-channel attention to highlight key features and suppress irrelevant ones.

Specifically, VAN employs shared convolution $f_{srd}(\cdot)$ (first three layers of ResNet18 [12]) to learn view-specific private latent sub-spaces. The property of weight sharing across views link multiple private latent sub-spaces. TCSA utilizes text guidance to encourage features from different views to have consistent semantic meanings within the same channel. Therefore, VAN aggregates features from multiple views by summing them within the same channel and uses layer normalization $ln(\cdot)$ to normalize aggregated result to v_1, which enhance the shared feature representation and capture complementary features across multiple views. This is summarized as follows:

$$v_1 = ln(f_{agg}(f_{srd}(x_v))), v_1 \in \mathbb{R}^{c_1 \times w \times h}, \tag{1}$$

where c_1 is the number of channels, w and h are the width and height, $f_{agg}(\cdot)$ denotes channel-wise aggregation. Next, we leverage a spatial-channel attention [15] to highlight critical features and suppress minor ones. In the channel attention module, we generate two different spatial context information by applying max pooling and average pooling. Then, the resulting feature is fed into a shared Multi-Layer Perceptron (MLP) to generate the channel attention map $M_c(v_1)$. This is summarized as follows:

$$M_c(v_1) = \sigma(MLP(AvgPool(v_1)) + MLP(MaxPool(v_1))), \tag{2}$$

where $\sigma(\cdot)$ denotes the sigmoid function. Spatial attention primarily emphasizes the spatial position information by weighting features from different locations. Specifically, we perform average pooling and max pooling along the channel dimension, and then concatenate the resulting feature maps. Finally, a convolution operation is applied to generate the spatial attention map:

$$M_s(v_1^c) = \sigma(f^{7 \times 7}([AvgPool(v_1^c); MaxPool(v_1^c)])), \tag{3}$$

where $v_1^c = M_c(v_1)$, σ denotes the sigmoid function and $f^{7 \times 7}$ represents convolution with the filter size of 7×7.

Vision Encoder. We use ReseNet50 [12] as the main visual encoder $f_v(\cdot)$ to map the representations of arbitrary multi-view x_v in private sub-spaces to a multimodal common sub-space z, aligning them with corresponding reports. We extract global representations v_g and local representations v_l from arbitrary multi-view and set a consistent multimodal dimension d_m to achieve alignment at both the global and local levels with report. We first project the output f_g of the global average pooling layer of ResNet50 via a learnable linear function $L_{gv}(\cdot)$ to d_m as the global representation. Then we extract and vectorize M feature maps from the intermediate layer of $f_v(\cdot)$ as local representation v_l and leverage a learnable linear function $L_{lv}(\cdot)$ to transform the feature dimension to d_m. This is summarized as follows:

$$v_g = L_{gv}(f_g), v_l = L_{lv}(f_l), v_g \in \mathbb{R}^{d_m}, v_l \in \mathbb{R}^{M \times d_m}. \tag{4}$$

2.3 Medical Report Encoding

We use BioClinicalBERT [16] as encoder $f_t(\cdot)$ to extract sentence embeddings and word embeddings as global and local representations from report x_t and project to the consistent multimodal dimension d_m, respectively. For a sentence with W words, assume that each word is tokenized to n_i sub-words. The tokenizer generate $N = \sum_{i=0}^{W} n_i$ word piece embeddings. Then $f_t(\cdot)$ extract features for each word piece embedding. The output $t_l' \in \mathbb{R}^{d_k \times N}$ of $f_t(\cdot)$ is the word embedding as the text local representations, where d_k is the encoding dimension of word embeddings. The sentence embedding is defined as the aggregation of all the word-embedding $t_g' = \sum_{i=0}^{N} t_l'$. Finally, we employ learnable linear functions L_{gt}, L_{lt} to project the word embedding and the sentence embedding dimension into d_m, respectively. As follows:

$$t_g = L_{gt}(l_g'), t_l = L_{lt}(t_l'), t_g \in \mathbb{R}^{d_m}, t_l \in \mathbb{R}^{N \times d_m}. \tag{5}$$

2.4 Cross-Modal Attention Alignment

We propose a cross-modal attention alignment method CMAA for word-based cross-view semantic alignment. CMAA matches the regions of interest across multiple views based on the given words by contrasting the words in the report with attention-weighted sub-regions of multiple views. Given a corresponding text with W words for x_t, an attention map M_{att} is constructed by computing the similarity between each word embedding and all sub-region representations. For each word, all sub-region representations are weighted by the attention map to aggregate as the attention-weighted local visual representation $c_i \in \mathbb{R}^{d_m}$. Firstly, we compute the dot-product similarity of W word embeddings $t_l \in \mathbb{R}^{W \times d_m}$ and M sub-region representations $v_t \in \mathbb{R}^{d_m \times M}$ to generate similarity matrix. Next, we normalize the similarity matrix twice to get attention map M_{att}. As follows:

$$M_{att} = S_{v-t}(S_{t-v}(v_t^T \cdot t_l)), M_{att} \in \mathbb{R}^{W \times M}, \tag{6}$$

where $S_{t-v}(\cdot)$ denotes similarity normalization of each sub-region to all words, $S_{t-v}(\cdot)$ denotes similarity normalization of each word to all subregions. Next we obtain the attention-weighted local visual representation c_i by aggregating all sub-regional representations according to attention weights $a_{ij} \in M_{att}$:

$$c_i = \sum_{j=1}^{M} a_{ij}v_j, v_{att} = \{c_1, c_2, ..., c_W\}, v_{att} \in \mathbb{R}^{W \times d_m}, \tag{7}$$

where v_{att} represents W attention-weighted local visual representations generated based on W word guidance, attention weight a_{ij} denotes the normalized similarity for a word across all sub-regions.

2.5   Loss

Local Loss: We use a feature matching function Z to aggregate similarity across all word embedding t_i and corresponding attention-weighted local visual representations c_i.

$$Z(x_v, x_t) = \log \left(\sum_{i=1}^{W} \exp\left(\langle c_i, t_i \rangle / \tau_3\right) \right)^{\tau_3}, \tag{8}$$

where τ_3 is the scaling temperature, $\langle c_i, t_i \rangle$ represents the cosine similarity between c_i and t_i. The matching function $Z(x_v, x_t)$ yields the similarity between $c_i \in v_{att}$ and $t_i \in t_l$. The local contrast loss can be defined as the posterior probability based on the matching function $Z(x_v, x_t)$. We also maximize the posterior probability of the word embedding t_i given its corresponding attention-weighted representations c_i. For N samples, the above can be defined as follows:

$$
\begin{aligned}
L_l^{(v|t)} &= \sum_{n=1}^{N} -\log \left(\frac{\exp\left(Z(x_v^n, x_t^n)/\tau_2\right)}{\sum_{k=1}^{N} \exp\left(Z(x_v^n, x_t^k)/\tau_2\right)} \right), \\
L_l^{(t|v)} &= \sum_{n=1}^{N} -\log \left(\frac{\exp\left(Z(x_t^n, x_v^n)/\tau_2\right)}{\sum_{k=1}^{N} \exp\left(Z(x_t^n, x_v^k)/\tau_2\right)} \right).
\end{aligned}
\tag{9}
$$

Global Loss: We apply contrastive loss functions following [5,7,9] to maximize the posterior probability of the global image representation v_g^n given its corresponding text representation t_g^n. We also maximize the posterior probability of the text. The above is summarized as follows:

$$
\begin{aligned}
L_g^{(v|t)} &= \sum_{n=1}^{N} -\log \left(\frac{\exp\left(\langle v_g^n, t_g^n \rangle / \tau_1\right)}{\sum_{k=1}^{N} \exp\left(\langle v_g^n, t_g^k \rangle / \tau_1\right)} \right), \\
L_g^{(t|v)} &= \sum_{n=1}^{N} -\log \left(\frac{\exp\left(\langle t_g^n, v_g^n \rangle / \tau_1\right)}{\sum_{k=1}^{N} \exp\left(\langle t_g^n, v_g^k \rangle / \tau_1\right)} \right),
\end{aligned}
\tag{10}
$$

where $\tau_1 \in \mathbb{R}$ is scaling temperature; $\langle v_g^n, t_g^k \rangle$ is the cosine similarity of the v_g^n and t_g^k; $\langle v_g^n, t_g^n \rangle$ is the cosine similarity of the v_g^n and t_g^n.

Total Loss: To achieve local and global contrastive losses jointly optimize the whole framework, the overall loss L is defined as follows:

$$L = L_g^{(t|v)} + L_g^{(v|t)} + L_l^{(t|v)} + L_l^{(v|t)}. \tag{11}$$

3 Experiments and Results

We perform zero shot, classification and image-text retrieval tasks on multiple datasets to compare with state of the art models to verify the effectiveness of TCSA. TCSA can input any number of views, while previous medical VL models only input one view (single-view models). For multi-view imaging classification task with $k > 1$ views, single-view models predicts k labels for k views, while TCSA predicts one label for all k views. For fair comparison, we provide two methods to calculate the evaluation metrics of TCSA. Method 1: for multi-view imaging with k views, TCSA predicts one label and assigns it to k views, resulting in k views having k consistent labels. Method 2: calculate evaluation metrics directly without assigning labels (results with $*$).

3.1 Datasets

MIMIC-CXR [14]: The MIMIC Chest X-ray (MIMIC-CXR) Database v2.0.0 is a large publicly available dataset of chest radiographs in DICOM format with free-text radiology reports. We pre-trained TCSA on the MIMIC-CXR 2.0.0 dataset, which contains 377,110 images corresponding to 227,835 radiographic cases.

CheXpert [3]: CheXpert dataset contains a total of 224,316 chest radiographs from 65,240 patients. According to the distribution of studies with varying numbers of views of CheXpert, we sampled 3,000 samples (5,000 chest radiographs) from CheXpert, comprising a mixture of imaging studies with varying number of views.

CheXpert 5×200 [7]: CheXpert 5×200 is a subset of CheXpert consisting of 1,000 chest radiographs from 1,000 studies.

3.2 Baseline

Random: ResNet50 [12] model with default random initialization parameters.

ImageNet [17]: ResNet50 model with weights pre-trained on the standard ImageNet ILSVRC-2012 task.

DSVE [18]: Previous methods require a pre-trained object detection model for local feature extraction, which is unsuitable for medical images. Therefore, we compared with DSVE on image and text retrieval task.

VSE++ [19]: VSE++ achieves the best performance for image-text retrieval by using only global representations.

Table 1. Image-text retrieval results on CheXpert 5×200 and top K Precision metrics are reported for $K = 5, 10, 100$.

Method	Prec@5	Prec@10	Prec@100
pre-trained on CheXpert			
DSVE [18]	40.6	32.8	24.7
VSE++ [19]	44.3	36.8	26.9
pre-trained on MIMIC-CXR			
ConVIRT [9]	37.8	36.6	30.8
GLoRIA(MIMIC) [7]	42.9	41.6	35.2
TCSA	**46.0**	**43.6**	**35.7**

ConVIRT [9]: ConVIRT proposed a framework based on contrast learning for medical images and reports, which only uses global representation. Since the code for ConVIRT [9] and report part of CheXpert is not publicly released, we replicated it to pre-trained on MIMIC-CXR based on their paper.

GLoRIA(MIMIC) [7]: GLoRIA proposed cross-modal local feature alignment to enable the model to extract local features and demonstrate state of the art performance. Since the report part of CheXpert is not publicly released, we implement GLoRIA based on the official source code[1] on the MIMIC-CXR and denoted as GLoRIA(MIMIC).

3.3   Image-Text Retrieval Result

In the image-text retrieval task, the closest matching text is located by leveraging a query image as the input and is decided by the similarity of their representations. The Precision@K metric was employed to evaluate whether the chosen reports belonged to the same category as the query image to calculate the precision of the top K reports retrieved. We reproduced ConVIRT and GLoRIA(MIMIC) under the same settings as TCSA. Table 1 shows the results of the image-text retrieval task on CheXpert 5×200. GLoRIA [7] observed that the performance of GLoRIA and ConVIRT for image-text retrieval in internal validation (pre-training and testing on the CheXpert) far outperforms DSVE and VSE++. Therefore, we compare the results of TCSA, ConVIRT, and GLoRIA on external datasets (pre-training on MIMIC-CXR and testing on CheXpert 5×200) with the internal validation results of DSVE and VSE++. The results in Table 1 show that even tested on an external dataset, TCSA, GLoRIA, and ConVIRT achieve comparable performance with DSVE, VSE++, with TCSA achieving the best results.

Due to the fact that medical reports are typically generated based on multiple views, while baseline models can only handle single views, there exists

---

[1] https://github.com/marshuang80/gloria.

Table 2. Zero-shot classification results (AUROC) †.

Method	CheXpert	CheXpert 5 × 200
ConVIRT [9]	61.8	62.8
GLoRIA(MIMIC) [7]	62.1	66.6
TCSA	**65.8 64.9***	**69.9**

[1] * Indicates AUROC obtained by Method 2. [2] † It is worth noting that we did not directly use the results reported by GLoRIA [7]. The detailed reasons are described in Sect. 3.4 Zero-shot Classification Results.

an information gap between reports and images, leading to sub-optimal performance in medical image-text retrieval task. TCSA leverages text-guided and VAN mechanisms to comprehensively analyze arbitrary multi-view, eliminating the information gap between reports and medical images. As a result, TCSA achieves superior results in image-text retrieval task.

3.4   Zero-Shot Classification Result

We followed GLoRIA [7] to perform zero-shot classification to evaluate TCSA on CheXpert and CheXpert 5 × 200. The prediction goal is to input image x_v and predict the label y of x_v; even the classifier is not trained with class labels y. We used the area under the ROC curve (AUROC) as the evaluation metric.

The results in Table 2 indicate that TCSA shows significant improvement compared to GLoRIA and ConVIRT, which can be attributed to the integration of multi-view information.

It is worth noting that we did not directly use the results reported by GLo-RIA [7] for the following two reasons: (1) As described in Sect. 3.2 Baseline for GLoRIA(MIMIC), GLoRIA [7] reported its model's performance from internal validation on CheXpert [3] by conducting pretraining and zero-shot classification. Internal validation (pretraining and validation are performed on the same dataset) tends to yield better performance. However, the report section of the CheXpert [3] dataset is not publicly available, preventing us from performing pretraining on CheXpert [3]. Consequently, we conducted pretraining on MIMIC-CXR and performed zero-shot classification on the external dataset CheXpert to validate the generalization performance of TCSA. External validation can result in performance lower than the internal validation reported by GLoRIA [7]. (2) As described in Sect. 3.1 Datasets, considering the distribution of single and multi-view in real medical scenarios, we resampled the CheXpert dataset as our validation dataset. However, the validation set used by GLoRIA [7] filters out a majority of multi-view samples. This leads to a higher inclusion of multi-view samples in our validation dataset. Since handling multi-view data poses a challenge for GLoRIA, this further contributes to the decrease in GLoRIA's performance under our experimental setup.

Table 3. CheXpert supervised classification result (AUROC) based on different portions of training data: 2%, 10%, 100%. We averaged the results from five individual runs to account for the variation in outcomes from randomly sampling training data †.

Method	Portion		
	2%	10%	100%
TCSA(no pretraining)	52.3	61.6	67.6
Random	52.9	57.2	68.1
ImageNet	61.3	64.4	69.8
ConVIRT [9]	63.7	64.7	74.1
GLoRIA(MIMIC) [7]	66.1	72.0	74.4
TCSA	**68.1 69.3***	**74.6 72.9***	**78.6 79.3***

[1] * Indicates AUROC obtained by Method 2. [2] † It is worth noting that we did not directly use the results reported by GLoRIA [7]. The detailed reasons are described in Sect. 3.5 Supervised Classification Result.

Table 4. CheXpert supervised classification ablation experiment result (AUROC) based on different portions of training data: 2%, 10%, 100%.

Pretraining	Spatial Channel Attention	VAN	2%	10%	100%
×	✓	✓	52.3	61.6	67.6
✓	×	✓	67.4	71.7	75.6
✓	✓	×	66.1	72.0	74.4
✓	✓	✓	**68.1**	**74.6**	**78.6**

3.5 Supervised Classification Result

We followed GLoRIA [7] to perform supervised image classification on CheXpert using fine-tuning and different amounts of training data (2%, 10%, and 100%). We matched GLoRIA's image count, using 3000 samples with 5000 images. The change was increasing from 1% to 2%. This adjustment was needed as, at 1%, TCSA could only sample 30 mixed multi-view and single-view training samples (compared to GLoRIA's 50 images at 1%). The limited training samples at 1% made it hard to cover all labels effectively, leading to training failure. We used AUROC as the evaluation metric.

As explained in Sect. 3.4, due to the unavailability of the report section of the CheXpert [3] dataset, we were unable to directly compare TCSA with GLoRIA in the same setting. Therefore, we resorted to external validation by pretraining on MIMIC [14] and validation on CheXpert [3]. This external validation approach resulted in lower performance compared to the reported results of GLoRIA [7]. Furthermore, considering the real medical scenario, our resampled validation set includes more multi-view samples, which poses a challenge for GLoRIA. In

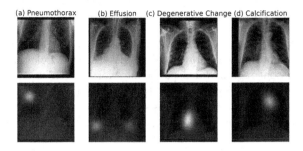

Fig. 3. Highlighted pixels represent higher activation weights by the corresponding word.

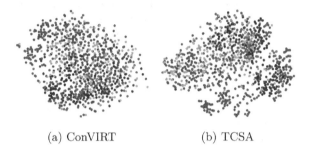

(a) ConVIRT   (b) TCSA

Fig. 4. t-SNE representation visualization of CheXpert 5×200 images by ConVIRT and TCSA.

conclusion, we present results that are inconsistent with GLoRIA [7] due to these reasons.

Table 3 shows that TCSA outperforms all baselines on training data of different sizes. Furthermore, TCSA trained with only 2% of the data achieves similar performance to imagenet-initialized trained with 100%. To validate the effectiveness of text-guided mechanisms, we employed randomly initialized VAN and ResNet50 as baselines for the supervised classification task, denoted as TCSA(no pretraining). TCSA(no pretraining) aligns with TCSA in its capability to accommodate arbitrarily multi-view inputs. However, TCSA (no pretraining) demonstrates notably inferior performance compared to pre-trained TCSA, achieving results akin to a randomly initialized ResNet50. This disparity suggests that TCSA(no pretraining), though capable of processing arbitrarily multi-view inputs, struggles to effectively exploit the information from multi-view. This suggests that the performance improvement of text-guided TCSA is not solely reliant on the implementation of arbitrary multi-view inputs, but rather benefits from the integration of the text-guided mechanism and arbitrary multi-view inputs.

3.6   Analysis of Our Framework

Visualization: As shown in Fig. 3, we visualize attention weights to evaluate the proposed method's representation qualitatively. TCSA can accurately identify important visual regions for a particular word. We also compare the representations of CheXpert 5×200 generated by ConVIRT and TCSA via t-SNE visualization. Due to the high inter-class similarity of medical images, clustering images in our setup is more challenging compared to clustering in typical object classification settings. As shown in the Fig. 4, TCSA demonstrates relatively better cluster separation compared to ConVIRT, indicating that TCSA presents superior clustering representation.

Ablation Studies: As shown in Table 4, we systematically remove each component one at a time while keeping the others fixed to observe the effect on the model's performance, thus demonstrating the necessity of each component.

4   Conclusion

Existing VL models in the medical domain have overlooked the varying numbers of views in real medical scenarios. To extend VL representation learning to image-text pairs with varying numbers of views, we propose a text-guided cross-view semantic alignment framework TCSA for adaptive multi-view visual representations learning. We evaluated TCSA on multiple tasks and datasets, demonstrating its effectiveness in various medical VL downstream tasks.

References

1. Litjens, G., et al.: A survey on deep learning in medical image analysis. Med. Image Anal. **42**, 60–88 (2017)
2. Wang, X., Peng, Y., Lu, L., Lu, Z., Bagheri, M., Summers, R.M.: Chestx-ray8: hospital-scale chest x-ray database and benchmarks on weakly-supervised classification and localization of common thorax diseases. In: Proceedings of the IEEE Conference on Computer Vision and Pattern Recognition, pp. 2097–2106 (2017)
3. Irvin, J., et al.: CheXpert: a large chest radiograph dataset with uncertainty labels and expert comparison. In: Proceedings of the AAAI Conference on Artificial Intelligence, vol. 33, pp. 590–597 (2019)
4. Melnyk, A., et al.: Clustering based identification of SARS-CoV-2 subtypes. In: Jha, S.K., Măndoiu, I., Rajasekaran, S., Skums, P., Zelikovsky, A. (eds.) ICCABS 2020. LNCS, vol. 12686, pp. 127–141. Springer, Cham (2021). https://doi.org/10.1007/978-3-030-79290-9_11
5. Radford, A., et al.: Learning transferable visual models from natural language supervision. In: International Conference on Machine Learning, pp. 8748–8763. PMLR (2021)
6. Zhou, H.-Y., Chen, X., Zhang, Y., Luo, R., Wang, L., Yu, Y.: Generalized radiograph representation learning via cross-supervision between images and free-text radiology reports. Nat. Mach. Intell. **40** (1), 32–40 (2022)

7. Huang, S.-C., Shen, L., Lungren, M.P., Yeung, S.: Gloria: a multimodal global-local representation learning framework for label-efficient medical image recognition. In: Proceedings of the IEEE/CVF International Conference on Computer Vision, pp. 3942–3951 (2021)
8. Wang, Z., Wu, Z., Agarwal, D., Sun, J.: Medclip: contrastive learning from unpaired medical images and text. arXiv preprint arXiv:2210.10163 (2022a)
9. Zhang, Y., Jiang, H., Miura, Y., Manning, C.D., Langlotz, C.P.: Contrastive learning of medical visual representations from paired images and text. arXiv preprint arXiv:2010.00747 (2020)
10. Yuan, J., Liao, H., Luo, R., Luo, J.: Automatic radiology report generation based on multi-view image fusion and medical concept enrichment. In: Shen, D., et al. (eds.) MICCAI 2019. LNCS, vol. 11769, pp. 721–729. Springer, Cham (2019). https://doi.org/10.1007/978-3-030-32226-7_80
11. Nooralahzadeh, F., Gonzalez, N.P., Frauenfelder, T., Fujimoto, K., Krauthammer, M.: Progressive transformer-based generation of radiology reports. arXiv preprint arXiv:2102.09777 (2021)
12. He, K., Zhang, X., Ren, S., Sun, J.: Deep residual learning for image recognition. In: Proceedings of the IEEE Conference on Computer Vision and Pattern Recognition, pp. 770–778 (2016)
13. Wang, F., Zhou, Y., Wang, S., Vardhanabhuti, V., Yu, L.: Multi-granularity cross-modal alignment for generalized medical visual representation learning. arXiv preprint arXiv:2210.06044 (2022b)
14. Johnson, A.E.W., et al.: Mimic-cxr-jpg, a large publicly available database of labeled chest radiographs. arXiv preprint arXiv:1901.07042 (2019)
15. Woo, S., Park, J., Lee, J.-Y., Kweon, I.S.: CBAM: convolutional block attention module. In: Ferrari, V., Hebert, M., Sminchisescu, C., Weiss, Y. (eds.) ECCV 2018. LNCS, vol. 11211, pp. 3–19. Springer, Cham (2018). https://doi.org/10.1007/978-3-030-01234-2_1
16. Alsentzer, E., et al.: Publicly available clinical BERT embeddings. arXiv preprint arXiv:1904.03323 (2019)
17. Deng, J., Dong, W., Socher, R., Li, L.-J., Li, K., Fei-Fei, L.: Imagenet: a large-scale hierarchical image database. In: Proceedings of the IEEE Conference on Computer Vision and Pattern Recognition, pp. 248–255. IEEE (2009)
18. Engilberge, M., Chevallier, L., Pérez, P., Cord, M.: Deep semantic-visual embedding with localization. In: RFIAP 2018-Congrès Reconnaissance des Formes, Image, Apprentissage et Perception (2018)
19. Faghri, F., Fleet, D.J., Kiros, J.R., Fidler, S.: Vse++: improving visual-semantic embeddings with hard negatives. arXiv preprint arXiv:1707.05612 (2017)

Multi-class Cancer Classification of Whole Slide Images Through Transformer and Multiple Instance Learning

Haijing Luan[1,2], Taiyuan Hu[1,2], Jifang Hu[1,2], Ruilin Li[1], Detao Ji[1,2], Jiayin He[1], Xiaohong Duan[3], Chunyan Yang[3], Yajun Gao[3], Fan Chen[3], and Beifang Niu[1,2(✉)]

[1] Computer Network Information Center,
Chinese Academy of Sciences, 100190 Beijing, China
luanhaijing@cnic.cn, bniu@sccas.cn
[2] University of Chinese Academy of Sciences, 100190 Beijing, China
[3] ChosenMed Technology (Beijing) Co., Ltd., 100176 Beijing, China

Abstract. Whole slide images (WSIs) are high-resolution and lack localized annotations, whose classification can be treated as a multiple instance learning (MIL) problem while slide-level labels are available. We introduce a approach for WSI classification that leverages the MIL and Transformer, effectively eliminating the requirement for localized annotations. Our method consists of three key components. Firstly, we use ResNet50, which has been pre-trained on ImageNet, as an instance feature extractor. Secondly, we present a Transformer-based MIL aggregator that adeptly captures contextual information within individual regions and correlation information among diverse regions within the WSI. Thirdly, we introduce the global average pooling (GAP) layer to increase the mapping relationship between WSI features and category features. To evaluate our model, we conducted experiments on the The Cancer Imaging Archive (TCIA) Clinical Proteomic Tumor Analysis Consortium (CPTAC) dataset. Our proposed method achieves a top-1 accuracy of 94.8% and an area under the curve (AUC) exceeding 0.996, establishing state-of-the-art performance in WSI classification without reliance on localized annotations. The results demonstrate the superiority of our approach compared to previous MIL-based methods.

Keywords: Multiple instance learning · Transformer · Whole slide image · Global average pooling

1 Introduction

Recently, with the emergence of digital scanners, it is capable of capturing biopsy slides as gigapixel whole slide images (WSIs) while preserving the original tissue structure. The advent of WSI, with its diverse nuclear, cytoplasmic, and

---

H. Luan and T. Hu—Equal contribution.

extracellular matrix features, has made up for the shortcomings of traditional biopsy slides that are prone to breakage, difficult retrieval, and poor diagnostic repeatability, driving pathology into a new stage of development and resulting in a substantial amount of clinical and research interest in the analysis of digital pathology.

Tasks related to pathological image analysis can be broadly categorized into classification, registration, detection, segmentation, localization, and generation [5]. Among the various tasks in pathological image analysis, WSI classification is widely regarded as a crucial task. Its importance primarily stems from its versatility, as it serves as the foundation for other tasks such as nuclei localization [22], mitosis detection [17], gland segmentation [3], pathological image retrieval [25], and pathological image registration. Furthermore, WSI classification can be employed to address intricate problems by transforming them into analogous classification tasks, as exemplified by the challenge of identifying the primary site of cancer of unknown primary (CUP).

Deep learning-based image classification methods have proven effective in the fields of natural and medical images. However, traditional deep learning methods are not suitable for the automatic classification of WSIs due to their high resolution, subtle differences in image features, overlapping cells, and color variations. An approach is to train models using high-resolution image patches and predict the label of a WSI based on the predictions made at the patch-level in supervised learning [23]. However, obtaining fine-grained labeled data, such as patch-level labels in WSI is an expensive and challenging task, and pathologists need to manually annotate region-of-interest (ROI) of WSI based on prior knowledge, which is not only time-consuming and labor-intensive but also requires extensive experience.

In the course of annotating the dataset, pathologists provide slide-level labels for WSI based on visual inspection. This procedure results in the absence of positional labels for the tumor-relevant regions, which poses a challenge in identifying the crucial local areas that significantly influence the slide-level WSI classification results. These characteristic highlights that the use of weakly supervised methods may be more suitable for addressing the classification tasks of WSIs. The pioneering research in weakly-supervised WSI classification has predominantly employed the framework of multiple instance learning (MIL), wherein a WSI is regarded as a collection, also referred to as a bag, of smaller image patches, known as instances. An approach is to extract and aggregate patch-level features or scores, followed by obtaining WSI-level labels through an embedding-level classifier or instance-level classifier. Recently, several MIL studies have leveraged convolutional neural networks (CNNs) for feature extraction and aggregation, leading to excellent performance in WSI classification.

Recently, a straightforward approach is to apply pooling operations to patch-level features or scores extracted by CNNs. MIL pooling methods include Max-pooling [7], Mean-pooling [16], Noisy-AND [12], and attention mechanisms. The first three methods are pre-defined and challenging to train, thus limiting the adaptability of MIL methods. In contrast, the attention mechanism offers a pliant pooling approach that can be trained concurrently with other model components. This mechanism provides high flexibility and interpretability, making it a promising alternative for MIL-based WSI classification. Based on the fundamental theorem of symmetric functions, Ilse et al. [9] outlined the usefulness of deep learning for modeling permutation-invariant bag scores and proposed a gating-based attention MIL pooling that assigns different trainable attention weights to each instance within a bag to calculate the WSI-level representation. The underlying idea of DSMIL [13] was to construct two different branches. The first branch of the proposed method utilizes an instance-based approach to compute scores for each instance, which are then aggregated using max-pooling to obtain the bag score. In the second branch, non-local attention is employed by computing the similarity between the key instance with the highest score in the first branch and the remaining instances, with each instance assigned a distinct attention score. The algorithm fuses the bag scores of the two branches for model training, effectively combining the advantages of instance-based and embedding-based methods. CLAM [15] employed attention-based MIL to automatically identify subregions which have high diagnostic values to accurately classify WSI. What sets CLAM apart from previous approaches is the use of instance-level clustering on a restricted set of representative regions. This clustering process serves to constrain and refine the feature space, further enhancing the classification performance.

The aforementioned studies have contributed to the advancement of WSI applications in routine clinical diagnosis, serving as valuable assistive tools for pathologists. However, all these methods assume that all instances are independent in the bag. A pertinent query arises as to whether pathologists can make valuable decisions for patients without considering the spatial information and morphological features of different instances in WSIs. In the context of MIL research, it is advantageous to consider the inter-instance relationships within WSI when constructing a model. In particular, the Transformer architecture [20] has achieved remarkable success in various visual recognition tasks, including image classification [6] and object detection [2]. Furthermore, the self-attention module in the Transformer architecture plays a crucial role in capturing the correlations among different instances, allowing for a linear combination of instance features within the bag. As a result, the transformer is well-suit for WSI classification, the results of which are closely related to the distribution of tissue regions in the WSIs. However, with higher magnification, the WSIs will generate a considerably larger number of patches, forming longer sequences that inevitably present significant challenges in terms of computational and memory requirements. In this study, we propose a novel model called TMG, which incorporates the Transformer-based Multiple instance learning with Global average pooling into histopathological images to achieve accurate WSI classification, with lower computational requirements and supporting for longer sequences. The main contributions of this study are as follows:

1. The traditional self-attention module can globally aggregate all instances of information when updating an instance; however, its time complexity is

$O(n^2)$. We integrated linear-attention into TMG, which can effectively reduce the model's time complexity to $O(n)$ without degrading the performance.

2. We propose the convolution-based multidimensional conditional position encoding (CMCPE), which can generate position information for WSIs of different resolution sizes. In comparison to traditional position encoding methods, the advantages of using CMCPE are summarised as follows: making the input sequence permutation-variant but translation-invariant; being capable of dealing with sequences longer than those during the training period; and providing the ability to encode absolute positions to a certain extent. What's more, it is simple and lightweight, which is easily plugged into transformer blocks.

3. To make the conversion between the feature map and the final classification result more natural, we propose using the global average pooling (GAP) layer in the vision transformer (ViT) to classify WSIs, leading to enhanced model robustness. Our experimental results demonstrate the effectiveness of GAP.

4. Experiments on histopathological images demonstrate the superiority of TMG in the The Cancer Imaging Archive (TCIA) Clinical Proteomic Tumor Analysis Consortium (CPTAC) datasets, compared to other state-of-the-art methods.

2   Methods

2.1   Transformer Network-Based Multiple Instance Learning

The MIL is a powerful tool for solving weakly supervised problems. It treats each WSI as a collection (called a bag) of smaller image patches (called instances). The entire WSI can be used as a research object without manually extracting the ROI. In our method, TMG, we define the ith patch obtained from gigapixel WSIs as an instance x_i, and the set of all patches in the WSI is taken as a bag ($B = \{x_i | i = 1, \ldots, n\}$). The classification result y for the entire WSI B is:

$$y(B) = g(h(f(x_1), f(x_2), \ldots f(x_i), \ldots, f(x_n))), \quad i = 1, \ldots, n \qquad (1)$$

where, $x_i \in R^D$ denotes the ith instance in bag B. As shown in Fig. 1, TMG contains $f(\cdot)$, $h(\cdot)$ and $g(\cdot)$, which are separately used as an instance feature extractor, a Transformer and MIL-based feature aggregator named TM, and GAP, respectively. In this study, the instance feature extractor consisted of a ResNet50 truncated after the third residual block, whose parameters were acquired through pretraining on ImageNet32.

To capture the long-term dependencies among different instances in WSIs, we employ the Transformer encoder structure in ViT as an aggregator for MIL and change the position where positional encoding is added. As shown in Fig. 1, each transformer encoder layer is composed of a multihead self-attention (MHSA) and a multi-layer perceptron (MLP) block. We follow ViT, adding the class token ($Class_0(B)$) and set of key features $M \subseteq f(x_1), \ldots f(x_i), \ldots, f(x_n)$ selected in

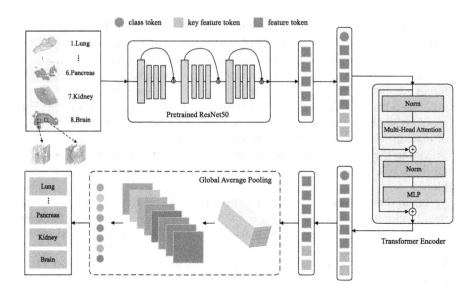

Fig. 1. Detailed overview of the TMG for the classification of WSI. The TMG consists of instance feature extractor, a feature aggregator TM, and GAP. The input histology image is divided into patches, which is feeding into the instance feature extractor to obtain the compact low-dimensional feature vector of length of 1024. The obtained features are the input into the feature aggregator TM to capture the local space information and morphological features in WSIs. The outputs of the TM are further input into the GAP to establish the association between the key information extracted by TM and the final predicted category information.

the feature extractor $f(\cdot)$ to the input instance token group. The input token embedding $z_0 \in R^{N \times D}$ can be represented as

$$z_0 = [Class_0(B), f(x_1), f(x_2), \dots f(x_i), \dots, f(x_n), M], \quad i = 1, \dots, n \quad (2)$$

In MHSA, we initially transform instance embedding into query Q, key K, and value V, following which we compute the similarity between the query and key vectors to obtain the attention matrix. Each value in the matrix records the correlation between a pair of instances. The output of the MHSA is the weighted average sum of all values, the weight of which is the attention matrix calculated by the query and key. The transformer encoder module is composed of L stacked layers, LN represents the layernorm, and MLP includes two fully connected layers with a nonlinear GELU activation function, which can be expressed as follows:

$$z'_l = MHSA(LN(z_{l-1})) + z_{l-1}, \quad l = 1, \dots, L \quad (3)$$

$$z_l = MLP(LN(z'_l)) + z'_l, \quad l = 1, \dots, L \quad (4)$$

The MHSA contains aggregation information that aggregates the contribution of each instance in the bag to the final WSI classification; however, its time complexity is $O(n^2)$. Methods such as Longformer [1], Linformer [21] and Reformer [11] can reduce the time complexity to $O(n)$ or $O(nlogn)$, which rely on additional constraints or assumptions to reduce the complexity of the model and do not provide rigorous theoretical proofs. Performer [4] provided fast attention via the positive orthogonal random features approach (FAVOR), thus achieving an unbiased estimation of the attention matrix under linear space-time complexity and not depending on sparsity or low-rankness assumptions. Specifically, the original attention formula can be expressed as:

$$Att(Q, K, V) = softmax(\frac{QK^T}{\sqrt{d}})V = D^{-1}AV \tag{5}$$

Here, $A = softmax(\frac{QK^T}{\sqrt{d}})$, $D = diag(A1_L)$, $1_L = (1)_{L \times L}$, and L is the length of embedding. In Performer, by applying a function mapping, matrix A can be decomposed into the product of two smaller matrices.

$$A = exp(QK^T) \approx \phi(Q)\phi(K)^T = Q'(K')^T \tag{6}$$

Here, $Q', K' \in R^{L \times r}$, r represents the dimension of the sequence after low-dimensional mapping. This mapping function preserves the inner product relationship between vectors and approximates the attention mechanism by performing dot product operations. The approximate self-attention Y of the Performer can be defined as [19]:

$$Y = D^{-1}(Q^{'}((K')^T V)) \tag{7}$$

Here, $D = diag(Q'((K')^T 1_L))$. Performer does not show the calculation of $A = exp(QK^T)$, which can reduce the time complexity of attention calculation to linear. In our method, we replace the self-attention module with Performer, which can effectively reduce the model complexity to $O(n)$ without degrading performance. In the multiple variants of ViT, the number of blocks used by the transformer encoder module is 12, 24, and 32. According to different model sizes, we use a block number of 2 to further reduce the complexity of the model.

For the self-attention module, the spatial position and edge information of the image is ignored. The absolute position encoding adopted by the transformer add the absolute positional encodings to the input embedding as new input embeddings. There are several choices for the absolute encoding in transformer, such as fixed encodings by sine and cosine with different frequencies and learnable encodings through training parameters [8]. In terms of classifying WSI, the number of patches contained in each WSI may vary, and the absolute position encoding of the ViT cannot provide position encoding information of the dynamic length, limiting the translation of the model and sequence length of the training images. Previous research findings have revealed that the enhancement of readout for absolute positional information is bolstered by the integration of larger receptive fields or the application of non-linear strategies to interpret positional nuances. Therefore, in this study, we transform the MLP into CMCPE, which can generate position information for WSIs of different resolution sizes. CMCPE can introduce the locality, two-dimensional neighbourhood structure, and translation invariance of convolution operations, thereby providing the absolute position of the image [10]. Given the different bag sizes (number of patches in each WSI), we proposed CMCPE to comprehensively consider the spatial information and morphological characteristics of different patches in the WSI. CMCPE operate on a bag of feature embeddings, denoted as $x \in R^{(N+1) \times D}$, where N is determined by the number of patches in a WSI and the number of key features after feature extraction, and it is dynamically changed. The purpose of this module is to enhance the positional encoding information of input features by utilizing convolutional layers with different kernel sizes. This intuitive operation aims to better capture the spatial relationships between features, thereby enhancing the model's understanding and representation of spatial information.

After feeding the input token embedding through L transformer encoder layers, we obtain the output token embedding $z_L \in R^{(N+1) \times D}$, where D represents the dimension of the token embedding. Prior to being fed into the GAP layer, the morphological features and local spatial location information extracted through the L transformer encoder layers are further consolidated using the self-attention module to obtain a comprehensive representation z_L. In subsequent experiments, $Class_L(B) = z_L[0]$ is the input token embedding of the MLP head in ViT, and $b_L = z_L[1, 2, \ldots, N]$ is the descriptive vector representation of the WSI, in which the instance embedding b_L can be used as the input of the GAP layer.

2.2   Global Average Pooling

In the ViT architecture, $Class_L(B) = z_L[0]$ is used as an input of the MLP Head for WSI classification. In the experiment, we constructed an association between the WSI feature map and the category feature map by introducing the GAP to achieve better classification results.

For image classification, the convolutional layer is typically used to extract two-dimensional feature information from images, which is converted into a one-dimensional vector and then fed into the fully connected layer to predict various categories. In WSI classification task, the WSI descriptive feature vector $b_L \in R^{N \times D}$ obtained by the transformer encoder is first reconstructed into a 2D image feature map $b_f \in R^{(\sqrt{N} \times \sqrt{N}) \times D}$. Then, convolution with a kernel size of 1 is applied to obtain a feature map with the same number of channels as the number of categories C, followed by applying the GAP, and finally rescaling the newly generated sequence to the serialised information format as the output. A schematic of GAP is shown in Fig. 2.

3   Results

3.1   Details of Implementation

Datasets. We acquired pathology slides and corresponding labels for WSIs containing tumour tissues corresponding to the eight grouping categories for WSI classification via TCIA CPTAC Pathology Portal. To ensure that the model

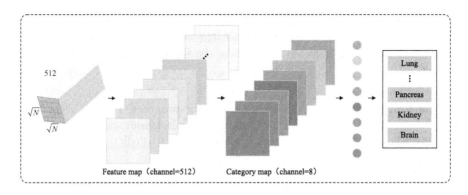

Fig. 2. Schematic diagram of global average pooling.

could effectively learn the distinctive features of tumours for precise classification of WSI, slides without tumour were excluded from the analysis, and only representative diagnostic slides of the tumour were included. In total, we obtain 6037 WSIs from the CPTAC study.

WSI Processing. Given the enormous bag sizes (numbers of patches in each WSI) in CPATC, we first used the CLAM library to segment tissues, which is publicly available over GitHub (https://github.com/mahmoodlab/CLAM), thereby obtaining a segmentation mask which contains enough tissue cells. CLAM mainly included the following processing steps: first, it converted the image from an RGB to HSV colour space and applied binary thresholding to the downsampled saturation channel of the image to calculate the binary mask of the tissue area. Second, 256×256 patches (without overlap) were cropped from the segmented tissue contours as instances. In the final step, we employed a truncated ResNet50 model that was pre-trained on the ImageNet dataset to convert each RGB image patch of size $256 \times 256 \times 3$ into a discriminative feature representation with a length of 1024. Significantly, we save each instance in the hierarchical data format hdf5 which contains an array of extracted features along with their corresponding patch coordinates.

Training. The model parameters were updated following each min-batch using the Adamax optimizer with an L2 weight decay of 1×10^{-5} and a learning rate of 2×10^{-4}. The running average of the first and second moments of the gradient was calculated with $\beta 1$ set to 0.9 and $\beta 2$ set to 0.999. Additionally, to ensure numerical stability, the ϵ term 1×10^{-8} was added to the denominator as the default setting.

Model Selection. The training process was limited to a maximum of 200 epochs, with continuous monitoring of the performance of the model on the validation set was monitored after each epoch. To prevent overfitting, early stopping was applied when the validation loss for the WSI classification did not decrease for 20 consecutive epochs beyond epoch 50. The best model, which achieved lowest validation loss was selected to evaluate the performance on the dependent test set.

Evaluation. To validate the performance of the proposed TMG in classifying WSI, we evaluated it using CPTAC dataset. The classification prediction of the network was the argmax of the class probability predicted by TMG. We evaluated the performance of the model using the following classification metrics: precision, recall, F1-score, mean average precision (mAP), AUC-ROC for each category, and aggregation for all classes by micro-averaging, macro-averaging, and weighted averaging (Fig. 3a, Fig. 3b, Table 1). In addition, with TMG, WSI classification yielded an AUC value over 0.99.

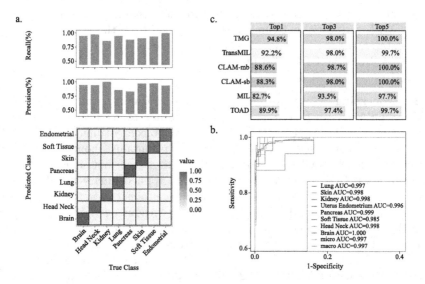

Fig. 3. Model performance of the TMG. a. Performance for the classification prediction of WSI on the CPTAC hold-out test set for 8 categories. Per precision and recall are plotted next to the confusion matrix. The columns of the confusion matrix represent the true category of the tumour and the rows represent the category predicted by the TMG. b. The AUC on the CPTAC hold-out test set of TMG in predicting eight tumours. c.Top-k model accuracies for WSI classification prediction on the CPTAC hold-out test set for $k \in 1, 3, 5$.

Computational Hardware and Software. All WSIs were processed on an Intel(R) Xeon(R) Gold 5218R CPU @ 2.10 GHz, equipped with two GeForce RTX 2080 Ti graphics processing units (GPUs). The WSI processing pipeline was implemented in Python (3.7.7) using the CLAM package. CLAM utilized

Table 1. Test performance for WSI classification prediction

category	Precision	Recall	F1-score	mAP	AUC-ROC
Lung	0.857	0.947	0.900	0.971	0.998
Skin	0.976	0.909	0.941	0.991	0.998
Kidney	1.000	0.857	0.923	0.965	0.997
Endometrial	0.939	1.000	0.969	0.994	0.999
Pancreas	0.833	0.882	0.857	0.910	0.985
Soft tissue	0.979	0.939	0.958	0.991	0.998
Head and neck	0.944	0.971	0.958	0.997	1.000
Brain	0.948	0.948	0.948	0.984	0.997
Micro average	0.936	0.936	0.934	0.977	0.996
Macro average	0.950	0.948	0.948	0.985	0.997
Weighted average	0.857	0.947	0.900	0.971	0.998

openslide-python (1.1.2) for WSI reading and opencv-python (4.1.1.26) for image processing. Each deep learning model was implemented on a GPU using the Pytorch (1.13.0) deep learning library. The scikit-learn (1.0.2), a scientific computing library, was used to calculate metrics for evaluating classification performance, including precision, recall, F1-score, and mAP. It was also utilized to evaluate the overall performance of the mode using the AUC-ROC metric. Additionally, numpy (1.18.1) was used for the numerical calculation and matplotlib (3.1.1) was used for plotting.

3.2   Performance Evaluation

Based on the CPTAC dataset, we compared TMG with state-of-the-art methods including MIL [9], CLAM [15], TOAD [14], and TransMIL [18]. To ensure comparability between the different methods, the feature extraction of the above method uses truncated ResNet50, whose parameters are pretrained on ImageNet32. Table 2 showcases the top-k accuracy of the model, where k = 1,3,5, which is a metric that evaluates how often the ground truth is found in the k-highest confidence predictions made by the model. It is also displayed the test error and ROC. In Fig. 3c, we can observe that the top-k accuracy of the TMG for k = 1,3,5 is 94.8%, 98.0%, and 100%. Our proposed TMG method achieved the best performance, outperforming the second-best method TransMIL by approximately 2.6% in top-1 accuracy, and had the lowest error rate of 5.21% on the hold-out test set.

3.3   Ablation Study

We conducted ablation experiments to study: (1) the selection of key features of the input token in the backbone network; (2) CMCPE captured the location-encoded neighbourhood information using convolution kernels of different sizes;

Table 2. Classification performance for each category. The top-k accuracy for k = 1,3,5, test error, and area under the curve (AUC) are shown for different methods, such as TOAD, MIL, CLAM-sb, CLAM-mb, TransMIL, and TMG.

method	top-1	top-3	top-5	test error	AUC
TOAD	0.899	0.974	0.997	0.10098	0.98838
MIL	0.827	0.935	0.977	0.17264	0.97262
CLAM-sb	0.883	0.980	1.000	0.11726	0.98865
CLAM-mb	0.886	0.987	1.000	0.11401	0.98859
TransMIL	0.922	0.980	0.997	0.07818	0.99370
TMG	0.948	0.980	1.000	0.05212	0.99643

and (3) where the GAP layer was added. To evaluate the impact of selecting key features in the input token for WSI classification prediction, we selected the first M or last M token features obtained from the feature extractor and spliced them into the original morphological features that combine with the class token. The experimental results of the ablation study were listed in Table 3. The results indicated that selecting the last M token features as key features could better classify WSI at various resolutions, and the top-1 accuracy and AUCs values had positive effects (Top-1 accuracy: 0.948 vs. 0.945; AUC:0.997 vs. 0.992). Next, to further verify the effectiveness of adding multidimensional position encoding and the superiority of the superposition of different dimensions of multidimensional position encoding, we captured the position information with different neighbourhood sizes by changing the size of the convolution kernel. The results showed that simultaneously capturing the position information with the neighbourhood of convolution kernel sizes of 1, 3, and 5 could significantly improve the top-1 accuracy of TMG and achieve an AUC value of 0.997 when achieving the WSI classification. Lastly, to evaluate the benefit of adding GAP, we additionally experimented with adding GAP to the class feature map or the WSI feature map. The experimental results demonstrate that the addition of the GAP layer to the class feature map could effectively enhance the correlation between the WSI feature map and the class feature map, leading to a noteworthy improvement of 2% in the top-1 accuracy performance (0.948 vs. 0.928). As shown in Table 3, all models used the same hyperparameter settings as those reported throughout the entire study. For all experiments, the cases were partitioned into 70:10:20 splits for training:validation:testing.

3.4   Model Convergence

In Fig. 4, we presented the training/validation accuracy and training/validation loss of TMG, as well as the latest MIL methods including TransMIL, TOAD, MIL, CLAM-mb, and CLAM-sb at different epochs. Unlike these methods, TMG leverages the morphological and spatial information among instances. Despite having a larger number of model parameters, TMG achieves comparable perfor-

Table 3. Ablation study. The top-k accuracy for k = 1,3,5 (represented by top-1, top-3, and top-5 in the below table), test error, and area under the curve (AUC) are shown for different experiments: 1) The key features are obtained from the first or last M-token features of the morphological features from the feature extractor, as shown in tests 1 and 2; 2) The location information in the neighbourhood of convolution kernel sizes of 1, 3, 5, or its combination, and without the location information, as shown in tests 2, 4, 5, 6, 7, 8, and 9; 3) Where the global average pool (GAP) layer is added-the class or the whole slide image (WSI) feature map, as shown in Tests 2 and 3.

Test	Key feature	Kernel size	GAP	top-1	top-3	top-5	Test error	AUC
1	first	1,3,5	✓	0.945	0.987	1.000	0.055375	0.995291
2	last	1,3,5	✓	0.948	0.980	1.000	0.052117	0.996643
3	last	1,3,5	×	0.928	0.987	1.000	0.071661	0.995544
4	last	×	✓	0.919	0.987	0.997	0.081433	0.994768
5	last	1	✓	0.932	0.977	0.997	0.068404	0.993394
6	last	3	✓	0.938	0.990	1.000	0.061889	0.997234
7	last	5	✓	0.925	0.977	0.993	0.074919	0.994470
8	last	7	✓	0.935	0.984	0.990	0.065147	0.992559
9	last	3,5,7	✓	0.938	0.984	0.997	0.061889	0.996415

mance in terms of training loss and accuracy to other MIL methods. Moreover, it outperforms existing MIL methods in terms of validation loss and accuracy.

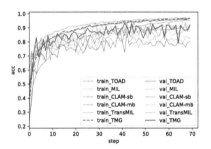
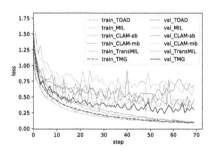

Fig. 4. The training/validation accuracy and training/validation loss of TMG, as well as the latest MIL methods.

3.5   t-SNE Visualization

To assess the effectiveness of adding the GAP layer, we conducted evaluations on the hold-out test dataset, specifically focusing on models that incorporated GAP into the category feature maps and WSI feature maps. We extracted features from eight types of tumors and utilized t-SNE, a non-linear dimensionality reduction method, to visualize the features in two dimensions. Figure 5 demonstrates that the inclusion of a GAP layer in the category feature map leads to improved WSI classification performance.

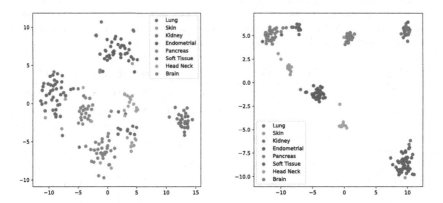

Fig. 5. The t-SNE visualization on the feature representation vector of WSI. The left figure illustrates the t-SNE visualization results when GAP is applied to the WSI feature maps, while the right figure demonstrates the t-SNE visualization results when GAP is applied to the category feature map.

4   Discussion and Conclusion

In this paper, we introduced TMG, a novel method based on Transformer and MIL, which shows promise in enhancing the classification performance of WSIs. The TMG is composed of three components: an instance feature extractor, a feature aggregator TM based on Transformer and MIL, and GAP. The instance feature extractor encodes each instance into a low-dimensional feature vector via a truncated ResNet50. TM is the backbone of the entire model, capturing the spatial information and morphological features in WSIs. GAP establishes the association between the key feature information extracted by TM and the final predicted category information. TMG demonstrated exceptional performance by achieving AUC values exceeding 0.996 in accurately predicting eight distinct tumour classes when testing on the hold-out CPTAC dataset. Using a single GeForce RTX 2080 Ti GPU, the average runtime (in seconds) for inference (including tissue segmentation, patching, feature extraction, and WSI classification prediction) on the dependent test set was 13.14 s (std, 6.06 s). Through extensive experiments, we demonstrate that our proposed method can achieve excellent performance, compared with previous methods, such as MIL, CLAM, TOAD, and TransMIL.

Although TMG has demonstrated excellent performance in predicting the type of cancer, further testing of its generalizability is needed. For example, we will assess the performance of TMG in subtype classification and compare it with novel methods such as GTP [24]. Furthermore, we are actively exploring the application of TMG to generalized classification problems, including the diagnosis of CUP and the prediction of tumor markers. Besides, we will persist in refining our method. Subsequently, we anticipate the integration of pathological images with multi-omics data in our model, thereby enhancing its performance.

Acknowledgments. This work was supported by the National Natural Science Foundation of China (grant numbers 92259101) and the Strategic Priority Research Program of the Chinese Academy of Sciences (grant number XDB38040100).

Availability. The pathology slides and corresponding labels for WSIs are available from the CPTAC Pathology Portal. All source code used in our study was implemented in Python using PyTorch learning library, which are available at https://github.com/Luan-zb/TMG.

References

1. Beltagy, I., Peters, M.E., Cohan, A.: Longformer: the long-document transformer. arXiv preprint arXiv:2004.05150 (2020)
2. Carion, N., Massa, F., Synnaeve, G., Usunier, N., Kirillov, A., Zagoruyko, S.: End-to-end object detection with transformers. In: Vedaldi, A., Bischof, H., Brox, T., Frahm, J.-M. (eds.) ECCV 2020. LNCS, vol. 12346, pp. 213–229. Springer, Cham (2020). https://doi.org/10.1007/978-3-030-58452-8_13
3. Chen, H., Qi, X., Yu, L., Heng, P.A.: Dcan: deep contour-aware networks for accurate gland segmentation. In: 2016 IEEE Conference on Computer Vision and Pattern Recognition (CVPR) (2016)
4. Choromanski, K., et al.: Rethinking attention with performers. arXiv preprint arXiv:2009.14794 (2020)
5. Deng, S., et al.: Deep learning in digital pathology image analysis: a survey. Front. Med. **14**(4), 18 (2020)
6. Dosovitskiy, A., et al.: An image is worth 16x16 words: transformers for image recognition at scale. In: International Conference on Learning Representations (2021). https://openreview.net/forum?id=YicbFdNTTy
7. Feng, J., Zhou, Z.H.: Deep miml network. In: Proceedings of the Thirty-First AAAI Conference on Artificial Intelligence. AAAI'17, pp. 1884–1890. AAAI Press (2017)
8. Gehring, J., Auli, M., Grangier, D., Yarats, D., Dauphin, Y.N.: Convolutional sequence to sequence learning. In: Precup, D., Teh, Y.W. (eds.) Proceedings of the 34th International Conference on Machine Learning. Proceedings of Machine Learning Research, vol. 70, pp. 1243–1252. PMLR, 06–11 August 2017. https://proceedings.mlr.press/v70/gehring17a.html
9. Ilse, M., Tomczak, J., Welling, M.: Attention-based deep multiple instance learning. In: Dy, J., Krause, A. (eds.) Proceedings of the 35th International Conference on Machine Learning. Proceedings of Machine Learning Research, vol. 80, pp. 2127–2136. PMLR, 10–15 July 2018. https://proceedings.mlr.press/v80/ilse18a.html
10. Islam, M.A., Jia, S., Bruce, N.D.: How much position information do convolutional neural networks encode. arXiv preprint arXiv:2001.08248 (2020)
11. Kitaev, N., Kaiser, Ł., Levskaya, A.: Reformer: the efficient transformer. arXiv preprint arXiv:2001.04451 (2020)

12. Kraus, O.Z., Ba, J.L., Frey, B.J.: Classifying and segmenting microscopy images with deep multiple instance learning. Bioinformatics **32**(12), i52–i59 (2016). https://doi.org/10.1093/bioinformatics/btw252
13. Li, B., Li, Y., Eliceiri, K.W.: Dual-stream multiple instance learning network for whole slide image classification with self-supervised contrastive learning. In: Conference on Computer Vision and Pattern Recognition Workshops. IEEE Computer Society Conference on Computer Vision and Pattern Recognition. Workshops 2021, pp. 14318–14328 (2021)
14. Lu, M.Y., et al.: AI-based pathology predicts origins for cancers of unknown primary. Nature **594**(7861), 106–110 (2021)
15. Lu, M.Y., Williamson, D.F.K., Chen, T.Y., Chen, R.J., Mahmood, F.: Data-efficient and weakly supervised computational pathology on whole-slide images. Nat. Biomed. Eng. **5**, 1–16 (2021)
16. Pinheiro, P.O., Collobert, R.: From image-level to pixel-level labeling with convolutional networks. In: 2015 IEEE Conference on Computer Vision and Pattern Recognition (CVPR), pp. 1713–1721 (2015). https://doi.org/10.1109/CVPR.2015.7298780
17. Sabeena Beevi, K., Nair, M.S., Bindu, G.: Automatic mitosis detection in breast histopathology images using convolutional neural network based deep transfer learning. Biocybern. Biomed. Eng. **39**(1), 214–223 (2019). https://doi.org/10.1016/j.bbe.2018.10.007, https://www.sciencedirect.com/science/article/pii/S0208521618302572
18. Shao, Z., et al.: Transmil: transformer based correlated multiple instance learning for whole slide image classification. In: Advances in Neural Information Processing Systems, vol. 34, pp. 2136–2147 (2021)
19. Tay, Y., Dehghani, M., Bahri, D., Metzler, D.: Efficient transformers: a survey. ACM Comput. Surv. **55**(6), 1–28 (2022)
20. Vaswani, A., et al.: Attention is all you need. In: Proceedings of the 31st International Conference on Neural Information Processing Systems. NIPS'17, pp. 6000–6010. Curran Associates Inc., Red Hook (2017)
21. Wang, S., Li, B.Z., Khabsa, M., Fang, H., Ma, H.: Linformer: self-attention with linear complexity. arXiv preprint arXiv:2006.04768 (2020)
22. Xing, F., Yang, L.: Robust nucleus/cell detection and segmentation in digital pathology and microscopy images: a comprehensive review. IEEE Rev. Biomed. Eng. **9**, 234–263 (2016). https://doi.org/10.1109/RBME.2016.2515127
23. Xu, Y., Jia, Z., Ai, Y., Zhang, F., Lai, M., Chang, E.I.C.: Deep convolutional activation features for large scale brain tumor histopathology image classification and segmentation. In: 2015 IEEE International Conference on Acoustics, Speech and Signal Processing (ICASSP), pp. 947–951 (2015). https://doi.org/10.1109/ICASSP.2015.7178109
24. Zheng, Y., et al.: A graph-transformer for whole slide image classification. IEEE Trans. Med. Imaging **41**(11), 3003–3015 (2022). https://doi.org/10.1109/TMI.2022.3176598
25. Zheng, Y., et al.: Diagnostic regions attention network (DRA-net) for histopathology WSI recommendation and retrieval. IEEE Trans. Med. Imaging **40**(3), 1090–1103 (2021). https://doi.org/10.1109/TMI.2020.3046636

ricME: Long-Read Based Mobile Element Variant Detection Using Sequence Realignment and Identity Calculation

Huidong Ma[1,2], Cheng Zhong[1,2(✉)], Hui Sun[3], Danyang Chen[1,2], and Haixiang Lin[4]

[1] School of Computer, Electronic and Information, Guangxi University, Nanning, Guangxi, China
chzhong@gxu.edu.cn

[2] Key Laboratory of Parallel, Distributed and Intelligent Computing in Guangxi Universities and Colleges, Nanning, Guangxi, China

[3] College of C.S., ICIC, Nankai-Orange D.T. Joint Lab, TMCC, SysNet, Nankai University, Tianjin, China

[4] Faculty of Electrical Engineering, Mathematics and Computer Science, Delft University of Technology, Delft, Netherlands

Abstract. The mobile element variant is a very important structural variant, accounting for a quarter of structural variants, and it is closely related to many issues such as genetic diseases and species diversity. However, few detection algorithms of mobile element variants have been developed on third-generation sequencing data. We propose an algorithm ricME that combines sequence realignment and identity calculation for detecting mobile element variants. The ricME first performs an initial detection to obtain the positions of insertions and deletions, and extracts the variant sequences; then applies sequence realignment and identity calculation to obtain the transposon classes related to the variant sequences; finally, adopts a multi-level judgment rule to achieve accurate detection of mobile element variants based on the transposon classes and identities. Compared with a representative long-read based mobile element variant detection algorithm rMETL, the ricME improves the F1-score by 11.5 and 21.7% on simulated datasets and real datasets, respectively.

Keywords: mobile element variants · sequence realignment · identity calculation · third-generation sequencing data

1 Introduction

Transposons are DNA sequences that can move autonomously across the genome. Transposons show an important component of the human genome, which occupy approximately half of the human genome [1]. The transposons that have been verified to remain active in the human genome include three classes, Alu, LINE-1 (L1) and SINE-VNTR-Alu (SVA) [1]. The variants caused by transposon position changes are called mobile

© The Author(s), under exclusive license to Springer Nature Singapore Pte Ltd. 2023
X. Guo et al. (Eds.): ISBRA 2023, LNBI 14248, pp. 165–177, 2023.
https://doi.org/10.1007/978-981-99-7074-2_13

element (ME) variants, which can be divided into mobile element insertion (MEI) variants and mobile element deletion (MED) variants. ME variants have demonstrated to be closely associated with various human genetic diseases such as hemophilia and neurofibromatosis [2, 3]. In addition, ME variants account for about a quarter of the overall structural variants [4]. Therefore, it is of great practical importance to carry out research on the detection algorithm of ME variants.

The representative algorithms for detecting ME variants on next-generation sequencing (NGS) data are Tea [3], MELT [4], Mobster [5], and Tangram [6]. Tea alignments the paired-end reads with the reference genome and the assembled sequences composed of ME sequences respectively, and extracts repeat-anchored mate reads and clipped reads based on the alignment information to determine the insertion mechanism of MEs [3]. MELT uses discordant read pairs from NGS data alignment information to detect MEI variants, filters ME variants based on proximity to known ME variants, sequencing depth, and mapping quality of reads; and finally uses discordant sequence pairs and split reads to determine the type of ME variants and precise breakpoints [4]. Mobster uses the discordant reads in the NGS data as the signals of MEI variants, and then extracts the variant sequences based on the signals and compares the variant sequences with the transposon consensus sequence (TCS) to determine the type of ME variations [5]. Tangram designs different ME variation features extraction methods for the read pair and split read in the NGS data alignment information respectively, and then determines the position and type of MEI variants according to the features [6].

Third-generation sequencing (TGS) data has great potential for structural variant detection. It has been shown that structural variant detection algorithms based on TGS data are better than those based on NGS data [7, 8]. rMETL [9] is a representative long-read based ME variant detection algorithm, which is divided into four steps. The first step extracts the candidate ME variant sequences from the long-read alignment file. The second step clusters the candidate variant sequences to determine the ME variant positions. The third step uses the read alignment tool NGMLR [10] to realign the candidate variant sequences with transposon consistency sequences. And the final step counts the transposon classes of the mapped sequences and selects the transposon class with the highest frequency as the type of the ME variant.

However, the existing long-read based ME variant detection algorithms encounter the following problems:

(1) Some insertions (INSs) or deletions (DELs) were not effectively detected, which will directly affect the recall of the final removable element variant detection;
(2) Some of the variant sequences could not be successfully realigned by NGMLR, which could not provide more judgment basis for the final ME variant detection to improve the recall and accuracy of the final detection results.
(3) If the maximum number of transposon class is zero, i.e., all the variant sequences are not mapped with the TCS, the variant may be missed as a false-negative mobile component variant. And if the transposon class with the highest number of occurrences is not unique, the variant may be misclassified as a false positive ME variant when randomly selecting a transposon as the final ME class.

To address the above-mentioned issues, we propose an ME variants detection algorithm ricME using sequence realignment and identity calculation to effectively improve the performance of detecting ME variants.

The remainder of this paper is organized as follows. Section 2 details the proposed ME variant detection algorithm ricME. Section 3 describes the experimental environment, dataset, results and analysis. Section 4 concludes the paper.

2   Method

Our proposed ricME algorithm comprises the following four steps.

Firstly, the long-read based structural variant detection algorithm cnnLSV [11] is executed on the read alignment file *Dset* to obtain the initial detection result *Cset* containing only INSs and DELs. Secondly, the ricME applies different sequence extraction methods according to the characteristics of INSs and DELs to extract variant sequences within *Cset*, and save the sequences into the set *S*. Thirdly, all variant sequences in the set *S* are realigned with TCS using the tool NGMLR. For the variant sequences that can be successfully aligned, the transposon class *te* mapped by the sequences is stored in the set *Tset*; for the variant sequences that are not aligned, the ricME calculates their identity with TCS, and the potential transposon class *pte* corresponding to the maximum identity is selected and deposited into the set *Tset*. Finally, the ricME uses a multilevel judgment rule to determine the final class of ME based on the distribution of *te*, *pte* and identity in the set *Tset*. Figure 1 shows the process of algorithm ricME.

2.1   Initial Variant Position Detection

Since ME variants are caused by transposons moving autonomously across the genome, ME variants are essentially special INSs and DELs. And the sequences of variants are highly similar to transposon families including Alu, L1 and SVA. Therefore, the first step of algorithm ricME is to initially identify INS and DEL position.

In our previous work, we proposed an algorithm called cnnLSV [11] to detect structural variants by encoding long-read alignment information and modeling convolutional neural network. Experiments have shown that the algorithm cnnLSV has a high overall F1-score compared to other existing algorithms. We use cnnLSV to detect the structural variants on sequence alignment file *Dset*, and save the INSs and DELs to the set *Cset*.

2.2   Variant Sequence Extraction

The algorithm ricME using the following method to extract the variant sequences for INS and DELs in *Cset*.

Variant Sequence Extraction for INSs. As shown in Fig. 2(a), the ricME extracts variant sequences using intra- and inter-alignment signatures.

(1) Extracting variant sequences based on intra-alignment signatures. The flag "I" in the CIGAR strings in the alignment information indicates the INS, while the number preceding the flag represents the length of the variant. The ricME searches all read

alignment information around the INSs in *Dset* to obtain the variant sequences with a length of more than 50 base pairs (bps) within the CIGAR strings, and saves the information about the variant sequences *seq* to the FA format file.

(2) Extracting variant sequences based on the inter-alignment signatures. When long-reads are aligned to the reference genome, long-reads that span structural variants may split into multiple segments. According to the characteristics of INSs, the distance between two segments from the same read will change before and after the alignment, and the redundant segments that cannot be aligned are exactly the INS variant sequence *seq*. In addition to the above case, a read alignment situation around the INS is also mentioned in the algorithm rMETL. When the sequencing reads cannot span the whole INS segment, there will be a fragment that can be aligned successfully, and the adjacent fragment is clipped off because it is located at the boundary of the variant region. The sequence is clipped off, and it is exactly the INS sequence *seq*. The ricME extracts the variant sequences according to the above two cases, and saves the information related to the variants to the FA format file.

Variant Sequence Extraction for DELs. Unlike the existing algorithm rMETL, which searches for variant features before detecting variant position, our algorithm ricME utilizes the detection results *Cset* of the existing structural variant detection algorithm cnnLSV to obtain the initial positions of DELs. Therefore, the ricME can directly intercept the variant sequence *seq* in the corresponding region of the reference genome based on the chromosome number *chr*, variant position *pos*, and variant length *svl* where the DEL occurs in the *Cset*, as shown in Fig. 2(b). And last, the ricME saves the information of the DEL to the FA format file.

2.3   Sequence Realignment and Identity Calculation

In algorithm rMETL, sequence realignment means that the variant sequences extracted from the long-read alignment information are realigned to TCS by NGMLR to obtain the transposon class *te*. The rMETL could judge the ME variant class according the distribution of the *te*.

However, some of the variant sequences were discarded because they could not be successfully aligned to TCS by NGMLR, i.e., they could not provide more basis for ME variant type judgment. In addition, relying only on the distribution of *te* to determine the ME class may lead to accidental errors. As shown in Fig. 3(a), three sequences seq_1, seq_3 and seq_6 are finally judged as the Alu class, and the other three sequences seq_4, seq_7 and seq_8 are aligned to the L1 class, i.e., the number of variant sequences supporting Alu and L1 classes are of equal sizes, which will lead to the difficulty for the algorithm rMETL to determine the final ME variant class. To solve the above problem, the algorithm ricME introduces the sequence identity calculation based on sequence realignment, as shown in Fig. 3(b).

Sequence identity reflects the degree of similarity between two sequences. The premise of sequence identity calculation is to align two sequences [12, 13]. The two-sequence alignment algorithms include global-based and local-based alignment. Due to the large differences between the lengths of the three transposons and the lengths of each variant sequence, if the global based approach is used to calculate the identity,

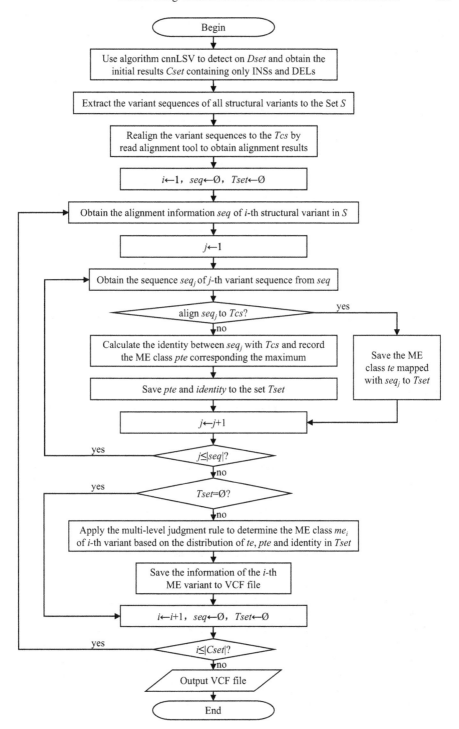

Fig. 1. Procedure of proposed mobile element variant detection algorithm ricME

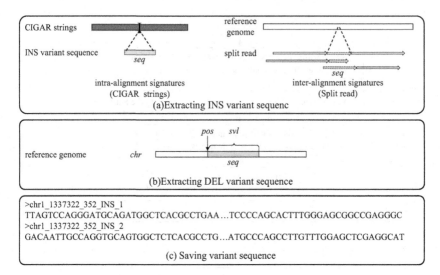

Fig. 2. Variant sequence extraction and storage in FA format

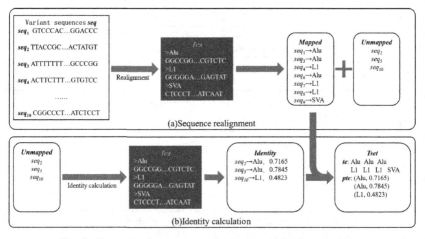

Fig. 3. Example of sequence realignment and identity calculation

the longer variant sequence will always get a higher identity with the longer transposon sequence. To avoid this bias caused by sequence length, the algorithm ricME uses the local-based alignment algorithm based on affine gap penalty to calculate the identity of two sequences. The ricME calculates the identity between unmapped variant sequences and the TCS to obtain the potential ME class *pte* and the corresponding identity *identity*, and stores the *te*, *pte*, and *identity* in the set *Tset*.

2.4 Mobile Element Variant Determination

In order to avoid the random selection error caused by the same frequency of transposon class *te*, the algorithm ricME introduces the potential variant class *pte* and identity *identity* as judgment factors, and proposes a multi-level judgment rule based on three factors of *te*, *pte* and *identity* to accurately detect the ME variant.

The process of the ME class judgment rule is as follows.

(1) To improve the accuracy of detection, the ricME defines the identity threshold $identity_0$ and eliminates the *pte* with *identity* < $identity_0$ within *Tset*.

(2) The ricME constructs 3 triples T_{Alu}, T_{L1} and T_{SVA} according to the distribution of *te*, *pte* and *identity* within *Tset*, where the triple $T_{me} = (n_{te}, n_{pte}, score)$, $me \in \{$Alu, L1, SVA$\}$, n_{te} denotes the number of *te* in *Tset* with class *me*, n_{pte} denotes the number of *pte*, and *score* denotes the *sum* of *identity* corresponding to the *pte* of class *me*.

(3) A multilevel judgment rule is used to determine the ME variant class. The ricME ranks the three tuples T_{Alu}, T_{L1} and T_{SVA} in the order of priority of n_{te}, $n_{pte,}$ and *score*.

(4) The ricME stores the ME variant information and transposon class to the VCF file as the final detection output.

3 Experiment

3.1 Experimental Environment and Data

Experimental Environment. The experiment was carried out on the computing node X580-G30 with CPU 2 × Intel Xeon Gold 6230, GPU 2 × Tesla T4, and main memory 192GB DDR4 of Sugon 7000A parallel computer cluster system at Guangxi University. The running operating system is CentOS 7.4. The proposed method was implemented by Python3.8 programming. The PyTorch was used to train and test the constructed network model.

Datasets

Simulated Datasets. Referring to the work of rMETL [9], the simulated datasets were generated as follows. Firstly, the 20,000 real ME variants of classes Alu, L1 and SVA, respectively, were collected from the database RepeatMasker [14]. And the sequences corresponding to the positions of the selected ME variants were deleted in chromosome 1 of the human reference genome. We extracted the sequence of chromosome 1 of the normal reference genome as seq_1, the sequence of chromosome 1 containing the MED variants as seq_2, and recorded the chromosomes, positions, lengths and classes of the real ME variants as the ground truth set. Secondly, we executed the tool PBSIM [15]to simulate sequencing reads for seq_1 to generate 4 PacBio CLR datasets with coverage of 50×, 30×, 20×, and 10×, respectively, and executed read alignment tool NGMLR to map the 4 datasets to the seq_2 to generate 4 long read alignment files for detecting MEI variants. Thirdly, we also used the PBSIM to simulate sequencing reads for seq_2 to generate 4 datasets with coverage of 50×, 30×, 20×, and 10×, respectively, and executed NGMLR to align the 4 datasets to the seq_1 to generate 4 read alignment files for detection MED variants. Finally, we used SAMtools [16] to sort and generate indexes for the 8 simulated datasets.

Real Datasets. The real dataset used is the HG002 CCS [17] dataset generated by the PacBio platform, which relates to the ground truth set that is a portion of the ME variants of HG002 dataset validated in [18]. This ground truth set contains 1353 Alu, 197 L1, and 90 SVA ME variants.

3.2 Detecting Performance Evaluation Metrics

The experiments use the precision *Pre*, recall *Rec* and F1-score *F*1 as the detection performance evaluation metric.

In the determination of true positive ME variant, when the detected ME variant *call* and the ME variant *base* of the ground truth set satisfy Eq. 1, then *call* is considered a true positive variant, otherwise it is a false positive variation:

$$
\begin{cases}
call_t = base_t \\
call_m = base_m \\
call_c = base_c \\
min(call_e + 1000, base_e) - max(call_s - 1000, base_s) \geq 0 \\
\frac{min(call_l, base_l)}{max(call_l, base_l)} \geq 0.7
\end{cases}
\tag{1}
$$

where $call_t$, $call_m$, $call_c$, $call_s$, $call_e$ and $call_l$ denote variant type, ME class, chromosome, start position, end position and the length of variant *call*, respectively, and $base_t$, $base_m$, $base_c$, $base_s$, $base_e$ and $base_l$ represent variant type, ME class, chromosome, start position, end position and the length of variant *base*, respectively, $call_m \in \{$Alu, L1, SVA$\}$ and $base_m \in \{$Alu, L1, SVA$\}$.

3.3 Experimental Results

Experiments on Simulated Datasets. The experiments were conducted on eight simulated datasets, including four datasets containing MEI variants with coverages of $50\times$, $30\times$, $20\times$, and $10\times$, respectively, and four datasets containing MED variants with coverages of $50\times$, $30\times$, $20\times$, and $10\times$, respectively.

Firstly, to verify the effectiveness of the algorithm ricME in detecting the initial variant position, we looked at the detection results of simple INSs and DELs. The results are shown in Table 1, where *TP-call* indicates the number of correctly detected variants in the detection results, and *TP-base* represents the number of correctly detected variants in the ground truth set.

From Table 1, it can be concluded that for the cases of INSs and DELs, the F1-score of the algorithm ricME is about 12–17% higher than that of the algorithm rMETL. For the INSs, the detection performance of algorithm rMETL, especially the recall, decreases significantly as the coverage of the dataset decreases. In contrast, algorithm ricME performed significantly better than algorithm rMETL in terms of recall *Rec*, especially at low coverage, and could detect more than 4000 INSs and had about 23%

higher recall than rMETL. Even though the *Pre* was slightly lower than that of rMETL under partial coverages, the larger recall by the ricME results in a significant higher $F1$ values. For the DELs, both algorithms rMETL and ricME achieved very high detection accuracy on the datasets with different coverages, especially rMETL could reach 99%. In terms of recall *Rec*, the algorithm ricME had a significant advantage over the algorithm rMETL, especially in the low coverage dataset, with a lead of about 13%. The $F1$ values of ricME were also higher than that of algorithm rMETL by about 6–8%. Compared with the similar algorithm rMETL, ricME had higher $F1$ values in the detection of INSs and DELs.

Table 1. Results of algorithms in detecting INSs and DELs on simulated datasets

Type	Coverage	Algorithm	TP-Call	TP-Base	FP	FN	Pre (%)	Rec (%)	F1 (%)
INS	50 ×	rMETL	13942	15013	1128	4987	92.515	75.065	82.881
		ricME	**18493**	**18692**	**694**	**1308**	**96.383**	**93.46**	**94.899**
	30 ×	rMETL	13100	14168	1046	5832	92.606	70.840	80.274
		ricME	**18218**	**18482**	**624**	**1518**	**96.688**	**92.410**	**94.501**
	20 ×	rMETL	12182	13235	916	6765	93.007	66.175	77.329
		ricME	**18038**	**18433**	**573**	**1567**	**96.921**	**92.165**	**94.483**
	10 ×	rMETL	9949	10944	636	9056	93.991	54.720	69.170
		ricME	**14980**	**15408**	**399**	**4592**	**97.406**	**77.040**	**86.034**
DEL	50 ×	rMETL	16472	16804	**129**	3196	**99.223**	84.02	90.991
		ricME	**18999**	**19017**	574	**983**	97.067	**95.085**	**96.066**
	30 ×	rMETL	16630	16919	**171**	3081	98.982	84.595	91.225
		ricME	**19026**	**19072**	619	**928**	96.849	**95.360**	**96.099**
	20 ×	rMETL	16318	16658	**122**	3342	**99.258**	83.290	90.576
		ricME	**18808**	**18913**	441	**1087**	97.709	**94.565**	**96.111**
	10 ×	rMETL	13500	13918	**60**	6082	**99.558**	69.590	81.919
		ricME	**16388**	**16583**	259	**3417**	98.444	**82.915**	**90.015**

Note that the values in bold represent the best results.

Next, the performance of two algorithms rMETL and ricME is compared for detecting ME variants. The experiment results are shown in Table 2.

From Table 2, we can see that the performance of the algorithm ricME was significantly better than that of the algorithm rMETL in detecting MEI variants and MED variants on datasets with different coverages, with 8–11% higher $F1$ values. For MEI variants, in terms of the detection precision *Pre*, the ricME slightly outperformed rMETL on high coverage datasets, while both performed comparably on low coverage datasets. In terms of recall *Rec*, the algorithm ricME was significantly higher than rMETL, especially about 15% higher on the low-coverage datasets. In addition, the value of metric

TP-base also shows that the ricME detected about 3000 more true positives than rMETL. For the overall performance metric F1-score, the ricME also obtained higher $F1$ values due to the high precision and recall. For the MED variants, in terms of the detection precision *Pre*, the rMETL achieved a high detection precision on all datasets, which was about 2% higher than that of ricME. However, in terms of recall *Rec*, algorithm ricME achieved significantly higher than rMETL on the datasets with each coverage, namely about 7–8% higher. In terms of the $F1$ values, the algorithm ricME obtained higher $F1$ values due to the overall high precision and recall, namely about 2–5% higher than the algorithm rMETL. The combined experiment results in Table 2 show that the algorithm ricME had a better performance in terms of recall *Rec* and was comparable to the algorithm rMETL in terms of detection precision *Pre*. This indicates that the proposed algorithm ricME is able to detect more ME variants than the algorithm rMETL, and has higher accuracy in the variant class judgment stage. The above results show that the proposed algorithm ricME use the sequence realignment and identity calculation to enhance the basis for variant class judgment, which improves the detection precision, recall and F1-score to achieve higher detection performance.

Table 2. Results of algorithms in detecting mobile element variants on simulated datasets

Type	Coverage	Algorithm	TP-call	TP-base	FP	FN	Pre (%)	Rec (%)	F1 (%)
MEI	50 ×	rMETL	13785	14837	1285	5163	91.473	74.185	81.927
		ricME	**16894**	**17432**	**1235**	**2568**	**93.188**	**87.16**	**90.073**
	30 ×	rMETL	12964	14017	**1182**	5983	91.644	70.085	79.428
		ricME	**16643**	**17150**	1294	**2850**	92.786	85.750	89.129
	20 ×	rMETL	12047	13085	**1051**	6915	**91.976**	65.425	76.461
		ricME	**16314**	**16861**	1428	**3139**	91.951	84.305	87.962
	10 ×	rMETL	9831	10807	**754**	9193	**92.877**	54.035	68.321
		ricME	**13397**	**13911**	1148	**6089**	92.107	69.555	79.258
MED	50 ×	rMETL	16388	16717	**213**	3283	**98.717**	83.585	90.523
		ricME	**17696**	**17875**	677	**2125**	96.315	89.375	92.715
	30 ×	rMETL	16542	16830	**259**	3170	**98.458**	84.150	90.744
		ricME	**17715**	**17914**	724	**2086**	96.074	89.570	92.708
	20 ×	rMETL	16238	16577	**202**	3423	**98.771**	82.885	90.133
		ricME	**17507**	**17742**	586	**2258**	96.761	88.710	92.561
	10 ×	rMETL	13437	13858	**123**	6142	**99.093**	69.290	81.554
		ricME	**15273**	**15562**	400	**4438**	97.448	77.810	86.529

Note that the values in bold represent the best results.

Experiments on Real Datasets. We used the tool SAMtools to downsample the long-read alignment file HG002 CCS 28 × to generate a new dataset with 10 × coverage, then executed the algorithms ricME and rMETL to detect ME variants on the two datasets, and compared the detection results with the ground truth set to evaluate the detection performance. The detection results of the two algorithms rMETL and ricME are shown in Table 3, where "\" indicates that the calculation of $F1$ is meaningless in the case that both metrics Pre and Rec are zero.

Table 3. Detection results of algorithms rMETL and ricME on dataset HG002

Type	Coverage	Algorithm	TP-call	TP-base	FP	FN	Pre (%)	Rec (%)	F1 (%)
28 ×	Alu	rMETL	4	4	**364**	1349	1.087	0.296	0.465
		ricME	**589**	**589**	1898	**764**	**23.683**	**43.533**	**30.677**
	L1	rMETL	14	14	**292**	183	4.575	7.107	5.567
		ricME	**59**	**59**	946	**138**	**5.871**	**29.949**	**9.817**
	SVA	rMETL	1	1	**101**	89	0.98	1.111	1.042
		ricME	**32**	**32**	1011	**58**	**3.068**	**35.556**	**5.649**
	All	rMETL	19	19	**757**	1621	2.448	1.159	1.573
		ricME	**680**	**680**	3855	**960**	**14.994**	**41.463**	**22.024**
10 ×	Alu	rMETL	2	2	**55**	1351	3.509	0.148	0.284
		ricME	**657**	**657**	2198	**696**	**23.012**	**48.559**	**31.226**
	L1	rMETL	1	1	**65**	196	1.515	0.508	0.76
		ricME	**62**	**62**	1053	**135**	**5.561**	**31.472**	**9.451**
	SVA	rMETL	0	0	**14**	90	0	0	\
		ricME	**26**	**26**	1133	**64**	**2.243**	**28.889**	**4.163**
	All	rMETL	3	3	**134**	1637	2.19	0.183	0.338
		ricME	**745**	**745**	4384	**895**	**14.525**	**45.427**	**22.012**

Note that the values in bold represent the best results.

As can be seen from Table 3, the overall detection performance of the algorithm ricME was significantly higher than that of the algorithm rMETL for the detection of the all classes of ME variants on the real datasets with coverages 28 × and 10 ×. The F1-scores of ricME was about 20% higher than that of rMETL. In terms of recall Rec, the ricME was much higher than rMETL, especially for the detection of Alu transposon class, which are 40% higher. In terms of precision Pre, although the Pre of the ricME was higher than that of rMETL in all classes, both performed less well. For the rMETL, the main reason is that the rMETL detected fewer true positive variants, i.e., insufficient detection of ME variants. For the algorithm ricME, the main reason for the low accuracy is the high number of false positive variants detected. However, it is worth noting that despite the high frequency and importance of ME variants, there are still few studies and annotations on ME variants [19, 20]. The ground truth set corresponding to the real dataset HG002 used in the experiments is the validated information of ME variants given in the work [18], and this ground truth set only contains some of the ME variants with a high confidence. This means that the false positives detected by the ricME may not actually mean that no variants have occurred.

4 Conclusion

Mobile element variant is a very import structural variant that is closely associated with a variety of genetic diseases. We propose the ricME, an algorithm for detecting the variation of movable components that integrates re-matching and sequence consistency calculation, to improve the existing representative ME variation detection algorithm rMETL. The ricME has the following features and innovations. First, the ricME use the detection results of algorithm cnnLSV to obtain the initial results with high recall. Secondly, the ricME extracts the variant sequences of all INSs and DELs in initial results. Thirdly, the ricME realigns and calculates identity between variant sequences with transposon consistency sequences to obtain the corresponding transposon classes and the identities. Finally, the ricME applies a multi-level judgment rule to determine the final ME class based on transposon classes, potential transposon classes and identities. The experiment results show that the proposed algorithm ricME outperforms the existing representative algorithm for ME variant detection in general. In the future, we will investigate algorithms for detecting more types of structural variants on more types of datasets.

Acknowledgement. This work is partly supported by the National Natural Science Foundation of China under Grant No. 61962004 and Guangxi Postgraduate Innovation Plan under Grant No. A30700211008.

References

1. Niu, Y., Teng, X., Zhou, H., et al.: Characterizing mobile element insertions in 5675 genomes. Nucleic Acids Res. **50**(5), 2493–2508 (2022)
2. Hancks, D.C., Kazazian, H.H.: Roles for retrotransposon insertions in human disease. Mob. DNA **7**(1), 1–28 (2016)
3. Lee, E., Iskow, R., Yang, L., et al.: Landscape of somatic retrotransposition in human cancers. Science **337**(6097), 967–971 (2012)
4. Gardner, E.J., Lam, V.K., Harris, D.N., et al.: The Mobile Element Locator Tool (MELT): population-scale mobile element discovery and biology. Genome Res. **27**(11), 1916–1929 (2017)
5. Thung, D.T., de Ligt, J., Vissers, L.E.M., et al.: Mobster: accurate detection of mobile element insertions in next generation sequencing data. Genome Biol. **15**(10), 1–11 (2014)
6. Wu, J., Lee, W.P., Ward, A., et al.: Tangram: a comprehensive toolbox for mobile element insertion detection. BMC Genom. **15**, 1–15 (2014)
7. Mahmoud, M., Gobet, N., Cruz-Dávalos, D.I., et al.: Structural variant calling: the long and the short of it. Genome Biol. **20**(1), 1–14 (2019)
8. Merker, J.D., Wenger, A.M., Sneddon, T., et al.: Long-read genome sequencing identifies causal structural variation in a Mendelian disease. Genet. Med. **20**(1), 159–163 (2018)

9. Jiang, T., Liu, B., Li, J., et al.: RMETL: sensitive mobile element insertion detection with long read realignment. Bioinformatics **35**(18), 3484–3486 (2019)

10. Sedlazeck, F.J., Rescheneder, P., Smolka, M., et al.: Accurate detection of complex structural variations using single-molecule sequencing. Nat. Methods **15**(6), 461–468 (2018)

11. Ma, H., Zhong, C., Chen, D., et al.: CnnLSV: detecting structural variants by encoding long-read alignment information and convolutional neural network. BMC Bioinform. **24**(1), 1–19 (2023)

12. Li, H.: Minimap2: pairwise alignment for nucleotide sequences. Bioinformatics **34**(18), 3094–3100 (2018)

13. Altschul, S.F., Erickson, B.W.: Optimal sequence alignment using affine gap costs. Bull. Math. Biol. **48**, 603–616 (1986)

14. Smit, A.F.A., Hubley, R., Green, P.: RepeatMasker Open-4.0. 2013–2015. http://www.repeat masker.org

15. Ono, Y., Asai, K., Hamada, M.: PBSIM: PacBio reads simulator—toward accurate genome assembly. Bioinformatics **29**(1), 119–121 (2013)

16. Danecek, P., Bonfield, J.K., Liddle, J., et al.: Twelve years of SAMtools and BCFtools. Gigascience **10**(2), giab008 (2021)

17. Zook, J.M., Catoe, D., McDaniel, J., et al.: Extensive sequencing of seven human genomes to characterize benchmark reference materials. Scientific Data **3**(1), 1–26 (2016)

18. Chu, C., Borges-Monroy, R., Viswanadham, V.V., et al.: Comprehensive identification of transposable element insertions using multiple sequencing technologies. Nat. Commun. **12**(1), 3836 (2021)

19. Hoen, D.R., Hickey, G., Bourque, G., et al.: A call for benchmarking transposable element annotation methods. Mob. DNA **6**, 1–9 (2015)

20. Ou, S., Su, W., Liao, Y., et al.: Benchmarking transposable element annotation methods for creation of a streamlined, comprehensive pipeline. Genome Biol. **20**(1), 1–18 (2019)

scGASI: A Graph Autoencoder-Based Single-Cell Integration Clustering Method

Tian-Jing Qiao, Feng Li, Shasha Yuan, Ling-Yun Dai, and Juan Wang[✉]

School of Computer Science, Qufu Normal University, Rizhao 276826, China
wangjuansdu@163.com

Abstract. Single-cell RNA sequencing (scRNA-seq) technology offers the opportunity to study biological issues at the cellular level. The identification of single-cell types by unsupervised clustering is a basic goal of scRNA-seq data analysis. Although there have been a number of recent proposals for single-cell clustering methods, only a few of these have considered both shallow and deep potential information. Therefore, we propose a graph autoencoder-based single-cell integration clustering method, scGASI. Based on multiple feature sets, scGASI unifies deep feature embedding and data affinity recovery in a uniform framework to learn a consensus affinity matrix between cells. scGASI first constructs multiple feature sets. Then, to extract the deep potential information embedded in the data, scGASI uses a graph autoencoder (GAEs) to learn the low-dimensional latent representation of the data. Next, to effectively fuse the deep potential information in the embedding space and the shallow information in the raw space, we design a multi-layer kernel self-expression integration strategy. This strategy uses a kernel self-expression model with multi-layer similarity fusion to learn a similarity matrix shared by the raw and embedding spaces of a given feature set, and a consensus learning mechanism to learn a consensus affinity matrix across all feature sets. Finally, the consensus affinity matrix is used for spectral clustering, visualization, and identification of gene markers. Large-scale validation on real datasets shows that scGASI has higher clustering accuracy than many popular clustering methods.

Keywords: scRNA-seq · Clustering · Graph Autoencoder · Multi-layer Kernel Self-expression Integration · Multi-layer Similarity Fusion · Consensus Learning

1 Introduction

Single-cell RNA sequencing (scRNA seq) technology can provide transcriptome profiles of individual cells. Therefore, scRNA-seq data processing enables researchers to trace the evolution of various cell lines, uncover complicated and uncommon cell subpopulations, and disclose various gene regulatory relationships between cells [1]. Unsupervised clustering is a key task in the analysis of scRNA-seq data. scRNA-seq data clustering aids in the discovery of novel cell types and clarifies intercellular heterogeneity.

Many traditional clustering techniques have been developed by researchers, such as K-means [2] and spectral clustering (SC) [3]. Some dimensionality reduction techniques

© The Author(s), under exclusive license to Springer Nature Singapore Pte Ltd. 2023
X. Guo et al. (Eds.): ISBRA 2023, LNBI 14248, pp. 178–189, 2023.
https://doi.org/10.1007/978-981-99-7074-2_14

have also been applied to reduce the dimensionality of single-cell data, mostly to select highly variable genes (HVGs) [4–7]. Indeed, scRNA data are high-dimensional, sparse, and noisy. Because of this, conventional analysis techniques frequently fail to yield the expected outcomes. Therefore, it is imperative to create novel clustering techniques to allow for precise single-cell type identification.

Recently, several methods have been developed for clustering single cells. Among them, most of the methods emphasize on learning the similarity between the cells in the original feature space. For example, Mei et al. proposed RCSL, which constructs a similarity matrix by measuring the global and local relationships between cells [8]. Since the multiple kernel learning technique tends to capture rich data structures, Wang et al. and Park et al. proposed SIMLR and MPSSC, respectively, by combining 55 different Gaussian kernels [9, 10]. To capture the complex relationships among cells, based on the low-rank self-expression model, the subspace clustering methods SinNLRR and NMFLRR are successively proposed [11, 12]. Furthermore, in order to achieve better clustering performance, some methods integrate different similarity or clustering results, such as SC3 and Seurat [13, 14]. Recently, SCENA, which learns a consistent affinity matrix for clustering by constructing multiple gene (feature) sets, was presented by Cui et al. [1]. The selection of multiple feature sets significantly improves clustering performance and stability compared to single feature set learning such as PCA.

However, the above methods usually only consider shallow information in scRNA-seq data. Recently, researchers have developed some deep learning-based clustering methods that are able to extract deep information hidden in scRNA-seq data. In particular, the autoencoder, which automatically learns the potential low-dimensional representation (embedding) of the data in an unsupervised manner, has received much attention. For example, scDeepCluster and scGMAI use an autoencoder to reconstruct the data and extract deep information [15, 16]. However, the autoencoder does not explicitly characterize the relationships between cells during the learning process and only considers gene expression patterns. The Graph Convolutional Networks (GCN) can take into account both content information (gene expression information) and structural information (cell-cell relationship information) of the data, and also learn how to represent nodes in the graph by propagating neighborhood information. The Graph Autoencoder (GAEs) model was created by Wang et al. by stacking multiple layers of GCN to integrate structural and content data in a deep learning framework [17]. Based on GAEs, Zhang et al. proposed scGAC, which combines multiple feature sets selection and consensus learning to learn the consensus affinity matrix for spectral clustering [18].

Although the above clustering methods have achieved good performance, the existing single-cell clustering methods lack a comprehensive consideration of the shallow information and the deep information. This tends to lead to suboptimal performance. Therefore, we propose a graph autoencoder-based single-cell integration clustering method scGASI. scGASI combines deep feature embedding and the multi-layer kernel self-expression integration strategy to learn an exact consensus affinity matrix between cells. And the consensus affinity matrix is used for spectral clustering, visualization, and identification of gene markers. To summarize, the main contributions made are as follows:

(1) scGASI selects multiple feature sets using the variance method. This preprocessing ensures that more accurate data information is captured and reduces the effect of noise on feature embedding and data affinity recovery. Moreover, we use GAEs to further mine deep information that comprehensively considers gene expression information and cell-cell relationship information in scRNA-seq data.
(2) To effectively extract and preserve shallow information in the original feature space and deep information in the embedding space, we propose a multi-layer kernel self-expression integration strategy. For a feature set, this strategy uses an enhanced kernel self-expression model to learn a common similarity matrix shared by the original and embedding spaces, and passes this matrix to the GAEs to guide feature learning.
(3) To improve the learning capability of the kernel self-expression model, we introduce a multi-layer similarity fusion term. The multi-layer similarity fusion term uses cosine similarity and Euclidean distance to characterize relationships among cells in the original and embedding spaces, and passes this information to the similarity matrix to guide its learning. Moreover, to enable interactive learning of multiple feature sets and to achieve learning of the consensus affinity matrix for multiple feature sets, we add a consensus learning term to the kernel self-expression model.

2 Related Work

2.1 Kernel Self-Expression Model

The task of self-expression learning is to approximate each data point in the union of subspaces as a linear combination of the other points located in the same subspace i.e. $x_i = \sum_j x_j c_{ij}$. The fundamental idea is that the weight c_{ij} should be high if x_i and x_j are similar. As a result, $\mathbf{C}$ is often referred to as the similarity matrix. The learning problem for the self-expression learning model is shown below:

$$\min_Z \frac{1}{2}\|\mathbf{X} - \mathbf{X}\mathbf{C}\|_F^2 + \beta\|\mathbf{C}\|_F^2 \ s.t. \ \mathbf{C} \geq 0. \tag{1}$$

Here, $\beta \geq 0$ is the regularization parameter, and $\mathbf{C} \geq 0$ ensures that samples are similar in a non-negative way.

However, the model (1) can only deal with linearly structured data and cannot capture the nonlinear relationships hidden in the data. Therefore, it is necessary to extend (1) to the kernel space to solve the nonlinear problem. The matrix $\mathbf{X}$ in the case of multiple kernel learning (MKL) can be represented as $\psi_\omega(\mathbf{X}) = [\sqrt{u_1}\psi_1(\mathbf{X}), \sqrt{u_2}\psi_2(\mathbf{X}), ..., \sqrt{u_Q}\psi_Q(\mathbf{X})]$, where $u = [u_1, ..., u_Q]^T$ is the coefficient of Q base kernel functions $\{\kappa_q(\cdot, \cdot)\}_{q=1}^{Q}$. As a result, the kernel self-expression model is obtained as shown below.

$$\min_{\mathbf{C},u} \frac{1}{2}Tr\left(\mathbf{K} - 2\mathbf{K}\mathbf{C} + \mathbf{C}^T\mathbf{K}\mathbf{C}\right) + \beta\|\mathbf{C}\|_F^2$$
$$s.t. \ \mathbf{C} \geq 0, \mathbf{K} = \sum_{q=1}^{Q} u_q\mathbf{K}_q, \ \sum_{q=1}^{Q} \sqrt{u_q} = 1, u_q \geq 0. \tag{2}$$

Here, $(\mathbf{K}_q)_{ij} = \kappa_q(x_i, x_j) = \psi_q(x_i)^T\psi_q(x_j)$, and $\mathbf{K}$ is the consensus kernel matrix.

3   Method

3.1   The Framework of scGASI

scGASI combines deep feature embedding and multi-layer kernel self-expression integration mechanism to learn a consensus cell-cell affinity matrix for downstream analysis such as clustering. Firstly, gene filtering and normalization are performed on the raw expression matrix to minimize the impact of noise on the clustering performance. Secondly, based on the preprocessed expression matrix $\mathbf{X} \in R^{m \times n}$ (m genes, n cells), to circumvent learning limitations of a single feature set, we use the variance method [1] to construct multiple feature sets $\mathbf{X}^v \in R^{T_v \times n}(T_v < m; \ v = 1, 2, ..., V)$. That is, all features are sorted by variance in descending order and instead of using a single feature set, we used V multiple feature sets (top T_1, top T_2, ..., top T_v). Meanwhile, we construct the KNN graph $\{ \mathbf{A}^v \in R^{n \times n} \}_{v=1}^{V}$ corresponding to the multiple feature set $\{\mathbf{X}^v\}_{v=1}^{V}$ as the initial input to the GAEs. Third, the GAEs is used to extract deep information containing gene expression information and cell-cell relationship information. Fourth, we use the multi-layer kernel self-expression integration mechanism to learn a common similarity matrix $\mathbf{C}^v \in R^{n \times n}$ for a single feature set, and a consensus affinity matrix $\mathbf{S} \in R^{n \times n}$ shared by multiple feature sets. Finally, iterating the above steps yields the final affinity matrix $\mathbf{S}$. And $\mathbf{S}$ is used for spectral clustering, visualizing and identifying gene markers. Figure 1 shows the framework of scGASI.

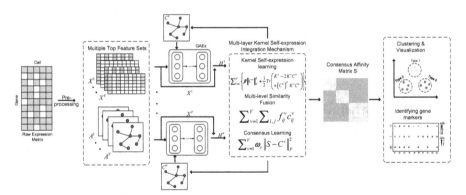

Fig. 1.   The framework of scGASI.

3.2   Graph Autoencoder

To construct our GAEs in this paper, we choose the stacked GCN proposed by Wang et al. [17]. For the data matrix $\mathbf{X}^v$, we first calculate the similarity matrix $\mathbf{W}^v$ between samples using the normalized Euclidean distance. We then construct an undirected K-nearest neighbor (KNN) graph by selecting the top K similar points of each sample as its neighbors. Thus, we obtain the adjacency matrix $\mathbf{A}^v$ from the non-graphical data $\mathbf{X}^v$.

Based on the data matrix $\mathbf{X} \in R^{T_v \times n}$ (T_v genes, n cells) and its corresponding adjacency matrix $\mathbf{A} \in R^{n \times n}$, GCN learns a embedding $\mathbf{H}^{l+1} \in R^{p \times n}(p \leq T_v)$ and a weight matrix $\mathbf{W}^l \in R^{n \times n}$. This process can be described as follows:

$$\mathbf{H}^{l+1} = f(\mathbf{H}^l, \mathbf{A}) = \sigma(\mathbf{A}/\mathbf{H}^l\mathbf{W}^l) = \sigma(\mathbf{D}^{-1/2}(\mathbf{A} + \mathbf{I}_n)\mathbf{D}^{-1/2}\mathbf{H}^l\mathbf{W}^l). \tag{3}$$

Here, $f(\mathbf{H}^l\mathbf{A})$ is the spectral convolution function. $l \in \{0, 1, ..., L\}$, L refers to the layer number of the GCN. $\mathbf{H}^0 = \mathbf{X}$ is the input of GCN, $\mathbf{H}^l$ is the input of the $l-th$ convolution, and $\mathbf{H}^{l+1}$ is the embedding from the $l - th$ layer. $\mathbf{W}^l$ Represents the weight matrix of the $l - th$ layer that will be optimized during the training process. $\sigma(\cdot)$ represents an activation function. $\mathbf{D}$ is degree matrix, and $\mathbf{D}_{ii} = \sum_k (\mathbf{A} + \mathbf{I}_n)_{ik}$ is the element within it. $\mathbf{I}_n \in R^{n \times n}$ is the identity matrix.

Thus, we obtain a single-layer GAE in scGASI as follows.

$$\min_{H,W} \|\mathbf{H} - \mathbf{A}/\mathbf{H}\mathbf{W}\|_F^2 + \lambda\|\mathbf{W}\|_F^2, \tag{4}$$

where $\lambda > 0$ is a parameter. In this paper, we choose the linear activation function.

Build a deep learning framework GAEs in scGASI after expanding the single-layer GAE to multiple layers, as follows.

$$\mathbf{H}^{l+1} = \mathbf{A}/\mathbf{H}^l\mathbf{W}^l. \tag{5}$$

By making the partial derivative of (4) with respect to $\mathbf{W}$ equal to 0, the optimization formula for $\mathbf{W}$ can be obtained as:

$$\mathbf{W} = \mathbf{H}^T\mathbf{A}/\mathbf{H}\left(\mathbf{H}^T\mathbf{A}/^T\mathbf{A}/\mathbf{H} + \lambda\mathbf{I}_n\right). \tag{6}$$

3.3 Multi-Layer Kernel Self-Expression Integration Mechanism

Both the raw data $\{\mathbf{X}^v\}_{v=1}^V$ and the embeddings $\{\mathbf{H}^v\}_{v=1}^V$ generated from GAE are used as input to this section. Based on (3), we can obtain the kernel self-expression model in scGASI:

$$\min_{\mathbf{C}^v,u_q^l} \sum_{v=1}^V \left\{\beta\|\mathbf{C}^v\|_F^2 + \frac{1}{2}Tr\left(\mathbf{K}^v - 2\mathbf{K}^v\mathbf{C}^v + (\mathbf{C}^v)^T\mathbf{K}^v\mathbf{C}^v\right)\right\} \tag{7}$$

$$s.t. \ \mathbf{C}^v \geq 0, \ \mathbf{K}^v = \sum_{l=1}^2\sum_{q=1}^Q u_q^l\mathbf{K}_q^l, \ \sum_{l=1}^2\sum_{q=1}^Q \sqrt{u_q^l} = 1, \ u_q^l \geq 0.$$

Here, $\mathbf{C}^v$ is the similarity matrix shared by the original data $\mathbf{X}^v$ and the embedding $\mathbf{H}^v$, $\mathbf{K}^v$ denotes the consensus kernel matrix formed by the $2Q$ kernel matrices of $\mathbf{X}^v$ and $\mathbf{H}^v$, and $\mathbf{K}_q^1$ (or $\mathbf{K}_q^2$) denotes the $q - th$ kernel matrix constructed from $\mathbf{X}^v$ (or $\mathbf{H}^v$).

Then, we further mine the similarity information between cells in raw and embedding spaces to improve the learning ability of model (7). In particular, we use the cosine measure and the normalized Euclidean distance to evaluate the non-linear and linear relationships between the cells. The similarity $f^v(h_i^v, h_j^v)$ between cells h_i^v and h_j^v is defined as follows.

$$F^v(h_i^v, h_j^v) = (1 - \alpha)\frac{(\sum_{q=1}^m |h_{iq}^v - h_{jq}^v|^2)^2}{\max_{i,j}(\sum_{q=1}^m |h_{iq}^v - h_{jq}^v|^2)^2} + \alpha \frac{<h_i^v, h_j^v>}{\|h_i^v\| \times \|h_i^v\|}, \tag{8}$$

where $\alpha \in (0, 1)$. The calculation of $f^v(x_i^v, x_j^v)$ is the same as $f^v(h_i^v, h_j^v)$. Thus, we obtain the fusion similarity $F_{ij}^v = 0.5 \times (f^v(h_i^v, h_j^v) + f^v(x_i^v, x_j^v))$. Then, we introduce the similarity constraint term $\sum_{i,j} F_{ij}^v c_{ij}^v$ into (7) and obtain the kernel self-expression model with multi-layer similarity fusion, as shown below.

$$\min_{C^v, u_q^l} \sum_{v=1}^V \left\{ \beta \|C^v\|_F^2 + \frac{1}{2}Tr\left(K^v - 2K^vC^v + (C^v)^T K^vC^v\right) + \sum_{i,j} F_{ij}^v c_{ij}^v \right\}$$

$$s.t. \ c_{ij}^v \geq 0, \ c_{ii}^v = 0, \ 1^T c_{ij}^v = 1, K^v = \sum_{l=1}^2 \sum_{q=1}^Q u_q^l K_q^l, \ \sum_{l=1}^2 \sum_{q=1}^Q \sqrt{u_q^l} = 1, \ u_q^l \geq 0. \tag{9}$$

However, in (9), information from different feature sets cannot be shared during the learning process. In order to guide the interaction across multiple feature sets and develop a consistent solution, we additionally add a consensus learning term $\sum_{v=1}^V \omega_v \|S - C^v\|_F^2$ to (9). And (9) is further rewritten as follows:

$$\min_{C^v, S, \omega_v, u_q^l} \sum_{v=1}^V \left\{ \beta \|C^v\|_F^2 + \frac{1}{2}Tr\left(K^v - 2K^vC^v + (C^v)^T K^vC^v\right) + \sum_{i,j} F_{ij}^v c_{ij}^v \right\}$$

$$+ \sum_{v=1}^V \omega_v \|S - C^v\|_F^2$$

$$s.t. \ c_{ij}^v \geq 0, \ c_{ii}^v = 0, \ 1^T c_{ij}^v = 1, \ s_{ij} \geq 0, \ 1^T s_{ij} = 1,$$

$$K^v = \sum_{l=1}^2 \sum_{q=1}^Q u_q^l K_q^l, \ \sum_{l=1}^2 \sum_{q=1}^Q \sqrt{u_q^l} = 1, \ u_q^l \geq 0. \tag{10}$$

where ω_v is the weight used for measuring the difference between S and C^v. In this paper, we refer to the strategy of obtaining the similarity matrix C^v and the affinity matrix S via (10) as the multi-layer kernel self-expression integration strategy.

Then, we apply the alternating direction method of multipliers to solve the objective function (10), since it is a non-convex optimization problem. The secondary variables $J^v \in R^{n \times n}$ and $Z^v \in R^{n \times n}$ are introduced and the following optimization problem is obtained:

$$\min_{C^v, Z^v, J^v, S, \omega_v, u_q^l} \sum_{v=1}^V \left\{ \begin{array}{l} \beta \|J^v\|_F^2 + \frac{1}{2}Tr\left(K^v - 2K^vC^v + (C^v)^T K^vC^v\right) + \sum_{i,j} F_{ij}^v z_{ij}^v \\ + \omega_v \|S - C^v\|_F^2 + \frac{\mu}{2}\|C^v - J^v + Y_1^v/\mu\|_F^2 + \frac{\mu}{2}\|C^v - Z^v + Y_2^v/\mu\|_F^2 \end{array} \right\} \tag{11}$$

Here, Y_1^v and Y_2^v are Lagrange multipliers, $\mu > 0$ is the regularization parameter. Then, we update each variable according to the alternating method. The rules for updating each variable can be obtained as follows:

$$\mathbf{J}^v = D_\eta(\mathbf{C}^v + \mathbf{Y}_1^v/\mu). \tag{12}$$

$$\mathbf{C}^v = [\mathbf{K}^v + 2\mu\mathbf{I}_n + 2\omega_v\mathbf{I}_n]^{-1}[\beta(\mathbf{K}^v)^T + \mu(\mathbf{Z}^v + \mathbf{J}^v) + 2\omega_v\mathbf{S} - \mathbf{Y}_1^v - \mathbf{Y}_2^v)]. \tag{13}$$

$$\omega_v = 1\big/2\sqrt{\|\mathbf{S} - \mathbf{C}^v\|_F^2}, \quad v = 1, 2, ..., V. \tag{14}$$

$$u_q^l = (h_q^l \sum_{j=1}^{L \times Q} 1/h_j)^{-2}, \quad h_q^l = Tr(\mathbf{K}_q^l - 2\beta\mathbf{K}_q^l\mathbf{C}^v + (\mathbf{C}^v)^T\mathbf{K}_q^l\mathbf{C}^v)\big/2. \tag{15}$$

Furthermore, the updates of $\mathbf{Z}^v$ and $\mathbf{S}$ can be solved by efficient iterative algorithms [19] and [18], respectively.

4   Results and Discussion

Table 1. Details of the seven scRNA-seq datasets.

Datasets	No. of cells	No. of genes	Class	Sparsity (%)	Species
Li_islet	60	4494	6	0.00	Homo sapiens
Goolam	124	40315	5	68.00	Mus musculus
Deng	135	12548	7	31.85	Mus musculus
Engel4	203	23337	4	80.46	Homo sapiens
Usoskin	622	17772	4	78.10	Mus musculus
Kolod	704	10684	3	27.87	Mus musculus
Tasic	1727	5832	48	32.70	Mus musculus

In the experiment, the evaluation criteria we used are the Normalized Mutual Information (NMI) [20] and the Adjusted Random Index (ARI) [21]. Seven different scRNA-seq datasets are used, including Li_islet [22], Goolam [23], Deng [24], Engel4 [25], Usoskin [26], Kolod [27] and Tasic [28]. Details are given in Table 1. In addition, we use 5 ($Q = 5$) different kernel functions, which are linear, polynomial, Gaussian, sigmoid, and inverse polynomial. And all kernel matrices are normalized to the range [0,1] to avoid inconsistent values.

4.1 Parameter Analysis

First, determine the parameters involved in the pre-processing step. In our model, 5 ($V = 5$) feature sets are selected empirically, denoted as $\{\mathbf{X}^v\}_{v=1}^V$ [1]. The size of the feature set is set automatically according to the total gene count m. The choice of parameter K in KNN is automatically set according to the cell count n.

Then, determine the parameters λ and L in the GAE, β and α in the model (10). The grid search technique is used to find the optimal parameters. To simplify the experiment, we fix $L = 3$, $\alpha = 0.4$, $log\lambda \in [-3, 1]$ and $log\beta \in [-3, 1]$.

Table 2. NMI of fourteen clustering methods on seven scRNA-seq datasets

Methods	Li_islet	Goolam	Deng	Engel4	Usoskin	Kolod	Tasic	Avg.R
PCA	0.5779	0.5547	0.7037	0.7178	0.5676	0.7935	0.4174	9.7
SC	0.9670	0.7253	0.6271	0.7082	0.6377	0.7784	0.4078	8.9
SC3	**1.0000**	0.7757	0.6459	0.8544	0.8126	**1.0000**	0.3856	5.3
Seurat	-	0.7597	0.6962	0.7589	0.7436	0.8407	0.4391	6.7
SIMLR	0.8000	0.5693	0.7422	0.7413	0.7476	**0.9915**	0.0731	8.0
MPSSC	0.8060	0.5639	**0.7554**	0.5465	0.5465	0.5130	**0.4657**	9.3
SinNLRR	**1.0000**	**0.8885**	0.7389	0.6932	0.8472	0.7856	0.4655	**4.9**
NMFLRR	**1.0000**	0.7253	0.7258	0.5372	0.5637	0.9849	0.4588	7.0
SCENA	**0.9735**	0.7977	0.7705	0.2275	0.6550	**1.0000**	0.3607	6.4
RCSL	0.5520	0.8371	0.7705	0.3189	0.6247	0.7238	0.2160	9.3
scDeepCluster	0.522	0.69	0.5277	0.552	0.5827	0.7403	0.4447	10.6
scGMAI	0.6735	0.6630	0.6664	0.8254	0.5156	0.8330	0.3882	9.7
scGAC	0.8390	0.7218	0.6989	**0.9689**	**0.8972**	**1.0000**	0.4264	5.1
scGASI	**1.0000**	**0.9065**	**0.7715**	**0.9640**	**0.9248**	**1.0000**	**0.4725**	**1.1**

4.2 Comparative Analysis of Clustering Results

The clustering performance of scGASI is verified in this section. We select 14 comparison methods, including two basic methods, PCA (average of 100 experiments) and SC (clustering affinity matrix by Pearson correlation coefficient), three integrated clustering methods, SC3, Seurat, and SCENA, five similarity learning-based clustering methods, SIMLR, MPSSC, SinNLRR, NMFLRR, and RCSL, and three deep learning-based clustering methods, scDeepCluster, scGMAI, and scGAC. For the sake of fairness, the main parameters are carefully adjusted and the data preprocessing instructions are followed for all comparison methods. In particular, the number of input clusters is set to the true number of clusters for all methods in order to make comparisons more fair.

The clustering results are shown in Tables 2 and 3, where "Avg.R" indicates the average ranking of each method, the optimal or suboptimal performance on each dataset is shown in bold, and "-" indicates datasets that could not be processed by this method. As shown in Tables 2 and 3, scGASI obtains the best performance on the six datasets and has the highest average ranking (Avg.R(NMI) = 1.1, Avg.R(ARI) = 1.3), significantly superior to SinNLRR with the sub-optimal ranking (Avg.R(NMI) = 4.9, Avg.R(ARI) = 4.7). Specifically, on the highly sparse Engel4 and Usoskin datasets, the clustering accuracy of scGASI is clearly better than other methods. What this suggests is that the clustering performance of scGASI is less susceptible to the effects of data sparsity. In conclusion, scGASI has more advantages than other clustering methods.

Table 3. ARI of fourteen clustering methods on seven scRNA-seq datasets.

Methods	Li_islet	Goolam	Deng	Engel4	Usoskin	Kolod	Tasic	Avg.R
PCA	0.2541	0.4070	0.4484	0.6299	0.5372	0.7694	0.1143	9.3
SC	0.9650	0.5439	0.3327	0.6741	0.6695	0.7273	0.1149	8.1
SC3	**1.0000**	0.6874	0.4221	0.7970	0.8453	**1.0000**	0.1063	5
Seurat	-	0.5821	**0.5497**	0.7087	0.6088	0.7232	0.1344	5.7
SIMLR	0.9020	0.2991	0.4565	0.6682	0.6830	0.9982	0.0010	8.6
MPSSC	0.6350	0.3046	0.4783	0.4377	0.0030	0.4957	0.1243	10.3
SinNLRR	**1.0000**	**0.9097**	0.4706	0.6533	0.8773	0.7291	0.1326	**4.7**
NMFLRR	**1.0000**	0.5440	0.4720	0.4436	0.4406	0.9920	0.1222	7.1
SCENA	**0.9678**	0.6132	0.5261	0.2057	0.4772	**1.0000**	0.0622	7.3
RCSL	0.3168	0.8600	0.5261	0.2455	0.5206	0.6191	0.0060	9.3
scDeepCluster	0.3554	0.5156	0.3379	0.4451	0.4462	0.5813	**0.1442**	9.9
scGMAI	0.5364	0.3518	0.4230	0.7332	0.4115	0.6993	0.1032	10.3
scGAC	0.7764	0.6296	0.4349	**0.9740**	**0.9374**	**1.0000**	0.1114	5.1
scGASI	**1.0000**	**0.9141**	**0.5504**	**0.9716**	**0.9503**	**1.0000**	**0.1405**	1.3

4.3 Visualization

Uniform manifold approximate and project (UMAP) is one of the most popular tools to visualize scRNA-seq data [6]. Here, we use the consensus affinity matrix S as input to UMAP to visually distinguish cell subpopulations. Figure 2 shows the visualization results of affinity matrices from Pearson correlation coefficient, SinNLRR, NMFLRR, RCSL, and scGASI on the Engel4 [25] and Kolod [27] datasets. As shown in Fig. 2, scGASI has larger interclass distances and smaller intraclass distances. scGASI completely distinguishes the three subpopulations on the Kolod dataset and partitions the majority of cells on the Engel4 dataset into their corresponding clusters. In conclusion, it can be said that scGASI has a greater potential to achieve the correct division of cell types.

4.4 Gene Markers

In this section, we apply the affinity matrix to the bootstrap Laplacian score [9] to select gene markers in each subpopulation. Figure 3 shows the top ten gene markers in the

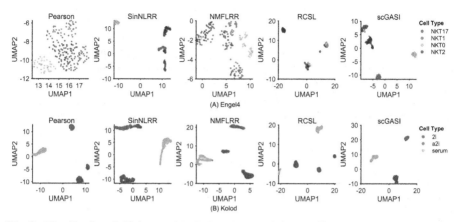

Fig. 2. Visualization of affinity matrices for Pearson correlation coefficient, SinNLRR, NMFLRR, RCSL and scGASI on (A) Engel4 and (B) Kolod datasets.

Engel4 dataset. In Fig. 3, eight of the ten genes we have detected are highly expressed genes in specific subpopulations, as demonstrated by Engel et al. [25]. Among them, Serpinb1a gene a peptidase inhibitor (Serpinb1a), which may be a negative feedback regulator of IL-17-producing T cells (including NKT17 cells) [25]; Tmem176a and Tmem176b are associated with the immature state of dendritic cells [29]; Blk encodes a kinase of the Src family that plays an important role in B-cell development [30]. In addition, 1300014I06Rik and Stmn1 genes are protein-coding genes that have not yet been identified and may be the focus of further research.

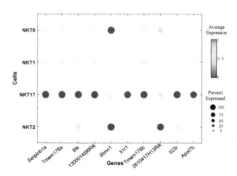

Fig. 3. The top ten gene markers in the Engel4 dataset.

5 Conclusion

Unsupervised clustering is one of the key issues in scRMA-seq data analysis, which can effectively differentiate cell subpopulations and help discover new cell subtypes. In this paper, we propose a new single-cell clustering method scGASI. scGASI combines multiple feature sets learning and the multi-layer kernel self-expression integration mechanism to learn an accurate consensus affinity matrix for downstream analysis. scGASI achieves superior clustering performance compared to fourteen downscaling or clustering methods on seven scRNA-seq datasets.

Although scGASI can effectively identify single cell types, GAE can be further developed. In the future, we plan to refine the GAE to better match the distribution of scRNA-seq data and enhance the scalability of scGASI.

Acknowledgment. This work is supported by the National Natural Science Foundation of China (No. 62172253), and jointly supported by the Program for Youth Innovative Research Team in the University of Shandong Province in China (No.2022KJ179).

References

1. Cui, Y., Zhang, S., Liang, Y., Wang, X., Ferraro, T.N., Chen, Y.: Consensus clustering of single-cell RNA-seq data by enhancing network affinity. Briefings Bioinform. **22**, bbab236 (2021)
2. Sinaga, K.P., Yang, M.-S.: Unsupervised K-means clustering algorithm. IEEE Access **8**, 80716–80727 (2020)
3. Von Luxburg, U.: A tutorial on spectral clustering. Stat. Comput. **17**, 395–416 (2007)
4. Wold, S., Esbensen, K., Geladi, P.: Principal component analysis. Chemom. Intell. Lab. Syst. **2**, 37–52 (1987)
5. Laurens, V.D.M., Hinton, G.: Visualizing data using t-SNE. J. Mach. Learn. Res. **9**, 2579–2605 (2008)
6. McInnes, L., Healy, J., Melville, J.: UMAP: uniform manifold approximation and projection for dimension reduction. arXiv preprint arXiv:1802.03426 (2018)
7. Wang, C.-Y., Gao, Y.-L., Kong, X.-Z., Liu, J.-X., Zheng, C.-H.: Unsupervised cluster analysis and gene marker extraction of scRNA-seq data based on non-negative matrix factorization. IEEE J. Biomed. Health Inform. **26**, 458–467 (2021)
8. Mei, Q., Li, G., Su, Z.: Clustering single-cell RNA-seq data by rank constrained similarity learning. Bioinformatics **37**, 3235–3242 (2021)
9. Wang, B., Zhu, J., Pierson, E., Ramazzotti, D., Batzoglou, S.: Visualization and analysis of single-cell RNA-seq data by kernel-based similarity learning. Nat. Methods **14**, 414–416 (2017)
10. Park, S., Zhao, H.: Spectral clustering based on learning similarity matrix. Bioinformatics **34**, 2069–2076 (2018)
11. Zheng, R., Li, M., Liang, Z., Wu, F.-X., Pan, Y., Wang, J.: SinNLRR: a robust subspace clustering method for cell type detection by non-negative and low-rank representation. Bioinformatics **35**, 3642–3650 (2019)
12. Zhang, W., Xue, X., Zheng, X., Fan, Z.: NMFLRR: clustering scRNA-seq data by integrating nonnegative matrix factorization with low rank representation. IEEE J. Biomed. Health Inform. **26**, 1394–1405 (2021)

13. Kiselev, V.Y., et al.: SC3: consensus clustering of single-cell RNA-seq data. Nat. Methods **14**, 483–486 (2017)
14. Stuart, T.: Comprehensive integration of single-cell data. Cell **177**, 1888–1902. e1821 (2019)
15. Tian, T., Wan, J., Song, Q., Wei, Z.: Clustering single-cell RNA-seq data with a model-based deep learning approach. Nat. Mach. Intell. **1**, 191–198 (2019)
16. Yu, B., et al.: scGMAI: a Gaussian mixture model for clustering single-cell RNA-Seq data based on deep autoencoder. Briefings Bioinform. **22**, bbaa316 (2021)
17. Wang, C., Pan, S., Long, G., Zhu, X., Jiang, J.: MGAE: marginalized graph autoencoder for graph clustering. In: Proceedings of the 2017 ACM on Conference on Information and Knowledge Management, pp. 889–898. (2017)
18. Zhang, D.-J., Gao, Y.-L., Zhao, J.-X., Zheng, C.-H., Liu, J.-X.: A new graph autoencoder-based consensus-guided model for scRNA-seq cell type detection. IEEE Trans. Neural Netw. Learn. Syst. (2022)
19. Huang, J., Nie, F., Huang, H.: A new simplex sparse learning model to measure data similarity for clustering. In: Twenty-Fourth International Joint Conference on Artificial Intelligence (2015)
20. Strehl, A., Ghosh, J.: Cluster ensembles—a knowledge reuse framework for combining multiple partitions. J. Mach. Learn. Res. **3**, 583–617 (2002)
21. Meilă, M.: Comparing clusterings—an information based distance. J. Multivar. Anal. **98**, 873–895 (2007)
22. Li, J., et al.: Single-cell transcriptomes reveal characteristic features of human pancreatic islet cell types. EMBO Rep. **17**, 178–187 (2016)
23. Goolam, M., et al.: Heterogeneity in Oct4 and Sox2 targets biases cell fate in 4-cell mouse embryos. Cell **165**, 61–74 (2016)
24. Deng, Q., Ramsköld, D., Reinius, B., Sandberg, R.: Single-cell RNA-seq reveals dynamic, random monoallelic gene expression in mammalian cells. Science **343**, 193–196 (2014)
25. Engel, I., et al.: Innate-like functions of natural killer T cell subsets result from highly divergent gene programs. Nat. Immunol. **17**, 728–739 (2016)
26. Usoskin, D., et al.: Unbiased classification of sensory neuron types by large-scale single-cell RNA sequencing. Nat. Neurosci. **18**, 145–153 (2015)
27. Kolodziejczyk, A.A., et al.: Single cell RNA-sequencing of pluripotent states unlocks modular transcriptional variation. Cell Stem Cell **17**, 471–485 (2015)
28. Tasic, B., et al.: Adult mouse cortical cell taxonomy revealed by single cell transcriptomics. Nat. Neurosci. **19**, 335–346 (2016)
29. Condamine, T., et al.: Tmem176B and Tmem176A are associated with the immature state of dendritic cells. J. Leukoc. Biol. **88**, 507–515 (2010)
30. Castillejo-López, C., et al.: Genetic and physical interaction of the B-cell systemic lupus erythematosus-associated genes BANK1 and BLK. Ann. Rheum. Dis. **71**, 136–142 (2012)

ABCAE: Artificial Bee Colony Algorithm with Adaptive Exploitation for Epistatic Interaction Detection

Qianqian Ren, Yahan Li, Feng Li, Jin-Xing Liu, and Junliang Shang[✉]

School of Computer Science, Qufu Normal University, Rizhao 276826, Shandong, China
shangjunliang110@163.com

Abstract. The detection of epistatic interactions among multiple single-nucleotide polymorphisms (SNPs) in complex diseases has posed a significant challenge in genome-wide association studies (GWAS). However, most existing methods still suffer from algorithmic limitations, such as high computational requirements and low detection ability. In the paper, we propose an artificial bee colony algorithm with adaptive exploitation (ABCAE) to address these issues in epistatic interaction detection for GWAS. An adaptive exploitation mechanism is designed and used in the onlooker stage of ABCAE. By using the adaptive exploitation mechanism, ABCAE can locally optimize the promising SNP combination area, thus effectively coping with the challenges brought by high-dimensional complex GWAS data. To demonstrate the detection ability of ABCAE, we compare it against four existing algorithms on eight epistatic models. The experimental results demonstrate that ABCAE outperforms the four existing methods in terms of detection ability.

Keywords: Adaptive exploitation · Artificial bee colony · Complex disease Epistatic interaction

1 Introduction

Genome-wide association studies (GWAS) play a crucial role in finding the genetic mechanisms behind complex diseases [1, 2]. The advent of high-throughput sequencing technologies has enabled the identification of millions of single-nucleotide polymorphisms (SNPs) associated with various diseases [3, 4]. These high-dimensional SNP datasets have provided valuable insights into the genetic mechanism of diseases, but they have also presented challenges for detecting epistatic interactions among SNPs [5–7].

In the field of GWAS, several algorithms have recently been tried to uncover epistatic interactions from SNP data [8, 9]. For example, Jing and Shen introduced a heuristic optimization framework that combines logistical regression and Bayesian network (MACOED) to detect epistatic interactions [10]. Tuo et al. proposed a niche harmony search algorithm (NHSA) that uses joint entropy as a heuristic factor to detect epistatic

X. Guo et al. (Eds.): ISBRA 2023, LNBI 14248, pp. 190–201, 2023.
https://doi.org/10.1007/978-981-99-7074-2_15

interactions [11]. Zhang et al. proposed a selectively informed particle swarm optimization algorithm (SIPSO) that employs mutual information as its fitness function to detect epistatic interactions [12]. Aflakparast presented a cuckoo search epistasis method (CSE) specifically designed for detecting epistatic interactions [13]. However, these algorithms often face limitations such as curse of dimensionality and insufficient detection ability detecting potential epistasis [14]. Moreover, these limitations make it difficult to develop powerful algorithms for detecting epistatic interactions in GWAS [15].

To tackle the aforementioned challenges, an artificial bee colony algorithm with adaptive exploitation (ABCAE) is proposed. Our algorithm is specifically used for detecting epistatic interactions in GWAS data. In this algorithm, an adaptive exploitation mechanism is designed and used in the onlooker stage of ABCAE. As a result, the proposed algorithm can locally optimize the promising SNP combination area. We conduct experiments using eight small epistatic models, providing empirical evidence to validate the effectiveness of the proposed algorithm. The results indicate that ABCAE outperforms other four epistatic interaction detection algorithms, highlighting its potential for advancing epistatic interaction detection in GWAS.

2 Materials

2.1 Artificial Bee Colony Algorithm

Inspired by the foraging behavior of honey bee swarms, the artificial bee colony (ABC) algorithm consists of three types of bees: employed bees, onlooker bees, and scout bees [16]. The colony is divided into two equal halves: employed bees and onlooker bees. If a food source is not optimized within a given limited number of times, the employed bee corresponding to the food source will become the scout bee, looking for a new food source. The employed bees are responsible for finding new food sources and then passing the information to onlooker bees. The onlooker bees choose high-quality food sources from the information they receive. The scout bees discard food sources and choose new food sources according to preset conditions.

2.2 Fitness Function

Mutual information (MI) is used to evaluate the association between epistatic interactions and disease phenotype [17], and MI is defined as

$$MI(A, B) = H(A) + H(B) - H(A, B) \qquad (1)$$

where A is an epistatic interaction, B is the disease phenotype, $H(A)$ is the entropy of A, $H(B)$ is the entropy of B, and $H(A, B)$ is the joint entropy of A and B. Here, $H(A)$ and $H(A, B)$ are defined as follow:

$$H(A) = -\sum_{a \in A} p(a) log(p(a)) \qquad (2)$$

$$H(A, B) = -\sum_{a \in A} \sum_{b \in B} p(a, b) log(p(a, b)) \qquad (3)$$

A high MI value indicates a high correlation between epistasis and disease.

3 Method

In this section, we put forward a novel and effective ABCAE algorithm to the complex task of detecting genetic interactions.

3.1 Overview Framework

At the onset of ABCAE, we initialize the process by creating a population of potential solutions within the feasible space. To ensure a diverse initial population, we employ a random generation process. Specifically, each food source contains K SNPs, and each SNP is generated by

$$x_{ij} = x_j^{min} + rand(0,\ 1)(x_j^{max} - x_j^{min}) \tag{4}$$

where $i = \{1,\ldots,M\}, j = \{1,\ldots,K\}$. M is the population size; K is the number of SNPs in an epistatic interaction; $rand(0, 1)$ is a random number uniformly distributed between 0 and 1. Additionally, x^{max} and x^{min} correspond to the upper and lower boundaries respectively for each dimension. The creation of the randomly generated population serves as a foundation for subsequent iterations and evolutionary processes. As the algorithm progresses, food sources within the population will undergo evaluation, selection, and modification, leading to the emergence of superior solutions over time.

Following the initialization phase, the employed bees take charge of searching for the food sources. Each employed bee generates a new position by incorporating the information from its previous position, leading to the discovery of a novel food source. This process can be succinctly described as

$$v_{ij} = x_{ij} + \delta_{ij}(x_{ij} - x_{kj}) \tag{5}$$

where δ_{ij} is a randomly generated number between 0 and 1. The index k, belonging to the set $\{1,\ldots,M\}$, is randomly selected.

When acquiring a new food source V_i, ABCAE performs a selection process to compare it with the previous food source X_i. If V_i demonstrates improved results over X_i, the employed bees adopt V_i and discard the previous one. Otherwise, X_i is retained within the population. The specific formula is described as

$$X_i(t) = \begin{cases} V_i(t) & fit(V_i(t)) > fit(X_i(t)) \\ X_i(t) & otherwise \end{cases} \tag{6}$$

Then, the onlooker bee searches for new food sources according to the probability value P_i corresponding to X_i. The detailed search process of onlooker bees and the calculation of P_i will be described in the next section.

Algorithm 1 ABCAE

Input: maximum number of function evaluations $neval_{max}$, population size M, parameter *limit*

Output: detected epistatic interactions

1. $Neval_{max} = 0$;

2. $t = 0$;

3. **for** $i = 1$ to M **then**

4.   Randomly initialize the food source $X_i(t)$;

5.   Calculate the fitness value of $X_i(t)$;

6. **end for**

7. Rank M food sources in descending order according to fitness values.

8. **while** $neval_{max}$ is not reached **do**

9.   *** **employed bee stage** ***

10.   **for** $i = 1$ to $M/2$ **then**

11.   **for** $j = 1$ to D **then**

12.   Generates a new SNP $v_{ij}(t)$;

13.   **end for**

14.   Calculate the fitness value of $V_i(t)$;

15.   **if** $f(V_i(t)) > f(X_i(t))$ **then**

16. $X_i(t) = V_i(t)$;

17.   **end if**

18.   **end for**

19.   *** **end employed bee stage** ***

20.   *** **onlooker bee stage** ***

21.   **for** $i = M/2$ to M **then**

22.   Calculate P_i related to X_i;

23.   **for** $j = 1$ to D **then**

24.   Generate a random number r;

25.   **if** $r < P_i$ **then**

26.   Generates a new SNP $x_{ij}(t)$;

27.   **end if**

28.   **end for**

29.   **end for**

30.   *** **end onlooker bee stage** ***

31.   *** **scout bee stage** ***

32.   **while** *limit* is not reached **do**

33.   Update $X_{best}(t)$;

34.   **end while**

35.   *** **end scout bee stage** ***

36. $t = t + 1$;

37. **end while**

Once all onlooker bees finish this process, a decision will be made regarding the retention of the best food source X_{best}. by examining the scout bee. The scout bee tries to change the SNPs in X_{best} *limit* times. If the best food source X_{best} cannot be improved, X_{best} is discarded.

3.2 Adaptive Exploitation Mechanism

ABCAE gives each food source X_i of the population a probability of being selected by the onlooker bees, and the probability corresponding P_i to X_i is calculated as

$$P_i = \left| \frac{fit(X_i) - fit_{avg}}{fit_{max} - fit_{min}} \right| \tag{7}$$

where $fit(X_i)$ is the fitness function value for X_i, fit_{avg} is the average fitness function value for all the food source in the population, fit_{max} is the fitness function value for the best food source, and fit_{min} is the fitness function value for the worst food source in the population.

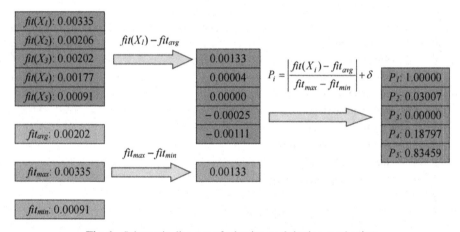

Fig. 1. Schematic diagram of adaptive exploitation mechanism

In order to understand adaptive exploitation mechanism easily, the schematic diagram of the calculation process of this mechanism is shown in Fig. 1. Suppose there are five food sources that can be selected by following bees: X_1, X_2, X_3, X_4 and X_5, and their fitness values are $fit(X_1) = 0.00335$, $fit(X_2) = 0.00206$, $fit(X_3) = 0.00202$, $fit(X_4) = 0.00177$, and $fit(X_5) = 0.00091$, respectively. fit_{avg}, fit_{max} and fit_{min} for the five food sources are 0.00202, 0.00335 and 0.00091. Thus, the difference between the fitness value of each food source and the average fitness value can be obtained by calculation, which are 0.00133, 0.00004, 0.00000, –0.00025, and –0.00111 respectively. Furthermore, the P_i associated with each X_i can be obtained, which are $P_1 = 1.00000$, $P_2 = 0.03007$, $P_3 = 0.00000$, $P_4 = 0.18797$, and $P_5 = 0.83459$, respectively. It can be observed that the food source with larger or smaller fitness value ($fit(X_1) = 0.00335$ and $fit(X_5) = 0.00091$) is more likely to be selected by the onlooker bees, while the other food sources ($fit(X_2) =$ 0.00206, $fit(X_3) = 0.00202$, and $fit(X_4) = 0.00177$) are less likely to be selected by the onlooker bees.

The adaptive exploitation mechanism is beneficial to improve the diversity of population, and the reasons are as follows. The solutions with low fitness values are easy to be chosen by onlooker bees to find better solutions, which gives them the opportunity to improve their own quality. The solutions with high fitness values are also easy to be selected by the onlooker bees, which can avoid the premature convergence caused by its large advantage in the population.

4 Experiments

4.1 Dataset

This paper utilizes eight different epistatic models to detect specific two-locus epistatic interactions. The detailed information of these models is shown in Table 1. Each model is generated by varying the strength of the marginal effect while maintaining a fixed interaction structure. The minor allele frequency (MAF) varies from 0.05 to 0.5. 100 different datasets are produced by each epistatic model, each of which consists of 1000 SNPs, 2000 cases and 2000 controls. These datasets are generated using the EpiSIM software [18].

Table 1. Details of eight different epistatic models

		AA	Aa	aa			AA	Aa	aa
Model 1	BB	0.087	0.087	0.087	Model 2	BB	0.078	0.078	0.078
	Bb	0.087	0.146	0.190		Bb	0.078	0.105	0.122
	bb	0.087	0.190	0.247		bb	0.078	0.122	0.142
Model 3	BB	0.084	0.084	0.084	Model 4	BB	0.092	0.092	0.092
	Bb	0.084	0.210	0.210		Bb	0.092	0.319	0.319
	bb	0.084	0.210	0.210		bb	0.092	0.319	0.319
Model 5	BB	0.072	0.164	0.164	Model 6	BB	0.067	0.155	0.155
	Bb	0.164	0.072	0.072		Bb	0.155	0.067	0.067
	bb	0.164	0.072	0.072		bb	0.155	0.067	0.067
Model 7	BB	0.000	0.000	0.100	Model 8	BB	0.000	0.020	0.000
	Bb	0.000	0.050	0.000		Bb	0.020	0.000	0.020
	bb	0.100	0.000	0.000		bb	0.000	0.020	0.000

4.2 Parameter Setting and Evaluation Metric

To evaluate the detection ability of ABCAE, we conduct experiments using eight different types of datasets. In the case of ABCAE, the initial population is randomly generated in the search space. We set the population size 200, which consists of 100 employed bees and 100 onlooker bees. Furthermore, the parameter *limit* is set to 10. With regard to the stop condition, the maximum number of fitness function evaluations $Neval_{max}$ is set to 50,000 for each dataset. Additionally, each algorithm is executed 10 times independently on each epistatic model. By averaging the results of these 10 independent runs, we compute the average values for performance comparisons.

For the convenience of comparing the ability of different algorithms to detect epistatic interactions, we utilize *Power* as the evaluation metric, which is defined as

$$Power = \frac{W}{Z} \tag{8}$$

where W denotes the number of datasets where epistatic interactions are identified, while Z denotes 100 datasets produced using the same model.

4.3 Detection Power Comparison

In this section, ABCAE compares its detection power with four existing algorithms for two-locus epistatic interaction detection. These methods include MACOED [10], NHSA [11], SIPSO [12] and CSE [13]. The detection power values of the five algorithms are summarized in Fig. 2. Based on the experimental results, several noteworthy observations can be made as follows.

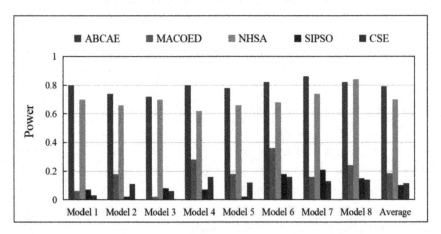

Fig. 2. Power of five algorithms on eight epistatic models

For Model 1, ABCAE obtains the best power value, whereas CSE performs the lowest power value. NHSA yields the best performance among the remaining algorithms. Additionally, MACOED and SIPSO yield almost the same power value for that epistatic model. For Model 2 and Model 4, ABCAE offers the best solutions, whereas SIPSO provides lowest power value. In Model 3, ABCA and MACOED provide almost the same power values, whereas SIPSO yields the worst results. Similarly, for Model 5 and

Model 6, ABCAE consistently outperforms other algorithms. In the case of Model 7, our proposed algorithm, ABCAE, continues to exhibit superior performance. Regarding Model 8, NHSA obtains the best power value. In this figure, the term Average means the overall performance of five algorithms on eight models. For the average power value, the ranking is as follows: ABCAE, NHSA, MACOED, SIPSO, and CSE. Therefore, ABCAE demonstrates superior performance compared to other algorithms, including MACOED, NHSA, SIPSO, and CSE.

4.4   Parameter Analysis

Population Size (M).   Population size is a significant aspect to consider in ABCAE. In this section, we aim to use eight models to analyze the impact of population size on our proposed algorithm. Each model comprises 1000 SNPs with 2000 cases and 2000 controls. In the experiment, ABCAE is executed with six different population sizes: $M = 50, M = 100, M = 150, M = 200, M = 250$, and $M = 300$. The remaining parameters are consistent with those introduced in Sect. 4.2. The experimental results are summarized in Fig. 3. For the average detection power, $M = 200$ yields the best solution, whereas $M = 50$ obtains the poorest among the other population settings. In summary, $M = 200$ brings ABCAE the best detection ability when considering $Neval_{max} = 50,000$.

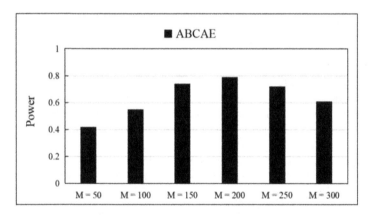

Fig. 3. Power of ABCAE with different population size

Control Parameter limit.   For ABCAE, *limit* plays a crucial role in determining which food source to give up. If the changed SNP fails to show further improvement beyond the *limit* threshold, the corresponding food source is discarded. Consequently, the scout bee discovers a new food source to replace the abandoned food source. The control parameter *limit* affects the detection ability of ABCAE. In this section, we vary *limit* from the set {5, 10, 15, 20, 25, 30} to compare the detection ability of the algorithm under different settings. The experimental results are summarized in Fig. 4. It is observed that the control parameter *limit* set to 10 produces superior models compared to other *limit* settings. Conversely, *limit* of 5 yields the poorest performance. Additionally, the control parameter *limit* set to 20 demonstrates similar average values to those obtained with *limit* set to 25. In summary, it can be concluded that selecting *limit* = 10 enhances the detection ability of ABCAE when considering $Neval_{max}$ = 50,000.

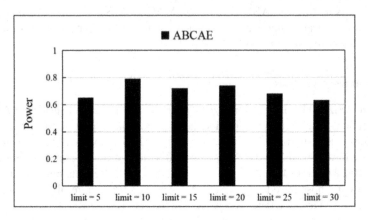

Fig. 4. Power of ABCAE with different *limit*.

4.5   Age-Related Macular Degeneration Studies

In this section, we utilize our proposed ABCAE algorithm to perform two-locus epistasis detection from the age-related macular degeneration (AMD) study [19]. The AMD dataset contains 103,611 SNPs from 96 case samples and 50 control samples. The parameter setting for ABCAE is described in Sect. 4.1. All the output epistasis generated by ABCAE, include the SNPs belonging to the CHF gene and other SNPs located on the respective genes. As shown in Table 2. The almost epistasis obtained by our algorithm involves rs380390 and rs1329428. The two SNPs belonging to the CHF gene and has been validated to be related AMD [20]. In the epistatic interaction (rs1329428, rs1822657), rs1822657 belonging to the NCAM2 gene related to Alzheimer's disease [21]. In the epistatic interaction (rs380390, rs2224762), SNP rs2224762 belongs to the KDM4C gene related to Glioblastoma [22]. For the epistatic interaction (rs1394608, rs3743175), SNP rs1394608 belongs to the SGCD gene which is reported to be related to the AMD disease [23]. SNP rs3743175 is located in the SCAPER gene, and SCAPER is related to syndromic intellectual disability [24]. In the epistatic interaction (rs1394608, rs9328536), SNP rs9328536 is located in the MED27 gene, and MED27 is associated with melanoma [25, 26]. For the epistatic interaction (rs1740752, rs943008), SNP rs943008 is located in the NEDD9 gene, and NEDD9 plays a role in cancer metastasis [27]. In the epistatic interaction (rs3775652, rs725518), SNP rs3775652 belongs to the INPP4B gene related to ovarian cancer [28], and SNP rs725518 belongs to the RRM1 gene related to lung cancer [29]. In addition, ABCAE also identifies some interactions contain SNPs in the non-coding region (N/A), such as rs1363688, rs1374431, rs10512937, rs7294734, rs2402053, and rs1740752.

Table 2. Epistatic interactions detected by ABCAE on AMD data

Epistatic interaction	Related genes	Associated disease
rs380390	CFH	Age-related macular degeneration
rs1363688	N/A	None
rs380390	CFH	Age-related macular degeneration
rs2224762	KDM4C	Glioblastoma
rs1329428	CFH	Age-related macular degeneration
rs1822657	NCAM2	Alzheimer's disease
rs380390	CFH	Age-related macular degeneration
rs1374431	N/A	None
rs1394608	SGCD	Age-related macular degeneration
rs3743175	SCAPER	Syndromic intellectual disability
rs380390	CFH	Age-related macular degeneration
rs2402053	N/A	None
rs1329428	CFH	Age-related macular degeneration
rs7294734	N/A	None
rs1329428	CFH	Age-related macular degeneration
rs9328536	MED27	Melanoma
rs1740752	N/A	None
rs943008	NEDD9	Cancer metastasis
rs3775652	INPP4B	Ovarian cancer
rs725518	RRM1	Lung Cancer

5   Conclusion

The paper addresses detecting epistatic interaction detection in GWAS through the application of an artificial bee colony algorithm with adaptive exploitation (ABCAE). By integrating the principles of artificial bee colony optimization, our algorithm takes advantage of the collective intelligence of bee colonies and their efficient foraging behavior. This enables us to efficiently explore the vast search space of epistatic interactions. To further enhance the performance of ABCAE, an adaptive exploitation mechanism is designed and used in the onlooker stage of ABCAE. By using the adaptive exploitation mechanism, ABCAE is able to optimize the promising SNP combination area, effectively tackling the challenges posed by high-dimensional and complex GWAS data. A comprehensive evaluation involving eight epistatic models is conducted. The detection ability of various algorithms, including MACOED, NHSA, SIPSO, and CSE is compared against ABCAE. Through extensive experiments and evaluations on different datasets, we demonstrate the efficacy and superiority of our proposed algorithm in detecting epistatic interactions.

Furthermore, through parameter analysis, the reason why the algorithm sets the parameter value is explained. Overall, ABCAE represents a significant advancement in the field of epistatic interaction detection. It offers a promising avenue for researchers to gain deeper insights into the complex genetic mechanisms underlying various diseases and opens up new opportunities for precision medicine and personalized treatments.

Acknowledgment. This work was supported by the National Natural Science Foundation of China (61972226 and 62172254).

References

1. Moore, J.H., Asselbergs, F.W., Williams, S.M.: Bioinformatics challenges for genome wide association studies. Bioinformatics **26**(4), 445–455 (2010)
2. Price, A.L., Patterson, N.J., Plenge, R.M., Weinblatt, M.E., Shadick, N.A., Reich, D.: Principal components analysis corrects for stratification in genome-wide association studies. Nat. Genet. **38**, 904–909 (2006)
3. Orliac, E.J., et al.: ImprovingGWAS discovery and genomic prediction accuracy in biobank data. Proc. Natl. Acad. Sci. **119**(31), e2121279119 (2022)
4. Price, A.L., Patterson, N.J., Plenge, R.M., Weinblatt, M.E., Shadick, N.A., Reich, D.: Principal components analysis corrects for stratification in genome-wide association studies. Nat. Genet. **38**(8), 904–909 (2006)
5. Jiang, X., Neapolitan, R.E., Barmada, M.M., Visweswaran, S.: Learning genetic epistasis using Bayesian network scoring criteria. BMC Bioinform. **12**(89) (2011)
6. Han, B., Chen, X.-W.: bNEAT: a Bayesian network method for detecting epistatic interactions in genome-wide association studies. BMC Genomics 1–8 (2011)
7. Upstill-Goddard, R., Eccles, D., Fliege, J., Collins, A.: Machine learning approaches for the discovery of gene–gene interactions in disease data. Brief. Bioinform. **14**, 251–260 (2013)
8. Cao, X., Yu, G., Ren, W., Guo, M., Wang, J.: DualWMDR: detecting epistatic interaction with dual screening and multifactor dimensionality reduction. Hum. Mutat. **41**, 719–734 (2020)
9. Guan, B., Xu, T., Zhao, Y., Li, Y., Dong, X.: A random grouping-based self-regulating artificial bee colony algorithm for interactive feature detection. Knowl. Based Syst. **243**, 108434 (2022)
10. Jing, P.-J., Shen, H.-B.: MACOED: a multi-objective ant colony optimization algorithm for SNP epistasis detection in genome-wide association studies. Bioinformatics **31**(5), 634–641 (2014)
11. Tuo, S., Zhang, J., Yuan, X., He, Z., Liu, Y., Liu, Z.: Niche harmony search algorithm for detecting complex disease associated highorder SNP combinations. Sci. Rep. **7**(1), 11529 (2017)
12. Zhang, W., Shang, J., Li, H., Sun, Y., Liu, J.-X.: SIPSO: selectively informed particle swarm optimization based on mutual information to determine SNP-SNP interactions. In: Huang, D.-S., Bevilacqua, V., Premaratne, P. (eds.) ICIC 2016. LNCS, vol. 9771, pp. 112–121. Springer, Cham (2016). https://doi.org/10.1007/978-3-319-42291-6_11
13. Aflakparast, M., Salimi, H., Gerami, A., Dubé, M.P., Visweswaran, S., Masoudi-Nejad, A.: Cuckoo search epistasis: a new method for exploring significant genetic interactions. Heredity **112**(6), 666 (2014)
14. Upstill-Goddard, R., Eccles, D., Fliege, J., Collins, A.: Machine learning approaches for the discovery of gene-gene interactions in disease data. Brief. Bioinform. **14**(2), 251–260 (2012)
15. Li, X., Zhang, S., Wong, K.-C.: Nature-inspired multiobjective epistasis elucidation from genome-wide association studies. IEEE/ACM Trans. Comput. Biol. Bioinform. **17**(1), 226–237 (2020)

16. Karaboga, D., Basturk, B.: A powerful and efficient algorithm for numerical function optimization: artificial bee colony (ABC). J. Global Optim. **39**(3), 459–471 (2007)
17. Zhao, J., Zhou, Y., Zhang, X., Chen, L.: Part mutual information for quantifying direct associations in networks. Proc. Natl. Acad. Sci. **113**(18), 5130–5135 (2016)
18. Shang, J., Zhang, J., Lei, X., Zhao, W., Dong, Y.: EpiSIM: simulation of multiple epistasis, linkage disequilibrium patterns and haplotype blocks for genome-wide interaction analysis. Genes Genom. **35**, 305–316 (2013)
19. Klein, R.J., et al.: Complement factor H polymorphism in age-related macular degeneration. Science **308**(5720), 385–389 (2005)
20. Tuo, S., Liu, H., Chen, H.: Multipopulation harmony search algorithm for the detection of high-order SNP interactions. Bioinformatics **36**(16), 4389–4398 (2020)
21. Leshchyns'ka,I., Liew, H.T., Shepherd, C., Halliday, G.M., Stevens, C.H., Ke, Y.D.: Aβ-dependent reduction of NCAM2-mediated synaptic adhesion contributes to synapse loss in Alzheimer's disease. Nat. Commun. **6**(1), 8836 (2015)
22. Chen, Y., Fang, R., Yue, C., Chang, G., Li, P., Guo, Q., et al.: Wnt-induced stabilization of KDM4C is required for Wnt/β-catenin target gene expression and glioblastoma tumorigenesis. Cancer Res. **80**(5), 1049–1063 (2020)
23. Tang, W., Wu, X., Jiang, R., Li, Y.: Epistatic module detection for case-control studies: a Bayesian model with a gibbs sampling strategy. PLoS Genet. **5**, e1000464 (2009)
24. Kahrizi, K., et al.: Homozygous variants in the gene SCAPER cause syndromic intellectual disability. Am. J. Med. Genet. A **179**(6), 1214–1225 (2019)
25. Sun, Y., Wang, X., Shang, J., Liu, J.-X., Zheng, C.H., Lei, X.: Introducing heuristic information into ant colony optimization algorithm for identifying epistasis. IEEE/ACM Trans. Comput. Biol. Bioinform. **17**(4), 1253–1261 (2020)
26. Tang, R., et al.: MED27 promotes melanoma growth by targeting AKT/MAPK and NF-kB/iNOS signaling pathways. Cancer Lett. **373**(1), 77–87 (2016)
27. Kim, M., Gans, J., Nogueira, C., Wang, A., Paik, J.H., Feng, B.: Comparative oncogenomics identifies NEDD9 as a melanoma metastasis gene. Cell **125**(07), 1269–1281 (2006)
28. Salmena, L., Shaw, P.A., Fans, I., Mclaughlin, J.R., Rosen, B., Risch, H.: Prognostic value of INPP4B protein immunohistochemistry in ovarian cancer. Eur. J. Gynaecol. Oncol. **36**(3), 260–267 (2015)
29. Lee, J.J., Maeng, C., Baek, S., Kim, G., Yoo, J.H., Choi, C.: The immunohistochemical over-expression of ribonucleotide reductase regulatory subunit M1 (RRM1) protein is a predictor of shorter survival to gemcitabine-based chemotherapy in advanced non-small cell lung cancer (NSCLC). Lung Cancer **70**(2), 205–210 (2010)

USTAR: Improved Compression of k-mer Sets with Counters Using de Bruijn Graphs

Enrico Rossignolo and Matteo Comin$^{(\boxtimes)}$

Department of Information Engineering, University of Padua, Padua, Italy
{enrico.rossignolo,matteo.comin}@unipd.it

Abstract. A fundamental operation in computational genomics is to reduce the input sequences to their constituent k-mers. Finding a space-efficient way to represent a set of k-mers is important for improving the scalability of bioinformatics analyses. One popular approach is to convert the set of k-mers into a de Bruijn graph and then find a compact representation of the graph through the smallest path cover.

In this paper, we present USTAR, a tool for compressing a set of k-mers and their counts. USTAR exploits the node connectivity and density of the de Bruijn graph enabling a more effective path selection for the construction of the path cover. We demonstrate the usefulness of USTAR in the compression of read datasets. USTAR can improve the compression of UST, the best algorithm, from 2.3% up to 26,4%, depending on the k-mer size.

The code of USTAR and the complete results are available at the repository https://github.com/enricorox/USTAR.

Keywords: k-mer set with counts · compression · smallest path cover

1 Introduction

The majority of bioinformatics analysis is performed by k-mer based tools that provide several advantages with respect to the ones that directly process reads or reads alignments. These tools operate primarily by transforming the input sequence data, which may be of various lengths depending on the technology used for sequencing, into a k-mer set that is a set of strings with fixed length and their multiplicities, called counts.

k-mers-based methods achieve better performance in many applications. In genome assembly, Spades [2] used k-mers-based methods to reconstruct the entire genome from reads obtaining efficiently highly accurate results. Also the assembly validation of Merqury [22] uses k-mer counts. In metagenomics, Kraken [27] is capable to classify and identify microorganisms in complex environmental samples using k-mers and it is 900 times faster than MegaBLAST. Since the introduction of Kraken, most of the tools for metagenomic classification are based on k-mers [1,5,19,24]. In genotyping, several tools [9,13,14,26] use k-mers instead of alignment to identify genetic variations in individuals or populations. In phylogenomics, Mash [15] uses k-mers to efficiently estimate genomes and metagenomes

X. Guo et al. (Eds.): ISBRA 2023, LNBI 14248, pp. 202–213, 2023.
https://doi.org/10.1007/978-981-99-7074-2_16

distances in order to reconstruct evolutionary relationships among organisms. In database searching many k-mers-based methods [3,10,12,16,25] have been proposed in order to search sequences efficiently.

Overall, k-mer-based methods have revolutionized many areas of bioinformatics and they have become an essential tool for analyzing large-scale genomic data. These tools often rely on specialized data structures for representing sets of k-mers (for a survey, see [6]). Since modern sequencing datasets are huge, the space used by such data structures is a bottleneck when attempting to scale up to large databases. Conway and Bromage [8] showed that at least $\log \binom{4^k}{n}$ bits are needed to losslessly store a set of n k-mers, in the worst case. However, a set of k-mers generated from a sequencing experiment typically exhibits the spectrum-like property [6] and contains a lot of redundant information. Therefore, in practice, most data structures can substantially improve on that bound [7].

Since storing a k-mer set requires non-negligible space, it's desirable to reduce the size that can be very large. For example, the dataset used to test the BIGSI [21] index takes approximately 12 TB to be stored in compressed form.

The best tool to compress a set of k-mers with counts is UST [21] (see Sect. 1.1) and it uses the De Bruijn graph representation of the input k-mer set. The problem of finding the smallest k-mer set representation is equivalent to finding the smallest path cover in a de Bruijn graph (Sect. 2). In this paper, we present USTAR (Unitig STitch Advanced constRuction), which follows a similar paradigm, but implements a better strategy for exploring De Bruijn graphs. The USTAR strategy leverages the density of the de Bruijn graph and node connectivity, enabling a more effective path selection for the construction of the path cover, and thus improving the compression. In Sect. 3 we reported a series of results on several real sequencing datasets. We showed that USTAR achieves the best compression ratio of k-mers and counts and it outperforms UST, and other tools.

1.1 Related Works

The problem of k-mer set compression has been addressed by several researchers, in this section we summarize the most recent findings. k-mers counters are tools that are designed to count and store distinct k-mers, a particularly hard challenge for large datasets. The most famous tools are Squeakr [17], KMC [11] and DSK [23]. Squeakr is an approximate and exact k-mers counting system that exploits Bloom filters, a probabilistic data structure, in order to efficiently store k-mers. KMC uses disk files as bins in which divide, sort, and count k-mers. Finally, DSK uses hash tables in order to update k-mers counters. The last two tools are not specifically designed for compression but they are still capable to reduce the size storing only distinct k-mers and their counts.

A k-mer set with counters can be represented by a de Bruijn graph (dBG) that can be exploited for efficient storage. BCALM2 [7] is a tool for the low-memory construction of dBGs that are compacted, meaning that maximal non-branching paths are merged in a single node labeled with k-mers glued together and the list of the counts. Compaction not only provides advantages in terms of memory used but also in terms of disk space. The idea, in order to compress k-mers, is to save $k-1$ characters per link. For example, given a dBG, the non-branching path (ACT, CTG, TGA) can be replaced by a single node $ACTGA$. The sequence represented by a non-branching path is called unitig and it is an attempt to compress k-mers using dBGs.

Another way to reduce the redundancy of a k-mer set is to exploit its spectrum-like property [6], i.e. the existence of long strings that "generate" all the k-mers. This idea has been developed in parallel by the authors of ProphAsm [4] and UST [21]. The authors of ProphAsm [4] refer to these long strings as simplitigs and they build them by linking overlapping unitigs and k-mers during the exploration of a dBG computed on the fly. They showed that simplitigs outperformed unitigs, the k-mers representation proposed by BCALM2, in terms of computational resources and compression rate. Also UST [21] links overlapping unitigs and k-mers, but it uses as input the compacted de Bruijn graph computed by BCALM2 and it considers also k-mers counts. They find a nearly tight lower bound for the optimal k-mers representation and they showed empirically that in most cases their greedy algorithm is within 3% of the lower bound.

2   USTAR: Unitig STitch Advanced ConstRuction

2.1   Definitions

For the purpose of this paper, we consider a string made up of characters in $\Sigma = \{A, C, T, G\}$. A string of length k is called k-mer and its *reverse complement* $rc(\cdot)$ is obtained by reversing the k-mer and replacing each character with its complement, that is $A \mapsto T$, $C \mapsto G$, $T \mapsto A$, $G \mapsto C$. Since we don't know from which DNA strand it is taken, we consider a k-mer and its reverse complement as the same k-mer. Given a string $s = \langle s_1, \ldots, s_{|s|} \rangle$, we denote the first i characters of s as $pref_i(s) = \langle s_1, \ldots, s_i \rangle$ and the last i characters of s as $suf_i(s) = \langle s_{|s|-i+1}, \ldots, s_{|s|} \rangle$. We define the *glue* operation between two strings u and v such that $suf_{k-1}(u) = pref_{k-1}(v)$ as the concatenation of u and the suffix of v:

$$u \odot^{k-1} v = u \cdot suf_{|v|-(k-1)}(v)$$

For example, given two 3-mers $u = CTG$ and $v = TGA$, their gluing is $u \odot^2 v = CTGA$.

A set of k-mers can be represented by a de Bruijn graph, of which we will give a node-centric definition, meaning that the arcs are implicitly given by the nodes. Thus we can refer to k-mers and dBG(K) interchangeably.

Given a k-mer set $K = \{m_1, \ldots m_{|K|}\}$, a de Bruijn graph of K is a directed graph dBG(K) = (V, A) in which:

1. $V = \{v_1, \ldots, v_{|K|}\}$
2. each node $v \in V$ has a label $lab(v_i) = m_i$

3. each node $v \in V$ has two different sides $s_v \in \{0, 1\}$, where $(v, 1)$ is graphically represented with a tip
4. a node side (v, s_v) is spelled as

$$spell(v, s_v) = \begin{cases} lab(v) & s_v = 0 \\ rc(lab(v)) & s_v = 1 \end{cases} \tag{1}$$

5. there is an arc between two node sides (v, s_v) and (u, s_u) if and only if there are spellings that share a $(k-1)$-mer. In particular, it must be

$$((v, s_v), (u, s_u)) \in A \iff suf_{k-1}(spell(v, 1 - s_v)) = pref_{k-1}(spell(u, s_u))$$

The right-hand condition is also known as (v, u)-oriented-overlap [21].

Note that nodes' sides allow treating a k-mer and its reverse complement as if they were the same k-mer. Furthermore, nodes can be associated with k-mer counts.

A path $p = \langle (v_1, s_1), \ldots, (v_l, s_l) \rangle$ is spelled by gluing the spelling of its node sides:

$$spell(p) = spell(v_1, s_1) \odot^{k-1} spell(v_1, s_1) \odot^{k-1} \cdots \odot^{k-1} spell(v_l, s_l)$$

The path p is said to be a *unitig* if its internal nodes have in-degree and out-degree equal to 1. A unitig is said to be *maximal* if it cannot be extended on either side. In order to decrease its memory footprint, a dBG(K) can be *compacted* by replacing maximal unitigs with single nodes labeled with the spellings of the unitigs.

An example of compacted dBG(K) with

$$K = \{ACT, CTG, TGA, CTT, TTG, TGC\}$$

is in Fig. 1. It has been compacted by replacing the maximal unitig (CTT, TTG) with the node $CTTG$.

2.2 Vertex-Disjoint Path Cover Problem

Compressing a k-mer set K can be achieved by finding a representation S of K made of strings of any length such that the set of its substrings of length k is equal to K.

A natural way to measure the size of a string set S is by computing its *cumulative length* defined as the sum of all the string lengths:

$$CL(S) = \sum_{s \in S} |s|$$

where $|s|$ is the length of the string s. It has been shown in [4,21] that, when S does not contain duplicate k-mers, the cumulative length of S is proportional to its cardinality, in particular, it holds

$$CL(S) = |K| + (k-1) \cdot |S|$$

where $|S|$ is the cardinality of the set S. Therefore our goal, finding the best representation of a k-mer set K, is equivalent to minimizing the number of strings in the string set S.

Consider again the example in Fig. 1. From the path $p = (ACT, CTG, TGA)$ we can compute its spell $spell(p) = ACTGA$ that contains all the 3-mers ACT, CTG and TGA in p. Thus from a set of paths P that contains all the nodes in dBG(K) we can derive a set S of compressed k-mers. By imposing that all the paths are vertex-disjoint, we guarantee that k-mers are represented only once. Therefore a vertex-disjoint path cover can be used in order to compute S for compression.

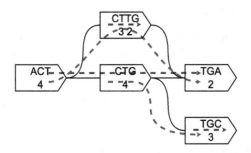

Fig. 1. An example of a compacted de Bruijn graph. Nodes are labeled with k-mers and their counts. Undirected arcs are used in place of two arcs with opposite directions. UST may choose the path cover in red while USTAR is forced to choose the path cover in green. (Color figure online)

Recalling that in order to minimize the cumulative length of a string set S that represents K we need to minimize the number of strings $|S|$ and that it corresponds to the number of paths $|P|$, we can solve the following problem.

Problem 1. Given a de Bruijn graph dBG(K) of a k-mer set K, the **minimum vertex-disjoint path cover problem** is to find the minimum number of vertex-disjoint paths that cover the graph.

For general graphs the problem above is known to be NP-hard [4,21] since it can be reduced from Hamilton: a graph has a Hamiltonian path if and only if it has a vertex-disjoint path cover of cardinality 1. However, it is not clear if the problem is still NP-hard for de Bruijn graphs.

Nevertheless, greedy and non-optimal algorithms have been proposed. ProphAsm [4] uses a simple heuristic that takes an arbitrary k-mer in the dBG(K), and it tries to extend it forward and backward as long as possible and it restarts until it consumes all the k-mers. Similarly, using as input the compacted dBG(K) constructed by BCALM2, UST [21] takes an arbitrary node, tries to extend it forward as long as possible, and restarts until there are available nodes. In the end, UST merges linked paths. Both methods perform a similar strategy by picking the first available k-mer, and without considering the graph structure. If we consider the example in Fig. 1, ProphAsm and UST, by choosing nodes arbitrarily, may build the path cover (in red) $P = \{(ACT, CTG, TGA), (CTT, TTG), (TGC)\}$ from which derives, by computing the spelling of each path, the set of strings $S = \{ACTGA, CTTG, TGC\}$. In this example, the cumulative length is $CL(S) = 12$.

In this work, we present USTAR (Unitig STitch Advanced constRuction) that, unlike previous algorithms, exploits the connectivity of the dBG graph and the values of counts. USTAR also implements a heuristic to compute simplitigs. As UST, also USTAR takes advantage of the compacted de Bruijn graph computed by BCALM2. Similarly to UST and ProphAsm, at each step, USTAR selects a seed node in the graph, and then it tries to compute a path starting from this node. A path is constructed by connecting adjacent nodes until the path cannot be further extended. The algorithm continues with the selection of a new seed node until all nodes have been covered by a path. The two key operations in this algorithm are how to select a good seed node, and how to extend a path among the available connections.

In USTAR the counts associated with each node and the topology properties can be used to determine the best seed and how to extend it. The distribution of counts is in general very skewed, with several low values and few very high values. Since the counts distribution is non-uniform, and skewed, it turns out that higher counts are harder to compress. For this reason, the exploration strategy of USTAR chooses as seed the node that has the highest average counts. In general neighboring nodes usually have similar counts, so that choosing the seed based on the highest average count might improve the compression of these high counts.

As for the path cover construction, we observe that UST and ProphAsm might choose a highly connected node, and since this node will not be available in the subsequent iterations, this selection may lead to isolated nodes, that will increase the cumulative length, like in the example in Fig. 1. Instead, in USTAR we try to avoid this scenario and, in fact, paths are extended by selecting the node with fewer connections so that highly connected nodes are still available for future iterations. These choices will help to have a lower CL since they create fewer and longer simplitigs.

Following the example in Fig. 1, USTAR guarantees that while constructing the first path, the most connected node CTG is avoided. This will produce a cover of the dBG with the paths (in green) $P' = \{(ACT, CTT, TTG, TGA), (CTG, TGC)\}$ and thus a set of strings $S' = \{ACTTGA, CTGC\}$. If we measure the cumulative length of S' we have that $CL(S') = 10$.

$$CL(S') = 10 < CL(S) = 12 < CL(K) = 18$$

Overall, in this example, the uncompressed k-mer set will require $CL(K) = 18$, with UST the k-mer set can be compressed with $CL(S) = 12$, whereas USTAR will produce a better compression with $CL(S') = 10$.

3   Results

In this section, we present a series of experiments in order to find the best tool that compresses k-mers and counts. In our evaluation, we compared USTAR with several other tools: Squeaker, KMC, DSK, BCALM2, and UST. We used for testing a set of real reads datasets taken from previous studies [4,7,11,17,23]. A summary of all datasets is reported in Table 1. For each dataset, we extracted all k-mers (see Table 1) and the corresponding counts, and use this information as input for all compression tools. For some tools, like UST and USTAR, it is required to build a compacted dBG with BCALM2 as a preprocessing.

Table 1. A summary of the read datasets used in the experiments. Datasets are downloaded from NCBI's Sequence Read Archive.

name	description	read length	#reads	size [GB]
SRR001665	Escherichia coli	36	20,816,448	9.304
SRR061958	Human Microbiome 1	101	53,588,068	3.007
SRR062379	Human Microbiome 2	100	64,491,564	2.348
SRR10260779	Musa balbisiana RNA-Seq	101	44,227,112	2.363
SRR11458718	Soybean RNA-seq	125	83,594,116	3.565
SRR13605073	Broiler chicken DNA	92	14,763,228	0.230
SRR14005143	Foodborne pathogens	211	1,713,786	0.261
SRR332538	Drosophila ananassae	75	18,365,926	0.683
SRR341725	Gut microbiota	90	2 5,479,128	1.254
SRR5853087	Danio rerio RNA-Seq	101	119,482,078	3.194
SRR957915	Human RNA-seq	101	49,459,840	3.671

In the first experiment, we ran all compression tools on all datasets for $k = 21$ and we reported the results in Table 2. In all cases, the stored data is additionally compressed using MFCompress [18] for nucleotide sequences or with bzip3 for binary data. In Table 2 are reported the dimensions of the files compressed by the different tools.

We can observe that USTAR is on average the best compressor, and it consistently outperforms the other tools on all datasets. As expected UST and BCALM are the second and third best methods, however, USTAR shrinks the representation by 76% over BCALM and 4.2% over UST.

Since it is clear that UST is the best competitor in the next tests we compare USTAR with UST. We used three different evaluation metrics:

- CL: the cumulative length as defined in Sect. 2, to test the quality before the k-mers compression with the dedicated compressor MFCompress;
- *counts*: the file size of counts after compression with a general-purpose compressor;

Table 2. Datasets (with $k = 21$) are processed with DSK, KMC, Squeaker, BCALM, UST, and USTAR. Nucleotide files are then compressed with MFCompress [18] while other text or binary files are compressed with bzip3. The average file size over all datasets is reported in the last row.

Dataset	DSK	KMC	Squeakr	BCALM	UST	USTAR
SRR001665_1	76,729,965	63,102,295	62,769,135	43,100,358	12,641,658	12,332,551
SRR001665_2	89,356,517	73,618,021	70,364,023	54,549,879	15,492,263	15,109,673
SRR061958_1	1,853,355,280	1,512,526,861	1,214,304,214	792,616,145	194,173,905	185,905,825
SRR061958_2	2,269,394,440	1,850,606,165	1,445,986,432	940,752,737	235,657,588	225,975,765
SRR062379_1	771,892,475	633,615,665	559,244,946	334,386,186	82,713,766	79,283,723
SRR062379_2	766,644,876	629,023,699	556,418,241	327,042,925	80,164,746	76,708,406
SRR10260779_1	594,043,132	489,620,438	459,620,233	272,605,742	64,644,700	61,724,139
SRR10260779_2	661,730,544	545,447,915	501,793,581	311,074,932	72,772,294	69,375,320
SRR11458718_1	660,336,575	547,192,385	515,587,875	278,247,157	64,694,925	61,236,404
SRR11458718_2	699,675,661	580,313,686	542,321,885	304,467,895	68,982,466	65,438,050
SRR13605073_1	286,147,403	236,056,529	244,522,615	110,324,321	25,833,347	24,546,244
SRR14005143_1	72,421,457	59,702,423	75,386,963	26,222,881	6,419,520	6,220,215
SRR14005143_2	148,413,200	121,547,826	126,063,532	51,976,493	13,117,896	12,655,430
SRR332538_1	61,647,503	50,466,675	65,343,140	21,192,576	5,737,778	5,599,034
SRR332538_2	125,336,255	100,440,228	116,698,603	77,057,667	14,410,775	13,528,977
SRR341725_1	972,617,730	799,134,833	700,565,933	262,398,076	80,436,678	78,193,253
SRR341725_2	1,005,087,513	825,643,578	719,993,731	277,159,709	84,250,689	81,877,574
SRR5853087_1	1,494,920,206	1,234,975,195	1,084,532,779	1,073,165,506	191,108,921	177,278,725
SRR957915_1	1,016,375,644	837,315,550	732,334,056	590,259,971	122,748,678	116,872,195
SRR957915_2	1,589,786,146	1,301,318,582	1,062,835,997	829,172,816	182,073,051	172,757,385
Average	760,795,626	624,583,427	542,834,396	348,888,699	80,903,782	77,130,944

– *overall*: the sum of compressed file sizes of k-mers and counts.

Given the metric M the improvement over UST is computed as

$$\Delta M = \frac{M_{UST} - M_{USTAR}}{M_{UST}}$$

where M can be CL, *counts* or *overall*.

In the next experiment, we compared USTAR and UST with these three metrics. In Table 3 we reported the average improvements of USTAR w.r.t. UST when tested on all the above datasets while varying the size of k-mers, using odd lengths as previously done by other authors [4,20,21].

We can observe that for all values of k, USTAR improves over UST, for both nucleotide and count compression. For large values of $k = 31$, the overall advantage of USTAR is 2.30%, and this improvement is mainly achieved with a better compression of counts, in fact, Δcounts $= 12.70\%$. We suspect that with higher values of k the compression might decrease, but more experiments are needed to confirm this hypothesis.

If smaller values of k are considered the overall improvement increases with $k = 21$ and $k = 17$ and it reaches the maximum value of 26.40% for $k = 15$.

Table 3. Average improvements of USTAR w.r.t. UST varying the k-mer size.

k-mer size	Δ CL [%]	Δ counts [%]	Δ overall [%]
15	33.64	12.07	26.40
17	13.61	15.85	13.92
21	2.10	14.17	4.20
31	0.97	12.70	2.30

A similar behavior can be observed for ΔCL, which increases from 0.97% with $k = 31$ to 33.64% with $k = 15$. We can note that the count improvement is roughly constant as k varies with an improvement of 12.07–15.85%. In general, Δoverall is mainly driven by ΔCL.

Since USTAR exploits the structure and connectivity of the dBG graph it is interesting to further study the behavior of USTAR w.r.t. to the number of arcs in the graph. Recall that nodes in a dBG have two sides and that each side can have four arcs, one per nucleotide, thus the maximum number of arcs is $2 \cdot 4 \cdot \#nodes$. Based on this observation we can define the graph density as the number of arcs over the maximum number of arcs:

$$density = \frac{\#arcs}{8 \cdot \#nodes}$$

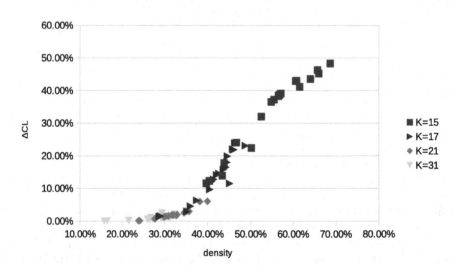

Fig. 2. Improvement on the cumulative length, w.r.t. UST, as a function of the dBG graph density.

In Fig. 2 are shown the improvements in the cumulative length for all datasets with different k-mer sizes, plotted against their density. Each point in the Figure represents a dataset. The plot has a well-defined curve that slowly raises until $density = 35\%$ and then increases almost linearly. ΔCL spans from 0.02% to 48.31% showing that the compression ratio strongly increases as the density increases. We can say that the graph density is very important in determining the improvement in k-mers compression because, with a denser graph, we obtained significantly higher improvements. We also considered other properties such as the number of k-mers, the number of unitigs, the number of isolated nodes, the read length, and the counts variance. However all these features are not as significant as the graph density, and so they are omitted.

Indeed, a higher density graph implies that USTAR has more nodes from which it can choose, unlike UST which picks the first connected k-mer, so that more connected nodes can be available to paths that would have been made up of single nodes. The higher compression ratio achieved by USTAR on denser graphs confirms that the strategy based on node connectivity works well. The fact that UST chooses almost randomly how to extend a path, it cannot guarantee to work well, even with a denser graph there exists the possibility that many single-node paths are generated, which substantially increases the number of paths and thus the cumulative length of compressed k-mers, worsening the compression ratio. If we decrease the k-mer size we observe a higher density: with small k there are more k-mers and it is more likely to have more connections between k-mers leading to an increase in density. In summary, USTAR can compress k-mers and counts with a better compression ratio w.r.t. to several other tools. USTAR is more effective on dense dBG graphs and for small k-mer sizes.

4   Conclusions

In this paper, we have presented USTAR, a tool for compressing k-mer sets with counters. Our approach utilizes a vertex-disjoint path cover to find a representation of the k-mer set that minimizes the cumulative length of the compressed data. By exploring de Bruijn graphs and making informed choices based on node connectivity and average counts, we have achieved better compression ratios compared to existing tools such as UST.

We have evaluated USTAR using various datasets and compared it with several other tools. The results demonstrate that our method consistently outperforms UST, and other tools, in terms of compression ratio, especially for smaller k-mer sizes. The improvements range from 0.97% to 33.64% in terms of cumulative length and it's almost constant for counts. The overall improvement in compressed file size is driven by the reduction in cumulative length.

Furthermore, we have observed that graph density plays a crucial role in determining the effectiveness of USTAR. Denser graphs yield higher improvements, and lowering the k-mer size contributes to increased density. These findings highlight the power of our method, particularly for datasets with higher graph density.

In conclusion, USTAR offers an effective solution for compressing k-mer sets with counters. By leveraging de Bruijn graphs and making informed choices in the construction of the vertex-disjoint path cover, we have achieved superior compression ratios compared to existing tools. Our method has the potential to enhance the storage and processing efficiency of analysis in bioinformatics, enabling more efficient analysis of large-scale genomic data using k-mer-based tools.

Acknowledgments. Authors are supported by the National Recovery and Resilience Plan (NRRP), National Biodiversity Future Center - NBFC, NextGenerationEU.

References

1. Andreace, F., Pizzi, C., Comin, M.: Metaprob 2: metagenomic reads binning based on assembly using minimizers and k-mers statistics. J. Comput. Biol. **28**(11), 1052–1062 (2021). https://doi.org/10.1089/cmb.2021.0270
2. Bankevich, A., et al.: Spades: a new genome assembly algorithm and its applications to single-cell sequencing. J. Comput. Biol. **19**(5), 455–477 (2012)
3. Bradley, P., Den Bakker, H.C., Rocha, E.P., McVean, G., Iqbal, Z.: Ultrafast search of all deposited bacterial and viral genomic data. Nat. Biotechnol. **37**(2), 152–159 (2019)
4. Břinda, K., Baym, M., Kucherov, G.: Simplitigs as an efficient and scalable representation of de Bruijn graphs. Genome Biol. **22**(1), 1–24 (2021)
5. Cavattoni, M., Comin, M.: Classgraph: improving metagenomic read classification with overlap graphs. J. Comput. Biol. **30**(6), 633–647 (2023). https://doi.org/10.1089/cmb.2022.0208, pMID: 37023405
6. Chikhi, R., Holub, J., Medvedev, P.: Data structures to represent a set of k-long DNA sequences. ACM Comput. Surv. (CSUR) **54**(1), 1–22 (2021)
7. Chikhi, R., Limasset, A., Medvedev, P.: Compacting de Bruijn graphs from sequencing data quickly and in low memory. Bioinformatics **32**(12), i201–i208 (2016)
8. Conway, T.C., Bromage, A.J.: Succinct data structures for assembling large genomes. Bioinformatics **27**(4), 479–486 (2011)
9. Denti, L., Previtali, M., Bernardini, G., Schönhuth, A., Bonizzoni, P.: Malva: genotyping by mapping-free allele detection of known variants. Iscience **18**, 20–27 (2019)
10. Harris, R.S., Medvedev, P.: Improved representation of sequence bloom trees. Bioinformatics **36**(3), 721–727 (2020)
11. Kokot, M., Długosz, M., Deorowicz, S.: KMC 3: counting and manipulating k-mer statistics. Bioinformatics **33**(17), 2759–2761 (2017)
12. Marchet, C., Iqbal, Z., Gautheret, D., Salson, M., Chikhi, R.: Reindeer: efficient indexing of k-mer presence and abundance in sequencing datasets. Bioinformatics **36**(Supplement_1), i177–i185 (2020)
13. Marcolin, M., Andreace, F., Comin, M.: Efficient k-mer indexing with application to mapping-free SNP genotyping. In: Lorenz, R., Fred, A.L.N., Gamboa, H. (eds.) Proceedings of the 15th International Joint Conference on Biomedical Engineering Systems and Technologies, BIOSTEC 2022, Volume 3: BIOINFORMATICS, 9–11 February 2022, pp. 62–70 (2022)
14. Monsu, M., Comin, M.: Fast alignment of reads to a variation graph with application to SNP detection. J. Integr. Bioinform. **18**(4), 20210032 (2021)
15. Ondov, B.D., et al.: Mash: fast genome and metagenome distance estimation using minhash. Genome Biol. **17**(1), 1–14 (2016)

16. Pandey, P., Almodaresi, F., Bender, M.A., Ferdman, M., Johnson, R., Patro, R.: Mantis: a fast, small, and exact large-scale sequence-search index. Cell Syst. **7**(2), 201–207 (2018)
17. Pandey, P., Bender, M.A., Johnson, R., Patro, R.: Squeakr: an exact and approximate k-mer counting system. Bioinformatics **34**(4), 568–575 (2018)
18. Pinho, A.J., Pratas, D.: Mfcompress: a compression tool for fasta and multi-fasta data. Bioinformatics **30**(1), 117–118 (2014)
19. Qian, J., Comin, M.: Metacon: unsupervised clustering of metagenomic contigs with probabilistic k-mers statistics and coverage. BMC Bioinform. **20**(367) (2019). https://doi.org/10.1186/s12859-019-2904-4
20. Rahman, A., Chikhi, R., Medvedev, P.: Disk compression of k-mer sets. Algorithms Mol. Biol. **16**(1), 1–14 (2021)
21. Rahman, A., Medvedev, P.: Representation of k-mer sets using spectrum-preserving string sets. In: Schwartz, R. (ed.) RECOMB 2020. LNCS, vol. 12074, pp. 152–168. Springer, Cham (2020). https://doi.org/10.1007/978-3-030-45257-5_10
22. Rhie, A., Walenz, B.P., Koren, S., Phillippy, A.M.: Merqury: reference-free quality, completeness, and phasing assessment for genome assemblies. Genome Biol. **21**, 245 (2020)
23. Rizk, G., Lavenier, D., Chikhi, R.: DSK: k-mer counting with very low memory usage. Bioinformatics **29**(5), 652–653 (2013)
24. Storato, D., Comin, M.: K2mem: discovering discriminative k-mers from sequencing data for metagenomic reads classification. IEEE/ACM Trans. Comput. Biol. Bioinf. **19**(1), 220–229 (2022). https://doi.org/10.1109/TCBB.2021.3117406
25. Sun, C., Harris, R.S., Chikhi, R., Medvedev, P.: Allsome sequence bloom trees. J. Comput. Biol. **25**(5), 467–479 (2018)
26. Sun, C., Medvedev, P.: Toward fast and accurate SNP genotyping from whole genome sequencing data for bedside diagnostics. Bioinformatics **35**(3), 415–420 (2019)
27. Wood, D.E., Salzberg, S.L.: Kraken: ultrafast metagenomic sequence classification using exact alignments. Genome Biol. **15**(3), 1–12 (2014)

Graph-Based Motif Discovery in Mimotope Profiles of Serum Antibody Repertoire

Hossein Saghaian[1]([✉]), Pavel Skums[1], Yurij Ionov[2], and Alex Zelikovsky[1]([✉])

[1] Department of Computer Science, Georgia State University, Atlanta, GA, USA
`ssaghaeiannejadesfa1@gsu.edu, pskums@gsu.edu, alexz@cs.gsu.edu`

[2] Department of Cancer Genetics, Roswell Park Comprehensive Cancer Center, Buffalo, NY, USA
`Yurij.Ionov@roswellpark.org`

Abstract. Phage display technique has a multitude of applications such as epitope mapping, organ targeting, therapeutic antibody engineering and vaccine design. One area of particular importance is the detection of cancers in early stages, where the discovery of binding motifs and epitopes is critical. While several techniques exist to characterize phages, Next Generation Sequencing (NGS) stands out for its ability to provide detailed insights into antibody binding sites on antigens. However, when dealing with NGS data, identifying regulatory motifs poses significant challenges. Existing methods often lack scalability for large datasets, rely on prior knowledge about the number of motifs, and exhibit low accuracy. In this paper, we present a novel approach for identifying regulatory motifs in NGS data. Our method leverages results from graph theory to overcome the limitations of existing techniques.

Keywords: Phage display · Next Generation Sequencing · Microbiome · Regulatory motif · Graph theory

1 Introduction

Phage display (or biopanning) is a technique for studying different protein interactions including protein-DNA, protein-peptide and protein-protein interactions using bacteriophage (a type of virus that only infects bacteria) [31]. In this technique, antibody genes are combined on a strand of DNA. The DNA is then packaged in a protein coat made from bacteriophage. The antibody genes make the antibody hat (receptor), which is attached to the top surface of the virus coat. The virus is called phage and the combination is called phage antibody. Each phage antibody hat is unique and binds to a specific target molecule (for example an antigen, an epitope, or a peptide). Target refers to the substance that is used to scan phage library and template is considered its natural partner. Only the antibody phage hat that fits the shape of a disease target will bind

© The Author(s), under exclusive license to Springer Nature Singapore Pte Ltd. 2023
X. Guo et al. (Eds.): ISRA 2023, LNBI 14248, pp. 214–226, 2023.
https://doi.org/10.1007/978-981-99-7074-2_17

to the target molecule. Changing the antibody genes will change the type of antibody hat and what it can bind to (antigen, or epitope). Many of these antibody phages have been made and the pool contains billions of unique antibody phages. Together all these antibody phages are called phage display library. This pool of antibody phages contains unique receptors for specific target binding and thus, can be screened to reveal specific disease targets. For example, cancer patients' serum can be incubated with phage library to reveal cancer specific epitopes. Once antibody phages bind with specific targets, they can be pulled out and further replicated using a host bacteria [22].

Geysen et al. [12] first identified peptide binding to target and mimicking the binding site on the template, which was called mimotope. Mimotope is useful in many applications such as epitope mapping [32], vaccines [16], therapeutics [21], defining drug target [27], protein network detection [33, 34]. As a result of exposure to antigenic proteins, patient's immune system produces antibodies, which could be used as biomarkers for cancer or viral infection detection [38]. Molecular interactions between antigen's epitope and antibody are often mediated by short linear motifs of the amino acid sequence of the antigen [9, 18]. Such interactions could be experimentally detected using random peptide phage display libraries.

Traditional phage display is laborious and prone to finding false positive hits. In recent years, many studies have been devoted to taking advantage of Next Generation Sequencing (NGS) technique in the analysis of phage display screens [7, 26]. NGS enables phage display screening to produce huge number of outputs (short peptides). Another contribution of NGS to the analysis of phage display screening is that it accelerates and improves selection process and therefore, avoids repetitive selections and restricts the number of false positive hits, in contrast with traditional phage display [36]. In effect, a library of all possible peptides of fixed length is generated, and peptides recognized by antibodies contained in the human serum are selected, amplified in bacteria and sequenced using NGS [5, 11]. Such methods produce data sets consisting of hundreds of thousands of peptide sequences. The computational problem consists in discovery of true binding motifs corresponding to epitopes related to the diagnosed disease.

The problem of detection of epitope-specific binding motifs from NGS data is computationally challenging for several reasons. The generated data is large and usually noisy, as a result of biopanning; mimotopes which are considered as desired signals are mixed with target unrelated peptides (TUPs) that are undesired signals. Thus, a significant portion of sequenced peptides is not related to the repertoires of antibody specificities, but produced by nonspecific binding and preferential amplification in bacteria [11]. High heterogeneity of antigen and antibody populations, as well as antibody-antigen recognition poly-specificity manifests itself in presence of multiple binding motifs of various lengths within the same data set [18, 35].

The development of regulatory motif finding algorithms began in the late 1980s and early 1990s, when researchers began using computational methods to identify and analyze patterns in DNA sequences. Early motif finding algorithms used sequence patterns to identify conserved motifs. One of the first algorithms was Gibbs Sampling, developed by Lawrence et al. in 1993 [19]. This algorithm used a probabilistic approach to identify common patterns in a set of DNA sequences. In the mid-1990s, researchers developed a new way to represent motifs called position weight matrices (PWMs). PWMs show the probability of each nucleotide at each position in a motif. This information can be used to develop algorithms like MEME [4] and AlignACE [29], which can find motifs in DNA sequences. In addition, machine learning techniques have been used to develop new motif finding algorithms that are based on classification and regression models. These models, such as support vector machines (SVMs), hidden Markov models (HMMs), and neural networks, have been used to discover motifs in DNA sequences. Some examples of machine learning-based algorithms include SVMotif [17], MDscan-Motif [20], and DeepBind [1].

The tools that have been developed to address the limitations of existing motif finding methods use a variety of algorithmic techniques, such as clustering [11,18], Gibbs sampling [2,24], artificial neural networks [23], and mixture model optimization [15]. However, these methods face serious challenges, when dealing with NGS data: many of them are not scalable for large data sets, require prior knowledge about the number of motifs to identify, and have low accuracy [11,18]. For instance, MEME [4] requires the number of motifs to be known as an input parameter, and it has low accuracy on random peptide phage display libraries (RPPDL).

We introduce a method for finding regulatory motifs, which relies on results from graph theory. In order to evaluate our method, we generated samples of sequences with different lengths. We planted predefined motifs into the samples and applied our algorithm to simulated data. We also applied our method to mouse microbiome data including two groups of 5 mice. Results indicate that our graph-based approach successfully identifies motifs in replicates samples.

In the following, we will introduce our method. We will then validate our approach on both simulated and real data, and analyze the results.

2   Method

Our graph-based method uses the concept of graphs in mathematics to formulate the problem of motif discovery. In this method, we construct a directed graph in which nodes are $k-$mers. We assume that $k = 4$ and thus, we deal with 4$-$mers in this research. The reason is that all peptides are of length 7 and according to [14] choosing $k = 4$ results in more statistically significant motifs.

For given set of peptides, we find all 4$-$mers of that set. By 4$-$mer we mean a string of amino acids of length 4 in which the first and the last positions occupied by amino acid and the second and the third positions either amino acid or gap (dash). In other words, at most, one deletion is allowed in the 2nd or 3rd positions. We build a graph in which vertices are 4$-$mers. Two vertices are connected by an edge, if there are enough M number of peptides supporting simultaneously both 4$-$mers. The value of M is a parameter that can be adjusted depending on how many peptides we expect to form a motif.

Let u, x be two 4−mers belonging to the same peptide P, and the first position of u strictly precedes the first position of x in P. Then there is a directed edge $u \to x$. Together u and x can make a 5−, 6− or 7−subset $k - ux$ inside P. This $k-$ subset forms an edge $(k - ux)$, if the first amino acid belongs to u and the last belongs to x. The support of an edge $(k - ux)$ is equal to the number of peptides containing $k-$ subset $k - ux$. We consider only edges with the support greater than M.

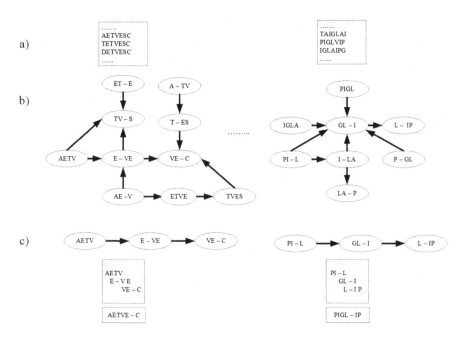

Fig. 1. Schematic of the Graph-based method. a) A sample directed graph made from 4−mers as vertices. Two vertices are connected if they both belong to a peptide. Direction of the edge is from the 4−mer that fills lower indexes to the 4−mer that occupies higher indexes when align to the peptide. b) Peptides that contribute to particular part of the graph. c) Examples of paths of length 2 (2− hops) in the graph. 4−mers are aligned and as a result, $k-$ subsets corresponding to paths are created.

After building the graph, we extract all paths of the graph with length equal to 2, i.e. we find all 2-hop paths of the graph (see Fig. 1−c). Each path uxv contains two directed edges with $(k1 - ux)$ and $(k2 - xv)$ subsets with enough number of M supports. Together $(k1-ux)$ and $(k2-xv)$ make a $(k-uxv)$ subset of the whole path. We assume that support of a path uxv is equal to minimum of {support (ux), support (xv)}. In the next step, we align all peptides that support the path and align them to the achieved $k-$subset $(k - uxv)$. As a result, for each path of length 2, we obtain a set of aligned peptides which accounts for a motif and we can represent this motif with a position probability matrix. The following formula is used to calculate probability $P(Aa)$ of each amino acid at an individual position of the motif

$$P(Aa) = \frac{N_A a + \frac{p}{n}}{\sum N + p} \tag{1}$$

where $N_A a$ is the number of amino acid at each position, N is the total number of peptides contributing in motif, p is the pseudo count which is added to the nominator and denominator of the fraction (in this study value of p is considered to be 1), and n is the total number of amino acids ($n = 20$). The reason for applying pseudo counts to the formula above is that in some positions, counts of one particular amino acid would be zero (in contrast with other positions), which results in probability of some significant motifs to be zero.

In studying motifs, some positions might be more important and subsequently, contain more information. To explore this, information content matrix is calculated based on Shannon entropy equation [3]. Alternatively, in some researches, information content is represented as relative entropy, Kullback-Leibler divergence (KL divergence). In contrast with Shannon entropy, KL divergence takes into account the non-uniform background frequencies. The relative entropy is calculated using the following equation:

$$IC(Aa) = P(Aa) \times \log_2 \frac{P(Aa)}{B_A a} \tag{2}$$

in which $B_A a$ is the background frequency of amino acid a. In this study, we accounted for the non-uniform background frequencies of amino acids (i.e., the probability of each amino acid is not equal). To do this, we calculated the frequency of each amino acid at each position of $7-$mers in each sample. In equation 2, IC can take negative values. In order to avoid negative values, we simply replace them with zero.

2.1   Motif Validation

Methods of quantifying similarity between motifs include (but not limited to) Pearson Correlation Coefficient (PCC) [25], Average Log-Likelihood Ration (ALLR) [37], Fisher-Irwin exact test (FIET) [30], Kullback-Leibler divergence (KLD) [28], Euclidean distance [6] and Tomtom (E value) [13]. We used Pearson Correlation Coefficient to measure if two motifs are identical. The Pearson correlation coefficient (PCC) is a measure of the linear relationship between two motifs. It is calculated by comparing the occurrence profiles of the two motifs across a set of sequences. A high positive PCC indicates that the two motifs are similar, while a low or negative PCC indicates that they are dissimilar. To this end, we selected a threshold correlation number *correlation* $\geq$ 0.75. Any two motifs with PCC greater than 0.75 are assumed identical motifs. Accordingly, we calculated the number of retrieved motifs. The reason we used the Pearson correlation coefficient over other alternative approaches in this study is that we represent motifs as position probability matrices. Employing the Pearson correlation coefficient for comparing two matrices aligns most effectively with our graph-based methodology.

In order to verify the discovered motifs obtained through our graph-based method, we conducted tests using simulated data. The simulated data was generated by creating several position probability matrices, which were used to generate sets of 7-mers. These 7-mers served as the intentionally placed motifs within our simulations. If any gaps were present in the motifs, we filled them with random amino acids. Subsequently, we applied a graph-based approach to analyze the simulated set of peptides. We utilized simulated data to evaluate the performance of our graph-based method in detecting the target motifs. By creating position probability matrices and generating corresponding peptides, we introduced noise and assessed the algorithm's ability to identify the desired motifs in the presence of such noise.

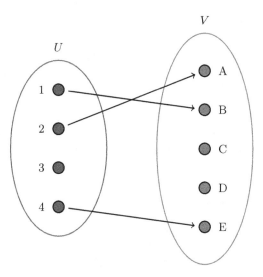

Fig. 2. Set U represents set of true planted motifs, set V represents set of retrieved motifs. When motifs are retrieved (in set V), Gale-Shapley algorithm [10] is used to do matching between set U and set V. Each motif in set U is matched with only one motif in set V with whom it has highest Pearson Correlation Coefficient

In Fig. 2, the U set represents the planted motifs, while the V set represents the collection of all retrieved motifs. To determine the correlation between the members of set U and set V, we employed the Pearson correlation coefficient. The matching between the two sets was accomplished using the Gale-Shapley algorithm [10] (Table 1).

Table 1. Confusion matrix built after matching related to Fig. 2

		Actual	
		Positive	Negative
Predicted	Positive	A, B, E	C, D
	Negative	3	*NotApplicable*

3 Results and Discussion

3.1 Data Set

Our mouse microbiome data is the mixture of the 3 libraries: $M2$, $L2$ and $M1$. M means that the IgM antibodies were used for analysis and the L means that antibody were isolated from the same serum samples by using protein L, which do not discriminate between classes of antibodies. With IgM antibodies the experiment was repeated with the same serum samples two times, this is why there are the $M2$ and the $M1$ libraries. L library include all antibodies (IgG, IgA, IgE) except the IgM. We expect that IgM repertoire is the most sensitive to the environment changes and we should find real differences in the IgM repertoire that is in the $M2$ and the $M1$ samples. So, totally we have 3 libraries $M2$, $L2$ and $M1$. Each library consist of 24 serum samples obtained from two groups of mice. Prior to time $t = 0$, all mice from both groups were maintained at 22 °C. At t = 0, serum samples were collected from mice in both groups, resulting in samples $S1$–$S6$ and $S7$–$S12$. Each group initially contained 5 mice, but the serum samples from mouse 5 in the first group and mouse 10 in the second group were duplicated as technical replicas. Thus, there were 6 profiles in each group.

After the initial bleeding, the first group of mice was kept at 22 °C, and the second bleeding occurred after 6 weeks ($t = 6w$) at the same temperature, producing samples $S13$–$S18$. Again, serum samples from mouse 5 were duplicated. Meanwhile, the second group of mice was kept at 30 °C, and after 6 weeks at $t = 6w$, samples $S19$–$S24$ were collected. Serum from mouse 10 was duplicated in this case (Fig. 3).

The phage DNA used for insert sequencing was derived from antibody-bound phages immediately after the initial incubation of the phage library with serum antibodies. This process was carried out without amplifying the isolated antibody-binding phages in bacteria. Consequently, the number of sequencing reads cannot serve as a quantitative measure of the antibody titer. This limitation arises because, in many cases, the number of corresponding antibodies far exceeds the number of their specific targets.

To create quantitative profiles, we propose quantifying the number of distinct peptide sequence variants or determining the size of the peptide family related to each motif. This approach allows us to calculate how frequently each motif appears in each profile, enabling the generation of a motif signature that can differentiate between housing at 22 °C and 30 °C. In summary, to discern

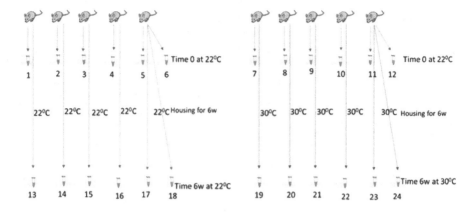

Fig. 3. Mouse microbiome data

the impact of temperature conditions, we focus on measuring motif occurrences and constructing signature profiles rather than relying on sequencing reads or antibody titers.

3.2   Results

Simulated Data: In order to confirm the motifs identified through our graph-based method, we conducted tests using simulated data. The simulated data was generated by using multiple position probability matrices, which were then utilized to generate sets of 7-mers. Essentially, we created a pool of diverse position probability matrices, each representing an identical motif with varying levels of information content associated with it. We did this by collecting all of the position probability matrices that were generated when we applied our method to real data. Subsequently, we randomly selected a specific number of matrices from this pool and generated a corresponding number of 7-mer peptides based on each selected matrix.

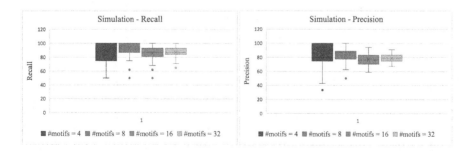

Fig. 4. Whisker bar plot of the simulated data. Number of planted motifs is on the x axis while the y axis shows recall and precision for each group.

To address any gaps within the motif, we inserted random amino acids, thereby introducing noise to the simulated data. Next, we applied a graph-based approach to analyze the set of generated peptides. The objective was to determine whether the graph-based algorithm was able to detect the intended target motifs or not. We conducted a series of experiments involving the insertion of sets of 4, 8, 16, and 32 motifs. Each set was tested 100 times, and the recall (sensitivity) was calculated for each experiment. Figure 4 visualizes the results of all simulations through a whisker bar plot.

Real Data: The primary objective of our research was to investigate any qualitative or quantitative differences in the profiles associated with the second bleeding of both the first and second groups of mice, which were respectively kept at 22 °C and 30 °C. The focus was to determine if variations in both time and temperature would lead to significant differences or patterns indicating an increase or decrease in motifs within the samples. However, our analysis did not reveal any significant differences or discernible patterns that would suggest an increase or decrease in motifs across the samples. Despite the variations in time and temperature, the motifs in the samples did not exhibit notable changes in their quantity or quality.

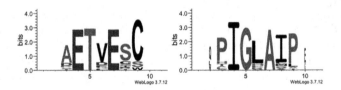

Fig. 5. Two technologically inserted motifs were successfully identified in all 24 samples with graph-based method

To further explore the data, we employed 4−mer sequences instead of motifs, as most motifs comprised consensus sequences of 4 amino acids. There were a total of 160, 000 distinct 4-mers, and for each profile, we calculated the number of occurrences of each tetra peptide. This involved determining how many different 7−mer sequences contained a particular tetra peptide in each profile, generating a signature of tetra peptides related to housing at 30 °C. The tetra peptides could exhibit either increased or decreased numbers with the temperature shift. However, we did not find a meaningful relationship between the samples in terms of increasing or decreasing tetra peptides.

It's important to note that although no significant differences were observed in this study, negative results are valuable as they contribute to the understanding of the regulatory motifs' behavior under specific conditions. These findings indicate that the specific motifs we investigated were not significantly affected by the time and temperature variations in the experiment. Additionally, it is worth highlighting that our method successfully identified two technologically inserted motifs ("IGLAEIP" and "AETVESC") in all of the samples (see Fig. 5). This discovery suggests that our method is valid and capable of detecting important motifs. Comparatively, we also applied a well-known motif discovery method called MEME [4] to analyze all 24 samples. However, MEME failed to identify these two motifs in all of the 24 samples.

In Fig. 6, we have summarized a number of discovered motif logos using our graph-based and MEME.

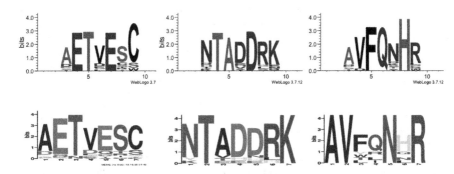

Fig. 6. The first row shows some sample motifs discovered by our graph-based method. The second row is assigned to motifs discovered by MEME. The logos of the first row was created using a version of WebLogo [8] modified to display aligned pairs of Logos

3.3   Discussion

To evaluate the results, we obtained a set of results for different values of three critical parameters: the number of peptides that support a two-hop path in the graph (which we call the "support path"), the information content of the planted motifs, and the correlation number. These parameters are all assumed to have a significant impact on the measurements.

We first hypothesized that increasing the path support would result in stronger motifs that were easier to retrieve. We started with a path support of 4 and increased it from there. We saw that the number of extracted motifs increased until we reached a certain point. However, the results remained unchanged when we chose higher values for the path support. This is because target motifs are mainly clustered in groups of a certain number of peptides. For example, if we start with a path support of 4, meaning that there are at least 4 peptides supporting the path, as we increase the path support, more target motifs will be detected by the graph-based method. However, from a certain number of path support onward, we get plenty of non-targeted motifs and the number of successfully detected target motifs will not increase.

We planted a variety of motifs with different information contents in our simulated data. As we expected, motifs with higher information contents were more likely to be retrieved than those with lower information contents.

Our graph-based motif discovery model has several advantages over existing methods. First, it is scalable to large datasets, making it possible to find motifs in hundreds of thousands of peptides in a reasonable amount of time. Second, it does not require prior knowledge of the number of motifs, unlike the MEME method. Third, it has been shown to be more accurate on simulated data than the MEME method.

In summary, our research did not yield significant differences or patterns in the quantity or quality of motifs or tetra peptides when comparing the serum samples collected after 6 weeks and between samples kept at 22 °C and 30 °C. These results suggest that the temperature shift did not have a noticeable impact on the identified motifs or tetra peptides in the samples.

References

1. Alipanahi, B., Delong, A., Weirauch, M.T., Frey, B.J.: Predicting the sequence specificities of dna-and rna-binding proteins by deep learning. Nat. Biotechnol. **33**(8), 831–838 (2015)
2. Andreatta, M., Lund, O., Nielsen, M.: Simultaneous alignment and clustering of peptide data using a gibbs sampling approach. Bioinformatics **29**(1), 8–14 (2013)
3. Ash, R.B.: Information Theory. Courier Corporation, North Chelmsford (2012)
4. Bailey, T.L., Elkan, C., et al.: Fitting a mixture model by expectation maximization to discover motifs in bipolymers (1994)
5. Bratkovič, T.: Progress in phage display: evolution of the technique and its applications. Cell. Mol. Life Sci. **67**(5), 749–767 (2010)
6. Choi, I.G., Kwon, J., Kim, S.H.: Local feature frequency profile: a method to measure structural similarity in proteins. Proc. Natl. Acad. Sci. **101**(11), 3797–3802 (2004)
7. Christiansen, A., et al.: High-throughput sequencing enhanced phage display enables the identification of patient-specific epitope motifs in serum. Sci. Rep. **5**(1), 1–13 (2015)
8. Crooks, G.E., Hon, G., Chandonia, J.M., Brenner, S.E.: Weblogo: a sequence logo generator. Genome Res. **14**(6), 1188–1190 (2004)
9. Dinkel, H., et al.: The eukaryotic linear motif resource elm: 10 years and counting. Nucleic Acids Res. **42**(D1), D259–D266 (2014)
10. Gale, D., Shapley, L.S.: College admissions and the stability of marriage. Am. Math. Mon. **69**(1), 9–15 (1962)
11. Gerasimov, E., Zelikovsky, A., Măndoiu, I., Ionov, Y.: Identification of cancer-specific motifs in mimotope profiles of serum antibody repertoire. BMC Bioinf. **18**(8), 1–6 (2017)
12. Geysen, H.M., Rodda, S.J., Mason, T.J.: A priori delineation of a peptide which mimics a discontinuous antigenic determinant. Mol. Immunol. **23**(7), 709–715 (1986)
13. Gupta, S., Stamatoyannopoulos, J.A., Bailey, T.L., Noble, W.S.: Quantifying similarity between motifs. Genome Biol. **8**, 1–9 (2007)
14. Ionov, Y., Rogovskyy, A.S.: Comparison of motif-based and whole-unique-sequence-based analyses of phage display library datasets generated by biopanning of anti-borrelia burgdorferi immune sera. PLoS ONE **15**(1), e0226378 (2020)

15. Kim, T., et al.: Musi: an integrated system for identifying multiple specificity from very large peptide or nucleic acid data sets. Nucleic Acids Res. **40**(6), e47–e47 (2012)
16. Knittelfelder, R., Riemer, A.B., Jensen-Jarolim, E.: Mimotope vaccination-from allergy to cancer. Expert Opin. Biol. Ther. **9**(4), 493–506 (2009)
17. Kon, M.A., Fan, Y., Holloway, D., DeLisi, C.: Svmotif: a machine learning motif algorithm. In: Sixth International Conference on Machine Learning and Applications (ICMLA 2007), pp. 573–580. IEEE (2007)
18. Krejci, A., Hupp, T.R., Lexa, M., Vojtesek, B., Muller, P.: Hammock: a hidden markov model-based peptide clustering algorithm to identify protein-interaction consensus motifs in large datasets. Bioinformatics **32**(1), 9–16 (2016)
19. Lawrence, C.E., Altschul, S.F., Boguski, M.S., Liu, J.S., Neuwald, A.F., Wootton, J.C.: Detecting subtle sequence signals: a gibbs sampling strategy for multiple alignment. Science **262**(5131), 208–214 (1993)
20. Liu, X.S., Brutlag, D.L., Liu, J.S.: An algorithm for finding protein-dna binding sites with applications to chromatin-immunoprecipitation microarray experiments. Nat. Biotechnol. **20**(8), 835–839 (2002)
21. Macdougall, I.C., et al.: A peptide-based erythropoietin-receptor agonist for pure red-cell aplasia. N. Engl. J. Med. **361**(19), 1848–1855 (2009)
22. Murphy, K., Weaver, C.: Janeway's Immunobiology. Garland Science, New York City (2016)
23. Nielsen, M., Lund, O.: Nn-align. an artificial neural network-based alignment algorithm for mhc class ii peptide binding prediction. BMC Bioinf. **10**(1), 296 (2009)
24. Nielsen, M., et al.: Improved prediction of mhc class i and class ii epitopes using a novel gibbs sampling approach. Bioinformatics **20**(9), 1388–1397 (2004)
25. Pietrokovski, S.: Searching databases of conserved sequence regions by aligning protein multiple-alignments. Nucleic Acids Res. **24**(19), 3836–3845 (1996)
26. Rentero Rebollo, I., Sabisz, M., Baeriswyl, V., Heinis, C.: Identification of target-binding peptide motifs by high-throughput sequencing of phage-selected peptides. Nucleic Acids Res. **42**(22), e169 (2014)
27. Rodi, D.J., Janes, R.W., Sanganee, H.J., Holton, R.A., Wallace, B., Makowski, L.: Screening of a library of phage-displayed peptides identifies human bcl-2 as a taxol-binding protein. J. Mol. Biol. **285**(1), 197–203 (1999)
28. Roepcke, S., Grossmann, S., Rahmann, S., Vingron, M.: T-reg comparator: an analysis tool for the comparison of position weight matrices. Nucleic Acids Res. **33**(suppl_2), W438–W441 (2005)
29. Roth, F.P., Hughes, J.D., Estep, P.W., Church, G.M.: Finding dna regulatory motifs within unaligned noncoding sequences clustered by whole-genome mrna quantitation. Nat. Biotechnol. **16**(10), 939–945 (1998)
30. Schones, D.E., Sumazin, P., Zhang, M.Q.: Similarity of position frequency matrices for transcription factor binding sites. Bioinformatics **21**(3), 307–313 (2005)
31. Smith, G.P.: Filamentous fusion phage: novel expression vectors that display cloned antigens on the virion surface. Science **228**(4705), 1315–1317 (1985)
32. Smith, G.P., Petrenko, V.A.: Phage display. Chem. Rev. **97**(2), 391–410 (1997)
33. Thom, G., et al.: Probing a protein-protein interaction by in vitro evolution. Proc. Natl. Acad. Sci. **103**(20), 7619–7624 (2006)
34. Tong, A.H.Y., et al.: A combined experimental and computational strategy to define protein interaction networks for peptide recognition modules. Science **295**(5553), 321–324 (2002)

35. Van Regenmortel, M.H.V.: Specificity, polyspecificity and heterospecificity of antibody-antigen recognition. In: HIV/AIDS: Immunochemistry, Reductionism and Vaccine Design, pp. 39–56. Springer, Cham (2019). https://doi.org/10.1007/978-3-030-32459-9_4

36. Wang, L.F., Yu, M.: Epitope identification and discovery using phage display libraries: applications in vaccine development and diagnostics. Curr. Drug Targets **5**(1), 1–15 (2004)

37. Wang, T., Stormo, G.D.: Combining phylogenetic data with co-regulated genes to identify regulatory motifs. Bioinformatics **19**(18), 2369–2380 (2003)

38. Zhong, L., Coe, S.P., Stromberg, A.J., Khattar, N.H., Jett, J.R., Hirschowitz, E.A.: Profiling tumor-associated antibodies for early detection of non-small cell lung cancer. J. Thorac. Oncol. **1**(6), 513–519 (2006)

Sequence-Based Nanobody-Antigen Binding Prediction

Usama Sardar[1], Sarwan Ali[2], Muhammad Sohaib Ayub[1], Muhammad Shoaib[1], Khurram Bashir[1], Imdad Ullah Khan[1], and Murray Patterson[2(✉)]

[1] Lahore University of Management Sciences, Lahore, Pakistan
usamasardar2022@gmail.com, {15030039,mshoaib,khurram.bashir,
imdad.khan}@lums.edu.pk
[2] Georgia State University, Atlanta, GA, USA
{sali85,mpatterson30}@gsu.edu

Abstract. Nanobodies (Nb) are monomeric heavy-chain fragments derived from heavy-chain only antibodies naturally found in Camelids and Sharks. Their considerably small size (∼3–4 nm; 13 kDa) and favorable biophysical properties make them attractive targets for recombinant production. Furthermore, their unique ability to bind selectively to specific antigens, such as toxins, chemicals, bacteria, and viruses, makes them powerful tools in cell biology, structural biology, medical diagnostics, and future therapeutic agents in treating cancer and other serious illnesses. However, a critical challenge in nanobodies production is the unavailability of nanobodies for a majority of antigens. Although some computational methods have been proposed to screen potential nanobodies for given target antigens, their practical application is highly restricted due to their reliance on 3D structures. Moreover, predicting nanobody-antigen interactions (binding) is a time-consuming and labor-intensive task. This study aims to develop a machine-learning method to predict Nanobody-Antigen binding solely based on the sequence data. We curated a comprehensive dataset of Nanobody-Antigen binding and non-binding data and devised an embedding method based on gapped k-mers to predict binding based only on sequences of nanobody and antigen. Our approach achieves up to 90% accuracy in binding prediction and is significantly more efficient compared to the widely-used computational docking technique.

Keywords: Nanobody · Antigen · Classification · k-mers · Binding Prediction

1 Introduction

Nanobodies (Nbs) are single-domain antibodies (sdAb), derived from heavy-chain only antibodies naturally occurring in Camelids and Sharks. They represent a unique class of proteins/antibodies having a molecular weight of 12–15

U. Sardar and S. Ali—Equal Contribution.

© The Author(s), under exclusive license to Springer Nature Singapore Pte Ltd. 2023
X. Guo et al. (Eds.): ISBRA 2023, LNBI 14248, pp. 227–240, 2023.
https://doi.org/10.1007/978-981-99-7074-2_18

kDa that combine the advantageous characteristics of conventional antibodies with desirable attributes of small-molecule drugs. Nbs are remarkably adaptable to various applications and offer several advantages over conventional antibodies [6]. Every Nb contains the distinct structural and functional properties found in naturally-occurring heavy-chain antibodies. They have a naturally low potential for causing immune responses and exhibit high similarity to variable regions of the heavy chain (VH) in human antibodies, making them excellently suited for therapeutic and diagnostic applications. Due to their small size, unique structure, and high stability, Nbs can access targets that are beyond the reach of conventional antibodies and small-molecule drugs [7,23]. Nbs Structure prediction and modeling are still challenging tasks [5,29]. Hundreds of Nb crystallographic structures have been deposited in the Protein Data Bank (PDB) [3,4]. Despite this, the current representation falls short of capturing the vast structural and sequence diversity observed in Nb hypervariable loops. Moreover, Nbs display a greater range of conformational variations, lengths, and sequence variability in their CDR3 compared to antibodies [17]. This makes modeling and prediction of their 3D structure more complex.

Machine Learning (ML) plays a crucial role in predicting nanobody-antigen(Nb-Ag) interactions. ML offers a powerful and effective approach to analyzing and comprehending complex patterns within extensive datasets [8]. Traditional non-computational methods for determining Nb-Ag interactions can be both costly and time-consuming. ML provides a faster and more cost-effective alternative, enabling scientists to prioritize potential nanobodies candidates for further research [28]. The large amount of training data that is easily accessible when utilizing sequence-based ML techniques is advantageous. Even though there is an increasing amount of data on protein structures, most Nb-Ag sequences still lack validated structural details, even though the number of protein sequence entries is still rising quickly [11,26].

ML algorithms are capable of handling a vast amount of data, encompassing nanobodies and antigen sequences, structural information, and experimental binding data. Through the examination of this data, ML algorithms can find complex relationships and patterns that might be hard for people to see on their own. These methods can automatically extract relevant details from raw data, such as structural information or amino acid sequences. This feature extraction method makes it easier to spot important molecular traits that influence antibody and Nb-Ag interaction [20,30]. ML models can gain knowledge from large databases of training samples and produce precise predictions. By being trained on known Nb-Ag binding data, these models may understand the underlying principles and patterns regulating binding interactions. This allows them to make accurate predictions for previously undiscovered Nb-Ag combinations. Traditional experimental techniques, on the other hand, need a lot of time and money to determine Nb-Ag binding. To arrange possible antibody candidates for further investigation, ML offers a quicker and more economical alternative [16,22,30,31].

Nanobody-antigen binding play a significant role in the immune response. By predicting and studying these bindings, researchers can gain insights into how nanobodies recognize and neutralize specific antigens. This understanding is fundamental for elucidating immune mechanisms and developing strategies for diagnostics to combat infectious diseases, autoimmune diseases, and cancer. Predicting Nb-ag binding can aid in discovering and engineering nanobodies with desired properties. Binding helps identify the essential antigens for vaccine development, effective vaccine formulations, and understanding the mechanisms of immune protection.

Predicting binding interactions enables the selection of highly specific nanobodies and antigens for accurate and sensitive diagnostic tests. Binding can guide the rational design of therapeutic nanobodies or nanobodies-based drugs and diagnostic tests. Scientists can modify or engineer nanobodies to improve their affinity, selectivity, and therapeutic potential by understanding the binding interactions between nanobodies and their target antigens. This approach can be applied in areas such as diagnostic tests and cancer immunotherapy, where nanobodies are designed to target specific tumor antigens [27].

The input in our study is nanobody sequences and antigen sequences and the output is binding/docking score (Yes/No).In this paper, we trained ML on nanobodies and antigen sequences extracted from the single-domain antibody (sdAb) database to determine binding. We make the following contributions.

- We have performed the comparison of various ML approaches for predicting nanobody-antigen binding from sequences only.
- We evaluated the impact of various sequence features (e.g., isoelectric point, hydrophilicity) on the prediction accuracy of ML models.
- We have curated a dataset of nanobody-antigen pairs for training and testing machine-learning models and made it publicly available for further research.

The rest of the paper is organized as follows: Sect. 2 explores the existing research and highlights the research gaps. Section 3 explains the data collection, feature extraction, and data visualization. The proposed embedding is discussed in Sect. 4. Section 5 describes the ML models and evaluation metrics. We discuss our results in Sect. 6. Finally, we provide the conclusion and future directions in Sect. 7.

2 Related Work

Antibodies (Abs) are crucial tools in biological research and the biopharmaceutical industry due to their exceptional binding specificity and strong affinity for target antigens. The effectiveness of the immune system directly reflects the diversity of antigens against which specific tightly binding 'B-lymphocyte antigen receptors (BCRs) can be generated. The vast range of binding specificity is achieved through sequence variations in the heavy chain (VH) and light chain (VL), resulting in an estimated diversity of BCRs in humans [18] that surpasses the population size of B-lymphocytes in an individual. However, it is still unclear how this immense sequence diversity translates into antigen specificity. Although not every unique combination of VH-VL sequences leads to a distinct binding specificity, predicting the number and positions of amino acid mutations required to change binding specificity has proven challenging [21].

A more manageable system is offered by heavy-chain antibodies found in camelid species like camels, llamas, and alpacas, where the light chain is absent. These antibodies, known as nanobodies (Nbs), consist of an isolated variable VHH domain that is about ten times smaller than conventional Abs but retains comparable binding specificity [19].

Both Nbs and Abs face the fundamental challenge of deciphering the molecular code that links amino acid sequence, particularly the choice of paratope residues, to the binding specificity of the folded molecule. In regular Abs, the paratope is situated at the interface of the VH and VL domains, typically comprising residues from up to six distinct hypervariable loop regions. The VH and VL domains can dock together in various ways, allowing the antibody to maximize the diversity of potential antigen-binding surfaces. In contrast, the Nb paratope is entirely contained within the VHH domain, significantly limiting the range of possible antigen-binding surfaces without seemingly affecting the diversity of resulting binding specificities. Indeed, Nbs typically bind their target antigens with affinities comparable to classical monoclonal Abs [18].

Several studies have been conducted to generate antibodies/nanobodies using non-computational methods. Experimental techniques such as hybridoma technology and phage display [25] are used to generate specific antibodies/nanobodies, but these have limitations and challenges. Hybridoma technology involves immunizing animals with an antigen and fusing B cells from the immunized animal with cancer cells to create hybridoma cells that produce specific antibodies. However, this method raises ethical concerns due to animal cruelty and is time-consuming, labor-intensive, and limited in antibody/nanobody diversity. On the other hand, phage display utilizes bacteriophages to display antibody fragments, but it also has time-consuming rounds of selection and amplification, labor-intensive requirements, and high costs.

Several ML methods are used to predict nanobody-antigen binding. Sequence-based methods utilize amino acid sequences, extracting features such as physio-chemical properties, sequence motifs, and sequence profiles [28]. These features are then used as input for machine learning algorithms like support vector machines (SVM), random forests, or neural networks. Structure-based methods employ three-dimensional structures obtained from experimental techniques like X-ray crystallography or homology modeling. Structural features like solvent accessibility, electrostatic potential, or shape complementarity are extracted and fed into machine-learning models. Hybrid methods combine sequence-based and structure-based features, integrating both sequence and structural information to capture a broader range of characteristics. Deep learning methods, such as convolutional neural networks (CNN) and recurrent neural networks (RNN), learn complex patterns and relationships from large datasets, including sequence and structural information, for accurate predictions [5]. Docking-based methods use molecular docking algorithms to predict the binding orientation and affinity by calculating a binding score based on the optimal spatial arrangement of the interacting molecules [22].

3   Proposed Approach

The proposed pipeline comprised different steps, including data collection, Nb-Ag sequence analysis, numerical embedding generation, and optimal feature extraction from the Nb and Ag sequences. We discuss each step in detail.

3.1   Data Collection

We collected 47 Ag sequences from UniProt[1], and for each, we collected all binding Nbs from Single Domain Antibody Database[2], which are total 365, as shown in Table 1 along with the basic statistics for the length of the antigen sequences including average, minimum, and maximum lengths, etc. A basic summary of the number of nanobodies binding to antigens is given in Table 2

Table 1. Sequence length statistics for antigen and nanobody sequences.

Type	Count	Sequence Length Statistics				
		Mean	Min	Max	Std. Dev.	Median
Antigens	47	671.51	158	1816	421.24	480
Nanobodies	365	122.84	104	175	8.87	123

Table 2. Statistics for nanobody sequences binding to each antigen.

Type	Mean	Min	Max	Std. Dev.	Median
Nanobodies in each antigen	7.77	1	36	9.28	4

3.2   Features Extracted from Sequences

We performed basic protein sequence analysis using the 'bioPython' package[3] on each nanobody and antigen sequence to determine their features. These features include charge at pH, Grand Average of Hydropathy (GRAVY), molecular weight, aromaticity, instability index, isoelectric point, secondary structure

---

[1] https://www.uniprot.org/.
[2] http://www.sdab-db.ca/.
[3] https://biopython.org/docs/dev/api/Bio.SeqUtils.ProtParam.html.

fraction (helix, turn, and sheet), and molar extinction coefficient (reduced and oxidized).

Charge at pH: The charge of a protein at a pH is determined by the presence of charged amino acids (aspartic acid, glutamic acid, lysine, arginine, histidine, and cysteine) and their ionization state. These amino acids gain or lose protons at different pH values, resulting in a net charge. The charge at pH affects the protein's solubility, interaction with other molecules, and biological function.

Grand Average of Hydropathy (GRAVY): GRAVY [14] measures the overall hydrophobicity or hydrophilicity of a protein, calculated by averaging the hydropathy values of its amino acids. Positive GRAVY values indicate hydrophobic, while negative values represent hydrophilic regions and provide insights into protein stability, membrane interactions, and protein-protein interactions.

Molecular Weight: Molecular weight refers to the sum of the atomic weights of all atoms in a protein molecule. It is calculated based on the amino acid composition of the protein sequence. Molecular weight impacts various protein properties, such as protein folding, thermal stability, and mobility, and is crucial for protein identification, characterization, and quantification.

Instability Index: The instability index [9] is a measure of the propensity of a protein to undergo degradation or unfold. Higher instability index values indicate increased susceptibility to degradation and decreased protein stability. The index is useful for evaluating protein expression, protein engineering, and predicting potential regions of protein instability.

Isoelectric Point: The isoelectric point (pI) is the pH at which a protein has a net charge of zero. It is determined by the presence of charged amino acids and their ionization states. The pI influences protein solubility, crystallization, and electrophoretic mobility. Knowledge of the pI is crucial for protein purification, protein characterization, and protein separation techniques based on charge.

Secondary Structure Fraction (Helix, Turn, and Sheet): The secondary structure fraction [10,12,13] refers to the proportions or percentages of different secondary structure elements (helices, turns, and sheets) in a protein sequence. These elements are determined by the pattern of hydrogen bonds between amino acids. The secondary structure fraction provides insights into the protein's folding, stability, and functional properties. Different secondary structure fractions contribute to the unique 3D structure and biological function of the protein.

Molar Extinction Coefficient (Reduced and Oxidized): The molar extinction coefficient is a measure of the ability of a molecule to absorb light at a specific wavelength. It quantifies the efficiency of light absorption by the molecule. The molar extinction coefficients can be different for reduced and oxidized forms of a protein due to changes in the chromophores. These coefficients are useful for protein quantification, monitoring protein folding/unfolding, and studying protein-protein interactions.

Remark 1. We compute classification results with and without these features for all embedding methods, including the proposed and baseline methods.

3.3   Obtaining Non-Binding Nb-Ag Pairs

We created the proximity matrix of these sequences using Clustal Omega[4] to evaluate the pairwise edit distance between antigens and nanobodies. This pairwise distance is used to identify further binding pairs and non-binding pairs of antigens and nanobodies. There are three pairs of antigens with very high similarity (distance $<$ .25), namely, antigen green fluorescent protein (GFP) and Superfolder green fluorescent protein (sfGFP), (pairwise distance $= 0.05042$), RAC-gamma serine/threonine-protein kinase pleckstrin homology domain (Akt3PH) and RAC-alpha serine/threonine-protein kinase pleckstrin homology domain (Akt1PH) (pairwise distance $= 0.19833$), Glioblastoma multiforme dihydropyrimidinase-related protein 2 (DPYSl2) and/or methylenetetrahydrofolate dehydrogenase 1 (MTHFD1) and Glioblastoma multiforme collapsin response mediator protein 1 (CRMP1) (pairwise distance $= 0.0.236014$). For these pairs, if a nanobody binds with one of them, we assume it also binds to the other. Thus, we add 1388 of additional binding pairs that bind with each other.

The non-binding pairs are obtained as follows: Suppose we have two binding Nb-Ag (n_i, g_j) and (n_ℓ, g_k). Then both the pairs (n_i, g_k) and (n_ℓ, g_j) are candidates for being declared as non-binding pairs if the distance between g_i and g_j is more than a certain threshold (we set the threshold $\in \{.8, .85, .9\}$. A random sample of such candidate pairs is added to the dataset as non-binding pairs, which consist of 1728 such pairs in total.

Data Visualization: In order to visually assess the proximity of similar points in the Spike2Vec-based embeddings, we employ the t-Distributed Stochastic Neighbor Embedding (t-SNE) technique to obtain two-dimensional representations of the embeddings. These representations are then plotted using a scatterplot [15]. The t-SNE plots for the Nanobody and Antigen-based embeddings are depicted in Fig. 1. The colored data points show the different antigen categories (47 in total). Although the data points are scattered in the whole plot, we can observe small grouping for different labels, which shows that the embedding captures the hidden hierarchical and structural information inherent in the protein sequences.

4   Representation Learning for Nb-Ag Binding Prediction

We learn various machine learning models on the training data of binding and non-binding pairs. To train the classifiers for the binary classification problem, we generate the feature vector (also called embeddings) for the nanobody and antibody in pairs. The corresponding feature vectors are a concatenation of the features extracted from the sequence as outlined above and feature vector embedding of the whole nanobody and antigen sequences, using state-of-the-art sequence2vec models discussed below.

---

[4] https://www.ebi.ac.uk/Tools/msa/.

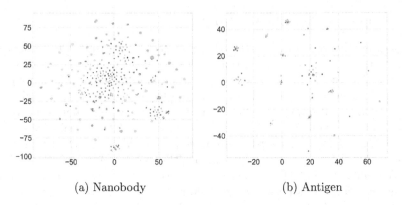

(a) Nanobody   (b) Antigen

Fig. 1. t-SNE plots for Nanobody and Antigen Embeddings using Spike2Vec approach. The figure is best seen in color. (Color figure online)

Embedding Generation: To generate the fixed length numerical embedding from the variable length protein sequences, we use the idea of gapped k-mers. The gapped k-mers are a representation of biological sequences, such as DNA or protein sequences, that capture patterns while allowing for flexibility in the positioning of the k-mer elements (i.e., nucleotides or amino acids). In gapped k-mers, there are gaps or missing positions between the elements of the k-mers. These gaps introduce variability and enable the capture of more diverse and complex patterns compared to traditional contiguous k-mers.

The spectrum generated from gapped k-mers refers to the collection of all possible (unique) gapped k-mers (along with their count) that can be formed from a given protein sequence. These k-mers counts are then added into the numerical vector, which is generated based on all possible k-mers for the given alphabet Σ that corresponds to $ACDEFGHIKLMNPQRSTVWXY$-. Note that in Σ, the character '$-$' is used to include the gap in the k-mers. For example, for a k-mer 'ACD', the gapped k-mers will comprise of '-CD', 'A-D', 'AC-', and 'ACD'. An important point to note here is that we generated 4 k-mers from just 1 original k-mer. This extra information helps us to preserve more information, which eventually helps in the downstream supervised analysis. The pseudocode for generating gapped k-mers-based spectrum is given in Algorithm 1. The spectrum provides a comprehensive representation of the sequence, taking into account both conserved elements and the flexibility introduced by the gaps. Each gapped k-mer in the spectrum represents a distinct pattern or subsequence present in the sequence, capturing variations and relationships between elements that may be important for various biological processes. Using gapped k-mers offers several advantages compared to other typical biological methods:

Increased Sensitivity: Gapped k-mers can capture more complex patterns and relationships compared to traditional contiguous k-mers. Gapped k-mers can capture conserved elements that are not necessarily adjacent in the sequence by

---

Algorithm 1. Gapped k-mers Spectrum

```
 1: Input: Set of sequences S, alphabet Σ, k is size of k-mer
 2: Output: Spectrum V
 3: combos = GenerateAllCombinations(Σ, k)              ▷ all possible combinations of k-mers
 4: V = [0] * |Σ|^k
 5: for s in S do
 6:     Gkmers = GappedKmers(s,k)                        ▷ compute all possible gapped k-mers
 7:     for i ← 1 to |Gkmers| do
 8:         idx = combos.index(Gkmers[i])
 9:         V[idx] ← V[idx] + 1
10: return(V)
```

---

allowing gaps. This increased sensitivity can be crucial for identifying motifs or regions of interest but may exhibit variable spacing.

Enhanced Flexibility: Gapped k-mers offer flexibility in terms of the spacing between elements. This flexibility allows for the inclusion of different variations and insertions, providing a more comprehensive representation of the sequence. Gapped k-mers can accommodate diverse patterns and handle insertions or deletions more effectively than traditional contiguous k-mers.

Comprehensive Motif Representation: The spectrum generated from gapped k-mers provides a comprehensive representation of the sequence by capturing a wide range of conserved patterns and variations. This allows for a more detailed analysis of complex motifs or functional regions that involve specific arrangements of elements.

Improved Specificity: Gapped k-mers can help improve specificity by reducing false-positive matches. By considering both conserved elements and gaps, gapped k-mers can differentiate between true motifs and random matches that may occur by chance in traditional k-mers.

After generating the embeddings using gapped k-mers spectrum, we use those embeddings as input to the machine learning classifiers for binary classification to predict the binding of Nb-Ag pairs binding.

5 Experimental Setup

In this section, we discuss the dataset statistics and evaluation metrics along with baseline models. All experiments are carried out using Python on a system equipped with a 2.4 GHz Core i5 processor, 32 GB of memory, and the Windows 10 operating system. For experiments, we randomly divide the data into 70–30% training and testing set and use 10% of the data from the training set as the validation data. The experiments are repeated 5 times, and we show average results to eliminate any biases in the random splits. For the sake of reproducibility, our code and preprocessed datasets can be accessed online[5].

---

[5] https://github.com/sarwanpasha/Nanobody_Antigen.

Baseline Models: To evaluate the proposed embedding, we compare their results with several popular baseline models from the literature. The baselines includes Spike2Vec [2], Minimizers [24], and PWM2Vec [1].

Spike2Vec [2]: The objective of this approach is to generate numerical embeddings for input protein sequences, facilitating the utilization of machine learning models. To achieve this, the method begins by creating k-mers from the given spike sequence. k-mers are employed due to their ability to retain the sequential information of the protein sequence.

Definition 1 *(k-mers).* *K-mers refer to sets of consecutive amino acids (called mers) of length k in a given sequence. The consecutive k-mers are computed using the sliding window approach, where the next k-mer is 1 window to the right of the previous k-mer. In the domain of natural language processing (NLP), they are referred to as n-grams.*

All possible k-mers that can be generated for a given sequence of length N are $N - k + 1$. Spike2Vec calculates the frequency vector based on k-mers to convert the alphabetical information of k-mers into a numerical representation. This vector captures the occurrence counts of each k-mer in the sequence, also called k-mers spectrum.

Minimizers [24]: The Minimizer-based feature vector is a method that involves computing a *minimizer* of length m (also called m-mer) for a given k-mer. In the case of m-mer, we have $m < k$. The m-mer is determined as the lexicographically smallest sequence in both the forward and reverse order of the k-mer. A fixed-length frequency vector is constructed from the set of minimizers, where each bin in the frequency vector represents the count of a specific minimizer. This method is also referred to as m-mers spectrum. The length of each vector is determined by the alphabet size Σ (where Σ contains all possible characters or amino acids in protein sequence i.e. $ACDEFGHIKLMNPQRSTVWXY$) and the length of the minimizers denoted as m (we set m to 3, which is decided using the standard validation set approach). Hence, the length of the vector is $|\Sigma|^m$.

PWM2Vec [1]: The PWM2Vec is a feature embedding method that transforms protein sequences into numerical representations. Instead of relying on the frequency of k-mers, PWM2Vec assigns weights to each amino acid within a k-mer. These weights are determined using a position weight matrix (PWM) associated with the k-mer. By considering the position-specific importance of amino acids, PWM2Vec captures both the relative significance of amino acids and preserves the ordering information within the sequences.

Evaluation Metrics: To assess the performance of embeddings, we employ several evaluation metrics, including accuracy, precision, recall, F1 score (weighted), F1 score (macro), Receiver Operating Characteristic (ROC) curve, Area Under the Curve (AUC), and training runtime. For metrics designed for binary classification tasks, we adopt the one-vs-rest approach for multi-class classification.

Machine Learning Classifiers: Supervised analysis entails the utilization of diverse linear and non-linear classifiers, such as Support Vector Machine (SVM),

Table 3. Classification results (averaged over 5 runs) for different evaluation metrics. The best values for each embedding are underlined, while the overall best values among all embeddings for different evaluation metrics are shown in bold.

	Embeddings	Algo.	Acc. ↑	Prec. ↑	Recall ↑	F1 (Weig.) ↑	F1 (Macro) ↑	ROC AUC ↑	Train Time (sec.) ↓
Without Sequence Features	Spike2Vec	SVM	0.818	0.824	0.818	0.818	0.818	0.819	5.662
		NB	0.813	0.815	0.813	0.813	0.813	0.813	<u>0.103</u>
		MLP	0.844	0.846	0.844	0.844	0.844	0.844	4.075
		KNN	0.892	0.893	0.892	0.892	0.892	0.892	1.290
		RF	<u>0.906</u>	<u>0.911</u>	<u>0.906</u>	**0.906**	**0.906**	<u>0.906</u>	3.725
		LR	0.813	0.815	0.813	0.813	0.813	0.814	2.417
		DT	0.878	0.878	0.878	0.878	0.877	0.878	1.293
	Minimizers	SVM	0.824	0.826	0.824	0.823	0.823	0.823	5.444
		NB	0.791	0.792	0.791	0.790	0.790	0.790	<u>0.091</u>
		MLP	0.844	0.845	0.844	0.844	0.844	0.844	2.997
		KNN	0.880	0.880	0.880	0.880	0.880	0.880	1.257
		RF	<u>0.892</u>	<u>0.898</u>	<u>0.892</u>	<u>0.892</u>	<u>0.892</u>	<u>0.893</u>	4.000
		LR	0.811	0.812	0.811	0.811	0.811	0.811	1.343
		DT	0.851	0.851	0.851	0.850	0.850	0.850	1.677
	PWM2Vec	SVM	0.810	0.812	0.810	0.809	0.809	0.809	5.732
		NB	0.792	0.793	0.792	0.792	0.792	0.792	<u>0.095</u>
		MLP	0.820	0.821	0.820	0.820	0.819	0.820	3.730
		KNN	0.875	0.875	0.875	0.875	0.875	0.875	1.232
		RF	<u>0.892</u>	<u>0.899</u>	<u>0.892</u>	<u>0.891</u>	<u>0.891</u>	<u>0.892</u>	3.746
		LR	0.804	0.805	0.804	0.804	0.804	0.804	7.137
		DT	0.866	0.866	0.866	0.866	0.866	0.866	1.692
	Gapped k-mers	SVM	0.814	0.816	0.814	0.813	0.813	0.812	5.740
		NB	0.798	0.798	0.798	0.797	0.797	0.796	<u>0.087</u>
		MLP	0.824	0.825	0.824	0.824	0.824	0.824	2.886
		KNN	0.885	0.886	0.885	0.885	0.885	0.885	0.995
		RF	**0.907**	**0.912**	**0.907**	<u>0.894</u>	<u>0.894</u>	**0.908**	3.755
		LR	0.812	0.813	0.812	0.812	0.812	0.812	4.395
		DT	0.872	0.872	0.872	0.872	0.871	0.872	1.777
With Sequence Features	Spike2Vec	SVM	0.791	0.796	0.791	0.790	0.790	0.790	8.804
		NB	0.695	0.737	0.695	0.680	0.678	0.691	<u>0.085</u>
		MLP	0.811	0.814	0.811	0.811	0.811	0.811	2.326
		KNN	0.844	0.845	0.844	0.844	0.844	0.844	0.953
		RF	<u>0.897</u>	<u>0.903</u>	<u>0.897</u>	<u>0.896</u>	<u>0.896</u>	<u>0.898</u>	3.890
		LR	0.827	0.827	0.827	0.827	0.826	0.827	1.183
		DT	0.847	0.848	0.847	0.847	0.847	0.847	1.246
	Minimizers	SVM	0.778	0.783	0.778	0.777	0.777	0.777	10.938
		NB	0.674	0.736	0.674	0.649	0.647	0.670	<u>0.094</u>
		MLP	0.801	0.806	0.801	0.800	0.800	0.800	3.228
		KNN	0.842	0.842	0.842	0.842	0.842	0.842	0.827
		RF	<u>0.896</u>	<u>0.902</u>	<u>0.896</u>	<u>0.896</u>	<u>0.896</u>	<u>0.897</u>	3.801
		LR	0.823	0.823	0.823	0.823	0.823	0.823	1.167
		DT	0.846	0.846	0.846	0.845	0.845	0.845	1.297
	PWM2Vec	SVM	0.766	0.770	0.766	0.765	0.765	0.766	9.569
		NB	0.679	0.726	0.679	0.659	0.657	0.674	<u>0.087</u>
		MLP	0.811	0.813	0.811	0.811	0.811	0.811	2.889
		KNN	0.828	0.828	0.828	0.827	0.827	0.827	0.768
		RF	<u>0.893</u>	<u>0.901</u>	<u>0.893</u>	<u>0.892</u>	<u>0.892</u>	<u>0.894</u>	3.765
		LR	0.819	0.819	0.819	0.819	0.819	0.819	1.495
		DT	0.851	0.851	0.851	0.851	0.851	0.850	1.279
	Gapped k-mers	SVM	0.785	0.792	0.785	0.784	0.783	0.784	9.270
		NB	0.720	0.745	0.720	0.712	0.711	0.718	<u>0.086</u>
		MLP	0.807	0.810	0.807	0.806	0.806	0.806	2.432
		KNN	0.839	0.839	0.839	0.839	0.838	0.838	0.753
		RF	<u>0.895</u>	<u>0.901</u>	<u>0.895</u>	<u>0.894</u>	<u>0.894</u>	<u>0.895</u>	3.468
		LR	0.823	0.823	0.823	0.823	0.823	0.823	1.123
		DT	0.860	0.861	0.860	0.860	0.860	0.860	0.955

Naive Bayes (NB), Multi-Layer Perceptron (MLP), K-Nearest Neighbors (KNN), Random Forest (RF), Logistic Regression (LR), and Decision Tree (DT).

6   Results and Discussion

The classification results for different evaluation metrics with and without the sequence features are reported in Table 3. We observe (bold values) that the gapped k-mers spectrum outperforms all embeddings in the case of average accuracy, precision, recall, and ROC-AUC using the random forest classifier. For Weighted and Macro F1, the baseline Spike2Vec performs better than other embeddings using the random forest classifier. One interesting observation is that the random forest classifier consistently outperforms different classifiers for all embeddings and evaluation metrics, as shown with underlined values in Table 3.

Comparing embeddings with and without sequence features, we observe that using these features degrades classifiers' accuracy. This degradation is due to redundancy and feature dimensionality as some features might capture similar information as the k-mers spectrum. For example, the charge at pH, aromaticity, or GRAVY may already encode certain aspects of protein sequence patterns. Including redundant features can lead to multicollinearity, where the features are highly correlated, making it difficult for the classifier to distinguish their contributions. In the case of feature dimensionality, adding new features increases the dimensionality of the input space. With more features, the classifier faces the curse of dimensionality. Insufficient training data or a limited number of samples in each class relative to the feature space can result in overfitting, reduced generalization performance, and decreased accuracy.

7   Conclusion

This study aimed to develop an ML approach to predict Nb-Ag binding solely based on sequences, thereby reducing the need for computationally intensive techniques such as docking. The proposed method utilized an embedding approach using gapped k-mers to generate a spectrum, which was then used for supervised analysis. Experimental evaluation of our approach demonstrates that the gapped k-mers spectrum outperformed competing embeddings. Our approach offers a more efficient and cost-effective alternative for screening potential Nbs and holds promise for facilitating the development of Nb-based diagnostics and therapeutics for various diseases, including cancer and other serious illnesses. Future research involves evaluating the proposed model on more extensive datasets and also working on the generalizability and robustness of the model. Additionally, exploring the integration of additional features and considering other machine learning algorithms could further enhance predictive performance.

References

1. Ali, S., Bello, B., Chourasia, P., Punathil, R.T., Zhou, Y., Patterson, M.: PWM2Vec: an efficient embedding approach for viral host specification from coronavirus spike sequences. Biology 11(3), 418 (2022)

2. Ali, S., Patterson, M.: Spike2vec: an efficient and scalable embedding approach for covid-19 spike sequences. In: IEEE International Conference on Big Data (Big Data), pp. 1533–1540 (2021)
3. Berman, H.M., et al.: The protein data bank. Nucleic Acids Res. **28**(1), 235–242 (2000)
4. Burley, S.K., et al.: Rcsb protein data bank: biological macromolecular structures enabling research and education in fundamental biology, biomedicine, biotechnology and energy. Nucleic Acids Res. **47**(D1), D464–D474 (2019)
5. Cohen, T., Halfon, M., Schneidman-Duhovny, D.: Nanonet: rapid and accurate end-to-end nanobody modeling by deep learning. Front. Immunol. **13**, 958584 (2022)
6. Cortez-Retamozo, V., et al.: Efficient cancer therapy with a nanobody-based conjugate. Can. Res. **64**(8), 2853–2857 (2004)
7. Deffar, K., Shi, H., Li, L., Wang, X., Zhu, X.: Nanobodies-the new concept in antibody engineering. Afr. J. Biotechnol. **8**(12), 2645–2652 (2009)
8. Farhan, M., Tariq, J., Zaman, A., Shabbir, M., Khan, I.: Efficient approximation algorithms for strings kernel based sequence classification. In: Advances in Neural Information Processing Systems (NeurIPS), pp. 6935–6945 (2017)
9. Guruprasad, K., Reddy, B.B., Pandit, M.W.: Correlation between stability of a protein and its dipeptide composition: a novel approach for predicting in vivo stability of a protein from its primary sequence. Protein Eng. Des. Sel. **4**(2), 155–161 (1990)
10. Haimov, B., Srebnik, S.: A closer look into the α-helix basin. Sci. Rep. **6**(1), 38341 (2016)
11. Hou, Q., et al.: Serendip-ce: sequence-based interface prediction for conformational epitopes. Bioinformatics **37**(20), 3421–3427 (2021)
12. Hutchinson, E.G., Thornton, J.M.: A revised set of potentials for β-turn formation in proteins. Protein Sci. **3**(12), 2207–2216 (1994)
13. Kim, C.A., Berg, J.M.: Thermodynamic β-sheet propensities measured using a zinc-finger host peptide. Nature **362**(6417), 267–270 (1993)
14. Kyte, J., Doolittle, R.F.: A simple method for displaying the hydropathic character of a protein. J. Mol. Biol. **157**(1), 105–132 (1982)
15. Van der M., L., Hinton, G.: Visualizing data using t-SNE. J. Mach. Learn. Res. (JMLR) **9**(11), 2579–2605 (2008)
16. Miller, N.L., Clark, T., Raman, R., Sasisekharan, R.: Learned features of antibody-antigen binding affinity. Front. Mol. Biosci. **10**, 1112738 (2023)
17. Mitchell, L.S., Colwell, L.J.: Analysis of nanobody paratopes reveals greater diversity than classical antibodies. Protein Eng. Des. Sel. **31**(7–8), 267–275 (2018)
18. Mitchell, L.S., Colwell, L.J.: Comparative analysis of nanobody sequence and structure data. Proteins Struct. Funct. Bioinf. **86**(7), 697–706 (2018)
19. Muyldermans, S.: Nanobodies: natural single-domain antibodies. Ann. Rev. Biochem. **82**, 775–797 (2013)
20. Myung, Y., Pires, D.E., Ascher, D.B.: Csm-ab: graph-based antibody-antigen binding affinity prediction and docking scoring function. Bioinformatics **38**(4), 1141–1143 (2022)
21. Peng, H.P., Lee, K.H., Jian, J.W., Yang, A.S.: Origins of specificity and affinity in antibody-protein interactions. Proc. Natl. Acad. Sci. **111**(26), E2656–E2665 (2014)
22. Ramon, A., Saturnino, A., Didi, K., Greenig, M., Sormanni, P.: Abnativ: vq-vae-based assessment of antibody and nanobody nativeness for engineering, selection, and computational design. In: bioRxiv, p. 2023-04 (2023)
23. Revets, H., De Baetselier, P., Muyldermans, S.: Nanobodies as novel agents for cancer therapy. Expert Opin. Biol. Ther. **5**(1), 111–124 (2005)

24. Roberts, M., Hayes, W., Hunt, B.R., Mount, S.M., Yorke, J.A.: Reducing storage requirements for biological sequence comparison. Bioinformatics **20**(18), 3363–3369 (2004)
25. Rossant, C.J., et al.: Phage display and hybridoma generation of antibodies to human cxcr2 yields antibodies with distinct mechanisms and epitopes. MAbs **6**(6), 1425–1438 (2014)
26. Schwede, T.: Protein modeling: what happened to the "protein structure gap"? Structure **21**(9), 1531–1540 (2013)
27. Sormanni, P., Aprile, F.A., Vendruscolo, M.: Rational design of antibodies targeting specific epitopes within intrinsically disordered proteins. Proc. Natl. Acad. Sci. **112**(32), 9902–9907 (2015)
28. Tam, C., Kumar, A., Zhang, K.Y.: Nbx: machine learning-guided re-ranking of nanobody-antigen binding poses. Pharmaceuticals **14**(10), 968 (2021)
29. Valdés-Tresanco, M.S., Valdés-Tresanco, M.E., Jiménez-Gutiérrez, D.E., Moreno, E.: Structural modeling of nanobodies: a benchmark of state-of-the-art artificial intelligence programs. Molecules **28**(10), 3991 (2023)
30. Yang, Y.X., Huang, J.Y., Wang, P., Zhu, B.T.: Area-affinity: a web server for machine learning-based prediction of protein-protein and antibody-protein antigen binding affinities. J. Chem. Inf. Model. **63**, 3230–3237 (2023)
31. Ye, C., Hu, W., Gaeta, B.: Prediction of antibody-antigen binding via machine learning: development of data sets and evaluation of methods. JMIR Bioinf. Biotechnol. **3**(1), e29404 (2022)

Approximating Rearrangement Distances with Replicas and Flexible Intergenic Regions

Gabriel Siqueira[1]([⊠])([iD]), Alexsandro Oliveira Alexandrino[1]([iD]),
Andre Rodrigues Oliveira[2]([iD]), Géraldine Jean[3]([iD]), Guillaume Fertin[3]([iD]),
and Zanoni Dias[1]([iD])

[1] Institute of Computing, University of Campinas, Campinas, Brazil
{gabriel.siqueira,alexsandro,zanoni}@ic.unicamp.br
[2] Computing and Informatics Department, Mackenzie Presbyterian University, São Paulo, Brazil
andre.rodrigues@mackenzie.br
[3] Nantes Université, École Centrale Nantes, CNRS, LS2N, UMR 6004, Nantes, France
{geraldine.jean,guillaume.fertin}@univ-nantes.fr

Abstract. Many tools from Computational Biology compute distances between genomes by counting the number of genome rearrangement events, such as reversals of a segment of genes. Most approaches to model these problems consider some simplifications such as ignoring nucleotides outside genes (the so-called intergenic regions), or assuming that just a single copy of each gene exists in the genomes. Recent works made advancements in more general models considering replicated genes and the number of nucleotides in intergenic regions. Our work aims at adapting those results by applying some flexibilization to match intergenic regions that do not have the same number of nucleotides. We propose the Signed Flexible Intergenic Reversal Distance problem, which seeks the minimum number of reversals necessary to transform one genome into the other and encodes the genomes using flexible intergenic region information while also allowing multiple copies of a gene. We show the relationship of this problem with the Signed Minimum Common Flexible Intergenic String Partition problem and use a $2k$-approximation to the partition problem to show a $8k$-approximation to the distance problem, where k is the maximum number of copies of a gene in the genomes.

Keywords: Rearrangement Distance · Intergenic Regions · Partition · Reversal

1 Introduction

In comparative genomics, the genetic changes caused by global mutations are the main features used to infer the distance between genomes of different species. These global mutations are called genome rearrangements, and they can change gene order or the genetic material of a genome. The most studied rearrangements are the reversals [7], which invert a segment of the genome, and the transpositions [1], which exchange the position of two adjacent segments.

The sequence of genes in a genome is often modeled as a string, in which each character corresponds to a gene. Besides, each character has a plus or minus sign that indicates gene orientation, when such information is available.

When comparing two genomes using rearrangements, we seek the minimum number of rearrangements necessary to transform the sequence of genes of the source genome into the sequence of genes of the target genome. This number is the so-called distance between the genomes. Considering the use of reversals, the problem of calculating such distance can be solved in polynomial time [7], if no gene has more than one copy. Otherwise, the problem is NP-hard [10] and the best known approximation for it has approximation factor $\Theta(k)$ [8], where k is the maximum number of copies of a gene in the genomes.

Recent studies started considering the distribution of intergenic regions in the genomes [6,9]. These studies only used sizes (number of nucleotides) of intergenic regions, because it is difficult to find correlations between intergenic regions of distinct genomes. Moreover, genome rearrangements often break intergenic regions [2,3]. To model intergenic regions, we use a list of non-negative integers, where each value is the size of an intergenic region. In this way, a genome is represented by a string, modeling gene order, and a list, modeling intergenic region sizes. In genome rearrangement distance problems considering intergenic regions, beyond transforming one string into the other, the rearrangements must be capable of transforming the intergenic region sizes from the source genome into the intergenic region sizes of the target genome.

Considering gene orientation and no gene repetitions, Oliveira et al. [9] showed that the Signed Intergenic Reversal Distance problem is NP-hard and introduced a 2-approximation algorithm. Later, Siqueira et al. [11] studied the problem allowing gene repetition. This version of the problem is also NP-hard and the authors presented a $\Theta(k)$-approximation algorithm, where k is the maximum number of copies of a gene in the genomes. Their work uses the Signed Minimum Common Intergenic String Partition problem to approximate the distance problem, but there is a flaw when they claimed a $2k$-approximation for the partition problem, which is fixed in the present work.

Recently, Brito et al. [4,5] introduced a new version of the problem with *flexible* intergenic regions in genomes without gene repetition. Each element of the target intergenic list in this version is a range of values. They presented an NP-hardness proof and a 2-approximation algorithm for the Signed Flexible Intergenic Reversal Distance problem, considering genomes without gene repetition.

In this work, we investigate the Signed Flexible Intergenic Reversal Distance (SFIRD) problem when more than one copy of each gene is allowed. In Sect. 2, we present the basic notations and definitions for the distance problem. In Sect. 3, we present an approximation algorithm for the SFIRD problem. And, in Sect. 4, we conclude the paper and we present directions for future works.

2 Definitions

For any set L of elements, we adopt the usual notation $|L|$ for the number of elements in L. If such set has some ordering of its elements, as in the case of a string or a sequence, the element in the i-th position of L is denoted by L_i.

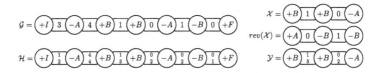

Fig. 1. On top, a rigid genome $\mathcal{G} = ([+I \ -A \ +B \ +B \ -A \ -B \ +F], [3\ 4\ 1\ 0\ 1\ 0])$ followed by one of its subgenomes $\mathcal{X}$ and the genome $rev(\mathcal{X})$. On the bottom, a flexible genome $\mathcal{H} = ([+I \ -A \ +B \ +B \ -A \ -B \ +F], [(1,2)\ (4,4)\ (1,3)\ (0,2)\ (0,2)\ (0,1)])$ followed by one of its subgenomes $\mathcal{Y}$.

In our genome representation, we include some information extracted from the intergenic regions. We will use two representations for intergenic regions: the rigid representation, which represents the intergenic regions by their sizes, and the flexible representation, which uses an interval of possible values for the intergenic region sizes instead of only one number. A *genome* $\mathcal{G} = (S, \breve{S})$ is a signed string S and a list of intergenic regions $\breve{S}$. Each character of the string S represents a gene, and a $+$ or $-$ sign is associated with the character to represent the gene orientation. If $\mathcal{G}$ is a *rigid genome* then $\breve{S}$ is a list of $|\breve{S}| = |S| - 1$ integers, such that $\breve{S}_i$ represents the size of the intergenic region between genes S_i and S_{i+1}. If $\mathcal{G}$ is a *flexible genome* then $\breve{S}$ is a list of $|\breve{S}| = |S| - 1$ pairs of integers, such that $\breve{S}_i = (\breve{S}_i^{min}, \breve{S}_i^{max})$ represents an interval $[\breve{S}_i^{min}, \breve{S}_i^{max}]$ that contains the size of the intergenic region between genes S_i and S_{i+1}. In Fig. 1, genome $\mathcal{G}$ is rigid, and genome $\mathcal{H}$ is flexible.

A *subgenome* $\mathcal{X} = (A, \breve{A})$ of $\mathcal{G} = (S, \breve{S})$ is a genome that appears at some position of $\mathcal{G}$, more precisely A is a substring of S and the intergenic regions in $\breve{S}$ between characters of S are equal to $\breve{A}$. We use the notation $\mathcal{X} \subset \mathcal{G}$ to represent the subgenome relation. In Fig. 1, $\mathcal{X}$ is a subgenome of $\mathcal{G}$ and $\mathcal{Y}$ is a subgenome of $\mathcal{H}$.

The *reverse* of a genome $\mathcal{G} = (S, \breve{S})$ is a genome $rev(\mathcal{G}) = (P, \breve{P})$, such that $P_i = -S_{|S|-i+1}, \forall 1 \leq i \leq |S|$ and $\breve{P}_i = \breve{S}_{|\breve{S}|-i+1}, \forall 1 \leq i \leq |\breve{S}|$.

We call two genomes $\mathcal{G} = (S, \breve{S})$ and $\mathcal{H} = (P, \breve{P})$ *co-tailed* if $S_1 = P_1$ and $S_{|S|} = P_{|P|}$. When sequencing a real genome $\mathcal{G} = (S, \breve{S})$, we have intergenic regions at the beginning and at the end of the genome, so for our representation we artificially insert two genes, corresponding to the initial gene $S_1 = +I$ and the final gene $S_{|S|} = +F$ in $\mathcal{G}$. Consequently, always choosing the same characters to insert, we ensure that any two genomes being compared are co-tailed. For that reason, henceforward we assume that all genomes, excluding subgenomes, are co-tailed. For example, in Fig. 1, any of the two genomes $\mathcal{G}$ and $\mathcal{H}$ is co-tailed, because both start with $+I$ and end with $+F$.

The *occurrence* of a gene (character) α in a genome $\mathcal{G} = (S, \breve{S})$, denoted by $occ(\alpha, \mathcal{G})$, is the number of characters in S that are equal to α disregarding the signs. We denote the largest value of $occ(\alpha, \mathcal{G})$ for any α by $occ\_max(\mathcal{G})$. In genome $\mathcal{G}$ of Fig. 1, we have $occ(I, \mathcal{G}) = occ(F, \mathcal{G}) = 1$, $occ(A, \mathcal{G}) = 2$ and $occ(B, \mathcal{H}) = 3$, so $occ\_max(\mathcal{G}) = 3$.

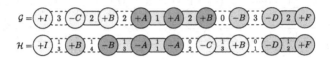

Fig. 2. A signed common flexible intergenic partition between two genomes $\mathcal{G} = (S, \breve{S})$ and $\mathcal{H}$. Note that the cost of the partition is 4 and the breakpoints of $\mathcal{G}$ are $\breve{S}_1 = 3$, $\breve{S}_3 = 2$, $\breve{S}_6 = 0$, and $\breve{S}_7 = 3$.

Let $\mathcal{G} = (S, \breve{S})$ be a rigid genome and $\mathcal{H} = (P, \breve{P})$ be a flexible genome. We say that $\mathcal{G}$ and $\mathcal{H}$ are *balanced* if for any gene in $\mathcal{G}$ or in $\mathcal{H}$, we have $occ(\alpha, \mathcal{G}) = occ(\alpha, \mathcal{H})$, and $\sum_{i=1}^{|\breve{P}|} \breve{P}_i^{min} \leq \sum_{i=1}^{|\breve{S}|} \breve{S}_i \leq \sum_{i=1}^{|\breve{P}|} \breve{P}_i^{max}$. We say that $\mathcal{G}$ and $\mathcal{H}$ are *fully compatible* if $S = P$ and $\breve{P}_i^{min} \leq \breve{S}_i \leq \breve{P}_i^{max}, \forall 1 \leq i \leq |\breve{S}|$. We say that $\mathcal{G}$ and $\mathcal{H}$ are *partially compatible* if they are fully compatible or $\mathcal{G}$ and $rev(\mathcal{H})$ are fully compatible. The genomes $\mathcal{G}$ and $\mathcal{H}$ of Fig. 1 are balanced but not fully compatible, because $\breve{S}_1 > \breve{P}_1^{max}$. On the other hand, the subgenomes $\mathcal{X}$ and $\mathcal{Y}$ are fully compatible and the subgenomes $rev(\mathcal{X})$ and $\mathcal{Y}$ are partially compatible.

Two rigid genomes $\mathcal{G} = (S, \breve{S})$ and $\mathcal{H} = (P, \breve{P})$ are *balanced* if $occ(\alpha, \mathcal{G}) = occ(\alpha, \mathcal{H})$ for every character α in $\mathcal{G}$ or in $\mathcal{H}$, and $\sum_{i=1}^{|\breve{S}|} \breve{S}_i = \sum_{i=1}^{|\breve{P}|} \breve{P}_i$.

A *partition* of a genome $\mathcal{G} = (S, \breve{S})$ is a sequence $\mathbb{S}$ composed of subgenomes of $\mathcal{G}$, such that $\mathcal{G}$ can be broken into the subgenomes in $\mathbb{S}$. Formally, there is a sequence $B = (b_1, b_2, \ldots, b_{|B|})$ of $|B| = |\mathbb{S}| - 1$ intergenic regions of $\breve{S}$, called *breakpoints*, such that, $\mathbb{S}_1$ is the subgenome to the left of breakpoint b_1, and $\mathbb{S}_i, \forall 2 \leq i \leq |B|$, is the subgenome between breakpoint b_{i-1} and b_i, and $\mathbb{S}_{|\mathbb{S}|}$ is the subgenome to the right of breakpoint $b_{|B|}$.

Given two balanced rigid genomes $\mathcal{G}$ and $\mathcal{H}$, a *signed common intergenic partition* of $\mathcal{G}$ and $\mathcal{H}$ is a pair $(\mathbb{S}, \mathbb{P})$, such that $\mathbb{S}$ is a partition of $\mathcal{G}$, $\mathbb{P}$ is a partition of $\mathcal{H}$, and we can reorder the elements of $\mathbb{S}$ to obtain $\mathbb{P}$.

Given a rigid genome $\mathcal{G}$ and a flexible genome $\mathcal{H}$, such that $\mathcal{G}$ and $\mathcal{H}$ are balanced, a *signed common flexible intergenic partition* of $\mathcal{G}$ and $\mathcal{H}$ is a pair $(\mathbb{S}, \mathbb{P})$, such that $\mathbb{S}$ is a partition of $\mathcal{G}$, $\mathbb{P}$ is a partition of $\mathcal{H}$, and we can reorder the elements of $\mathbb{S}$ to obtain a sequence $\mathbb{P}'$ such that $\mathbb{P}'_i$ is partially compatible with $\mathbb{P}_i$, for all $1 \leq i \leq |\mathbb{P}|$. Figure 2 shows an example of such common partition.

We use only the term common partition, when the type of common partition is clear by the context or the text applies to either type of partition. The cost of a common partition $\mathcal{P} = (\mathbb{S}, \mathbb{P})$, denoted by $cost(\mathcal{P})$ (or $cost((\mathbb{S}, \mathbb{P}))$) is the number of breakpoints in $\mathbb{S}$, which is equal to the number of breakpoints in $\mathbb{P}$.

The *Signed Minimum Common Intergenic String Partition* (SMCISP) problem seeks a signed common intergenic partition between two rigid genomes with minimum cost. Similarly, the *Signed Minimum Common Flexible Intergenic String Partition* (SMCFISP) problem seeks a signed common flexible intergenic partition between a rigid genome and a flexible genome with minimum cost.

Given a rigid genome $\mathcal{G} = (S, \breve{S})$ and the integers i, j, x, y, with $2 \leq i \leq j \leq |S| - 1$, $0 \leq x \leq \breve{S}_{i-1}$, and $0 \leq y \leq \breve{S}_j$. The *reversal* $\rho_{(x,y)}^{(i,j)}$ is an operation that transforms $\mathcal{G}$ into a genome $\mathcal{G}.\rho_{(x,y)}^{(i,j)} = (S', \breve{S}')$, where $S' = [S_1 \ldots S_{i-1} \ \underline{-S_j \ldots -S_i} \ S_{j+1} \ldots S_{|S|}]$ and $\breve{S}' = [\breve{S}_1 \ldots \breve{S}_{i-2} \ x + y \ \underline{\breve{S}_{j-1} \ldots \breve{S}_i} \ x' + y' \ \breve{S}_{j+1} \ldots \breve{S}_{|\breve{S}|}]$, with $x' = \breve{S}_{i-1} - x$ and $y' = \breve{S}_j - y$.

Given a rigid genome $\mathcal{G}$ and a flexible genome $\mathcal{H}$. The *Signed Flexible Intergenic Reversal Distance* problem seeks the minimum number of reversals (denoted by $d_{FR}(\mathcal{G}, \mathcal{H})$) necessary to transform $\mathcal{G}$ into a genome $\mathcal{H}'$ fully compatible with $\mathcal{H}$.

We also need some definitions from graph theory. Given a graph $G = (V, E)$, with vertex set V and edge set E. We say that G is *bipartite* if we can separate V into two sets V_1 and V_2, such that there is no edge of E connecting two vertices of the same set. A bipartite graph G admits a *perfect match* if there is a bijective function f from V_1 to V_2, such that v and $f(v)$ are connected by an edge.

A *vertex cover* of a graph $G = (V, E)$ is a subset V' of V, such that any edge in E is incident to at least one vertex of V'. The *Minimum Vertex Cover* problem seeks a vertex cover with minimum size. Finding such cover is a known NP-hard problem that admits a simple 2-approximation, which consists of repeating the process of picking an uncovered edge and including both vertices incident to it in V' until a cover is found.

3   Approximating the Distance Problems

In this section, we propose an algorithm with approximation factor $2k$ for the SMCFISP poblem and a proof that it can also be used to approximate the SFIRD problem. The algorithm adapts ideas from the algorithm for rigid intergenic regions [11]. Some of these ideas were originally proposed in a context without intergenic regions by Kolman and Waleń [8].

For the algorithm, we need a structure that codifies the compatibility between subgenomes. Given a rigid genome $\mathcal{G}$ and a flexible genome $\mathcal{H}$, such that $\mathcal{G}$ and $\mathcal{H}$ are balanced, the *block compatibility graph* $\mathcal{B}(\mathcal{G}, \mathcal{H}) = (V, E)$ is defined by the vertex set V and edge set E. The vertices from V correspond to each possible subgenome of $\mathcal{G}$ and $\mathcal{H}$. For the set E, there is an edge $e_{v,u} \in E$ between vertices v, corresponding to a subgenome $\mathcal{X} \subset \mathcal{G}$, and u, corresponding to a subgenome $\mathcal{Y} \subset \mathcal{H}$, if $\mathcal{X}$ and $\mathcal{Y}$ are partially compatible. Figure 3 shows some connected components of the block compatibility graph between two genomes.

We can generalize the block compatibility graph to work with partitions of genomes. Let $\mathbb{S}$ be a partition of $\mathcal{G}$ and $\mathbb{P}$ a partition of $\mathcal{H}$. The graph $\mathcal{B}(\mathbb{S}, \mathbb{P})$ is constructed by removing from $\mathcal{B}(\mathcal{G},\mathcal{H})$ all the subgenomes that contain a break-point of $\mathbb{S}$ or $\mathbb{P}$. Note that $\mathcal{B}(\mathcal{G},\mathcal{H}) = \mathcal{B}([\mathcal{G}],[\mathcal{H}])$ (here $[\mathcal{G}]$ denotes a partition with a single sub-genome $[\mathcal{G}]$).

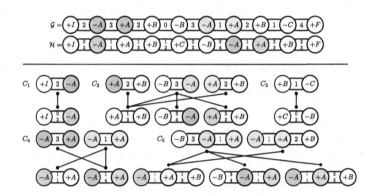

Fig. 3. A rigid genome $\mathcal{G}$ and a flexible genome $\mathcal{H}$ followed by some components of the block compatibility graph $\mathcal{B}(\mathcal{G},\mathcal{H})$. We include colors in some genes to help the identification of the positions of subgenomes. We are showing all components that are not a single element (as these components do not admit a perfect match) and that do not have vertices with a single gene (as these components admit a perfect matching because the genomes are balanced). From the shown components C_1, C_3, and C_4 admit a perfect matching while C_2 and C_5 do not. The propagation of the intergenic region $(1,1)$ in the component C_5 is shown in red. (Color figure online)

One important property is that $\mathcal{B}(\mathbb{S},\mathbb{P})$ is a bipartite graph, because two vertices corresponding to subgenomes of $\mathcal{G}$, or two vertices corresponding to subgenomes of $\mathcal{H}$, are never connected with edges. Consequently, each connected component of $\mathcal{B}(\mathbb{S},\mathbb{P})$ is also a bipartite graph. We say that a subgenome belongs to a connected component if it corresponds to a vertex in that component. The following lemma relates the connected components with the possible common partitions.

Lemma 1. *Let $\mathcal{G}$ be a rigid genome and $\mathcal{H}$ be a flexible genome, such that $\mathcal{G}$ and $\mathcal{H}$ are balanced. Given a partition $\mathbb{S}$ of $\mathcal{G}$ and a partition $\mathbb{P}$ of $\mathcal{H}$. Then $\mathcal{P} = (\mathbb{S},\mathbb{P})$ is a common signed flexible partition of $\mathcal{G}$ and $\mathcal{H}$, if and only if all components of $\mathcal{B}(\mathbb{S},\mathbb{P})$ admit a perfect matching.*

Proof. First, we prove that if there is a common signed flexible partition $\mathcal{P} = (\mathbb{S},\mathbb{P})$ of $\mathcal{G}$ and $\mathcal{H}$, then a component C of $\mathcal{B}(\mathbb{S},\mathbb{P})$ must admit a perfect matching.

Consider a correspondence between the blocks of $\mathbb{S}$ and $\mathbb{P}$ and a subgenome $\mathcal{X}$ of $\mathcal{G}$ that belongs to C, $\mathcal{X}$ must be a subgenome of a block from $\mathbb{S}$. So there exists a unique subgenome $\mathcal{Y}$ from C in a block of $\mathbb{P}$ corresponding to the block of $\mathcal{X}$ whose vertex is connected to the vertex of $\mathcal{X}$. We are going to construct the perfect matching by matching each $\mathcal{X}$ with its correspondent $\mathcal{Y}$.

Let us show that for any two subgenomes $\mathcal{X}_1$ and $\mathcal{X}_2$ of $\mathcal{G}$ that belong to C the matching subgenomes $\mathcal{Y}_1$ and $\mathcal{Y}_2$ are distinct, which ensures that our matching is possible. Let B_1, B_2, B'_1, and B'_2 be the blocks of $\mathcal{P}$ containing $\mathcal{X}_1$, $\mathcal{X}_2$, $\mathcal{Y}_1$, and $\mathcal{Y}_2$, respectively. If $B_1 \neq B_2$, then $B'_1 \neq B'_2$, and consequently $\mathcal{Y}_1 \neq \mathcal{Y}_2$. If $B_1 = B_2$, then $B'_1 = B'_2$. In that case, $\mathcal{X}_1$ and $\mathcal{X}_2$ are in two distinct positions of B_1, which means $\mathcal{Y}_1$ and $\mathcal{Y}_2$ are in two distinct positions of B'_1, so $\mathcal{Y}_1 \neq \mathcal{Y}_2$.

We now prove that if all connected components of $\mathcal{B}(\mathbb{S}, \mathbb{P})$ admit a perfect matching, then $\mathcal{P}$ is a common signed flexible partition. To ensure that $\mathcal{P}$ is a common partition, we must show that there is a one-to-one correspondence between the blocks of $\mathbb{S}$ and the blocks of $\mathbb{P}$.

The proof is by induction on the size of the sequences $\mathbb{S}$ and $\mathbb{P}$. If both sequences have a single block, then the blocks of both sequences must be equal, otherwise the largest of the two blocks would be the only one in its component, and that component would not have a perfect matching.

Now consider that one of the sequences has more than one block. In that case, both sequences must have more than one block, otherwise either the component of the block in the sequence with one block or the component of some block of the other sequence would not admit a perfect matching. Take the largest block in any one of the two sequences. Without loss of generality, assume that this is a block B from $\mathbb{S}$. Once the connected component of B admits a perfect matching, there must be a substring B' from $\mathcal{H}$ that is partially compatible with B, and does not have a breakpoint from $\mathbb{P}$. The substring B' must be a block of $\mathbb{P}$, because it does not have a breakpoint and there is no block in $\mathbb{P}$ bigger than B'. If we remove the block B from $\mathbb{S}$ and the block B' from $\mathbb{P}$, we have two sequences $\mathbb{S}'$ and $\mathbb{P}'$ such that $\mathcal{B}(\mathbb{S}', \mathbb{P}')$ admits a perfect matching, because after removing B and B' we only removed pairs of vertices from the matches of the components from $\mathcal{B}(\mathbb{S}, \mathbb{P})$. By induction hypothesis, there is a one-to-one correspondence between the blocks of $\mathbb{S}'$ and the blocks of $\mathbb{P}'$. By including B and B' we have a one-to-one correspondence between the blocks of $\mathbb{S}$ and the blocks of $\mathbb{P}$. $\qquad\square$

Let us define some ideas to work with connected components of the block compatibility graph. For any subgenome $\mathcal{X}$ in a block of $\mathbb{S}$ or $\mathbb{P}$, let $\mathcal{B}^C(\mathbb{S}, \mathbb{P}, \mathcal{X})$ be the connected component of $\mathcal{B}(\mathbb{S}, \mathbb{P})$ containing $\mathcal{X}$. We recursively define the *propagation* of an intergenic region $\breve{A}_i$ in a subgenome $\mathcal{X} = (A, \breve{A})$ from the component $C = \mathcal{B}^C(\mathbb{S}, \mathbb{P}, \mathcal{X})$ as the set $\mathtt{prop}(\breve{A}_i, C)$ of intergenic regions, such that $\breve{A}_i \in \mathtt{prop}(\breve{A}_i, C)$ and, for a genome $(B, \breve{B}) \in \mathcal{B}^C(\mathbb{S}, \mathbb{P}, \mathcal{X})$ corresponding to a vertex u, the intergenic region $\breve{B}_j$ is in $\mathtt{prop}(\breve{A}_i, C)$ if there is a genome $(C, \breve{C}) \in \mathcal{B}^C(\mathbb{S}, \mathbb{P}, \mathcal{X})$ corresponding to a vertex v, such that there is an edge $e_{v,u}$ in $\mathcal{B}(\mathbb{S}, \mathbb{P})$, and there is a intergenic region $\breve{C}_k \in \mathtt{prop}(\breve{A}_i, C)$, such that $k = j$ and $(B, \breve{B})$ and $(C, \breve{C})$ are fully compatible or $k = |\breve{B}| - j + 1$ and $(B, \breve{B})$ and $rev((C, \breve{C}))$ are fully compatible. In Fig. 3, we indicate the propagation of one intergenic region as an example.

We say that two connected components C and C' of $\mathcal{B}(\mathbb{S}, \mathbb{P})$ *intersect* if there are two subgenomes $\mathcal{X}$ and $\mathcal{Y}$ such that $C = \mathcal{B}^C(\mathbb{S}, \mathbb{P}, \mathcal{X})$, $C' = \mathcal{B}^C(\mathbb{S}, \mathbb{P}, \mathcal{Y})$, and $\mathcal{X}$ has at least one intergenic region in common with $\mathcal{Y}$. Any such intergenic region and intergenic regions in their propagations are said to be in the intersection between C and C'. Let us extend the notion of subgenomes to connected components of $\mathcal{B}(\mathbb{S}, \mathbb{P})$. We say that $C \subset \mathcal{B}^C(\mathbb{S}, \mathbb{P}, \mathcal{X})$ if $\mathcal{Y} \subset \mathcal{X}$ for some $\mathcal{Y}$, such that $C = \mathcal{B}^C(\mathbb{S}, \mathbb{P}, \mathcal{Y})$. In Fig. 3, we have $C_2 \subset C_5$ and $C_4 \subset C_5$. If for two components C and C' the relation $C \subset C'$ is not true, we use the notation $C \not\subset C'$. The next lemma shows that the definition of $\subset$ for components is not dependent on the choice of $\mathcal{X}$.

Lemma 2. *Given a rigid genome $\mathcal{G}$ and a flexible genome $\mathcal{H}$, such that $\mathcal{G}$ and $\mathcal{H}$ are balanced. Let $\mathbb{S}$ be a partition of $\mathcal{G}$ and $\mathbb{P}$ be a partition of $\mathcal{H}$. Given two subgenomes $\mathcal{X}$ and $\mathcal{Y}$ from blocks of $\mathbb{S}$ or $\mathbb{P}$, if $\mathcal{Y} \subset \mathcal{X}$ then for every $\mathcal{X}'$ corresponding to a vertex of $\mathcal{B}^C(\mathbb{S}, \mathbb{P}, \mathcal{X})$ there is a genome $\mathcal{Y}'$ corresponding to a vertex of $\mathcal{B}^C(\mathbb{S}, \mathbb{P}, \mathcal{Y})$, such that $\mathcal{Y}' \subset \mathcal{X}'$.*

Proof. The proof is by induction on the number of vertices in the connected component $\mathcal{B}^C(\mathbb{S}, \mathbb{P}, \mathcal{X})$. For the base case, if $\mathcal{B}^C(\mathbb{S}, \mathbb{P}, \mathcal{X})$ has a single vertex, then $\mathcal{X}'$ must be $\mathcal{X}$ and we can take $\mathcal{Y}' = \mathcal{Y}$.

Now consider a connected component $\mathcal{B}^C(\mathbb{S}, \mathbb{P}, \mathcal{X})$ with $k \geq 2$ vertices. Let v be a vertex of $\mathcal{B}^C(\mathbb{S}, \mathbb{P}, \mathcal{X})$ corresponding to a genome $\mathcal{X}'$. If $\mathcal{X}' = \mathcal{X}$, we can take $\mathcal{Y}' = \mathcal{Y}$. Otherwise, let u be a vertex from $\mathcal{B}^C(\mathbb{S}, \mathbb{P}, \mathcal{X})$ corresponding to a genome $\mathcal{X}''$, such that there is an edge $e_{u,v}$ in $\mathcal{B}(\mathbb{S}, \mathbb{P})$.

If we remove v, by induction hypothesis, there is a genome $\mathcal{Y}''$ corresponding to a vertex of $\mathcal{B}^C(\mathbb{S}, \mathbb{P}, \mathcal{Y})$, such that $\mathcal{Y}'' \subset \mathcal{X}''$. Once $\mathcal{X}''$ and $\mathcal{X}'$ are partially compatible, there must be a subgenome $\mathcal{Y}'$ of $\mathcal{X}'$ that is partially compatible with $\mathcal{Y}''$ and consequently corresponding to a vertex of $\mathcal{B}^C(\mathbb{S}, \mathbb{P}, \mathcal{Y})$. $\square$

Let $\mathbf{T}_{\mathbb{S}, \mathbb{P}}$ be the set of all connected components of $\mathcal{B}(\mathbb{S}, \mathbb{P})$ that do not admit a perfect matching, and consider the subset $\mathbf{T}_{\mathbb{S}, \mathbb{P}}^{\min} = \{C \in \mathbf{T}_{\mathbb{S}, \mathbb{P}} | C' \not\subset C, \forall C' \in \mathbf{T}_{\mathbb{S}, \mathbb{P}} \text{ and } C' \neq C\}$. By Lemma 1, $(\mathbb{S}, \mathbb{P})$ is a common partition if and only if $\mathbf{T}_{\mathbb{S}, \mathbb{P}}$ is empty. If that is not the case, we must add at least one breakpoint in one genome from each component of $\mathbf{T}_{\mathbb{S}, \mathbb{P}}$. Note that, to include a breakpoint in some genome corresponding to a vertex on a component $C' \in \mathbf{T}_{\mathbb{S}, \mathbb{P}} \setminus \mathbf{T}_{\mathbb{S}, \mathbb{P}}^{\min}$, it suffices to include a breakpoint in some genome of a component $C \in \mathbf{T}_{\mathbb{S}, \mathbb{P}}^{\min}$. Considering the genomes of Fig. 3, in $\mathbf{T}_{[\mathcal{G}], [\mathcal{H}]}^{\min}$ we have the component C_2 and the components corresponding to the following subgenomes (each with a single vertex): $([+I \ -A \ +A], [2 \ 3])$, $([+B \ -B], [0])$, $([-C \ +F], [4])$, $([+I \ -A \ +A], [(0, 2) \ (1, 1)])$, $([+B \ +C], [(1, 2)])$, and $([+B \ +F], [(2, 3)])$.

There is a useful property of the intersection of components in $\mathbf{T}_{[\mathcal{G}], [\mathcal{H}]}^{\min}$. We describe it in the following lemma.

Lemma 3. *Given a rigid genome $\mathcal{G}$ and a flexible genome $\mathcal{H}$, such that $\mathcal{G}$ and $\mathcal{H}$ are balanced. Let $C, C' \in \mathbf{T}_{[\mathcal{G}], [\mathcal{H}]}^{\min}$ be two components, such that C and C' intersect. Then there is a component C'' of $\mathcal{B}(\mathcal{G}, \mathcal{H})$ that admits a perfect matching and whose genomes contain every intergenic region from the intersection between C and C'.*

Proof. By the definition of intersection, there are two subgenomes $\mathcal{X}$ and $\mathcal{Y}$ such that $C = \mathcal{B}^C([\mathcal{G}], [\mathcal{H}], \mathcal{X})$, $C' = \mathcal{B}^C([\mathcal{G}], [\mathcal{H}], \mathcal{Y})$, and $\mathcal{X}$ has at least one intergenic region in common with $\mathcal{Y}$. The subgenome $\mathcal{W}$ in the intersection of the genomes $\mathcal{X}$ and $\mathcal{Y}$ has at least one intergenic region, and it is a subgenome of $\mathcal{X}$ and $\mathcal{Y}$. Consequently, the component $C'' = \mathcal{B}^C([\mathcal{G}], [\mathcal{H}], \mathcal{W})$ is a subcomponent of C and C'. By definition of $\mathbf{T}^{\min}_{[\mathcal{G}],[\mathcal{H}]}$, C'' must admit a perfect matching. Furthermore, by definition of a subcomponent and by Lemma 2, the genomes of C'' contain every intergenic region from the intersection between C and C'. $\square$

The next lemma describes the possible intersections of elements from $\mathbf{T}^{\min}_{[\mathcal{G}],[\mathcal{H}]}$.

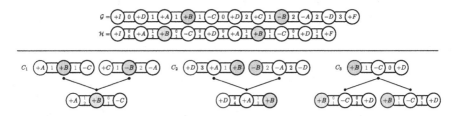

Fig. 4. A rigid genome $\mathcal{G}$ and a flexible genome $\mathcal{H}$ followed by some components of the block compatibility graph $\mathcal{B}(\mathcal{G}, \mathcal{H})$. We include colors in some genes to help the identification of the positions of subgenomes. Component C_1 intersects component C_2 (the intergenic regions of that intersection are shown in red) and component C_3 (the intergenic regions of that intersection are shown in blue). (Color figure online)

Lemma 4. *Given a rigid genome $\mathcal{G}$ and a flexible genome $\mathcal{H}$, such that $\mathcal{G}$ and $\mathcal{H}$ are balanced. For every component $C \in \mathbf{T}^{\min}_{[\mathcal{G}],[\mathcal{H}]}$, one of the following cases is true.*

1. *C does not intersect any other component of $\mathbf{T}^{\min}_{[\mathcal{G}],[\mathcal{H}]}$.*
2. *There is at least one intergenic region $\breve{A}_i$ in every genome $(A, \breve{A})$ of C, such that genomes of every component of $\mathbf{T}^{\min}_{[\mathcal{G}],[\mathcal{H}]}$ that intersects C contain an intergenic region in $\mathbf{prop}(\breve{A}_i, C)$.*
3. *There are two intergenic regions $\breve{A}_i$ and $\breve{A}_j$ in every genome $(A, \breve{A})$ of C, such that genomes of every component of $\mathbf{T}^{\min}_{[\mathcal{G}],[\mathcal{H}]}$ that intersects C contain an intergenic region in $\mathbf{prop}(\breve{A}_i, C)$ or in $\mathbf{prop}(\breve{A}_j, C)$.*

Proof. We are going to prove the lemma by showing that if cases 1 or 2 do not occur, then case 3 is true. Note that, for any component $C' \in \mathbf{T}^{\min}_{[\mathcal{G}],[\mathbb{P}]}$ that intersects C' we have $C' \not\subset C$ and $C \not\subset C'$. So the genomes $\mathcal{X} = (A, \breve{A})$ from C and $\mathcal{Y}$ from C' that intersect must include either $\breve{A}_1$ or $\breve{A}_{|A|}$. Note that the set $\mathbf{prop}(\breve{A}_1, C) \cup \mathbf{prop}(\breve{A}_{|A|}, C)$ contains every first and last intergenic regions in the genomes of C, and the lemma follows. $\square$

The original algorithm for rigid intergenic regions [11] does not consider the intersections from case 3, which is a flaw in the (claimed) proof of the $2k$-approximation. We fix such flaw using the *block intersection graph* $\mathcal{I}(\mathcal{G}, \mathcal{H})$ to describe such intersections. The vertices of this graph are the intersection from components of $\mathbf{T}^{\min}_{[\mathcal{G}],[\mathcal{H}]}$ including at least one component from case 3, and two vertices are connected with an edge if their corresponding intersections include the same connected component. Note that we need to add a breakpoint in the elements of a vertex cover of this graph to be able to add a breakpoint in all components from case 3. For example, the components from Fig. 4 result in two vertices (the intersections of $\{C_1, C_2\}$ and of $\{C_1, C_3\}$) connected by one edge (resulting from the component C_1).

For every genome $\mathcal{X}$ such that $\mathcal{B}^C([\mathcal{G}], [\mathcal{H}], \mathcal{X})$ is in $\mathbf{T}^{\min}_{[\mathcal{G}],[\mathcal{H}]}$, let us associate to $\mathcal{X}$ a set $\mathtt{break}(\mathcal{X}, \mathcal{G}, \mathcal{H})$ with one or two intergenic regions, that will become breakpoints. The association depends on the cases in Lemma 4. In case 1, $\mathtt{break}(\mathcal{X}, \mathcal{G}, \mathcal{H})$ has only one arbitrary intergenic region of $\mathcal{X}$. In cases 2 and 3, $\mathtt{break}(\mathcal{X}, \mathcal{G}, \mathcal{H})$ has the intergenic regions of $\mathcal{X}$ described in the respective cases.

Let us extend this definition to include the new possibilities that appear when we start to build the partitions of $\mathcal{G}$ and $\mathcal{H}$. Consider the breakpoints from a partition $\mathbb{S}$ of $\mathcal{G}$ and a partition $\mathbb{P}$ of $\mathcal{H}$. For a subgenome $\mathcal{X} = (A, \breve{A})$ belonging to a component $C \in \mathbf{T}^{\min}_{\mathbb{S},\mathbb{P}}$, we will associate a set with one or two intergenic regions, denoted by $\mathtt{break}(\mathcal{X}, \mathbb{S}, \mathbb{P})$, defined as follows. If $\mathcal{X}$ is in a component of $\mathbf{T}^{\min}_{\mathcal{G},\mathcal{H}}$ then $\mathtt{break}(\mathcal{X}, \mathbb{S}, \mathbb{P}) = \mathtt{break}(\mathcal{X}, \mathcal{G}, \mathcal{H})$. Otherwise, there must be a breakpoint in some other subgenome from $\mathcal{B}^C([\mathcal{G}], [\mathcal{H}], \mathcal{X})$ that made this component part of $\mathbf{T}^{\min}_{\mathbb{S},\mathbb{P}}$. So $\mathtt{break}(\mathcal{X}, \mathbb{S}, \mathbb{P})$ is an intergenic region $\breve{A}_i \in \mathtt{prop}(\breve{B}_j, \mathcal{B}^C([\mathcal{G}], [\mathcal{H}], \mathcal{X}))$, such that $\breve{B}_j$ is a breakpoint from a genome $(B, \breve{B}) \in \mathcal{B}^C([\mathcal{G}], [\mathcal{H}], \mathcal{X})$.

Now consider Algorithm 1, that inserts breakpoints in the genomes according to the elements of $\mathbf{T}^{\min}_{\mathbb{S},\mathbb{P}}$ and the values of $\mathtt{break}$. Let us prove that this strategy for the construction of a common partition ensures an approximation.

Theorem 1. *Algorithm 1 has an approximation factor of $2k$ for the SMC-FISP problem between a rigid genome $\mathcal{G}$ and a flexible genome $\mathcal{H}$, where $k = occ\_max(\mathcal{G})$.*

Proof. By Lemma 1, if $\mathbf{T}^{\min}_{\mathbb{S}^{i-1},\mathbb{P}^{i-1}} = \emptyset$ then $(\mathbb{S}, \mathbb{P}) = (\mathbb{S}^{i-1}, \mathbb{P}^{i-1})$ is a common signed flexible intergenic partition between $\mathcal{G}$ and $\mathcal{H}$. So we just have to ensure that the algorithm inserts at most $2kR^*$ breakpoints, where R^* is the number of breakpoints in an optimal common partition.

Let us look at the breakpoints in an optimal common partition $(\mathbb{S}^*, \mathbb{P}^*)$ of $\mathcal{G}$ and $\mathcal{H}$. We know that there must be at least one breakpoint in a genome of each component of $C \in \mathbf{T}^{\min}_{[\mathcal{G}],[\mathcal{H}]}$, we will denote the set of breakpoints in C by $\mathtt{break}^*(C, \mathcal{G}, \mathcal{H})$.

Consider the breakpoints $\mathtt{break}(\mathcal{X}, \mathbb{S}^{i-1}, \mathbb{P}^{i-1})$ inserted in the subgenome $\mathcal{X} = (A, \breve{A})$ during the i-th iteration of the algorithm. Note that we have two cases: (i) $\mathcal{X}$ is in a component $C \in \mathbf{T}^{\min}_{[\mathcal{G}],[\mathcal{H}]}$ that does not yet have any breakpoint in its genomes; (ii) during some previous iteration j, we included a breakpoint

---

Algorithm 1: $2k$-approximation for SMCFISP

In: A rigid genome $\mathcal{G}$ and a flexible genome $\mathcal{H}$, such that $\mathcal{G}$ and $\mathcal{H}$ are balanced
Out: A signed common flexible intergenic partition between $\mathcal{G}$ and $\mathcal{H}$

1 $(\mathbb{S}^0, \mathbb{P}^0) \leftarrow ([\mathcal{G}], [\mathcal{H}])$
2 $i \leftarrow 0$
3 **while** $\mathbf{T}^{\min}_{\mathbb{S}^{i-1}, \mathbb{P}^{i-1}} \neq \emptyset$ **do**
4 $\quad$ $i \leftarrow i+1$
5 $\quad$ Choose a subgenome $\mathcal{X}$ of $\mathcal{G}$ or $\mathcal{H}$ that belongs to a component in $\mathbf{T}^{\min}_{\mathbb{S}^{i-1}, \mathbb{P}^{i-1}}$
6 $\quad$ $(\mathbb{S}^i, \mathbb{P}^i) \leftarrow \mathbb{S}^{i-1}$ and $\mathbb{P}^{i-1}$ with the breakpoints in $\mathtt{break}(\mathcal{X}, \mathbb{S}^{i-1}, \mathbb{P}^{i-1})$.
7 **return** $(\mathbb{S}^i, \mathbb{P}^i)$

$\breve{B}_m$ in some subgenome $\mathcal{Y} = (B, \breve{B})$ from the component $\mathcal{B}^\mathcal{C}([\mathcal{G}], [\mathcal{H}], \mathcal{X})$ that caused that component to no longer admit a perfect matching.

Let us count the number R of breakpoints inserted in all instances of case (i) and compare it with R^*. For that, we look at the different cases from Lemma 4. In case 1, C does not intersect any other component of $\mathbf{T}^{\min}_{[\mathcal{G}], [\mathcal{H}]}$, so there is no other component of $C' \in \mathbf{T}^{\min}_{[\mathcal{G}], [\mathcal{H}]}$, such that $\mathtt{break}^*(C, \mathcal{G}, \mathcal{H}) = \mathtt{break}^*(C', \mathcal{G}, \mathcal{H})$. In this case, we count one breakpoint for both R and R^*. In case 2, $\mathtt{break}(\mathcal{X}, \mathbb{S}^{i-1}, \mathbb{P}^{i-1})$ has only the intergenic region $\breve{A}_\ell$, such that every component $C' \in \mathbf{T}^{\min}_{[\mathcal{G}], [\mathbb{P}]}$, with $\mathtt{break}^*(C, \mathcal{G}, \mathcal{H}) = \mathtt{break}^*(C', \mathcal{G}, \mathcal{H})$, contains an intergenic region in $\mathtt{prop}(\breve{A}_\ell, C)$. In this case, we also count one breakpoint for R and one breakpoint for R^*. All other components C' with $\mathtt{break}^*(C, \mathcal{G}, \mathcal{H}) = \mathtt{break}^*(C', \mathcal{G}, \mathcal{H})$ will no longer appear in case (i).

Let us consider all occurrences of case 3, by looking at the graph $\mathcal{I}(\mathcal{G}, \mathcal{H})$. Let VC^* be the size of a minimum vertex cover of $\mathcal{I}(\mathcal{G}, \mathcal{H})$. We know that the optimal common partition needs at least VC^* breakpoints to include all components of case 3. Besides, by Lemma 3, for each such breakpoints we are removing a genome from at least one component with perfect matching from $\mathcal{B}(\mathcal{G}, \mathcal{H})$, so the optimal common partition will have at least one other breakpoint in these components of $\mathbf{T}^{\min}_{[\mathcal{G}], [\mathcal{H}]}$. By definition of $\mathtt{break}(\mathcal{X}, \mathbb{S}^{i-1}, \mathbb{P}^{i-1})$, we are including breakpoints in the two vertices from the edge corresponding to the component C in $\mathcal{I}(\mathcal{G}, \mathcal{H})$. That process corresponds to the classical 2-approximation algorithm for the minimum vertex cover problem, so we are including at most $2VC^*$ vertices. In all occurrences of this case, we count at most $2VC^*$ breakpoints for R and $2VC^*$ breakpoints for R^*. Considering all three cases, we can see that $R \leq R^*$.

Now, we look at the breakpoints inserted in case (ii). By definition of $\mathtt{break}$, we include the breakpoint in $\mathtt{prop}(\breve{B}_m, \mathcal{B}^\mathcal{C}(\mathbb{S}^j, \mathbb{P}^j, \mathcal{X}))$, so it will be between genes equal to B_m and B_{m+1}. As we only have k copies of each gene in $\mathcal{G}$, k copies of each gene in $\mathcal{H}$, and in case (i) we add less than R^* breakpoints, we will have at most $2kR^*$ breakpoints in the resulting common partition. $\qquad\square$

It is worth noting that, by setting the intervals corresponding to every flexible intergenic region to a single number, our algorithm also ensures a $2k$-

approximation for the SMCISP problem. Besides by setting the intergenic regions to zero, our algorithm ensures a $2k$-approximation to the variant of the problem that does not consider intergenic regions.

Now, we present a relation between SMCFISP and the distance problems that allows us to adapt an approximation for the SMCFISP problem to obtain an approximation for the SFIRD problem.

Theorem 2. *Let $\mathcal{G}$ be a rigid genome and $\mathcal{H}$ be a flexible genome, such that $\mathcal{G}$ and $\mathcal{H}$ are balanced. Let $(\mathbb{S}, \mathbb{P})$ be the signed common flexible intergenic partition between $\mathcal{G}$ and $\mathcal{H}$ returned by an ℓ-approximation for the SMCISP problem. If there is an algorithm that produces a sequence of d reversals, with $d \leq r \cdot cost((\mathbb{S}, \mathbb{P}))$, capable of transforming $\mathcal{G}$ into a genome fully compatible with $\mathcal{H}$, then it is a $2r\ell$-approximation for the SFIRD problem.*

Proof. Let $(\mathbb{S}^*, \mathbb{P}^*)$ be a minimum cost signed common flexible intergenic partition between $\mathcal{G}$ and $\mathcal{H}$. An ℓ-approximation algorithm for the SMCFISP problem returns a common signed flexible intergenic partition $(\mathbb{S}, \mathbb{P})$, such that $cost(\mathbb{S}^*, \mathbb{P}^*) \leq cost(\mathbb{S}, \mathbb{P}) \leq \ell cost(\mathbb{S}^*, \mathbb{P}^*)$.

Consider a sequence of reversals that turns $\mathcal{G}$ into a genome $\mathcal{H}'$ compatible with $\mathcal{H}$. We know that if there is a common intergenic partition $\mathcal{P}$ of minimum cost between $\mathcal{G}$ and $\mathcal{H}'$, then any sequence of reversals that transforms $\mathcal{G}$ into $\mathcal{H}'$ must have size at least $\frac{cost(\mathcal{P})}{2}$ [11, Lemma 5]. As $\mathcal{H}'$ is fully compatible with $\mathcal{H}$, there must also be a common flexible intergenic partition between $\mathcal{G}$ and $\mathcal{H}$ with cost $cost(\mathcal{P})$. Once $(\mathbb{S}^*, \mathbb{P}^*)$ has minimum cost, we have that $cost((\mathbb{S}^*, \mathbb{P}^*)) \leq cost(\mathcal{P})$. Consequently, any sequence of reversals that transforms $\mathcal{G}$ into a genome fully compatible with $\mathcal{H}$ must have size at least $\frac{cost((\mathbb{S}^*, \mathbb{P}^*))}{2}$ (in other words, $d_{FR}(\mathcal{G}, \mathcal{H}) \geq \frac{cost((\mathbb{S}^*, \mathbb{P}^*))}{2})$. As we can turn $\mathcal{G}$ into a genome fully compatible with $\mathcal{H}$ using $k \leq r \cdot cost((\mathbb{S}, \mathbb{P}))$ reversals, we have $d_{FR}(\mathcal{G}, \mathcal{H}) \leq k \leq 2r\ell d_{FR}(\mathcal{G}, \mathcal{H})$.   □

Corollary 1. *An ℓ-approximation for the SMCFISP problem ensures a 4ℓ-approximation for the SFIRD problem.*

Proof. Let $\mathcal{G} = (A, \breve{A})$ be a rigid genome and $\mathcal{H}$ be a flexible genome, such that $\mathcal{G}$ and $\mathcal{H}$ are balanced. Let $(\mathbb{S}, \mathbb{P})$ be the signed common flexible intergenic partition between $\mathcal{G}$ and $\mathcal{H}$ returned by an ℓ-approximation for the SMCFISP problem.

We can use the correspondence between the blocks in $(\mathbb{S}, \mathbb{P})$ to derive a correspondence between the genes, such that we can treat $\mathcal{G}$ and $\mathcal{H}$ as if they did not have replicas.

In that scenario, Brito *et al.* [4, Theorem 3.12] showed an algorithm that find a sequence R of $2(|\breve{S}| - f(\mathcal{G}, \mathcal{H}))$ reversals that turns $\mathcal{G}$ into a genome fully compatible with $\mathcal{H}$, such that $f(\mathcal{G}, \mathcal{H})$ is the number of free cycles from a structure called Flexible Weighted Cycle Graph. What is important to our proof is that there is one such cycle for each intergenic region of $\breve{A}$ that is not a breakpoint in $\mathbb{S}$. Consequently, $2(|\breve{S}| - f(\mathcal{G}, \mathcal{H})) < 2cost((\mathbb{S}, \mathbb{P}))$, and the theorem follows by setting $r = 2$ in Theorem 2.   □

4 Conclusion

In this work, we proposed a $2k$-approximation for the Signed Minimum Common Flexible Intergenic String Partition problem and showed that it can also be used to approximate the Signed Flexible Intergenic Reversal Distance (SFIRD) problem with a factor of $8k$. The relationship between the problems can be used as well to directly convert any improvement in the partition problem approximation (resp. in the distance problem approximation) without gene repetition, into an improvement in the approximation for the distance problem with gene repetition.

Future works in problems involving flexible intergenic regions could establish similar relationships considering other rearrangement events such as transposition, deletion, and insertion. Another path is to study other approaches to the partition problem, such as heuristics or exact algorithms for particular sets of instances.

Acknowledgment. This work was supported by the Coordenação de Aperfeiçoamento de Pessoal de Nível Superior - Brasil (CAPES) - Finance Code 001 and the São Paulo Research Foundation, FAPESP (grants 2013/08293-7 , 2021/13824-8 , and 2022/13555-0).

References

1. Bafna, V., Pevzner, P.A.: Sorting by transpositions. SIAM J. Disc. Math. **11**(2), 224–240 (1998)
2. Biller, P., Guéguen, L., Knibbe, C., Tannier, E.: Breaking good: accounting for fragility of genomic regions in rearrangement distance estimation. Genome Biol. Evol. **8**(5), 1427–1439 (2016)
3. Biller, P., Knibbe, C., Beslon, G., Tannier, E.: Comparative genomics on artificial life. In: Beckmann, A., Bienvenu, L., Jonoska, N. (eds.) CiE 2016. LNCS, vol. 9709, pp. 35–44. Springer, Cham (2016). https://doi.org/10.1007/978-3-319-40189-8_4
4. Brito, K.L., Alexandrino, A.O., Oliveira, A.R., Dias, U., Dias, Z.: Genome rearrangement distance with a flexible intergenic regions aspect. IEEE/ACM Trans. Comput. Biol. Bioinf. **20**(3), 1641–1653 (2023)
5. Brito, K.L., Oliveira, A.R., Alexandrino, A.O., Dias, U., Dias, Z.: Reversal and transposition distance of genomes considering flexible intergenic regions. Procedia Comput. Sci. **195**, 21–29 (2021)
6. Fertin, G., Jean, G., Tannier, E.: Algorithms for computing the double cut and join distance on both gene order and intergenic sizes. Algor. Molec. Biol. **12**(1), 16 (2017)
7. Hannenhalli, S., Pevzner, P.A.: Transforming cabbage into turnip: polynomial algorithm for sorting signed permutations by reversals. J. ACM **46**(1), 1–27 (1999)
8. Kolman, P., Waleń, T.: Reversal distance for strings with duplicates: linear time approximation using hitting set. In: Erlebach, T., Kaklamanis, C. (eds.) WAOA 2006. LNCS, vol. 4368, pp. 279–289. Springer, Heidelberg (2007). https://doi.org/10.1007/11970125_22
9. Oliveira, A.R., et al.: Sorting signed permutations by intergenic reversals. IEEE/ACM Trans. Comput. Biol. Bioinf. **18**(6), 2870–2876 (2021)

10. Radcliffe, A.J., Scott, A.D., Wilmer, E.L.: Reversals and transpositions over finite alphabets. SIAM J. Disc. Math. **19**(1), 224–244 (2005)
11. Siqueira, G., Alexandrino, A.O., Dias, Z.: Signed rearrangement distances considering repeated genes and intergenic regions. In: Proceedings of 14th International Conference on Bioinformatics and Computational Biology (BICoB 2022), vol. 83, pp. 31–42 (2022)

The Ordered Covering Problem
in Distance Geometry

Michael Souza[1]($\boxtimes$) , Nilton Maia[1], and Carlile Lavor[2]

[1] Federal University of Ceará, Fortaleza 60440-900, Brazil
{michael,nilton}@ufc.br
[2] IMECC, University of Campinas, Campinas 13081-970, Brazil
clavor@unicamp.br

Abstract. This study is motivated by the Discretizable Molecular Distance Geometry Problem (DMDGP), a specific category in Distance Geometry, where the search space is discrete. We address the challenge of ordering the DMDGP constraints, a critical factor in the performance of the state-of-the-art SBBU algorithm. To this end, we formalize the constraint ordering problem as a vertex cover problem, which diverges from traditional covering problems due to the substantial importance of the sequence of vertices in the covering. In order to solve the covering problem, we propose a greedy heuristic and compare it to the ordering of the SBBU. The computational results indicate that the greedy heuristic outperforms the SBBU ordering by an average factor of 1,300×.

1 Introduction

The Discretizable Molecular Distance Geometry Problem (DMDGP) is a notable combinatorial optimization problem encountered in the determination of three-dimensional molecular structures using a set of interatomic distances [3]. This problem has attracted growing attention in computational chemistry and structural biology due to its extensive applications in molecular modeling [5].

The primary goal of the DMDGP is to identify a protein conformation that adheres to a collection of distance constraints, which are usually obtained from experimental data such as Nuclear Magnetic Resonance (NMR) spectroscopy [8]. The computational complexity of the DMDGP persists as a challenge, especially for large and flexible molecules.

The graph representation of the DMDGP offers valuable insights into the problem's structure and has proven useful in developing efficient solution techniques [3]. There is considerable research on vertex ordering in DMDGP instances [1]. However, research on the importance of edge ordering is still incipient. Recently, researchers have examined the connections between the DMDGP and the edge ordering of the associated graph, discovering that the problem becomes more manageable when the graph edges are ordered in a particular manner [2].

The formal definition of the DMDGP can be given as follows [3].

X. Guo et al. (Eds.): ISBRA 2023, LNBI 14248, pp. 255–266, 2023.
https://doi.org/10.1007/978-981-99-7074-2_20

Definition 1 (DMDGP). *Given a simple, undirected, weighted graph $G = (V, E, d)$, $|V| = n$, with weight function $d : E \to (0, \infty)$ and a vertex order $v_1, \ldots, v_n \in V$, such that*

- *$\{v_1, v_2, v_3\}$ is a clique;*
- *For every $i > 3$, v_i is adjacent to $v_{i-3}, v_{i-2}, v_{i-1}$ and*

$$d(v_{i-1}, v_{i-3}) < d(v_{i-1}, v_{i-2}) + d(v_{i-2}, v_{i-3}),$$

the DMDGP consists in finding a realization $x : V \to \mathbb{R}^3$, such that

$$\forall \{u, v\} \in E, \|x_u - x_v\| = d(u, v), \tag{1}$$

where $\| \cdot \|$ denotes the Euclidean norm, $x_v := x(v)$, and $d(u, v) := d(\{u, v\})$ (each equation in (1) is called a distance constraint).

Assuming the vertex ordering of the DMDGP definition and denoting the edge $e = \{v_i, v_j\}$ by $e = \{i, j\}$, we define the *discretization edges* by

$$E_d = \{\{i, j\} \in E : |i - j| \leq 3\}$$

and the *pruning edges* by

$$E_p = E - E_d.$$

The origin of the adjectives "pruning" and "discretization" in edge classification is linked to the Branch-and-Prune (BP) method, the first algorithm proposed for solving the DMDGP [4]. In the BP, the discretization edges are used to represent the search space as a binary tree, and the pruning edges, on the other hand, are used as pruning in a depth-first search for viable realizations.

For many years, BP was the most efficient approach for solving DMDPG, but recently the SBBU algorithm has taken over this position [2]. However, its performance is strongly dependent on the order given to the pruning edges and this issue is an open problem [2].

Given a permutation $\pi = (e_1, \ldots, e_m)$ of the pruning edges, the central idea of the SBBU is to reduce the DMDGP to a sequence $(P(e_1), \ldots, P(e_m))$ of feasibility subproblems, each of them associated with a different pruning edge. The set of (binary) variables of each subproblem $P(e)$ is given by $[e] = [\{i, j\}] = \{b_{i+3}, \ldots, b_j\}$ and its computational cost grows exponentially with the number of variables. The subproblems may share variables and the efficiency of the SBBU comes from the fact that the variables of each solved subproblem can be removed from all subsequent subproblems.

In the following, we give an example that will be used throughout the paper to facilitate understanding of the concepts presented.

Example 1: Let $G = (V, E, d)$ be a DMDGP instance given by

$$V = \{v_1, v_2, \ldots, v_{15}\} \text{ and } E_p = \{e_1, e_2, e_3, e_4\},$$

where

$$e_1 = \{v_1, v_8\},$$
$$e_2 = \{v_5, v_{15}\},$$
$$e_3 = \{v_6, v_{14}\},$$
$$e_4 = \{v_{11}, v_{15}\}.$$

If we take the permutation $\pi = (e_1, e_2, e_3, e_4)$, the SBBU will solve the sequence of subproblems $(P(e_1), P(e_2), P(e_3), P(e_4))$, whose the sets of variables are $\{b_4, b_5, \ldots, b_8\}$, $\{b_8, b_9, \ldots, b_{15}\}$, $\{b_9, b_{10}, \ldots, b_{14}\}$, and $\{b_{14}, b_{15}\}$, respectively.

After solving the first subproblem $P(e_1)$, built from the edge $e_1 = \{1, 8\}$, we can remove the variable b_8 from the set of variables of remaining subproblems. For the same reason, after solving the second subproblem, i.e., $P(e_2)$, there will be no more available variables and the remaining subproblems are already solved.

In the worst case scenario, the cost of solving the sequence of subproblems $(P(e_1), P(e_2), P(e_3), P(e_4))$ will be

$$F(\pi) = 2^{|\{b_4, b_5, b_6, b_7, b_8\}|} + 2^{|\{b_9, b_{10}, b_{11}, b_{12}, b_{13}, b_{14}, b_{15}\}|}$$
$$= 2^5 + 2^7$$
$$= 160.$$

Another permutation, given by $\hat{\pi} = (e_4, e_3, e_2, e_1)$, has cost

$$F(\hat{\pi}) = 2^{|\{b_{14}, b_{15}\}|} + 2^{|\{b_9, b_{10}, b_{11}, b_{12}, b_{13}\}|} + 2^{|\{b_8\}|} + 2^{|\{b_4, b_5, b_6, b_7\}|}$$
$$= 2^2 + 2^5 + 2^1 + 2^4$$
$$= 54,$$

since the related subproblems P_4, P_3, P_2, P_1 would have, respectively, the variables $\{b_{14}, b_{15}\}$, $\{b_9, b_{10}, \ldots, b_{13}\}$, $\{b_8\}$, and $\{b_4, b_5, b_6, b_7\}$.

Our primary contribution is to formulate the SBBU pruning edge ordering problem as a graph covering problem. In addition to that, we propose a greedy heuristic to find such ordering and compare it to the ordering of the SBBU algorithm.

2   Preliminary Definitions

This section provides the definitions necessary to formalize the SBBU pruning edge ordering problem, called the *Ordered Covering Problem* (OCP).

Definition 2 (Edge Interval).
Given a graph $G = (V, E)$, for each $e = \{i, j\} \in E$, we define the edge interval

$$\llbracket e \rrbracket = \{i + 3, \ldots, j\}$$

and

$$\llbracket E \rrbracket = \bigcup_{e \in E} \llbracket e \rrbracket.$$

Definition 3 (Equivalence Relation in $\llbracket E \rrbracket$).
Given $i, j \in \llbracket E \rrbracket$, we say that

$$i \sim j \iff \{e \in E_p : i \in \llbracket e \rrbracket\} = \{e \in E_p : j \in \llbracket e \rrbracket\}.$$

In the Example 1, the vertex $u = 9$ is equivalent to vertex $v = 10$, because both belong to the same edge intervals, namely $\llbracket e_2 \rrbracket$ and $\llbracket e_3 \rrbracket$. However, $w = 14$ is not equivalent to $u = 9$, because $u = 9 \notin \llbracket e_4 \rrbracket$, but $w = 14 \in \llbracket e_4 \rrbracket$.

Definition 4 (Segment).
Let $S = \{\sigma_1, \ldots, \sigma_k\}$ be the partition of $\llbracket E \rrbracket$ induced by the equivalence relation of Definition 3. A segment is any element of the partition S.

In the Example 1, we have the partition $S = \{\sigma_1, \sigma_2, \sigma_3, \sigma_4, \sigma_5\}$ (see Fig. 1), where

$$\sigma_1 = \{4, 5, 6, 7\},$$
$$\sigma_2 = \{8\},$$
$$\sigma_3 = \{9, 10, 11, 12, 13\},$$
$$\sigma_4 = \{14\},$$
$$\sigma_5 = \{15\}.$$

Definition 5 (Pruning Edge Hypergraph). *Given a DMDGP instance with pruning edges $E_P = \{e_1, \ldots, e_m\}$ and a segment partition $S = \{\sigma_1, \ldots, \sigma_k\}$, we define the hypergraph of the pruning edges by $H = (E_P, T)$, where the set of vertices is E_P and, for each segment σ_i in S, there is an hyperedge $\tau_i \in T$ given by $\tau_i = \{e \in E_P : \sigma_i \subset \llbracket e \rrbracket\}$.*

In other words, the vertices of the pruning edge hypergraph are the pruning edges of the DMDGP graph and each of its hyperedge τ_i is the set of pruning edges whose intervals contain the segment σ_i. Also, note that there is a bijection between the set T of hyperedges in H and the set of segments in S. For simplicity, we will replace τ_i by σ_i in all representations and $H = (E_p, T)$ by $H = (E_p, S)$. Figure 2 illustrates the concepts given in Definition 5 associated to Example 1.

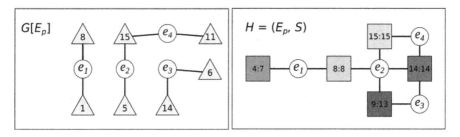

E_p	1	2	3	4	5	6	7	8	9	10	11	12	13	14	15
$e_1 = \{\,1, 8\,\}$	▓	▓	▓	▓	▓	▓	▓	░							
$e_2 = \{\,5, 15\,\}$					▓	▓		░		▓	▓	▓	▓	▓	░
$e_3 = \{\,6, 14\,\}$						▓	▓	▓	▓	▓	▓	▓	▓	░	
$e_4 = \{\,11, 15\,\}$											▓	▓	▓	▓	░
S				▓	▓	▓	▓	░	▓	▓	▓	▓	▓	░	
				σ_1		σ_2				σ_3			σ_4	σ_5	

Fig. 1. Segments of the graph defined in Example 1.

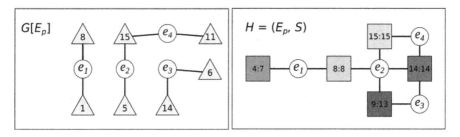

Fig. 2. Graph of Example 1 (restricted to the pruning edges) and the related hypergraph $H = (E_P, S)$.

3   The Ordered Covering Problem (OCP)

Before the definition of the OCP, we need additional concepts.

Definition 6 (Vertex Cover).
A hypergraph $H = (E_P, S)$ is said to be covered by $W \subset E_p$ if every hyperedge in S has at least one element in W.

In a conventional context, a vertex cover is a subset of vertices. However, in our application, the sequential arrangement of elements within the cover holds significance. Therefore, rather than a subset of vertices, our interest lies in an ordered list.

Definition 7 (Ordered Vertex Cover).
Given a pruning edge hypergraph $H = (E_p, S)$, a tuple $\pi = (\pi_1, \ldots, \pi_k)$ is an ordered vertex cover if H is covered by π.

For instance, considering the graph $G = (V, E)$ of Example 1, possible ordered covers are (e_1, e_2) or (e_1, e_3, e_4). However, (e_1, e_3) is not a cover, since the segment $\sigma_5 = \{15\}$ remains uncovered (see Fig. 2).

In the SBBU algorithm, every pruning edge e_i gives rise to a subproblem on binary variables. Each binary variable within this subproblem is uniquely associated with an element from $[\![e_i]\!]$. Leveraging DMDGP symmetries [7], binary variables from resolved subproblems can be eliminated, thereby retaining only those associated with vertices not included in earlier subproblems.

So, given a tuple $\pi = (\pi_1, \ldots, \pi_k)$ of pruning edges, we define

$$\Gamma(\pi_i) = [\![\pi_i]\!] - \bigcup_{j<i} [\![\pi_j]\!] \tag{2}$$

and the size of the search space for the subproblem corresponding to edge π_i is given by:

$$f(\pi_i) = \{2^{|\Gamma(\pi_i)|}, \text{ if } \Gamma(\pi_i) \neq \{\} \text{ and } 0, \text{ otherwise}\}. \tag{3}$$

Revisiting the Example 1, for the tuple $\pi = (\pi_1, \pi_2) = (e_1, e_2)$, the cost associated with each edge is computed as follows:

$$f(\pi_1) = 2^{|\{4,5,6,7,8\}|} = 2^5,$$
$$f(\pi_2) = 2^{|\{9,10,11,12,13,14,15\}|} = 2^7.$$

Likewise, the size of the search space for $\hat{\pi} = (e_4, e_3, e_2, e_1)$ is given by

$$f(\hat{\pi}_1) = 2^{|\{14,15\}|} = 2^2,$$
$$f(\hat{\pi}_2) = 2^{|\{9,10,11,12,13\}|} = 2^5,$$
$$f(\hat{\pi}_3) = 2^{|\{8\}|} = 2^1,$$
$$f(\hat{\pi}_4) = 2^{|\{4,5,6,7\}|} = 2^4.$$

The example above suggests the following definition.

Definition 8 (Ordered Covering Cost).
Given a segment hypergraph $H = (E_p, S)$, the total cost associated with the tuple $\pi = (\pi_1, \ldots, \pi_k)$ of pruning edges is calculated as:

$$F(\pi) = \sum_{i=1}^{k} f(\pi_i), \tag{4}$$

where f is the partial cost function defined in Eq. (3).

Finally, we can now define the Minimum Ordered Covering Problem.

Definition 9 (Ordered Covering Problem (OCP)).
Given a DMDGP instance $G = (V, E)$, with a pruning edge set E_P and a segment hypergraph $H = (E_P, S)$, the goal is to find:

$$\pi^* = \arg \min_{\pi \in \Pi(H)} F(\pi), \tag{5}$$

where $\Pi(H)$ represents all possible ordered vertex covers in H.

In the context of the graph represented in Fig. 2, the optimal solution is $\pi^* = (e_4, e_3, e_2, e_1)$, with a total cost $F(\pi^*) = 2^2 + 2^5 + 2^1 + 2^4 = 54$.

4   A Greedy Heuristic for the OCP

In this section, we propose a greedy heuristic (GD) to the OCP. GD is a sequential algorithm which begins by calculating the costs associated with each available pruning edge. The pruning edge with the lowest cost is selected, and its incident hyperedges are removed. The costs of the remaining available pruning edges are updated and the process repeats, each time selecting the pruning edge with the lowest cost until no edges remain. The pseudo-code for the GD heuristic is provided in Algorithm 1.

Algorithm 1. GD heuristic

1: **procedure** GD(E_p)
2:     Let $W = E_p$, $\pi = ()$, $i = 1$
3:     **while** $|W| > 0$ **do**   #*While W is not empty*
4:         $\bar{c}_i = \infty$
5:         **for** $e \in W$ **do**   #*Select the edge with the minimal cost*
6:             $\pi_i = e$, $c_i = f(\pi_i)$
7:             **if** $c_i < \bar{c}_i$ **then**
8:                 $\bar{e} = e$, $\bar{c}_i = c_i$
9:             **end if**
10:        **end for**
11:        $W = W - \{\bar{e}\}$
12:        $\pi_i = \bar{e}$, $i = i + 1$
13:    **end while**
14:    **return** π #*π is a permutation of E_p*
15: **end procedure**

Figure 3 offers a visual representation of the GD heuristic in action. Initially, the algorithm picks the pruning edge with the lowest current cost, specifically edge e_4 with a cost of 4. After removing the segments $\sigma_5 = \{15\}$ and $\sigma_4 = \{14\}$, and updating the costs, edges e_1 and e_3 are now the cheapest, both carrying a cost of 32. The algorithm, adhering to its greedy strategy, selects edge e_1, as it comes first among the lowest-cost options. Upon removing the segments $\sigma_1 = \{4, 5, 6, 7\}$ and $\sigma_2 = \{8\}$, the remaining edges, namely e_2 and e_3, now bear a cost of 32. The algorithm sticks to its strategy and selects the pruning edge e_2. Since all segments are now covered, the remaining edge costs nothing. Consequently, the sequence generated by the GD heuristic is (e_4, e_1, e_2, e_3) with a total cost of 68, which equals the sum of 4, 32, 32, and 0.

5   Analyzing Results and Discussion

This section presents a comparative analysis of the GD heuristic (see Algoritm 1) and the pruning edge ordering implemented in the SBBU algorithm as introduced by [2]. To provide a comprehensive comparison, we also include the exact

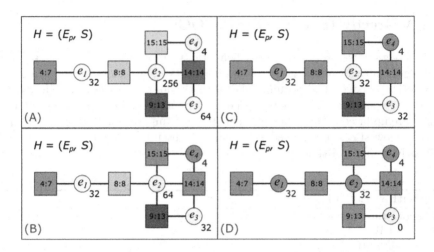

Fig. 3. Edge selection of GD heuristic, with steps (A), (B), (C), and (D).

solution obtained by brute force (BF), *i.e.*, derived from scrutinizing all conceivable permutations of pruning edges.

These experiments were performed on a system equipped with an Intel Core i7-3770 CPU running at a clock speed of 3.40 GHz, supported by 8 GB of RAM, and utilizing Ubuntu 20.04.6 LTS. The heuristic was implemented in Python 3.10.6.

Our tests involved 5,000 randomly generated instances, each with 30 vertices and 5 pruning edges, where one of them was $\{1, 30\}$ and the remaining four edges were randomly generated. For each random pruning edge $\{i, j\}$, we selected vertex i from the set $\{1, 2, \ldots, 26\}$ and j from the range $\{i+4, \ldots, 30\}$ at random. The edge $\{1, 30\}$ is the hardest constraint, since it has the largest search space [6].

We assessed the algorithms' efficiency using the following metric:

$$gap(\pi) = \frac{F(\pi) - F(\pi^\star)}{F(\pi^\star)},$$

where π represents a permutation of pruning edges and $\pi^\star$ indicates the optimal permutation of pruning edges computed via brute force (BF).

Table 1 illustrates the results for all instances, summarizing details about the number of vertices, pruning edges, and segments of each instance, alongside the cost of the optimal solution and the gaps related to the GD heuristic (gapGD) and the SSBU ordering (gapSB).

These results highlight the sub-optimality of GD and the SBBU ordering, showing their potential to produce solutions inferior to the optimal ones. The GD heuristic generated results that were, in the worst scenario, about 1.5× the optimal solution's cost (gapGD = 0.5). The performance of the SBBU ordering was much worse, with over a quarter of its results costing more than twice as much as the optimal solution, with some even reaching nearly 6,500×.

Table 1. Algorithmic cost comparison.

	—V—	—Ep—	—S—	costBF	gapGD	gapSB
mean	30	5	7.62	4.6E+4	1.0E−3	1.3E+1
std	0	0	0.94	2.9E+5	1.5E−2	1.4E+2
min	30	5	5	2.6E+2	0	0
25%	30	5	7	9.0E+2	0	0
50%	30	5	8	2.2E+3	0	1.0E−3
75%	30	5	8	8.5E+3	0	9.9E−1
max	30	5	9	8.4E+6	5.3E−1	6.5E+3

Despite the clear GD heuristic's average advantage, it only outperformed the SBBU ordering slightly over half of the 5K instances (in exact 2742 instances). That is, the SBBU ordering low average performance is the result of extreme gaps such as its maximum value of 6.5K. So, we can say that the great advantage of GD heuristic is its robustness.

Table 2 showcases the four instances where the GD heuristic produced the worst results. The instance labeled as test650 is noteworthy because the GD heuristic's gap was larger than the SBBU ordering gap. Figure 4 represents the hypergraph $H(E_p, S)$ for each of the instances highlighted in Table 2.

Table 2. Worst results of the GD heuristic.

ID	—V—	—Ep—	—S—	costBF	gapGD	gapSB
test374	30	5	8	720	0.525	0.525
test3007	30	5	7	2310	0.345	0.345
test3267	30	5	9	624	0.314	0.314
test650	30	5	9	2374	0.301	0.000

Table 3 outlines the four instances in which the SBBU ordering yielded its worst results. Figure 5 provides a graphical representation of the hypergraph $H(E_p, S)$ for each of the instances in Table 3.

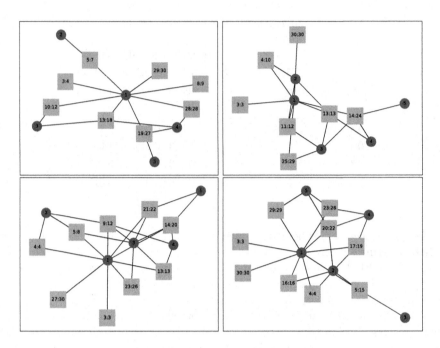

Fig. 4. Hypergraphs $H(E_p, S)$ for the instances test374, test3007, test3267, and test650 of Table 2.

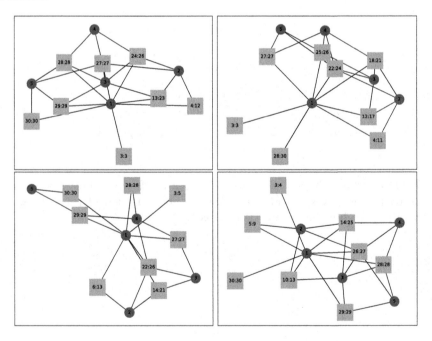

Fig. 5. Hypergraphs $H(E_p, S)$ for the instances test1675, test4351, test213, and test4562.

Table 3. Worst results of the SBBU ordering.

ID	—V—	—Ep—	—S—	costBF	gapGD	gapSB
test1675	30	5	8	2586	0.000	6486.71
test4351	30	5	8	416	0.000	5040.28
test213	30	5	8	652	0.000	3215.51
test4562	30	5	8	4168	0.000	2011.62

6   Conclusion

Despite the SBBU marked superiority over the BP algorithm, it is strongly dependent on the ordering of pruning edges [2]. We have formalized the pruning edge ordering problem associated with the SBBU algorithm, framing it as a vertex cover problem.

Distinct from traditional covering problems, the sequence of vertices in our cover problem holds significant relevance. To address this, we have introduced a greedy heuristic specifically designed to tackle this pruning edge ordering problem.

Through a series of computational experiments conducted on over 5,000 instances, we evaluated the efficiency of the proposed heuristic in comparison with the ordering used in the SBBU algorithm. The results were promising: on average, the cost of solutions derived from our heuristic was a mere 0.1% higher than the optimal solutions, with a standard deviation of 1.5%. Conversely, the average cost of the solutions derived from the SBBU ordering was staggering: 1,300× higher than the optimal cost with an alarming deviation of 14,200%.

These results are highly encouraging, suggesting the considerable potential of our proposed heuristic to enhance the performance of the SBBU algorithm. As we move forward, our future work will focus on examining the performance of the SBBU algorithm using our proposed heuristic in scenarios with larger and more intricate instances.

Acknowledgements. We thank the Brazilian research agencies FAPESP and CNPq, and the comments made by the reviewers.

Conflict of interest. No conflict of interest.

References

1. Cassioli, A., Günlük, O., Lavor, C., Liberti, L.: Discretization vertex orders in distance geometry. Disc. Appl. Math. **197**, 27–41 (2015)
2. Gonçalves, D.S., Lavor, C., Liberti, L., Souza, M.: A new algorithm for the k dmdgp subclass of distance geometry problems with exact distances. Algorithmica **83**(8), 2400–2426 (2021)
3. Lavor, C., Liberti, L., Maculan, N., Mucherino, A.: The discretizable molecular distance geometry problem. Comput. Optim. Appl. **52**, 115–146 (2012)

4. Liberti, L., Lavor, C., Maculan, N.: A branch-and-prune algorithm for the molecular distance geometry problem. Int. Trans. Oper. Res. **15**(1), 1–17 (2008)
5. Liberti, L., Lavor, C., Maculan, N., Mucherino, A.: Euclidean distance geometry and applications. SIAM Rev. **56**(1), 3–69 (2014)
6. Liberti, L., Masson, B., Lee, J., Lavor, C., Mucherino, A.: On the number of realizations of certain henneberg graphs arising in protein conformation. Disc. Appl. Math. **165**, 213–232 (2014)
7. Mucherino, A., Lavor, C., Liberti, L.: Exploiting symmetry properties of the discretizable molecular distance geometry problem. J. Bioinf. Comput. Biol. **10**(03), 1242009 (2012)
8. Wüthrich, K.: Protein structure determination in solution by nuclear magnetic resonance spectroscopy. Science **243**, 4887 (1989)

Phylogenetic Information as Soft Constraints in RNA Secondary Structure Prediction

Sarah von Löhneysen[1], Thomas Spicher[2,6], Yuliia Varenyk[2,7],
Hua-Ting Yao[2], Ronny Lorenz[2], Ivo Hofacker[2],
and Peter F. Stadler[1,2,3,4,5(✉)]

[1] Bioinformatics Group, Department of Computer Science, and Interdisciplinary Center for Bioinformatics, Universität Leipzig, Härtelstrasse 16-18, 04107 Leipzig, Germany
{lsarah,studla}@bioinf.uni-leipzig.de
[2] Institute for Theoretical Chemistry, University of Vienna, Vienna, Austria
{tspicher,yvarenyk,htyao,ronny,ivo}@tbi.univie.ac.at
[3] Max Planck Institute for Mathematics in the Sciences, Leipzig, Germany
[4] Facultad de Ciencias, Universidad Nacional de Colombia, Bogotá, Colombia
[5] Santa Fe Institute, Santa Fe, NM, USA
[6] UniVie Doctoral School Computer Science (DoCS), University of Vienna, Währinger Str. 29, 1090 Vienna, Austria
[7] Vienna BioCenter PhD Program, Doctoral School of the University of Vienna and Medical University of Vienna, 1030 Vienna, Austria

Abstract. Pseudo-energies are a generic method to incorporate extrinsic information into energy-directed RNA secondary structure predictions. Consensus structures of RNA families, usually predicted from multiple sequence alignments, can be treated as soft constraints in this manner. In this contribution we first revisit the theoretical framework and then show that pseudo-energies for the centroid base pairs of the consensus structure result in a substantial increase in folding accuracy. In contrast, only a moderate improvement can be achieved if only the information that a base is predominantly paired is utilized.

Keywords: RNA secondary structure · consensus structure · pseudo-energies

1 Introduction

Families of functional RNAs typically feature secondary structures that are very well conserved over large evolutionary time-scales, giving rise to a consensus structure that can be determined from alignments of the family members [1,14,

This work was funded by the Deutsche Forschungsgemeinschaft (DFG grant number STA 850/48-1) and by the Austrian Science Fund (FWF grant number F-80 and I-4520).

X. Guo et al. (Eds.): ISBRA 2023, LNBI 14248, pp. 267–279, 2023.
https://doi.org/10.1007/978-981-99-7074-2_21

30,35]. Individual family members nevertheless may exhibit structural variations that leave few conserved base pairs. In response, the `Rfam` database reports multiple families collected in a common "clan" [7]. On the other hand, it is well known that accuracy of secondary structure prediction for single sequences is limited [12] but can be improved drastically by comparative methods that enforce a consensus structure [8]. In order to accommodate the variability within such a family of related structures, one may use the consensus structure predicted from an alignment as constraint for the structure prediction of an individual sequence. A typical workflow for predicting structures for a set of related sequences thus comprises the following steps:

1. construction of a multiple sequence alignment (MSA) $\mathbb{A}$ of related sequences that includes the sequence X of interest.
2. computation of the consensus structure C for the alignment $\mathbb{A}$.
3. projection of the consensus structure C onto the sequence X by removing from C all positions in which X has a gap in $\mathbb{A}$. A base pair is removed from C if either paired position corresponds to a gap of X in $\mathbb{A}$.
4. prediction of the secondary structure of X using the projection C_X of the consensus structure C onto X as a hard constraint.

A prototypical implementation for the second step is `RNAalifold`, which extends the thermodynamic model from single sequences to alignments by averaging energy contributions over alignment columns [1,14]. In the same vein, `Pfold` [17,27] uses aligned sequences and phylogeny to compute a global consensus structure with the help of stochastic context free grammars. The first two steps can also be combined to computing a structure-based alignment using e.g. `locarna` [35,36]. Instead of using an MSA `cmalign` starts from a pre-computed covariance model (CM) for an RNA family and aligns the sequence of interest X to this CM. The CM in turn is obtained from an (often manually curated) multiple sequence alignment annotated by a consensus secondary structure C. The result is thus again a projection of the consensus structure C onto the sequence X. This workflow is used e.g. by the `R2DT` pipeline [29] for predicting and visualizing RNA secondary structures in `RNAcentral`. A prototypical implementation of the last step is the script `refold.pl`, which is part of the utilities distributed through the `ViennaRNA` github site[1].

A shortcoming of this approach is that it enforces the consensus structure even if individual structures may not fit well to the consensus, or if the consensus only applied to part of the structure. `TurboFold` [13] takes a different approach and uses position-specific modifications of the standard energy model [32] to include extrinsic information that summarizes structural information on related sequences. The "proclivity" for a base pair (i, j) is obtained by aggregating the probabilities that sequences positions potential homologous to i and j in other sequences form a base pair. The "proclivity" is then converted into a pseudo-energy added to every secondary structure of x that contains the base pair (i, j).

---

[1] https://github.com/ViennaRNA/ViennaRNA/blob/master/src/Utils/refold.pl.

Pseudo-energies were used already in early implementations of mfold [38] and the ViennaRNA Package [15] as a means of forcing or excluding base pairs and to encode exceptions to general energy rules. An example for the latter were bonus energies for extra-stable tetraloops [22]. RNAalifold uses optional bonus energies for sequence covariation [14], and TurboFold uses such terms to reward conservation of local structures. The use of pseudo-energies in RNA folding has become common-place with the advent of large-scale chemical probing data. For instance, SHAPE reactivities have been converted to position-specific stabilizing energies for unpaired bases using an empirical fit [3,11,37]. The same approach has been used for other types of chemical probing e.g. using DMS [2], for lead probing [18] and for enzymatic probing such as PARS [16,33]. Bonus energies can also be determined indirectly from the optimal balance of experimental signal and the thermodynamic folding model [34]. Instead of pseudo-energies, SCFG-based folding algorithms such as PPfold [27] and ProbFold [26] modify the emission probabilities to incorporate external evidence including probing data. A fully probabilistic model approach has been suggested in [5].

The effectivity of pseudo-energy contributions derives from the fact that the standard energy model for RNA secondary structures is a very good approximation [32]. As a consequence, it is unlikely that the biologically relevant RNA structure deviates dramatically from the predicted groundstate as far as its predicted free energy is concerned, even if the predicted structure may be very different. A moderate "bonus" thus is sufficient to nudge the ensemble of alternative structures towards features that are expected to be present according to the empirically determined evidence.

Evidence for evolutionary conserved RNA features can also be extracted from sequence alignments. Given sufficient data, covariance and mutual information measures can detect consensus base pairs. MIfold, for instance, uses column-wise scores of this type as a replacement for an energy model [6]. Direct evidence for the presence of individual helices is computed in ShapeSorter using a probabilistic model [31]. It is also possible, however, to convert mutual information values directly into pseudo-energies for base pairs.

In this contribution we consider the use of pseudo-energy contributions for extrinsic information on conserved secondary structures in detail. Such approaches are attractive because the ViennaRNA package features a generic interface for the position- and base pair-specific pseudo-energies [21], making them easy to implement. Section 2 revisits the pertinent theory. We then proceed to showing that pseudo-energies derived from an alignment-based consensus structure substantially improve the predictions of individual structures and investigate the influence of alignment quality.

2   Theory

Secondary Structures and Their Features. Throughout this contribution we fix an RNA sequence of interest X and a (pairwise or multiple) sequence alignment $\mathbb{A}$ that we assume to contain X. A secondary structure s (on X or

A) is a set of base pair such that every base or alignment position is contained in at most one base pair. Two base pairs (i, j) and (k, l) are said to be crossing if $k < i < l < j$ or $i < k < j < l$. Throughout, we consider only crossing-free secondary structures.

A secondary structure s is *compatible* with X, if every base pair adheres to certain "base pairing rules". In the standard RNA model, only GC, CG, AU, UA, GU, and UG pairs are allowed, all other conditions of nucleotides are forbidden. Analogously, s is compatible with A if, for each base pair $(i, j) \in s$, the corresponding alignment columns A_i and A_j are allowed to pair. In the case of the RNAalifold model, for instance, A_i and A_j are allowed to pair if the nucleotides $A_i(S)$ and $A_i(S)$ in row (sequence) S form one of the six possible base pairs. (The current implementation of RNAalifold tolerates a small number of exceptions, see [1]). The set of all secondary structures that are compatible with X will be denoted by $\Omega = \Omega(X)$. Analogously, we write $\Omega(A)$ for the set of all structures compatible with the alignment A.

A "feature" μ in an RNA secondary structure is most generally defined simply as a subset of the set of secondary structures. More intuitively, a feature comprises a pattern of base pairs and/or unpaired bases that is present in a secondary structure s. Simple examples include paired or unpaired single positions, specific base pairs, and the loops that appear in the recursions for secondary structure prediction. More complex examples are entire helices or abstract shapes (in sense of [9]). Here, we will be interested in features that are individual base pairs (i, j), or the pairing status of single positions. We will write (i) to designate the paired position i and $\neg(i)$ for an unpaired position. Furthermore, we write $s \in \mu$ if the secondary structure s has the feature μ, i.e., we formally treat μ simply as a set of structures.

Definition 1. *Two features μ' and μ'' are* incompatible *if $\mu' \cap \mu'' = \emptyset$.*

For example, crossing base pairs are incompatible features. Similarly, the base pair (i, j) is incompatible with each of the unpaired positions $\neg(i)$ and $\neg(j)$.

Pseudo-Energies. In their most general form, a pseudo-energy Γ_μ for a feature μ is an additive contribution to the energy $G(x)$ of every secondary structures x with feature μ. Thus $G(x) = G_0(x) + \Gamma_\mu$ if $x \in \mu$, while $G(x) = G_0(x)$ for $x \notin \mu$. Here, $G_0(x)$ denotes the energy of secondary structure x according to the standard energy model [32]. Additivity ensures that pseudo-energies Γ_μ may be considered simultaneously for an arbitrary set M of features:

$$G(x) = G_0(x) + \sum_{\mu \in M} \Gamma_\mu \tag{1}$$

For example, TurboFold considers pseudo-energies of the form $\Gamma_{(i,j)} = -a \ln \Pi_{(i,j)}$ derived from pairing "proclivities" $\Pi_{(i,j)}$ that in turn are computed from the base pair probabilities of related sequences [13]. The recursions of the dynamic programming algorithms underlying RNA folding readily accommodate certain pseudo-energy contributions. In particular, contributions for unpaired positions (i) and base pairs (i, j), but also entire hairpins enclosed by pair (i, j)

or interior loops between the pairs (i, j) and (k, l) are consistent with the folding algorithms [21]. A general issue is the conversion of quantities that measure extrinsic information a into energies. In general, empirical expressions are employed [3,13].

A more principled approach is available if the extrinsic information can be quantified as probability $p[\mu]$ or $p[\neg\mu]$ that a feature μ is present or absent, respectively. Following [2], the pseudo-energy for feature μ is then given by the scaled odds-ratio

$$\Gamma_\mu = -RT \ln \frac{p[\mu]}{p[\neg\mu]} \tag{2}$$

In the case of chemical probing data, $p[\mu]$ can be estimated by comparing empirically determined probing signals with known secondary structures [19,28]. For consensus structures, however, it is possible to obtain probabilities $p[(i, j)]$ of base pairs from a probabilistic model of the consensus structure for an alignment. In this contribution we use the partition function version of RNAalifold [1]. Pfold [17] or LocaRNA-P [35] could be used in the same manner.

In principle it is also possible to convert empirical base pair propensities into feature probabilities. To this end, one first converts the data to energy-like base pair propensities ε_{ij}. These can then be interpreted as energy model for a maximum matching model [24], whose partition functions over the structures on intervals are readily computed recursively as $Z_{ij} = Z_{i+1,j} + \sum_{i<k\leq j} Z_{i+1,k-1} Z_{k+1,j} \exp(-\beta\varepsilon_{ij})$, with the inverse temperature $\beta = -1/RT$. The backward recursion of McCaskill's algorithm [23] then yields base pair probabilities $p[(i, j)]$. As in the case RNAalifold -P, Pfold or LocaRNA-P, these derive from an ensemble of crossing-free RNA secondary structures Ω.

Feature Probabilities from Ensembles of Secondary Structures. Let Ω be an ensemble of (non-crossing) secondary structures and denote by $p(x)$ the probability of secondary structure x. Then the probability of feature μ is $p[\mu] := \sum_{x\in\mu} p(x)$. The probabilities of not observing feature μ is then $p[\neg\mu] = p[\Omega\setminus\mu] = 1 - p[\mu]$. Feature probabilities of this form are of interest because they have some mathematically appealing and practically useful properties.

We call a feature *dominating* if $p[\mu] > p[\neg\mu]$, i.e., if $p[\mu] > 1/2$.

Lemma 1. *Let $\mathcal{D}$ be a set of dominating features. Then $\mathcal{D}$ does not contain a pair of incompatible features.*

Proof. Suppose μ' and μ'' are incompatible and $\mu' \in \mathcal{D}$, i.e., $p[\mu'] > 1/2$. Then $\mu'' \subseteq \Omega \setminus \mu'$ and thus $p[\mu''] \leq p[\neg\mu] = 1 - p[\mu] < 1/2$ and thus $\mu'' \notin \mathcal{D}$.

The notion of incompatible features provides a simple general argument for the centroid of an ensemble of secondary structures being crossing-free, because base pairs of a centroid structure satisfy $p[(i, j)] > 1/2$ [4].

Corollary 1. *The dominating set of base pairs $\mathcal{C} = \{(i, j) \mid p[(i, j)] > 1/2\}$ is crossing-free.*

Now consider the special case that $p[\mu]$ is obtained also from a *Boltzmann ensemble* of secondary structures $\{s\}$ with corresponding energies $\varepsilon(s)$. Note that this model is **not** the energy model for RNA folding but instead incorporates the external information. Here, we use the energy model of `RNAalifold` for this purpose. Using the partition functions $Z_\varepsilon := \sum_s \exp(-\varepsilon(s)/RT)$ and $Z_\varepsilon[\mu] := \sum_{s \in \mu} \exp(-\varepsilon(s)/RT)$ yields $p[\mu] = Z_\varepsilon[\mu]/Z_\varepsilon$. A short computation then yields the identity

$$\Gamma_\mu = -RT \ln \frac{p[\mu]}{p[\neg\mu]} = G_\varepsilon[\mu] - G_\varepsilon[\neg\mu], \tag{3}$$

where $G_\varepsilon[\mu]$ and $G_\varepsilon[\neg\mu]$ are the free energies of ensembles w.r.t. the model ε constrained to structures containing and not containing μ, respectively. The difference $G_\varepsilon[\mu] - G_\varepsilon[\neg\mu]$ thus quantifies the evidence for the presence of μ and serves as bonus energy contribution.

Restrictions on Bonus Terms for Consensus Structures. In the setting of global consensus structures, a position i appears as unpaired both if it is unpaired in all contributing structures and if it is paired with different, and therefore incompatible pairing partners. That is, an unpaired position in a consensus structure may simply reflect the absence of a conserved base pair. This will always be the case if there is a certain part of the RNA that does not support a consensus structure. A large value of $p[\neg(i)]$, therefore, is not evidence of conserved unpairedness. In particular, alignments of many random sequences will not result in the prediction of consensus base pairs in `RNAalifold` and similar models, see Fig. 1.

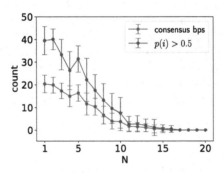

Fig. 1. Consensus base pairs in sets of random sequences. The number of base pairs predicted by `RNAfold` and the number of bases with a base pair probability > 0.5 was computed for $N = 1...20$ sequences with 80% similarity to a randomly generated reference sequence of length 80 nt. As expected, larger alignments yield consensus structures without significant base pairs.

This reasoning has also served as motivation for the definition of the "structure conservation index" (SCI) [10] as the ratio of the consensus folding energy and the average energy of the (unconstrained) secondary structure of the individual sequences. The SCI is in essence an estimate for the fraction of the folding energies that is explained by the consensus structure. Since the open structure without base pairs has energy 0, it quantifies consensus base pairs. We argue, therefore, that only the information on base pairs in the consensus structure can be used, while it remains open whether an unpaired position in the consensus is unpaired in the related sequences, or whether it is merely involved in non-conserved base pairs.

In a general setting, we assume that it is possible to determine the presence of a feature μ in the consensus, while it is not possible to distinguish the absence of μ from missing data. Clearly, there is support for μ only if $p[\mu] > p[\neg\mu]$, i.e., if μ is a dominating feature. The bonus energy for μ therefore takes the form

$$\Gamma_\mu = \min\left\{-RT \ln \frac{p[\mu]}{p[\neg\mu]},\ 0\right\} \tag{4}$$

Here we have used that $\Gamma_\mu < 0$ is equivalent to requiring $p[\mu] > p[\neg\mu] = 1 - p[\mu]$ and thus $p[\mu] > 1/2$. In the case of the RNAalifold model, therefore, we consider only the set $\mathcal{C}$ of dominating base pairs, which as argued above, coincides with the centroid structure in the RNAalifold model.

3   Implementation and Evaluation

In order to demonstrate that the inclusion of consensus structure information is useful we implemented the following simple workflow.

(i) We selected nine seed alignments $\mathbb{A}$ from Rfam that contained the largest number of seed sequences with known 3D structures and were free of pseudoknots. Alignments were augmented by the focal sequence X using mafft unless X was already included in $\mathbb{A}$. Then sequences with a similarity of more than 80% to other sequences were excluded because very similar sequences are expected to have very similar secondary structures and thus a single reprensative sequence suffices. The lengths of the focal sequences X varied between 74 and 184 nucleotides.

(ii) The consensus structure ensemble for $\mathbb{A}$ is computed using RNAalifold -p. From the corresponding base pair probability matrix, the set $\mathcal{C}$ of centroid base pairs and the corresponding probabilities $p(i', j')$ for the alignment columns i' and j' are retrieved. From these, we compute probability $p_{i'} := \sum_{j'<i'} p(j', i') + \sum_{j'>i'} p(i', j')$.

(iii) The columns i' and j' of the alignment $\mathbb{A}$ are translated to the corresponding sequence positions i and j in X. If row X of $\mathbb{A}$ shows a gap in alignment column i' or j', then the consensus base pair (i', j') is discarded. Similarly, if the nucleotides X_i and X_j at positions i and j cannot form a Watson-Crick or wobble pair, we ignore the constraint. As a result we obtain a set $\mathcal{C}_X$ of base pairs. The corresponding consensus probabilities $p[(i, j)]$ are converted into bonus energies $\Gamma_{(i,j)}$. Analogously, we convert the probability $p_{i'}$ that the alignment column containing position i of X is paired in the consensus into a bonus energy Γ_i for position i being paired. In the following we refer to these bonus energies as *phylogenetic soft constraints*.

(iv) We use RNAfold to compute the secondary structure of X using $\Gamma_{(i,j)}$ for $(i,j) \in \mathcal{C}_X$ as soft constraint. Alternatively, the $\Gamma_i < 0$ for paired positions are used as soft constraints. The code and implementation details can be accessed by visiting www.github.com/ViennaRNA/softconsensus.

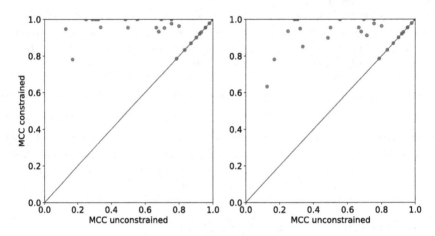

Fig. 2. Inclusion of phylogenetic soft constraints improves the accuracy of structure predictions. L.h.s.: Bonus energies $\Gamma_{(i,j)}$ RNAalifold centroid base pairs. R.h.s.: Bonus energies Γ_i for paired position with $p_i > 0.5$ in the RNAalifold ensemble of consensus structures. Bonus energies were obtained from Rfam seed alignments, see Electronic Supplement for details on the data set.

For benchmarking, we retrieved the reference structure of the focal sequences X from RNAcentral. Details can be found in the Electronic Supplement[2]. The accuracy of a predicted secondary structures was quantified by Matthews's correlation coefficient (MCC) based on the comparison of predicted base pairs and the base pairs of the reference structures.

In Fig. 2 we compare the predictions with and without the phylogenetic bonus energies. Using consensus base pairs (l.h.s.), we observe that for the majority of cases there is an improvement in accuracy, consistently reaching a level of at least 80% and sometimes perfect predictions, upon inclusion of phylogenetic information. The average improvement in MCC is 0.29 ($SD = 0.28$) overall and 0.45 ($SD = 0.23$) for sequences where unconstrained folding yields an MCC < 0.8. We have not encountered a case where the phylogenetic information is misleading, i.e., where the prediction deteriorates upon inclusion of the bonus energy terms. Using only the probability p_i that position i is paired in the consensus also improves the predicted structures. However, the improvement is more moderate with an average improvement of 0.26 ($SD = 0.25$) overall and 0.41 ($SD = 0.20$)

---

[2] www.bioinf.uni-leipzig.de/publications/supplements/23-002.

where MCC unconstrained < 0.8. For both features, no improvement is observed when the accuracy of the unconstrained prediction is more than 80%.

Figure 3 compares the distribution of MCC differences between unconstrained and constrained folding. We find that using hard constraints (i.e., enforcing or forbidding a base pair or unpaired position [21]) instead of bonus energy occasionally enforces incorrect base pairs. Furthermore, we observe that the pairing status of individual nucleotides conveys nearly as much information as the centroid base pairs of the consensus. We suspect, however, that the good performance in particular of paired nucleotides is in part the consequence of highly accurate, manually curated Rfam seed alignments.

In order to test the influence of size and quality of the alignment $\mathbb{A}$ we first sub-sampled the Rfam seed alignments by retaining only a number N of rows. Alternatively, we randomly selected N rows, removed the gaps and re-aligned the sequences with mafft. In both cases the focal sequence X is included in every subset. The accuracy of secondary structure predictions in dependency on the number N of sequences in the MSA $\mathbb{A}$ is illustrated in Fig. 4. Pseudo-energies derived from sub-sampled Rfam alignments already result in nearly the maximal improvement for two or three sequences. However, if sequences are realigned, larger samples are needed to reach the same accuracy, emphasizing the importance of high-quality alignments for the inference of consensus secondary structures. Pairwise alignments convey little improvement in this situation, presumably due to biases in the gap patterns and ambiguities in the alignment of more divergent regions.

We tested whether the MSA can be replaced by the superposition of base pair probabilities computed from individual sequences X_i of pairwise alignments (X, X_i). Averaging the base pair probabilities after projection to sequence X, however, yields few base pairs with $p[(i, j)] > 0.5$, and thus little improvement. The comparably poor performance of pairwise alignments may explain why the use of averaged base pair probabilities from pairwise sequence alignments of X with another related sequence does not improved structure prediction.

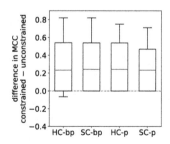

Fig. 3. Quantitative comparison of the effect of consensus information. Consensus information can be incorporated as hard (HC) or soft constraint (SC). Hard constraints can occasionally enforce misleading base pairs. Using the pairing status of individual nucleotides instead of the centroid base pairs occasionally leads to a smaller improvement of the MCC.

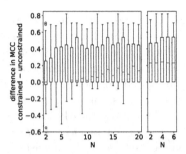

Fig. 4. Sequence-based `mafft` alignments without manual curation require 15–20 sequences to fully exploit the consensus structure information. Subsampling of `Rfam` seed alignments, on the other hand, already leverage the full information contained in the consensus structure with about 4 sequences; larger alignments do not seem to provide a further gain accuracy.

4 Discussion

Consensus base pairs inferred from phylogenetically related sequences are a powerful source of information to guide the prediction of accurate secondary structure. We have shown here that consensus structures computed from MSAs yield pseudo-energies that provide a theoretically well-founded and computationally efficient framework for this purpose. Moreover, we have seen that only "positive information", i.e., paired position and base pairs in the consensus structure can be used, since unpaired positions may arise simply from the lack of a consensus. It is also possible to use MSAs to superimpose independent secondary structure predictions. Preliminary data indicate, however, that this requires a sophisticated integration procedure, e.g. as implemented in the iterative scheme of `TurboFold`. The beneficial effect of consensus base pairs, finally, depends strongly on the accuracy of the alignment.

The approach taken here still assumes a single consensus structure. Some structured RNAs, however, function through structural transitions. A prime example are riboswitches relying on the alternative formation of a terminator to an anti-terminator hairpin. To handle such cases, alternative, usually incompatible, structures need to be considered. As noted e.g. in [25], conserved alternative conformations sometimes can be found as suboptimal local minima in the Boltzmann ensemble, and it may be possible to obtain direct evidence for the conservation of helices that are incompatible with the most stable structure [31]. An interesting generic approach to handle such cases is the ensemble tree proposed in [20].

References

1. Bernhart, S.H., Hofacker, I.L., Will, S., Gruber, A.R., Stadler, P.F.: RNAalifold: improved consensus structure prediction for RNA alignments. BMC Bioinf. **9**, 474 (2008). https://doi.org/10.1142/s0219720008003886
2. Cordero, P., Kladwang, W., VanLang, C.C., Das, R.: Quantitative dimethyl sulfate mapping for automated RNA secondary structure inference. Biochemistry **51**, 7037–7039 (2012). https://doi.org/10.1021/bi3008802
3. Deigan, K.E., Li, T.W., Mathews, D.H., Weeks, K.M.: Accurate SHAPE-directed RNA structure determination. Proc. Natl. Acad. Sci. USA **106**, 97–102 (2009). https://doi.org/10.1073/pnas.080692910
4. Ding, Y., Chan, C.Y., Lawrence, C.E.: RNA secondary structure prediction by centroids in a Boltzmann weighted ensemble. RNA **11**, 1157–1166 (2005). https://doi.org/10.1261/rna.2500605
5. Eddy, S.R.: Computational analysis of conserved RNA secondary structure in transcriptomes and genomes. Ann. Rev. Biophys. **43**, 433–456 (2014). https://doi.org/10.1146/annurev-biophys-051013-022950
6. Freyhult, E., Moulton, V., Gardner, P.: Predicting RNA structure using mutual information. Appl. Bioinf. **4**, 53–59 (2005). https://doi.org/10.2165/00822942-200504010-00006
7. Gardner, P.P., et al.: Rfam: wikipedia, clans and the "decimal" release. Nucleic Acids Res. **39**, D141–D145 (2011). https://doi.org/10.1093/nar/gkq1129
8. Gardner, P.P., Giegerich, R.: A comprehensive comparison of comparative RNA structure prediction approaches. BMC Bioinf. **5**, 140 (2004). https://doi.org/10.1186/1471-2105-5-140
9. Giegerich, R., Voß, B., Rehmsmeier, M.: Abstract shapes of RNA. Nucleic Acids Res. **32**, 4843–4851 (2004). https://doi.org/10.1093/nar/gkh779
10. Gruber, A.R., Bernhart, S.H., Hofacker, I.L., Washietl, S.: Strategies for measuring evolutionary conservation of RNA secondary structures. BMC Bioinf. **9**, 122 (2008). https://doi.org/10.1186/1471-2105-9-122
11. Hajdin, C.E., Bellaousov, S., Huggins, W., Leonard, C.W., Mathews, D.H., Weeks, K.M.: Accurate SHAPE-directed RNA secondary structure modeling, including pseudoknots. Proc. Natl. Acad. Sci. **110**(14), 5498–5503 (2013). https://doi.org/10.1073/pnas.1219988110
12. Hajiaghayi, M., Condon, A., Hoos, H.H.: Analysis of energy-based algorithms for RNA secondary structure prediction. BMC Bioinf. **13**, 22 (2012). https://doi.org/10.1186/1471-2105-13-22
13. Harmanci, A.O., Sharma, G., Mathews, D.H.: TurboFold: iterative probabilistic estimation of secondary structures for multiple RNA sequences. BMC Bioinf. **12**, 108 (2011). https://doi.org/10.1186/1471-2105-12-108
14. Hofacker, I.L., Fekete, M., Stadler, P.F.: Secondary structure prediction for aligned RNA sequences. J. Mol. Biol. **319**, 1059–1066 (2002). https://doi.org/10.1016/S0022-2836(02)00308-X
15. Hofacker, I.L., Fontana, W., Stadler, P.F., Bonhoeffer, L.S., Tacker, M., Schuster, P.: Fast folding and comparison of RNA secondary structures. Chem. Monthly **125**, 167–188 (1994). https://doi.org/10.1007/BF00818163
16. Kertesz, M., et al.: Genome-wide measurement of RNA secondary structure in yeast. Nature **467**(7311), 103–107 (2010). https://doi.org/10.1038/nature09322
17. Knudsen, B., Hein, J.: Pfold: RNA secondary structure prediction using stochastic context-free grammars. Nucleic Acids Res. **31**, 3423–3428 (2003). https://doi.org/10.1093/nar/gkg614

18. Kolberg, T., et al.: Led-seq - ligation- enhanced double-end sequence-based structure analysis of RNA. Nucleic Acids Res. (2013). https://doi.org/10.1093/nar/gkad312

19. Kolberg, T., et al.: Led-seq - ligation- enhanced double-end sequence-based structure analysis of rna. Nucleic Acids Res. (2023). https://doi.org/10.1093/nar/gkad312

20. Li, T.J.X., Reidys, C.M.: On an enhancement of RNA probing data using information theory. Alg. Mol. Biol. **15**, 15 (2020). https://doi.org/10.1186/s13015-020-00176-z

21. Lorenz, R., Hofacker, I.L., Stadler, P.F.: RNA folding with hard and soft constraints. Alg. Mol. Biol. **11**, 8 (2016). https://doi.org/10.1186/s13015-016-0070-z

22. Mathews, D.H., Sabina, J., Zuker, M., Turner, D.H.: Expanded sequence dependence of thermodynamic parameters improves prediction of RNA secondary structure. J. Mol. Biol. **288**, 911–940 (1999). https://doi.org/10.1006/jmbi.1999.2700

23. McCaskill, J.S.: The equilibrium partition function and base pariring probabilities for RNA secondary structures. Biopolmers **29**(6–7), 1105–1119 (1990). https://doi.org/10.1002/bip.360290621

24. Nussinov, R., Jacobson, A.B.: Fast algorithm for predicting the secondary structure of single stranded RNA. Proc. Natl. Acad. Sci. USA **77**, 6309–6313 (1980). https://doi.org/10.1073/pnas.77.11.6309

25. Ritz, J., Martin, J.S., Laederach, A.: Evolutionary evidence for alternative structure in RNA sequence co-variation. PLoS Comput. Biol. **9**, e1003152 (2013). https://doi.org/10.1371/journal.pcbi.1003152

26. Sahoo, S., Świtnicki, J.M.P., Pedersen, J.S.: ProbFold: a probabilistic method for integration of probing data in RNA secondary structure prediction. Bioinformatics **32**, 2626–2635 (2016). https://doi.org/10.1093/bioinformatics/btw175

27. Sükösd, Z., Knudsen, B., Kjems, J., Pedersen, C.N.S.: PPfold 3.0: fast RNA secondary structure prediction using phylogeny and auxiliary data. Bioinformatics **28**, 2691–2692 (2012). https://doi.org/10.1093/bioinformatics/bts488

28. Sükösd, Z., Swenson, M.S., Kjems, J., Heitsch, C.E.: Evaluating the accuracy of SHAPE-directed RNA secondary structure predictions. Nucleic Acids Res. **41**, 2807–2816 (2013). https://doi.org/10.1093/nar/gks1283

29. Sweeney, B.A., et al.: R2DT is a framework for predicting and visualising RNA secondary structure using templates. Nat. Commun. **12**, 3494 (2021)

30. Tagashira, M., Asai, K.: ConsAlifold: considering RNA structural alignments improves prediction accuracy of RNA consensus secondary structures. Bioinformatics **38**(3), 710–719 (2022). https://doi.org/10.1093/bioinformatics/btab738

31. Tsybulskyi, V., Meyer, I.M.: ShapeSorter: a fully probabilistic method for detecting conserved RNA structure features supported by SHAPE evidence. Nucleic Acids Res. **50**, e85 (2022). https://doi.org/10.1093/nar/gkac405

32. Turner, D.H., Mathews, D.H.: NNDB: the nearest neighbor parameter database for predicting stability of nucleic acid secondary structure. Nucleic Acids Res. **38**, D280–D282 (2010). https://doi.org/10.1093/nar/gkp892

33. Wan, Y., et al.: Landscape and variation of RNA secondary structure across the human transcriptome. Nature **505**, 706–709 (2014). https://doi.org/10.1038/nature12946

34. Washietl, S., Hofacker, I.L., Stadler, P.F., Kellis, M.: RNA folding with soft constraints: reconciliation of probing data and thermodynamic secondary structure prediction. Nucleic Acids Res. **40**, 4261–4272 (2012). https://doi.org/10.1093/nar/gks009

35. Will, S., Joshi, T., Hofacker, I.L., Stadler, P.F., Backofen, R.: LocARNA-P: accurate boundary prediction and improved detection of structured RNAs for genome-wide screens. RNA **18**, 900–914 (2012). https://doi.org/10.1261/rna.029041.111
36. Will, S., Missal, K., Hofacker, I.L., Stadler, P.F., Backofen, R.: Inferring non-coding RNA families and classes by means of genome-scale structure-based clustering. PLoS Comput. Biol. **3**, e65 (2007). https://doi.org/10.1371/journal.pcbi.0030065
37. Zarringhalam, K., Meyer, M.M., Dotu, I., Chuang, J.H., Clote, P.: Integrating chemical footprinting data into RNA secondary structure prediction. PLOS ONE **7**(10) (2012). https://doi.org/10.1371/journal.pone.0045160
38. Zuker, M., Jaeger, J.A., Turner, D.H.: A comparison of optimal and suboptimal RNA secondary structures predicted by free energy minimization with structures determined by phylogenetic comparison. Nucleic Acids Res. **19**, 2707–2714 (1991). https://doi.org/10.1093/nar/19.10.2707

NeoMS: Identification of Novel MHC-I Peptides with Tandem Mass Spectrometry

Shaokai Wang[1], Ming Zhu[2], and Bin Ma[1,3(✉)]

[1] David R. Cheriton School of Computer Science, University of Waterloo, Ontario, Canada
{s853wang,binma}@uwaterloo.ca
[2] Kuaixu Biotechnology, Shanghai, China
[3] Rapid Novor Inc., Ontario, Canada

Abstract. The study of immunopeptidomics requires the identification of both regular and mutated MHC-I peptides from mass spectrometry data. For the efficient identification of MHC-I peptides with either one or no mutation from a sequence database, we propose a novel workflow: NeoMS. It employs three main modules: generating an expanded sequence database with a tagging algorithm, a machine learning-based scoring function to maximize the search sensitivity, and a careful target-decoy implementation to control the false discovery rates (FDR) of both the regular and mutated peptides. Experimental results demonstrate that NeoMS both improved the identification rate of the regular peptides over other database search software and identified hundreds of mutated peptides that have not been identified by any current methods. Further study shows the validity of these new novel peptides.

Keywords: Neoepitope · MHC · Mass Spectrometry · Machine Learning

1 Introduction

In adaptive immunity, the major histocompatibility complex (MHC) presents a class of short peptides (also known as MHC peptides or HLA peptides) on the cell surface for T-cell surveillance. The systematic study of the MHC peptides is also referred to as immunopeptidomics. Two major classes of MHC exist: MHC-I and MHC-II. MHC-I molecules are expressed on all nucleated cells and MHC-II molecules are expressed on antigen-presenting cells. Normally, peptides presented by MHC-I are derived from endogenous proteins which are neglected by the cytotoxic T lymphocytes. However, the MHC-I of infected or tumor cells may present exogenous or mutated peptides (neoantigens) derived from either the viral proteome or cancer-related mutations, leading to the activation of specific cytotoxic T lymphocytes to eliminate the neoantigen-presenting cells. These abnormal MHC-I peptides also serve as excellent targets for immunotherapy, such as TCR-T [1] and cancer vaccines [2]. For these reasons, a method that

© The Author(s), under exclusive license to Springer Nature Singapore Pte Ltd. 2023
X. Guo et al. (Eds.): ISBRA 2023, LNBI 14248, pp. 280–291, 2023.
https://doi.org/10.1007/978-981-99-7074-2_22

can systematically determine all the MHC-I peptides becomes extremely useful for studying infectious diseases, developing novel cancer immunotherapy, and choosing the right immunotherapy for individual patients. Currently there are two main approaches to identifying abnormal MHC-I peptides: genomics and proteogenomics. In genomics, the identification of somatic mutations on neoepitopes is often performed using whole exome sequencing (WES) or transcriptome sequencing data [3]. On the other hand, the proteogenomics approach involves analyzing tissue samples using liquid chromatography tandem mass spectrometry (LC-MS/MS) and searching the obtained spectra against a personalized protein database constructed from exome sequencing or RNA sequencing data [4]. However, since there's no evidence or statistical validation, the false positives of both methods can be high.

Proteomics approaches that rely solely on MS spectra for peptide identification provide direct experimental evidence, reducing false discoveries and allowing for statistical validation of the False Discovery Rate (FDR). Traditional database search methods, including Peaks [5], Comet [6], and MaxQuant [7], are widely used for peptide identification, especially for spectra generated from tryptic-digested peptides. Tryptic digestion occurs when proteins are cleaved at the C-terminal of lysine (K) and arginine (R) residues. However, these methods often exhibit limited performance when applied to more complex MHC peptides, which undergo non-tryptic digestion. Non-tryptic digestion refers to protein cleavage occurring at various sites, not restricted to specific amino acid residues. To enhance the identification rates for non-tryptic peptides, current efforts focus on rescoring the database search results. Percolator [8] is a widely used tool that employs an SVM-based semi-supervised machine learning approach to rescore Peptide-Spectrum Matches (PSMs), thereby enhancing sensitivity in peptide identification. MHCquant [9] integrates Percolator into an immunopeptidomics data analysis workflow for MHC peptide identification, benefiting from improved sensitivity and accuracy. Recent advances in peptide property prediction, such as retention time (RT), MS/MS spectrum, and collisional cross sections, have enabled innovative workflows like DeepRescore [10], Prosit [11], MS2Rescore [12], and AlphaPeptDeep [13]. These workflows utilize prediction tools to generate PSM features and leverage Percolator for improved performance. However, a significant limitation of these tools is their inability to identify mutated peptides not present in the sequence database. This hinders their applicability in detecting novel peptides with mutations. De novo sequencing methods [14,15] can identify mutated peptides as they do not rely on a reference database. However, these methods suffer from a higher error rate and the lack of a universally accepted result validation method, which undermines confidence in the identified sequences. Open-search methods such as MSFragger [16], Open-pFind [17], TagGraph [18], and PROMISE [19] provide techniques for identifying peptides with post-translational modifications (PTMs), but not for mutation.

To confidently identify MHC peptides, we have developed a novel workflow called NeoMS that combines de novo sequencing and PSM rescoring techniques. NeoMS generates a candidate neoepitope database using de novo sequencing and k-mer tagging. To enhance peptide identification, we employ a trained light-GBM model instead of using semi-supervised learning to avoid overfitting. To ensure the accuracy of identified mutated peptides, NeoMS implements rigorous false discovery rate (FDR) control measures. The performance of NeoMS is evaluated using publicly available mass spectrometry data for MHC peptides. In comparison to other tested methods, NeoMS outperforms them by identifying a greater number of peptides. Importantly, NeoMS confidently identifies hundreds of mutated MHC-I peptides, providing high-confidence results. By combining de novo sequencing, PSM rescoring, and stringent FDR control, NeoMS offers a powerful and accurate approach for MHC peptide identification, including the detection of mutated peptides. This workflow has the potential to advance research in immunopeptidomics and facilitate the development of personalized cancer immunotherapy.

2   Methods

2.1   Datasets

Two LC-MS/MS datasets of human HLA I peptides: PXD000394 and PXD004894 were downloaded from the proteome change repository. The first dataset is used for training our lightGBM model and the second dataset is used for testing. The two datasets are derived from separate experiments, resulting in no overlap between them.

- Pride PXD000394 was acquired from a Thermo Q-Exactive instrument and contained 41 MS raw files [20]. It is a collection of six cell lines: JY, SupB15WT, HCC1143, HCC1937, Fib, and HCT116. All 41 MS raw files (12.6 million MS/MS spectra) were used for training our machine learning model.
- Pride PXD004894 was acquired from a Thermo Q-Exactive HF instrument [4]. This is a survey conducted on tissue samples associated with melanoma-associated tumors. A total of 25 melanoma patients were included in the study, and we specifically focused on three patients: Mel5, Mel8, and Mel15. To test our software, we utilized 24 MS raw files, which consisted of 1.12 million MS/MS spectra associated with these three patients. Among these files, 16 were associated with Mel15, containing 720,557 spectra. There were 4 raw files associated with Mel5, comprising 118,003 spectra, and another 4 raw files associated with Mel8, comprising 132,297 spectra.

2.2   Overall Workflow

The whole workflow of NeoMS is shown in Fig. 1. The input of the workflow is the peptide MS/MS spectra and a reference protein sequence database. The raw MS file is converted to the mgf format using msConvert [21] before the analysis. There are four main steps of the analysis:

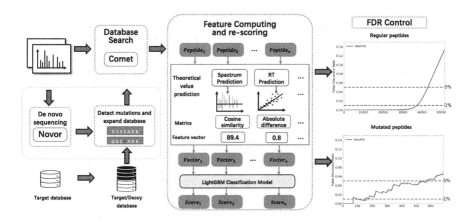

Fig. 1. The overall workflow of NeoMS.

1. Expanded database generation: Generate an expanded sequence database that consists of both the original sequences and possibly mutated sequences. To generate the mutated sequences, de novo sequence generated by Novor [22] is used to locate the possible point mutations.
2. Database search: Conduct database search analysis with Comet [6] by using the input spectra and the expanded sequence database.
3. PSM Rescoring: Rescore the PSMs found by the database search with newly computed scoring features and a machine-learning scoring function.
4. Result analysis: Control the FDR of the identified regular and mutated peptides.

These steps are further elaborated on in the following sections. Each step is dockerized, and the whole workflow is compiled in Nextflow [23]. The code is available on GitHub (https://github.com/waterlooms/NeoMS).

2.3   Generation of Expanded Database

The MS/MS spectra are de novo sequenced by Novor [22]. For each spectrum, Novor computes a peptide sequence and a positional confidence score for each amino acid of the peptide. Within the de novo sequences, a confident sequence tag is defined as a length-k substring (continuous subsequence) where each amino acid has a confidence score above a threshold t. By default, NeoMS sets $k = 7$ and $t = 60$, which were selected empirically.

These tags are searched in a target/decoy database to find approximate matches. This target sequence database is human protein database downloaded from uniprot [24]. Considering the non-tryptic manner of searching, the decoy is generated by random shuffling target protein sequences. Hence, the target/decoy database is the concatenation of these two database. A hit to a tag is a length-k substring in the database that approximately matches the tag with exactly one amino acid mutation. For each hit, a mutated sequence is constructed by concatenating the at-most n amino acids immediately before the hit, the mutated amino acids, and the at-most n amino acids immediately after the hit. By default, NeoMS sets $n = 12$. Since MHC-I peptides have lengths up to 13, this choice of n allows the inclusion of every mutated MHC-I peptide that has a confident de novo tag covering the point mutation.

The newly generated mutated sequences are appended to the target/decoy database to form an expanded database. Note that a hit can be either from a target or a decoy sequence. The target and decoy hits are treated equally throughout the analysis until the FDR control step. In the end, the resulting expanded database contains the target sequences and the decoy sequences, as well as the mutated sequences generated from the target and decoy sequences.

2.4   Database Search in the Expanded Database

Comet [6] is then used for the database search analysis by using the input MS/MS spectra and the expanded sequence database. We used the default Comet high-high parameter and made three adjustments:

1. Set *num_output_lines* to 10. Up to 10 candidate peptides are computed for each spectrum. These candidate peptides are further evaluated in the downstream rescoring analysis to choose the optimal one for each spectrum. For a certain spectrum, some true peptides with lower Comet scores could have higher rankings after rescoring. Meanwhile, the statistical relations of peptides for one spectrum produce important features for rescoring. It is noticed that in practice, any other database search tool that allows the output of multiple candidates per spectrum can be used in lieu of Comet as the base engine of NeoMS.
2. Set enzyme to *cut_everywhere*. By default, Comet searches peptides in tryptic digestion that only cut cleaves the C-terminal to lysine (K) and arginine (R). We set it as *cut_everywhere* to search in a non-tryptic manner.
3. Set *mass_tolerance* to 0.02 Dalton. To narrow down the search space and filters out the incorrect PSMs.

2.5   Rescoring

For each spectrum S, the top 10 peptide candidates $P_1, P_2, \ldots, P_n$. Five comet computed values: *xcorr*, *delta_cn*, *sp_score*, *mass_error*, and *e_value* are taken as features. Here the *e_value* score is converted to $log(e\_value)$ before using it in machine learning. Besides, the following set of peptide features is computed for each P_i:

1. The absolute difference between predicted RT for P_i and the experimental RT of the spectrum S. AutoRT [25] was used to make the prediction.
2. The similarity between S and the MS/MS spectrum predicted for P_i by pDeep2 [26]. The feature is computed by the Pearson correlation coefficient between the predicted b and y-ion intensities and their experimental intensities.

3. The similarity between S and the MS/MS spectrum predicted for P_i by Pred-Full [27]. PredFull predicts a sparse vector of length 20,000, where each dimension represents the maximum peak intensity in an m/z bin of width 0.1 mass units. The experimental spectrum is also converted to such a sparse vector. The feature is the Cosine Similarity of two vectors.

This list comprises eight peptide features. In addition, for the $log(e\_value)$ and the 3 peptide features above, three spectrum features are computed: maximum, mean, and variance of the feature's values on the top 10 peptides for the spectrum. In total, there are $4 \times 3 = 12$ features, consisting of 8 peptide features and 12 spectrum features. A machine learning model based on LightGBM is used to calculate a numeric score based on the 20 features.

2.6   FDR Control

The target-decoy approach [28] is adapted for controlling the FDR. The target and decoy databases are combined and analyzed together. After the NeoMS search, the regular and mutated peptides are separated, and their FDRs are also controlled separately. The score thresholds of the regular and mutated peptides are usually different because of their different distributions. In addition, a user can choose to use different FDR thresholds for the regular and mutated peptides, respectively. In our experiment, due to the higher complexity of identifying mutated peptides compared to regular peptides, we established distinct thresholds for each category: 1% for regular peptides and 5% for mutated peptides.

2.7   Training of the Scoring Function

The scoring function training is performed as a separate step of the scoring process itself. In contrast to Percolator's semi-supervised learning approach, we employ supervised learning, eliminating the need to train on testing data. The training is conducted only once using training data and remains unchanged for future analyses. The Lightgbm package is utilized to support the training process with specific parameter settings: max_depth as 9 and num_leaves as 51. The training is performed iteratively over several iterations to optimize the model's performance.

Before our training, we do not have precise positive and negative labels for PSMs. In the first iteration, the target peptides in Comet's search results with 1% FDR comprise the initial positive set, and the same amount of top-ranked decoy peptides constitute the initial negative set. Lightgbm is used to learn the initial GBM model. In each of the following iterations, the GBM model learned from the previous iteration is used to conduct the search. Then the target peptides with 1% FDR are added to the current positive set with deduplication, and the same amount of top-ranked decoy peptides are added to the current negative set with deduplication as well. Finally, the model is learned by using the accumulated positive and negative sets. The process is repeated a few times until the performance does not improve anymore.

3   Results

3.1   NeoMS Identified More Regular Peptides Than Other Methods

The performance of NeoMS was benchmarked against four other database search methods: Comet [6], MaxQuant [7], PeaksX [5], and DeepRescore [10]. Three patients' data (Mel5, Mel8, and Mel15) in PXD004894 and the human sequence database (UniProt UPID: UP000005640) were used to test the performance. For Comet, we use the default parameter setting with 3 changes mentioned in 2.4. DeepRescore is set to use the default parameter for Comet. PeaksX's database search results provided from the database searching part of an individual immunopeptidomes [14] on the same datasets were used, while MaxQuant's results provided by dataset paper [4] were used. As the other software does not search for mutated peptides, the mutation finding function was turned off in NeoMS here for a fair comparison. The number of PSMs and the number of peptides identified at 1% FDR by each software are plotted in Fig. 1. From both the perspective of peptide number and PSM number, NeoMS outperformed the other methods. For the three samples, our NeoMS have improved the identification number 2 to 3 times than our base database search method Comet. In comparison with other rescoring methods, NeoMS identifies 5% to 10% more than the second-best methods. To further analyze the authenticity of the identified peptides as MHC-bound peptides, we utilized MHCflurry [29] for binding affinity prediction. Following established methodologies, peptides are considered to bind to MHC I if their predicted affinity falls within the top 2% of the strongest binding peptides for any allele in the sample. In the Mel 15 experiment, NeoMS identified a total of 37,135 peptides, of which 33,810 (91.05%) were predicted to be MHC-bound. Comparatively, DeepRescore identified 32,600 MHC-bound peptides. It is important to note that while there is a significant overlap between the results of NeoMS and DeepRescore, each search engine also identified a considerable number of novel peptides. Among the 2,316 novel peptides exclusively identified by NeoMS, 1,955 (84.41%) were predicted to be MHC-bound, which reinforces the reliability of NeoMS in accurately identifying MHC-associated peptides.

3.2   NeoMS Identified Mutated MHC I Peptides

NeoMS demonstrates the capability to identify both MHC-I peptides with and without mutations in a unified search. In Fig. 3(A) and (B), the number of mutated peptides and PSMs identified by NeoMS on the Mel15 dataset (from PXD004894) is presented, respectively. The search was conducted on 16 MS files, comprising a total of 164,844 spectra, using the Uniprot human sequence database. The identification of mutated and unmutated peptides was performed independently, with separate FDR controls. NeoMS identified 37,042 peptides (393,871 PSMs) without any mutations at 1% FDR and 544 mutated peptides (4,028 PSMs) with exactly one amino acid mutation at a 5% FDR. The other two samples, Mel5 and Mel8, yielded the identification of 191 and 86 mutated

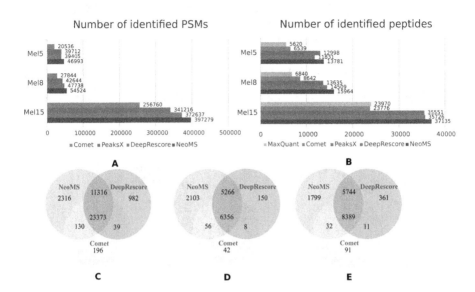

Fig. 2. A, B: Performance comparison in terms of identifying MHC-I peptides without mutations. Three patients' samples (Mel5, Mel8, Mel15). The x-axis indicates the number of PSM (A) and unique peptide (B) identifications at 1% FDR. NeoMS' mutation finding function is turned off for a fair comparison. We compared our identified peptides with four methods: Comet, MaxQuant, PeaksX, and DeepRescore. Since MaxQuant's PSM number is not provided, we compare our PSM number with the other 3 methods. C, D, E: Venn diagram of the unique peptides identified at 1% FDR on Mel15(C), Mel5(D), and Mel8(E) by the three search engines, NeoMS, DeepRescore, and Comet, respectively. The number in each area indicates the number of identified peptides.

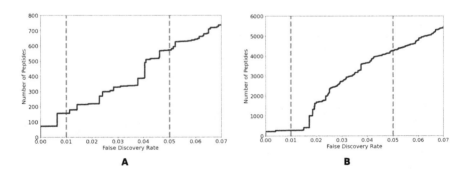

Fig. 3. The number of mutated peptides/PSMs identified by NeoMS from the Mel15 sample in dataset PXD004894. Panel (A) is the number of peptides and (B) is for the number of PSMs. The x-axis is the FDR threshold and the y-axis is the number of peptides and PSMs, respectively

peptides, respectively, at a 5% FDR. These results highlight NeoMS's ability to efficiently identify MHC-I peptides, both with and without mutations, in large-scale datasets. The accurate and comprehensive identification of mutated peptides opens up new avenues for investigating their associations with tumor cells and their potential as targets for T lymphocytes in cancer immunology research.

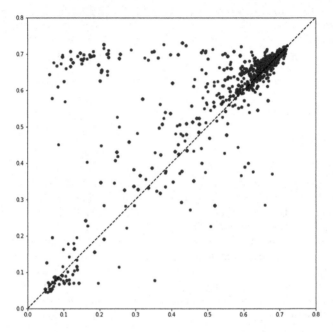

Fig. 4. Affinity comparison between the mutated peptides and their corresponding original peptides in the sequence database. Each data point corresponds to a pair of mutated and original peptides. The y and x-axes are their predicted affinity scores, respectively. The upper left area contains the peptides with both a high score and a high score increase caused by the single amino acid mutation.

3.3   The Mutation Increases Binding Affinity

For each mutated peptide identified with a 5% FDR threshold by NeoMS, we retrieved its corresponding original peptide in the sequence database. The pair of peptides differ by only one amino acid. Their MHC-I binding affinities were predicted by MHCflurry, using the six alleles provided together with the dataset in the paper [4]. The maximum affinity of a peptide from the six alleles was used as the affinity for the peptide. Figure 4 shows the scatter plot of the predicted affinity scores of the identified peptide pairs. Among these, there were 81 mutated target peptides with binding affinity scores increased by at least 0.1.

In contrast, only 8 have binding affinity scores decreased by at least 0.1. These clearly show that the single amino acid mutation in these mutated peptides generally improved the binding affinity.

4   Discussion

Immunopeptidomics studies require the identification of MHC-I peptides containing amino acid mutations with high confidence from a sequence database and MS data. The lack of protease specificity with the consideration of mutations together increased the search space as well as the spectrum complexity. Both a better scoring function and rigorous result validation are required. In this work, we proposed a novel computational workflow, NeoMS to meet the needs for MHC-I peptide identification. Based on a de novo-based approach to detect mutations and to expand the sequence database, NeoMS could identify both the regular peptides that do not contain mutations and the mutated peptides containing exactly one amino acid mutation in a unified and efficient search workflow. For the identification of regular peptides, NeoMS outperformed all other search engines: under the same FDR constraints, NeoMS identifies most PSMs and peptides. These peptides are further examined by MHC-Flurry and more than 90% of our peptides are bound to MHC. A distinct advantage of NeoMS is that it can identify MHC-I peptides with a single amino acid mutation. In our melanoma dataset, NeoMS identifies 544 mutated peptides. We examine the binding affinities of these peptides before mutation and after mutation. Most of the peptides binding affinity are increased by the mutation.

Indeed, there are certain limitations in the current NeoMS workflow. One of the limitations is that it can only identify peptides with a single mutation. While MHC-I peptides are typically short, there is still research value in identifying peptides with multiple mutations, as they can provide valuable insights into tumor heterogeneity and immune response. Another limitation is that the current NeoMS method is specifically designed for MHC-I peptides and does not address the identification of MHC-II peptides. MHC-II molecules primarily interact with immune cells, and their peptide length ranges from 13 to 25 amino acids. The longer length of MHC-II peptides adds complexity to the identification of mutated peptides. Future work should focus on improving NeoMS to enable the identification of mutated peptides in the context of MHC-II.

Acknowledgements. The authors would like to thank other people working in the same group for their help and guidance in this study.

References

1. He, Q., Jiang, X., Zhou, X., Weng, J.: Targeting cancers through tcr-peptide/mhc interactions. J. Hematol. Oncol. **12**(1), 1–17 (2019)
2. Peng, M., et al.: Neoantigen vaccine: an emerging tumor immunotherapy. Mol. Cancer **18**(1), 1–14 (2019)

3. Karasaki, T., et al.: Prediction and prioritization of neoantigens: integration of RNA sequencing data with whole-exome sequencing. Cancer Sci. **108**(2), 170–177 (2017)
4. Bassani-Sternberg, M., et al.: Direct identification of clinically relevant neoepitopes presented on native human melanoma tissue by mass spectrometry. Nat. Commun. **7**(1), 13404 (2016)
5. Zhang, J., et al. Peaks db: de novo sequencing assisted database search for sensitive and accurate peptide identification. Molecular Cellular Proteomics **11**(4) (2012)
6. Eng, J.K., Jahan, T.A., Hoopmann, M.R.: Comet: an open-source ms/ms sequence database search tool. Proteomics **13**(1), 22–24 (2013)
7. Cox, J., Mann, M.: Maxquant enables high peptide identification rates, individualized ppb-range mass accuracies and proteome-wide protein quantification. Nat. Biotechnol. **26**(12), 1367–1372 (2008)
8. The, M., MacCoss, M.J., Noble, W.S., Käll, L.: Fast and accurate protein false discovery rates on large-scale proteomics data sets with percolator 3.0. J. Am. Soc. Mass Spectrometry **27**, 1719–1727 (2016)
9. Bichmann, L., et al.: Mhcquant: automated and reproducible data analysis for immunopeptidomics. J. Proteome Res. **18**(11), 3876–3884 (2019)
10. Li, K., Jain, A., Malovannaya, A., Wen, B., Zhang, B.: Deepredcore: leveraging deep learning to improve peptide identification in immunopeptidomics. Proteomics **20**(21–22), 1900334 (2020)
11. Wilhelm, M., et al.: Deep learning boosts sensitivity of mass spectrometry-based immunopeptidomics. Nature Commun. **12**(1), 3346 (2021)
12. Declercq, A.: Ms2rescore: data-driven rescoring dramatically boosts immunopeptide identification rates. Molecular Cell. Proteomics **21**(8) (2022)
13. Zeng, W.-F.: Alphapeptdeep: a modular deep learning framework to predict peptide properties for proteomics. Nat. Commun. **13**(1), 7238 (2022)
14. Tran, N.H., Qiao, R., Xin, L., Chen, X., Shan, B., Li, M.: Personalized deep learning of individual immunopeptidomes to identify neoantigens for cancer vaccines. Nat. Mach. Intell. **2**(12), 764–771 (2020)
15. Qiao, R., et al.: Computationally instrument-resolution-independent de novo peptide sequencing for high-resolution devices. Nat. Mach. Intell. **3**(5), 420–425 (2021)
16. Kong, A.T., Leprevost, F.V., Avtonomov, D.M., Mellacheruvu, D., Nesvizhskii, A.I.: Msfragger: ultrafast and comprehensive peptide identification in mass spectrometry-based proteomics. Nat. Methods **14**(5), 513–520 (2017)
17. Chi, H., et al.: Comprehensive identification of peptides in tandem mass spectra using an efficient open search engine. Nat. Biotechnol. **36**(11), 1059–1061 (2018)
18. Devabhaktuni, A., et al.: Taggraph reveals vast protein modification landscapes from large tandem mass spectrometry datasets. Nat. Biotechnol. **37**(4), 469–479 (2019)
19. Kacen, A., et al.: Post-translational modifications reshape the antigenic landscape of the mhc i immunopeptidome in tumors. Nat. Biotechnol. **41**(2), 239–251 (2023)
20. Bassani-Sternberg, M., Pletscher-Frankild, S., Jensen, L.J., Mann, M.: Mass spectrometry of human leukocyte antigen class i peptidomes reveals strong effects of protein abundance and turnover on antigen presentation*[s]. Mole. Cell. Proteomics **14**(3), 658–673 (2015)
21. Adusumilli, R., Mallick, P.: Data conversion with proteowizard msconvert. Proteomics: methods and protocols, pp. 339–368 (2017)
22. Ma, B.: Novor: real-time peptide de novo sequencing software. J. Am. Soc. Mass Spectrom. **26**(11), 1885–1894 (2015)

23. Di Tommaso, P., Chatzou, M., Floden, E.W., Barja, P.P., Palumbo, E., Notredame, C.: Nextflow enables reproducible computational workflows. Nat. Biotechnol. **35**(4), 316–319 (2017)

24. UniProt Consortium: Uniprot: a hub for protein information. Nucleic Acids Res. **43**(D1), D204–D212 (2015)

25. Wen, B., Li, K., Zhang, Y., Zhang, B.: Cancer neoantigen prioritization through sensitive and reliable proteogenomics analysis. Nat. Commun. **11**(1), 1759 (2020)

26. Zeng, W.-F., Zhou, X.-X., Zhou, W.-J., Chi, H., Zhan, J., He, S.-M.: Ms/ms spectrum prediction for modified peptides using pdeep2 trained by transfer learning. Anal. Chem. **91**(15), 9724–9731 (2019)

27. Liu, K., Li, S., Wang, L., Ye, Y., Tang, H.: Full-spectrum prediction of peptides tandem mass spectra using deep neural network. Anal. Chem. **92**(6), 4275–4283 (2020)

28. Elias, J.E., Gygi, S.P.: Target-decoy search strategy for increased confidence in large-scale protein identifications by mass spectrometry. Nat. Methods **4**(3), 207–214 (2007)

29. O'Donnell, T.J., Rubinsteyn, A., Laserson, U.: Mhcflurry 2.0: improved pan-allele prediction of mhc class i-presented peptides by incorporating antigen processing. Cell Syst. **11**(1), 42–48 (2020)

On Sorting by Flanked Transpositions

Huixiu Xu[1], Xin Tong[1], Haitao Jiang[1(✉)], Lusheng Wang[2(✉)], Binhai Zhu[3(✉)], and Daming Zhu[1(✉)]

[1] School of Computer Science and Technology, Shandong University, Qingdao, China
{xuhuixiu,xtong}@mail.sdu.edu.cn, {htjiang,dmzhu}@sdu.edu.cn
[2] Department of Computer Science, City University of Hong Kong, Kowloon, Hong Kong, China
cswangl@cityu.edu.hk
[3] Gianforte School of Computing, Montana State University, Bozeman, MT 59717, USA
bhz@montana.edu

Abstract. Transposition is a well-known genome rearrangement event that switches two consecutive segments on a genome. The problem of sorting permutations by transpositions has attracted a great amount of interest since it was introduced by Bafna and Pevzner in 1995. However, empirical evidence has reported that, in many genomes, the participation of repeat segments is inevitable during genome evolution and the breakpoints where a transposition occurs are most likely accompanied by a triple of repeated segments. For example, a transposition will transform $r\ x\ r\ y\ z\ r$ into $r\ y\ z\ r\ x\ r$, where r is a relative short repeat appearing three times and x and y are long segments involved in the transposition. For this transposition event, the neighbors of segments x and y remain the same before and after the transposition. This type of transposition is called flanked transposition.

In this paper, we investigate the problem of sorting by flanked transpositions, which requires a series of flanked transpositions to transform one genome into another. First, we present an $O(n)$ expected running time algorithm to determine if a genome can be transformed into the other genome by a series of flanked transposition for a special case, where each adjacency (roughly two neighbors of two element in the genome) appears once in both input genomes. We then extend the decision algorithm to work for the general case with the same expected running time $O(n)$. Finally, we show that the new version, sorting by minimum number of flanked transpositions is also NP-hard.

Keywords: Genome rearrangement · flanked transpositions · decision algorithm · NP-hard

1 Introduction

Genome rearrangements involve global changes of genomes and reveamol the evolutionary events between species. The discovery of genome rearrangements

X. Guo et al. (Eds.): ISBRA 2023, LNBI 14248, pp. 292–311, 2023.
https://doi.org/10.1007/978-981-99-7074-2_23

dates back to the 1930 s, when Sturtevant and Dobzhansky found inversions on the genome of drosophila [1,2]. With the development of DNA sequencing, more and more genome rearrangement phenomena were discovered [3–5]. Sankoff et al. described the genome rearrangement events with some basic operations on the genomes, e.g., reversals, transpositions, block-interchanges and translocations [6], where reversals and transpositions occur the most frequently. Since then, the problems of transforming one genome into another by some rearrangement operations have been attracting a lot of attention in computational biology.

The problem of sorting by reversals has been investigated extensively. Watterson et al. pioneered the research on sorting by reversals [7]. For signed genomes, Hannenhalli and Pevzner proposed an $O(n^4)$ time exact algorithm, where n is the number of genes in the given permutation [8]. Sorting unsigned genomes by reversals was shown to be NP-hard by Caprara [9], and APX-hard by Berman et al. [10]. Other algorithmic improvements on sorting by reversals can be found in [11–15,31].

The progress on sorting by transpositions is relatively slow. Before the establishment of its NP-hardness, Bafna and Pevzner presented the first approximation algorithm for this problem, which guarantees an approximation ratio of 1.5 and runs in $O(n^2)$ time, where n is the length of the input genomes [16]. Hartman and Shamir simplified their algorithm, and also improved the running time to $O(n^{1.5}\sqrt{\log n})$ [17]. Consequently, Feng and Zhu proposed a data structure called "permutation tree", resulting in a running time of $O(n \log n)$ [18]. So far as we know, Elias and Hartman's algorithm holds the best approximation ratio of 1.375 [19]. In 2012, Bulteau et al. proved that sorting by transpositions is NP-hard, settling the complexity of this long standing open problem [20]. Another interesting aspect of sorting by transpositions is the "transposition diameter", which is to find the genome of length n that consumes the largest number of transpositions. Eriksson et al. presented an upper bound of the corresponding diameter [21].

The above model assumes that the rearrangement events could occur at any position of a genome. Statistics analysis showed that breakpoints, where rearrangements occur, are often associated with repetitive segments [22,23]. In fact, this phenomenon has been independently discovered by some previous works since 1997 [24–27]. Recently, studies on Pseudomonas aeruginosa, Escherichia coli, Mycobacterium tuberculosis and Shewanella further verified that rearrangement events are associated with repeats and that repeat neighbors of rearrangement segments remain the same before and after the rearrangement events [28,29]. An example in [29] revealed that the transposition event swaps the two consecutive segments $47 \sim 60DS1$ and $18 \sim 46DS2$ of two scaffolds of Pseudomonas aeruginosa strains, where the repeats +R appear at the three breakpoints. That is, one strains contains $+R47 \sim 60DS1 + R18 \sim 46DS2 + R$ and the other contains $+R18 \sim 46DS2 + R47 \sim 60DS1 + R$. Therefore, whether there exist repeats at the breakpoints of rearrangement events such that the repeat neighbors of rearrangement segments remain the same before and after the rearrangement event may give us a clue on whether the calculated rearrangement scenarios are biologically meaningful. Other rearrangement events also follows the same rule. For example, the reversal transforming $+y + B + x - B + z$ into $+y + B - x - B + z$ also have repeats +B and -B at the ends of x. The neighbors of x remain the same before and after the reversal. Similar examples were also found for block-interchanges, etc [28,29].

In this paper, we investigate the problem of sorting by flanked transpositions, which requires a series of flanked transpositions to transform one genome into another. For example, a transposition will transform r x r y z r into r y z r x r, where r is a relative short repeat appearing three times and x and y are long segments involved in the transposition. For this transposition event, the neighbors of segments x and y remain the same before and after the transposition. This type of transposition is called flanked transposition.

First, we present an $O(n)$ expected running time algorithm to determine if a genome can be transformed into the other genome by a series of flanked transposition for a special case, where each adjacency appears once in both input genomes. We then extend the decision algorithm to work for the general case with the same expected running time $O(n)$. Finally, we show that the new version, sorting by minimum number of flanked transpositions is also NP-hard.

This paper is organized as follows. In Sect. 2, we give some preliminary definitions. In Sect. 3, we present the decision algorithm for the simple case. In Sect. 4, we extend the decision algorithm for the general case. The NP-hardness results for the optimization version of the new model is given in Sect. 5. Section 6 gives the conclusion. Due to space restriction, we put the proofs of this lemma and some of the following lemmas and theorems in the appendix.

2   Preliminaries

In the literature on transposition, a genome is a permutation of integers from $\mathbb{N} = \{1, 2, \ldots, m\}$, where each integer stands for a gene (or a long segment of DNA sequence).

For flanked transposition, let $\mathbb{N} = \{1, 2, \ldots, m\}$ be the set of genes and $\mathbb{R} = \{x1, x2, \ldots, x_k\}$ be a set of distinct repeats, where each repeat represents a (relatively) short DNA segment. A genome π is a string of length n over $\mathbb{N} \cup \mathbb{R}$, where each $x \in \mathbb{N}$ appears once and every $x \in \mathbb{R}$ appears at least once. Each repeat $x \in \mathbb{R}$ may appear more than once and the appearance of x is referred to as its *occurrence*. The number of occurrences of the repeat x on π is its *duplicate number* on the genome, denoted by $dp[x, \pi]$. The duplicate number of a genome π, denoted by $dp[\pi]$, is the maximum duplicate number of the repeats on the genome. For example, the genome $\pi = [x, y, x, 1, y, x]$ contains one gene and two repeats x and y, where $dp[1, \pi] = 1$, $dp[x, \pi] = 3$, and $dp[y, \pi] = 2$. Thus, then $dp[\pi] = 3$.

Two genomes π and τ over $\mathbb{N} \cup \mathbb{R}$ are *related* if the duplicate number of each element is identical, i.e., for each element $x \in \mathbb{N} \cup \mathbb{R}$, $dp[x, \pi] = dp[x, \tau]$.

For simplicity, we set $\mathbb{G} = \mathbb{N} \cup \mathbb{R}$, and we can also assume that a genome π is a sequence of length n over $\mathbb{G}$, where every *element* $x \in \mathbb{G}$ with $dp[x, \pi] = 1$ is a *gene* and every element $x \in \mathbb{G}$ with $dp[x, \pi] > 1$ is a *repeat*.

A *transposition* swaps two consecutive segments on a genome. A transposition $\rho(i, j, k)$ will transform the genome $\pi = [x_1, \ldots, x_{i-1}, \underline{x_i, \ldots, x_{j-1}}, \underline{x_j, \ldots, x_{k-1}}, x_k, \ldots, x_n]$ into $\pi' = [x_1, \ldots, x_{i-1}, \underline{x_j, \ldots, x_{k-1}, x_i, \ldots, x_{j-1}}, x_k, \ldots, x_n]$, where $1 \leq i < j < k \leq n$. For convenience, we use $\rho(x_i, x_j, x_k)$ or $\rho(i, j, k)$ interchangeably to denote π. A transposition is called a *flanked transposition* if $x_i = x_j = x_k$. Note that, the right neighbor of x_i, x_j and x_k remain the same before and after the transposition.

Let $\pi = [x_1, x_2, \ldots, x_n]$ be a genome, each element in π, say x_i ($1 \leq i \leq n$), can be represented by a pair of ordered nodes, x_i^l and x_i^r. Consequently, π can also be described as $[x_1^l, x_1^r, x_2^l, x_2^r, \ldots, x_n^l, x_n^r]$. Moreover, x_i^r and x_{i+1}^l form an *adjacency*, denoted by $\langle x_i^r, x_{i+1}^l \rangle$, for $1 \leq i \leq n-1$. Let $\mathbb{A}[\pi]$ represent the set of $n-1$ adjacencies of π. A genome π is *simple* if the $n-1$ adjacencies in $\mathbb{A}[\pi]$ are distinct.

For the sake of unification, we add "0" to the two ends of every genome. Let's take the genome $\pi = [0, 2, 3, 2, 1, 2, 3, 0]$ as an example to illustrate the above notations. The set of adjacencies is $\mathbb{A}[\pi] = \{\langle 0^r, 2^l \rangle, \langle 2^r, 3^l \rangle, \langle 3^r, 2^l \rangle, \langle 2^r, 1^l \rangle, \langle 1^r, 2^l \rangle, \langle 2^r, 3^l \rangle, \langle 3^r, 0^l \rangle\}$. π can also be viewed as $[0^l, 0^r, 2^l, 2^r, 3^l, 3^r, 2^l, 2^r, 1^l, 1^r, 2^l, 2^r, 3^l, 3^r, 0^l, 0^r]$. π is not simple since the adjacency $\langle 2^r, 3^l \rangle$ appears twice.

Now, we formally define the problems investigated in this paper.

Definition 1. *Sorting by flanked transpositions (**SFT**). **Instance:** Two related genomes π and τ, such that $dp[\pi] = dp[\tau] \geq 3$. **Question:** Is there a sequence of flanked transpositions that transforms π into τ?*

Definition 2. *Sorting by the minimum number of flanked transpositions (**SMFT**). **Instance:** Two related genomes π and τ, such that $dp[\pi] = dp[\tau] \geq 3$. **Question:** A sequence of flanked transpositions $\rho_1, \rho_2, \ldots, \rho_k$ that transforms π into τ, such that k is minimized.*

Hereafter, we assume that $\pi = [x_0, x_1, \ldots, x_n, x_{n+1}]$ and $\tau = [y_0, y_1, \ldots, y_n, y_{n+1}]$ are two related genomes, where $x_0 = x_{n+1} = y_0 = y_{n+1} = 0$. Note that, each x_i is an occurrence of an element in π and each y_j is an occurrence of an element in τ. The indices i and j indicate their location in π and τ, respectively.

Since a flanked transposition always contains three occurrences x_i, x_j, and x_k of the same element(repeat), the following lemma is obvious.

Lemma 1. *Let π be a genome and π' be the genome obtained from π after a flanked transposition. $\mathbb{A}[\pi] = \mathbb{A}[\pi']$.*

Lemma 1 implies a necessary condition for answering **SFT**.

Theorem 1. *π cannot be transformed into τ by a series of flanked transpositions if $\mathbb{A}[\pi] \neq \mathbb{A}[\tau]$.*

3   An $O(n)$ Expected Time Decision Algorithm for the Simple Case

As a warm-up, in this section, we solve the problem of **SFT** in the case that π and τ are simple and $\mathbb{A}[\pi] = \mathbb{A}[\tau]$.

3.1   The 2-Color and 3-Color Cycle Graphs

If an adjacency $\langle x_{i-1}^r, x_i^l \rangle$ of π is identical to an adjacency $\langle y_{j-1}^r, y_j^l \rangle$ of τ, i.e., $x_{i-1} = y_{j-1}$ and $x_i = y_j$, we define a mapping f such that $f(\langle x_{i-1}^r, x_i^l \rangle) = \langle y_{j-1}^r, y_j^l \rangle$. Since both π and τ are simple and $\mathbb{A}[\pi] = \mathbb{A}[\tau]$, there is a unique bijection mapping between $\mathbb{A}[\pi]$ and $\mathbb{A}[\tau]$.

Let $f : \mathbb{A}(\pi) \to \mathbb{A}(\tau)$ be the bijection mapping. The adjacency $\langle x_{i-1}^r, x_i^l \rangle$ in π is matched to an adjacency $\langle y_{j-1}^r, y_j^l \rangle$ in τ, where x_{i-1} and y_{j-1} (x_i and y_j) are occurrences of the same element. That is,

$$f(\langle x_{i-1}^r, x_i^l \rangle) = \langle y_{j-1}^r, y_j^l \rangle. \tag{1}$$

Note that each adjacency $\langle x_{i-1}^r, x_i^l \rangle$ in π is represented by the occurrences of elements in π, e.g., x_{i-1}^r and x_i^l, while each adjacency in τ is represented by the occurrences of elements in τ, say, y_{j-1}^r and y_j^l. The equality (1) holds, when x_{i-1}, x_i, y_{j-1} and y_j are represented by the names of their elements. (instead of names of occurrences of elements).

Based on this bi-jection mapping f, we can construct the 2-color cycle graph $\mathcal{G}(\pi, \tau, f)$ as follows. For each x_i ($0 \le i \le n+1$) in π, we construct a red edge (x_i^l, x_i^r), labelled by x_i. For each y_k ($0 \le k \le n+1$) in τ, if $f(\langle x_{i-1}^r, x_i^l \rangle) = \langle y_{k-1}^r, y_k^l \rangle$, and $f(\langle x_j^r, x_{j+1}^l \rangle) = \langle y_k^r, y_{k+1}^l \rangle$, then we connect x_i^l and x_j^r with a blue edge labeled by y_k. Specially, x_0^l and x_0^r (as well as x_{n+1}^l and x_{n+1}^r) are connected by a blue edge and a red edge. An example is shown in Fig. 1.

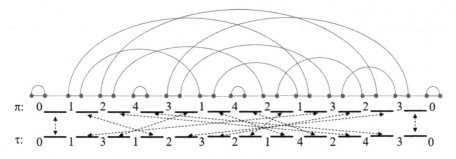

Fig. 1. The cycle graph $\mathcal{G}(\pi, \tau, f)$. (Color figure online)

In fact, each occurrence of an element in π corresponds to a red edge, and each occurrence of an element in τ corresponds to a blue edge in $\mathcal{G}(\pi, \tau, f)$.

Note that each node in $\mathcal{G}(\pi, \tau, f)$ is incident to one red edge and one blue edge. Thus, $\mathcal{G}(\pi, \tau, f)$ is composed of disjoint cycles, on which the red edges and blue edges appear alternatively. Let s be an integer. A cycle composed of s blue edges (also s red edges) is called an s-cycle, s is called the length of the cycle. An s-cycle is an even (resp. odd) cycle if s is even (resp. odd). An s-cycle is a long cycle when $s \ge 4$.

The cycle graph constructed here for flanked transposition are quite different from the original version.

Lemma 2. *All the nodes in the same cycle of $\mathcal{G}(\pi, \tau, f)$ are from the occurrences of the same element (repeat).*

Thus, we will say that a cycle in the 2-color graph $\mathcal{G}(\pi, \tau, f)$ corresponds to an element. Now, we add the third color edges to a 2-color graph by connecting each pairs of nodes x_i^r and x_{i+1}^l (for all $0 \le i \le n$) with a directed green edge (x_i^r, x_{i+1}^l) in $\mathcal{G}(\pi, \tau, f)$ to form a 3-colored graph.

Lemma 3. *If we connect each pairs of nodes x_i^r and x_{i+1}^l (for all $0 \le i \le n$) with a directed green edge (x_i^r, x_{i+1}^l) in $\mathcal{G}(\pi, \tau, f)$, then all the blue edges and green edges form a directed path P, where the directions of all green edges are the same in P and the directions of blue edges are determined by the direction of P accordingly, and the sequence of elements along the path is identical to $\tau = (y_0, y_1, \ldots, y_{n+1})$.*

Hereafter, we use $\mathcal{G}(\pi, \tau, f)$ to represent the 3-color graph and simply call it cycle graph. A path with alternative blue and green edges is a directed path and a cycle with alternative blue and red edges are undirected. Through the cycle graph, we can obtain the condition that two related simple genomes are identical.

Lemma 4. *Given two related genomes ν and μ with $\mathbb{A}[\nu] = \mathbb{A}[\mu]$. then ν and μ are identical, if and only if there exists a bijection f between $\mathbb{A}[\nu]$ and $\mathbb{A}[\mu]$, such that the 2-color graph $\mathcal{G}(\nu, \mu, f)$ is composed of 1-cycles.*

Lemma 4 provides an alternative away to transform π into τ, i.e., performing a series of flanked transpositions to transform π into π^*, and find a bijection f^* between π^* and τ, till $\mathcal{G}(\pi^*, \tau, f^*)$ only contains 1-cycles. Next, we show that how a flanked transposition impacts $\mathcal{G}(\pi, \tau, f)$.

Lemma 5. *Let x_i, x_j and x_k be three occurrences of the same element in π. Let $(x_i^l, x_{i'}^r)$, $(x_j^l, x_{j'}^r)$, and $(x_k^l, x_{k'}^r)$ be blue edges labeled by y_a, y_b, and y_c, respectively in $\mathcal{G}(\pi, \tau, f)$. Let $\rho(i, j, k)$ be a flanked transposition that transforms π into π'. Then, there exists a bijection f' between $\mathcal{A}[\pi']$ and $\mathcal{A}[\tau]$, such that $\mathcal{G}(\pi', \tau, f')$ is different from $\mathcal{G}(\pi, \tau, f)$ with blue edges $(x_i^l, x_{i'}^r)$, $(x_j^l, x_{j'}^r)$ and $(x_k^l, x_{k'}^r)$ being replaced by blue edges $(x_j^l, x_{i'}^r)$, $(x_k^l, x_{j'}^r)$, and $(x_i^l, x_{k'}^r)$, and green edges (x_{i-1}^r, x_i^l), (x_{j-1}^r, x_j^l) and (x_{k-1}^r, x_k^l) replaced by green edges (x_{i-1}^r, x_j^l), (x_{j-1}^r, x_k^l), and (x_{k-1}^r, x_i^l), respectively. Moreover, other edges remain the same in $\mathcal{G}(\pi', \tau, f')$.*

Lemma 5 also shows that once we perform a flanked transposition, we will also get a new bijection.

From Lemma 2, each cycle in $\mathcal{G}(\pi, \tau, f)$ corresponds to exactly one element. Since a flanked transposition always contains three occurrences of the same element, a flanked transposition can only affect cycles corresponding to the same element.

In the traditional transposition model, each gene appears exact once and a transposition can affect more than one cycle, where each cycle may contain more than one element. Consequently, a genome can always be transformed into the other by a series of transpositions. For the sorting by flanked transposition problem, it is not always possible to transform one genome into the other by using a series of flanked transpositions. Thus, the decision problem needs to be solved first.

Suppose that ρ is a flanked transposition that transforms π into π', and the bijection changes from f to f' accordingly. Let $c(\pi, \tau, f)$ (resp. $c(\pi', \tau, f')$), $c_o(\pi, \tau, f)$ (resp. $c_o(\pi', \tau, f')$) and $c_e(\pi, \tau, f)$ (resp. $c_e(\pi', \tau, f')$) denote the number of cycles, odd cycles, and even cycles in the 2-color graph $\mathcal{G}(\pi, \tau, f)$ (resp. $\mathcal{G}(\pi', \tau, f')$). We define $\Delta_o(\rho) = c_o(\pi', \tau, f') - c_o(\pi, \tau, f)$, $\Delta(\rho) = c(\pi', \tau, f') - c(\pi, \tau, f)$. Bafna and Pevzner showed the following important property of transpositions, which still holds for flanked transpositions.

Lemma 6. [16] *For every (flanked) transposition ρ, $\Delta(\rho) \in \{2, 0, -2\}$ and $\Delta_o(\rho) \in \{2, 0, -2\}$.*

Now, we present another necessary condition for **SFT**.

Theorem 2. *Let π and τ be two related simple genomes with $\mathbb{A}(\pi) = \mathbb{A}(\tau)$, and f be the bijection between $\mathbb{A}(\pi)$ and $\mathbb{A}(\tau)$. π cannot be transformed into τ by a series of flanked transpositions if there exists some element x such that the number of even cycles corresponding to x in $\mathcal{G}(\pi, \tau, f)$ is odd.*

Hereafter, we assume that the number of even cycles corresponding to every element/repeat is even. In the following part of this section, we will show that the condition is also sufficient by modifying all the cycles into 1-cycles. See Lemma 7.

The outline to convert all cycles into 1-cycles is the same as [16], where we first repeatedly split long cycles so that there are only 2-cycles and 3-cycles left. We then remove 2-cycles. After that we remove 3-cycles. The way to remove 2-cycles is simple and identical to that in [16]. The ways to split long cycles and remove 3-cycles contain brand new techniques and are totally different here.

3.2   Splitting Long (Red and Blue) Cycles

Let us consider an arbitrary long (red and blue) cycle, $C = (x_p^l, x_p^r, x_i^l, x_i^r, x_j^l,$ $x_j^r, x_k^l, x_k^r, x_q^l, x_q^r, \ldots, x_p^l)$ with length s such that $s \geq 4$ in $\mathcal{G}(\pi, \tau, f)$, where $q = p$ when $s = 4$. From Lemma 2, all the x_i in C are occurrences of the same element, say, x. Recall that, each blue edge is labeled with an occurrence of the element in τ. We assume that the blue edge (x_i^r, x_j^l) in C is labeled with y_b and the blue edge (x_k^r, x_q^l) in C is labeled with y_a, where y_a and y_b are both occurrences of x in τ.

By adding an occurrence of x at proper positions of π and τ, respectively, We will obtain π' and τ' such that in $\mathcal{G}(\pi', \tau', f')$(where f' is a bijection between $\mathbb{A}(\pi')$ and $\mathbb{A}(\tau')$), C is split into an $(s - 2)$ cycle and a 3-cycle. Repeating the process, we can obtain a pair of new genomes π' and τ', as well as a bijection

f' between $\mathbb{A}(\pi')$ and $\mathbb{A}(\tau')$, such that all the cycles in $\mathcal{G}(\pi', \tau', f')$ are of length $s \leq 3$.

Consider an arbitrary occurrence, say, x_i, of the element x in the long cycle C. By adding a new occurrence $x_{i'}$ of the common element x in C at the position right to x_i in π, we get a new genome π'.

Starting from x_i^r, we can through nodes/edges in the long cycles C in the order x_j^l, x_j^r, x_k^l, x_k^r, x_q^l, and x_q^r. Consider the blue edge (x_k^r, x_q^l) in $\mathcal{G}(\pi, \tau)$ that is labeled with y_a, where y_a is an occurrence of x in τ. Let f be the bijection between $\mathbb{A}(\pi)$ and $\mathbb{A}(\tau)$. We have $f(\langle x_k^r, x_{k+1}^l \rangle) = \langle y_a^r, y_{a+1}^l \rangle$, where x_{k+1} and y_{a+1} are occurrences of some elements on the right of x_k and y_a in π and τ, respectively. Similarly, we have $f(\langle x_{q-1}^r, x_q^l \rangle) = \langle y_{a-1}^r, y_a^l \rangle$. Recall that, the blue edge (x_i^r, x_j^l) is labeled with y_b. Then we have, $f(\langle x_{j-1}^r, x_j^l \rangle) = \langle y_{b-1}^r, y_b^l \rangle$ and $f(\langle x_i^r, x_{i+1}^l \rangle) = \langle y_b^r, y_{b+1}^l \rangle$.

Finally, by adding a new occurrence $y_{a'}$ of the common element x in C at the position right to y_a in τ, we get a new genome τ'.

It is easy to see that $\mathbb{A}(\pi') = \mathbb{A}(\tau')$. In fact, $\mathbb{A}[\pi'] = \mathbb{A}[\pi] - \{\langle x_i^r, x_{i+1}^l \rangle\} + \{\langle x_i^r, x_{i'}^l \rangle, \langle x_{i'}^r, x_{i+1}^l \rangle\}$, and $\mathbb{A}[\tau'] = \mathbb{A}[\tau] - \{\langle y_a^r, y_{a+1}^l \rangle\} + \{\langle y_a^r, y_{a'}^l \rangle, \langle y_{a'}^r, y_{a+1}^l \rangle\}$. Though π' and τ' could no longer be simple, we can still obtain a bijection f' between $\mathbb{A}(\pi')$ and $\mathbb{A}(\tau')$ by changing f as follows:

- Remove 2 the mapping $f(\langle x_k^r, x_{k+1}^l \rangle) = \langle y_a^r, y_{a+1}^l \rangle$ and $f(\langle x_i^r, x_{i+1}^l \rangle) = \langle y_b^r, y_{b+1}^l \rangle$ from f.
- Add 3 new matches, (1) $f'(\langle x_k^r, x_{k+1}^l \rangle) = \langle y_{a'}^r, y_{a+1}^l \rangle$, (2) $f'(\langle x_i^r, x_{i'}^l \rangle) = \langle y_a^r, y_{a'}^l \rangle$, and (3) $f'(\langle x_{i'}^r, x_{i+1}^l \rangle) = \langle y_b^r, y_{b+1}^l \rangle$ in f'.

Using f', we can obtain $\mathcal{G}(\pi', \tau', f')$. The only reason that we define f' in the above way is that $\mathcal{G}(\pi', \tau', f')$ is constructed from f', the long cycle C in $\mathcal{G}(\pi, \tau, f)$ is decomposed into a $(s-2)$-cycle and a 3-cycle in $\mathcal{G}(\pi', \tau', f')$. In fact, $\mathcal{G}(\pi, \tau, f)$ is changed into $\mathcal{G}(\pi', \tau', f')$ (see Fig. 2) as follows:

(a) Add two new nodes $x_{i'}^l$ and $x_{i'}^r$, which are connected by a red edge.
(b) Remove the two blue edges (x_k^r, x_q^l) and (x_i^r, x_j^l), since the two matches $f(\langle x_k^r, x_{k+1}^l \rangle) = \langle y_a^r, y_{a+1}^l \rangle$ and $f(\langle x_i^r, x_{i+1}^l \rangle) = \langle y_b^r, y_{b+1}^l \rangle$ don't belonging to f', .
(c) Add a blue edge $(x_{i'}^l, x_k^r)$, which is labeled by $y_{a'}$, due to the two new matches (1) and (2). Add a blue edge (x_q^l, x_i^r), which is labeled by y_a, due to the match $f(\langle x_{q-1}^r, x_q^l \rangle) = \langle y_{a-1}^r, y_a^l \rangle$, which still exists in f', and (2). Add a blue edge $(x_{i'}^r, x_j^l)$, which is labeled by y_b, due to the match $f(\langle x_{j-1}^r, x_j^l \rangle) = \langle y_{b-1}^r, y_b^l \rangle$, which still exists in f', and (3).

As a result, in $\mathcal{G}(\pi', \tau', f')$, $(x_{i'}^l, x_{i'}^r, x_j^l, x_j^r, x_k^l, x_k^r, x_{i'}^l)$ is a 3-cycle, while $(x_p^l, x_p^r, x_i^l, x_i^r, x_q^l, x_q^r, \ldots, x_p^l)$ becomes an $(s-2)$-cycle.

Lemma 7. *If π can be transformed into τ by m flanked transpositions, then π' can be transformed into τ' by at most $m+1$ flanked transpositions. If π' can be transformed into τ' by m' flanked transpositions, then π can be transformed into τ by at most m' flanked transpositions.*

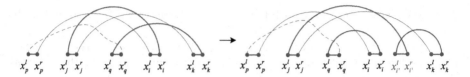

$$x_p^l \ x_p^r \quad x_j^l \ x_j^r \quad x_q^l \ x_q^r \quad x_i^l \ x_i^r \quad x_k^l \ x_k^r \qquad \rightarrow \qquad x_p^l \ x_p^r \quad x_j^l \ x_j^r \quad x_q^l \ x_q^r \quad x_i^l \ x_i^r \ x_{i'}^l \ x_{i'}^r \quad x_k^l \ x_k^r$$

Fig. 2. Splitting a long cycle: We illustrate the general case of a long cycle, where dashed lines represents the nodes/edges after x_q^r in C. In order to draw the figure, we arbitrarily fixed the order of the relative positions of the occurrences of the same element x (i.e., x_i, x_j, x_k, x_p, and x_q)in π. This will not affect the general scenario if we just focus on the order of edges in the cycles. (Color figure online)

After long cycle splitting, let the resulting genomes be $\hat{\pi}$ and $\hat{\tau}$, and the bijection be $\hat{f}$. Then $\mathcal{G}(\hat{\pi}, \hat{\tau}, \hat{f})$ only contains 1-cycles, 2-cycles and 3-cycles. Moreover, the numbers of even cycles in $\mathcal{G}(\hat{\pi}, \hat{\tau}, \hat{f})$ is still even.

The way that we handle long red and blue cycle is different from that in [16] due to the fact that each flanked transposition corresponds to three occurrences of the same element.

3.3   Removing (Red and Blue) 2-Cycles

From Theorem 2, for each element x, the number of even (red and blue) cycles must be even.

Lemma 8. *In $\mathcal{G}(\hat{\pi}, \hat{\tau}, \hat{f})$, any two red and blue 2-cycles corresponding to the same element can be transformed into a 1-cycle and a 3-cycle by a flank transposition.*

By use of Lemma 11, all the 2-cycles can be transformed into 3-cycles and 1-cycles, provided that the number of even cycles corresponding to each element in $\mathcal{G}(\hat{\pi}, \hat{\tau})$ is even. Let the resulting genome be $\bar{\pi}$, and the bijection becomes $\bar{f}$, then there are only 1-cycles and 3-cycles in $\mathcal{G}(\bar{\pi}, \hat{\tau}, \bar{f})$.

The above way to handle 2-cycle is similar to that in [16] except that two 2-cycles correspond to the same element.

3.4   Handling (Red and Blue) 3-Cycles

Given a (red and blue) 3-cycle $\mathcal{C} = (x_i^r, x_j^l, x_j^r, x_k^l, x_k^r, x_i^l)$ in $\mathcal{G}(\bar{\pi}, \hat{\tau}, \bar{f})$, there are 6 permutations of x_i, x_j and x_k. They are classified into two groups. $\mathcal{C}$ is *oriented* if $i < k < j$ or $j < i < k$ or $k < j < i$, and $\mathcal{C}$ is *unoriented* if $i < j < k$ or $j < k < i$ or $k < i < j$. As shown in Fig. 3, the 3-cycle in (a) is oriented, and the 3-cycle in (b) is unoriented.

The following observation is well-known in the literature of genome rearrangement, which was first proposed as *Lemma 3.3* in [16].

Observation 1. *Performing a flanked transposition on an oriented 3-cycle will transform it into three 1-cycles, and performing a flanked transposition on an unoriented 3-cycle will transform it into another unoriented 3-cycle.*

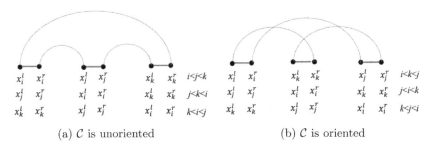

(a) $\mathcal{C}$ is unoriented   (b) $\mathcal{C}$ is oriented

Fig. 3. Two types of a 3-cycle $\mathcal{C} = (x_i^r, x_j^l, x_j^r, x_k^l, x_k^r, x_i^l)$, where (x_i^r, x_j^l), (x_j^r, x_k^l), and (x_k^r, x_i^l) are blue edges, and (x_i^l, x_i^r), (x_j^l, x_j^r), and (x_k^l, x_k^r) are red edges. (Color figure online)

In the previous studies on sorting genomes by transposition [16,17], once there is an unoriented 3-cycle (assume the indices of the three occurrences are i, j and k, where $i < j < k$), then a transposition $\rho(a, b, c)$, such that $i < a < j < b < k < c$ or $a < i < b < j < c < k$, will transform this unoriented 3-cycle into an oriented 3-cycle. In other words, it only needs one additional transposition to generate an oriented 3-cycle. For the flanked transposition model, this method does not work, since the three occurrences involved by a flanked transposition must be of the same element. In the following, we present a new method to generate an oriented 3-cycle by a series of flanked transpositions.

Let (x_i^α, x_j^β) and $(x_{i'}^{\alpha'}, x_{j'}^{\beta'})$ be two blue edges in $\mathcal{G}(\bar{\pi}, \hat{\tau})$, where $i < j$, $i' < j'$ and $\{\alpha, \beta\} = \{\alpha', \beta'\} = \{l, r\}$. We say that (x_i^α, x_j^β) and $(x_{i'}^{\alpha'}, x_{j'}^{\beta'})$ *cross* if $i < i' < j < j'$ or $i' < i < j' < j$. Note that the blue edge of 1-cycle does not cross any other blue edge, the three blue edge of an oriented 3-cycle cross with each other, and any two of the three blue edges of an unoriented 3-cycle cross. Two cycles $\mathcal{C}$ and $\mathcal{C}'$ *cross* with each other if there exist two blue edges $e \in \mathcal{C}$ and $e' \in \mathcal{C}'$, such that e and e' cross.

Lemma 9. *In $\mathcal{G}(\bar{\pi}, \hat{\tau}, \bar{f})$, an unoriented 3-cycle must cross with at least one other 3-cycle. Equivalently, a 3-cycle that does not cross with any other cycle must be oriented.*

An unoriented 3-cycle can be represented as $\mathcal{C} = (x_i^l, x_i^r, x_j^l, x_j^r, x_k^l, x_k^r, x_i^l)$ with $i < j < k$. (See Fig. 4.) For an unoriented 3-cycle $\mathcal{C}$, we obtain a graph $\mathbb{G}(\bar{\pi}, \hat{\tau}, \bar{f})(\mathcal{C})$ by deleting all the nodes and edges in $\mathcal{C}$ from $\mathcal{G}(\bar{\pi}, \hat{\tau}, \bar{f})$, and connecting x_i^r and x_j^l, where x_i and j are the two adjacent nodes of any two consecutive occurrences in π (after deleting) via a green edge. If the blue edges and green edges still form a path connection all the nodes in $\mathbb{G}(\bar{\pi}, \hat{\tau}, \bar{f})(\mathcal{C})$, we say that $\mathcal{C}$ is *reducible*, and the path is called the *remaining path* of $\mathcal{C}$; otherwise, $\mathcal{C}$ is *irreducible*. We give an example to show reducible and irreducible 3-cycles in Fig. 4.

Constructing a Pair of New Genomes by Deleting a Reducible 3-Cycle $\mathcal{C}$:

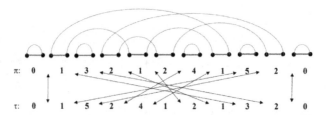

(a) The cycle graph $\mathcal{G}(\pi, \tau, f)$ of $\pi = [0,1,3,2,1,2,4,1,5,2,0]$ and $\tau = [0,1,5,2,4,1,2,1,3,2,0]$, both the 3-cycles corresponds to the element '1' and '2' are irreducible.

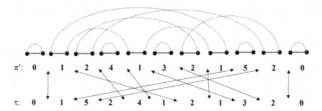

(b) $\pi' = \pi \bullet \rho(1,4,7) = [0,1,2,4,1,3,2,1,5,2,0]$, in the cycle graph $\mathcal{G}(\pi', \tau, f')$ of π, the 3-cycle corresponds to the element '1' becomes reducible.

Fig. 4. An example to illustrate how an irreducible 3-cycle becomes recducible by a flanked transposition. (Color figure online)

Let $\mathcal{C} = (x_i^l, x_i^r, x_j^l, x_j^r, x_k^l, x_k^r, x_i^l)$ with $i < j < k$ be a (unoriented) reducible 3-cycle in $\mathcal{G}(\pi, \tau, f)$, where there are only 3-cycles and 1-cycles in $\mathcal{G}(\pi, \tau, f)$. Let $\mathbb{G}(\pi, \tau, f)(\mathcal{C})$ be the graph obtained from $\mathcal{G}(\pi, \tau, f)$ by deleting all the nodes and edges in $\mathcal{C}$ and adding green edges as in the definition of reducible 3-cycle. Since $\mathcal{C}$ is reducible, there exists a blue and green path connecting all the nodes in $\mathbb{G}(\pi, \tau, f)(\mathcal{C})$.

Let the blue edges (x_k^r, x_i^l), (x_i^r, x_j^l) and (x_j^r, x_k^l) be labeled by y_p, y_q and y_r in τ, respectively. Let π' be the genome obtained by deleting x_i, x_j and x_k from π. Let τ' be the sequence of labels (y_a') of blue edges along the blue and green path in $\mathbb{G}(\pi, \tau, f)(\mathcal{C})$.

Lemma 10. *There is a bijection f' between $\mathcal{A}[\pi']$ and $\mathcal{A}[\tau']$, such that $\mathcal{G}(\pi', \tau', f')$ is the same as $\mathbb{G}(\pi, \tau, f)(\mathcal{C})$.*

Any irreducible unoriented 3-cycle can be represented as $\mathcal{C} = (x_i^l, x_i^r, x_j^l, x_j^r, x_k^l, x_k^r, x_i^l)$, with $i < j < k$.

Lemma 11. *An irreducible unoriented 3-cycle $\mathcal{C} = (x_i^l, x_i^r, x_j^l, x_j^r, x_k^l, x_k^r, x_i^l)$ in $\mathcal{G}(\pi, \tau, f)$ will be transformed into a reducible unoriented 3-cycles by the flanked transposition $\rho(i, j, k)$.*

Proof. Let the blue edges (x_i^l, x_k^r), (x_j^l, x_i^r) and $(x_k^l, x_j^r)x_k^l)$ are labeled by y_p, y_q and y_r in τ, respectively. From Lemma 3, the 3-colored graph $\mathcal{G}(\pi, \tau, f)$ contains a blue and green path P corresponding to τ, where the left end of P is y_0^l and the right end of P is y_{n+1}^r. Along this direction, $\tau = [y_0, \ldots, y_{n+1}]$. Let (x_t^l, x_{i-1}^r) be the blue edge incident to x_{i-1}^r, where x_t in π and x_{i-1}^r are occurrence of the same element. So the path must be of the form $[\ldots, x_t^l, x_{i-1}^r, x_i^l, x_k^r, \ldots]$ according to the direction, then the directed blue edge (x_t^l, x_{i-1}^r) must be labeled by y_{p-1}.

Note that, each x_i^r / x_i^l is incident to an unique blue edge and an unique green edge as well. Similarly, the blue edges incident to x_{j-1}^r, x_{j+1}^l, x_{k-1}^r, and x_{k+1}^l are labeled by y_{q-1}, y_{r+1}, y_{r-1}, and y_{p+1}, respectively (Fig. 4-(a)). Next, we show that $\mathcal{C}$ is reducible if and only if $p < q < r$ or $q < r < p$ or $r < p < q$.

We use a, b and c to indicate the minimum, medium, and maximum number in $\{p, q, r\}$, respectively. Then, $\tau = [y_0, \ldots, y_{a-1}, y_a, y_{a+1}, \ldots, y_{b-1}, y_b, y_{b+1}, \ldots, y_{c-1}, y_c, y_{c+1}, \ldots, y_{n+1}]$. After deleting y_a, y_b, and y_c, τ is partitioned into four sub-sequences: (P1) $[y_0, \ldots, y_{a-1}]$, (P2) $[y_{a+1}, \ldots, y_{b-1}]$, (P3) $[y_{b+1}, \ldots, y_{c-1}]$, and (P4) $[y_{c+1}, \ldots, y_{n+1}]$. In $\mathbb{G}(\pi, \tau, f)(\mathcal{C})$, the three new green edges (x_{i-1}^r, x_{i+1}^l), (x_{j-1}^r, x_{j+1}^l) and (x_{k-1}^r, x_{k+1}^l) connect the three pairs y_{p-1} and y_{q+1}, y_{q-1} and y_{r+1}, and y_{r-1} and y_{p+1}, respectively. Based on the order of p, q, r in τ, we have the following cases.

Case (1): $p < q < r$ (i.e., $a = p$, $b = q$, $c = r$), $q < r < p$ (i.e., $a = q$, $b = r$, $c = p$), or $r < p < q$ (i.e., $a = r$, $b = p$, $c = q$). In each of the three cases, the three pairs y_{a-1} and y_{b+1}, y_{b-1} and y_{b+1}, and y_{c-1} and y_{a+1} are connected by the three new green edges (x_{i-1}^r, x_{i+1}^l), (x_{j-1}^r, x_{j+1}^l) and (x_{k-1}^r, x_{k+1}^l), connecting the three pairs y_{a-1} and y_{b+1}, y_{b-1} and y_{b+1}, and y_{c-1} and y_{a+1}. Then the four sub-sequences can form a new genome $\tau' = [y_0, \ldots, y_{a-1}, \underline{y_{b+1}, \ldots, y_{c-1}}, y_{a+1}, \ldots, y_{b-1}, y_{c+1}, \ldots, y_{n+1}]$. Thus, in this case, $\mathcal{C}$ is reducible.

Case (2): $r < q < p$ (i.e., $a = r$, $b = q$, $c = p$), $p < r < q$ (i.e., $a = p$, $b = r$, $c = q$), or $q < p < r$ (i.e., $a = q$, $b = p$, $c = r$). In each of the three cases, the three pairs y_{a-1} and y_{b+1}, y_{b-1} and y_{b+1}, and y_{c-1} and y_{a+1} are connected by the three new green edges (x_{i-1}^r, x_{i+1}^l), (x_{j-1}^r, x_{j+1}^l) and (x_{k-1}^r, x_{k+1}^l), connecting the three pairs y_{a-1} and y_{b+1}, y_{b-1} and y_{b+1}, and y_{c-1} and y_{a+1}. Then the four sub-sequences can not form a new genome. Instead, they form a linear sequence $[y_0, \ldots, y_{a-1}, y_{c+1}, \ldots, y_{n+1}]$, and two circular sequences $[y_{a+1}, \ldots, y_{b-1}]$, and $[y_{b+1}, \ldots, y_{c-1}]$, where y_{b-1} connects y_{a+1} and y_{c-1} connects y_{b+1}, respectively. Thus, in this case, $\mathcal{C}$ is irreducible.

Since $\mathcal{C}$ is irreducible, we only have to consider **Case(2)**.

By performing $\rho(i, j, k)$, $\pi' = \pi \bullet \rho(i, j, k)$. From Lemma 8, blue edges $(x_i^l, x_k^r)($ labeled by $y_p)$, (x_j^l, x_i^r) (labeled by $y_q)$ and (x_k^l, x_j^r)(labeled by $y_r)$ are replaced by (x_j^l, x_k^r) (labeled by $y_p)$, (x_k^l, x_i^r) (labeled by $y_q)$, and (x_i^l, x_j^r) (labeled by $y_r)$, respectively. Also we can obtain a bijection f' as in the proof of Lemma 8. Thus, $\mathcal{C} = (x_i^l, x_i^r, x_j^l, x_j^r, x_k^l, x_k^r, x_i^l)$ in $\mathcal{G}(\pi, \tau, f)$ becomes $\mathcal{C}' = (x_j^l, x_j^r, x_i^l, x_i^r, x_k^l, x_k^r, x_j^l)$ in $\mathcal{G}(\pi', \tau, f')$. Other edges remain the same in $\mathcal{G}(\pi', \tau, f')$. Again, the blue and green path P' in $\mathcal{G}(\pi', \tau, f')$ also lead to $\tau = y_0, y_1, \ldots, y_{n+1}$ according to the order of their blue edges in P'. Again, by deleting $\mathcal{C}'$, τ is partitioned into four sub-sequences: (P1) $[y_0, \ldots, y_{a-1}]$, (P2) $[y_{a+1}, \ldots, y_{b-1}]$, (P3) $[y_{b+1}, \ldots, y_{c-1}]$, and (P4) $[y_{c+1}, \ldots, y_{n+1}]$. Moreover, the three new green edges (x_{i-1}^r, x_{j+1}^l), (x_{j-1}^r, x_{k+1}^l) and (x_{k-1}^r, x_{i+1}^l) (after applying $\rho(i, j, k)$ on π) connect the three

---

Algorithm 1. decision algorithm for the simple case

Input: two related simple genomes π and τ.
Output: Yes/No (whether π can be transformed into τ by a series of flanked transpositions).

1: Set up an arbitrary bijection f between $\mathbb{A}[\pi] = \mathbb{A}[\tau]$.
2: **if** there exist an unmatched adjacency **then**
3:    **return** No
4: **end if**
5: Construct the cycle graph $\mathcal{G}(\pi, \tau)$ based on the bijection f.
6: **for** each element x **do**
7:    Count the number $c_e(x, \pi, \tau, f)$ of even cycles corresponding to x in $\mathcal{G}(\pi, \tau, f)$.
8: **end for**
9: **if** there exist an element x, such that $c_e(x, \pi, \tau, f)$ is odd. **then**
10:    **return** No.
11: **else**
12:    **return** Yes.
13: **end if**

---

pairs y_{p-1} and y_{r+1}, y_{r-1} and y_{q+1}, and y_{q-1} and y_{p+1} in $\mathbb{G}(\pi', \tau, f')(\mathcal{C}')$, respectively.

In each of the three cases ($r < q < p$, $p < r < q$, or $q < p < r$), the three pairs y_{a-1} and y_{b+1}, y_{b-1} and y_{b+1}, and y_{c-1} and y_{a+1} are connected by the three new green edges (x_{i-1}^r, x_{j+1}^l), (x_{j-1}^r, x_{k+1}^l) and (x_{k-1}^r, x_{i+1}^l), connecting the three pairs y_{p-1} and y_{r+1}, y_{r-1} and y_{q+1}, and y_{q-1} and y_{p+1}. Then the four sub-sequences can form a new genome $\tau' = [y_0, \ldots, y_{a-1}, \underline{y_{b+1}, \ldots, y_{c-1}}, \underline{y_{a+1}, \ldots, y_{b-1}}, \underline{y_{c+1}, \ldots, y_{n+1}}]$. Thus, $\mathcal{C}'$ is reducible.

Lemma 12. *For any pair of genomes π and τ with f being a bijection between $\mathcal{A}[\pi]$ and $\mathcal{A}[\tau]$, if there are only 3-cycles and 1-cycles in $\mathcal{G}(\pi, \tau, f)$, then there exist a series of flanked transpositions to convert all the 3-cycles into 1-cycles and thus transform π into τ.*

Theorem 3. *Given two related simple genomes π and τ, π can be transformed into τ by a series of flanked transpositions if and only if $\mathbb{A}[\pi] = \mathbb{A}[\tau]$ and the number of even cycles corresponding to each repeat in $\mathcal{G}(\pi, \tau, f)$ is even, where f is the bijection between $\mathbb{A}[\pi]$ and $\mathbb{A}[\tau]$.*

The pseudo-code are shown in **Algorithm 1**.

Analysis of the Running Time of Algorithm 1: While setting up the bijection mapping, we can put the adjacencies of $\mathbb{A}[\pi]$ into a hash table, then search for each adjacency of $\mathbb{A}[\tau]$ in this hash table, if the search fails, then $\mathbb{A}[\pi] \neq \mathbb{A}[\tau]$, we will return "No". The expected running time of this process is $O(n)$. Once we have obtained the bijection mapping f, the nodes and red edges of the cycle graph $\mathcal{G}(\pi, \tau, f)$ can be constructed in linear time, the blue edges can also be constructed by traversing the occurrences in τ one by one in linear time. By traversing all the blue edges and red edges, we can figure out the number $c_e(x, \pi, \tau, f)$ of even cycles corresponding to each element x, which takes $O(n)$ time. Therefore, **Algorithm** 1 has an expected running time $O(n)$.

4   Extension to the General Case

In this section, we handle the general case when π and τ might not be simple, i.e., $\mathbb{A}[\pi]$ is no longer a set but a multi-set, and so is $\mathbb{A}[\tau]$. We assume that $\mathbb{A}[\pi] = \mathbb{A}[\tau]$.

Let x_i and x_j be two different occurrences of element u in π and x_{i+1} and x_{j+1} be two different occurrences of element v in π. Then the adjacency between x_i and x_{i+1} is the same as the adjacency between x_j and x_{j+1}. and they appear twice in $\mathbb{A}(\pi)$. In the general case, an adjacency may appear in $\mathbb{A}(\pi)$ more than once. To handle this case, for any consecutive occurrences x_i, and x_{i+1} of element x, we will add a new element z between x_i and x_{i+1} in π. Here each z appears once in the genome. If there are a total number of q consecutive occurrences of different elements in π, We will introduce q new letters.

Lemma 13. *Let* $\pi = [x_0, \ldots, x_{i-1}, x_i, \ldots, x_{n+1}]$, $\tau = [y_0, \ldots, y_{j-1}, y_j, \ldots, y_{n+1}]$, *where* x_{i-1}, x_i, y_{j-1}, *and* y_j *are all occurrences of some element. Let* z *be a new element not appearing in* π *and* τ, $\pi' = [x_0, \ldots, x_{i-1}, z, x_i, \ldots, x_{n+1}]$, $\tau' = [y_0, \ldots, y_{j-1}, z, y_j, \ldots, y_{n+1}]$. *Then,* π *can be transformed into* τ *by a series of flanked transpositions, if and only if* π' *can be transformed into* τ' *by a series of flanked transpositions.*

From Lemma 13, We can scan π from left to right, once there exist two consecutive occurrences of the same element, we insert an occurrence of a new element between them in π, then find two consecutive occurrence of the same element in τ, also insert an occurrence of the same new element between them in τ. We will obtain two new genomes, such that there is no consecutive occurrences of the same element in both. We still use π and τ to denote the two genomes.

As $\mathbb{A}[\pi] = \mathbb{A}[\tau]$, there are a lot of bijections between $\mathbb{A}[\pi]$ and $\mathbb{A}[\tau]$. According to an arbitrary bijection f, we can construct the cycle graph $\mathcal{G}(\pi, \tau, f)$. From Theorem 3, we would be lucky if the number of even cycles corresponding to each element is even. Otherwise, we have to try some other bijections. An element is *safe* under a bijection f if there are an even number of even cycles corresponding to it in $\mathcal{G}(\pi, \tau, f)$; otherwise, it is unsafe. The *safety* of an element refers to whether it is safe or not.

Intuitively, it is nearly impossible to find a proper bijection, under which all the elements are safe, since there could be an exponential number of bijections, but the following lemma gives us a critical clue to a proper bijection.

Let $\langle x_i^r, x_{i+1}^l \rangle$ and $\langle x_j^r, x_{j+1}^l \rangle$ be two identical adjacencies in $\mathbb{A}[\pi]$, and according to the bijection f, $\langle x_i^r, x_{i+1}^l \rangle$ is matched to $\langle y_p^r, y_{p+1}^l \rangle$ and $\langle x_j^r, x_{j+1}^l \rangle$ is matched to $\langle y_q^r, y_{q+1}^l \rangle$. Let f' be another bijection, in which $\langle x_i^r, x_{i+1}^l \rangle$ is matched to $\langle y_q^r, y_{q+1}^l \rangle$, $\langle x_j^r, x_{j+1}^l \rangle$ is matched to $\langle y_p^r, y_{p+1}^l \rangle$, and all the other matches in f' are the same as f. Assume that x_i, x_j, y_p, and y_q are occurrences of the element u, and x_{i+1} x_{j+1}, y_{p+1}, y_{q+1} are occurrences of the element v. Let $c_e^{f'}(u)$ and $c_e^{f}(u)$ denote the number of even (blue and red) cycles corresponding to the element u in $\mathcal{G}(\pi, \tau, f)$ and $\mathcal{G}(\pi, \tau, f')$ respectively. Then, we have,

Lemma 14. $c_e^{f'}(u) - c_e^{f}(u) \in \{-1, 1\}$, and $c_e^{f'}(v) - c_e^{f}(v) \in \{-1, 1\}$.

Proof. There must be two occurrences of u, say x_a and x_b, such that (x_a^l, x_i^r) (labeled by y_p) and (x_b^l, x_j^r) (labeled by y_q) are blue edges in $\mathcal{G}(\pi, \tau, f)$. Then, we have

$$f((x_{a-1}^r, x_a^l)) = \langle y_{p-1}^r, y_p^l \rangle, f((x_i^r, x_{i+1}^l)) = \langle y_p^r, y_{p+1}^l \rangle,$$

$$f((x_{b-1}^r, x_b^l)) = \langle y_{q-1}^r, y_q^l \rangle, f((x_j^r, x_{j+1}^l)) = \langle y_q^r, y_{q+1}^l \rangle.$$

Based on f, there are two blue edges (x_a^l, x_i^r) (labeled by y_p) and (x_b^l, x_j^r) (labeled by y_q) in $\mathcal{G}(\pi, \tau, f')$. When the bijection changes from f to f', we have,

$$f'((x_{a-1}^r, x_a^l)) = \langle y_{p-1}^r, y_p^l \rangle, f'((x_j^r, x_{j+1}^l)) = \langle y_p^r, y_{p+1}^l \rangle,$$

$$f'((x_{b-1}^r, x_b^l)) = \langle y_{q-1}^r, y_q^l \rangle, f'((x_i^r, x_{i+1}^l)) = \langle y_q^r, y_{q+1}^l \rangle.$$

The rest part of f' is the same as f. Based on f', the two new blue edges (x_a^l, x_j^r) (labeled by y_p) and (x_b^l, x_i^r) (labeled by y_q) in $\mathcal{G}(\pi, \tau, f')$ replace two blue edges (x_a^l, x_i^r) (labeled by y_p) and (x_b^l, x_j^r) (labeled by y_q) in $\mathcal{G}(\pi, \tau, f')$. Now, we consider two cases:

Case (I): x_i and x_j are in the same cycle, say $\mathcal{C}$, in $\mathcal{G}(\pi, \tau, f)$: Since both x_i^l and x_i^r are in $\mathcal{C}$, let us start with node x_i^l.

There is a path P_1 from x_i^l to x_b^r not including x_a^l in $\mathcal{C}$. Similarly, there is another path P_2 from x_j^l to x_a^r not including x_b^l in $\mathcal{C}$, where the P_1 and P_2 are node disjoint. See the dashed lines in Fig. 5 (a), where x_a^r and x_i^l are connected via two red edges with the blue edge (x_a^l, x_i^r) (labeled by y_p) in the middle, and x_j^r and $x_b =^l$ are connected via two red edges with the blue edge (x_b^r, x_j^l) (labeled by y_q) in the middle.

When the bijection is changed from f to f', the paths P_1 and P_2 still exist in $\mathcal{G}(\pi, \tau, f')$. P_1, the blue edge (x_b^l, x_i^r), and the two red edges (x_i^l, x_i^r) and (x_b^l, x_b^r) form a cycle $\mathcal{C}' = (x_b^l, x_b^r, P_1, x_i^l, x_i^r, x_b^l)$. Also, P_2, the blue edge (x_a^l, x_j^r), and the two red edges (x_j^l, x_j^r) and (x_a^l, x_a^r) form a cycle $\mathcal{C}'' = (x_a^l, x_a^r, P_2, x_j^l, x_j^r, x_a^l)$. See Fig. 5 (b).

Assume that $\mathcal{C}$ is an s-cycle, $\mathcal{C}'$ is an s'-cycle and $\mathcal{C}''$ is an s''-cycle, then $s = s' + s''$. If s is even, s' and s'' are both even or both odd. Thus, $c_e^{f'}(u) - c_e^{f}(u) \in \{-1, 1\}$. If S is odd, one of s' and s'' is even and the other is odd. Thus, $c_e^{f'}(u) - c_e^{f}(u) \in \{-1, 1\}$. Therefore, $c_e^{f'}(u) - c_e^{f}(u) \in \{-1, 1\}$ in any case.

Case (II): x_i and x_j are in two different cycles, say $\mathcal{C}'$ and $\mathcal{C}''$, in $\mathcal{G}(\pi, \tau, f)$.

In $\mathcal{C}'$, x_a^r and x_i^l are connected via two red edges with the blue edge (x_a^l, x_i^r) (labeled by y_p) in the middle. Thus there must be a path P_1 from x_i^l to x_a^r on $\mathcal{C}'$. In $\mathcal{C}''$, x_b^r and x_j^l are connected via two red edges with the blue edge (x_b^l, x_j^r) (labeled by y_q) in the middle. Thus, there must be a path P_2 from x_j^l to x_b^r on $\mathcal{C}''$. See the dashed lines in Fig. 5 (c)

Algorithm 2. decision algorithm for the general case

Input: two related simple genomes π and τ.
Output: Yes/No (whether π can be transformed into τ by a series of flanked transpositions).

1: Set up the bijection f between $\mathbb{A}[\pi] = \mathbb{A}[\tau]$.
2: **if** there exist an unmatched adjacency **then return**No
3: **end if**
4: Construct the cycle graph $\mathcal{G}(\pi, \tau, f)$ based on the bijection f
5: **for** each element x **do**
6:　count the number $c_e(x, \pi, \tau, f)$ of even cycles corresponding to x in $\mathcal{G}(\pi, \tau, f)$.
7:　**if** $c_e(x, \pi, \tau, f)$ is odd **then**
8:　　mark x as unsafe.
9:　**end if**
10: **end for**
11: Construct the multi-common adjacency graph $MCA(\pi)$ of π.
12: **for** each connected component in $MCA(\pi)$ **do**
13:　**if** the number of unsafe vertices is odd. **then return**No.
14:　**else return**Yes.
15:　**end if**
16: **end for**

When the matching is changed from f to f', the paths P_1 and P_2 still exist, and form a cycle $\mathcal{C}$ together with the two blue edges (x_b^l, x_i^r) and (x_a^l, x_j^r), and the four red edges (x_a^l, x_a^r), (x_b^l, x_b^r), (x_i^l, x_i^r) and (x_j^l, x_j^r), in $\mathcal{G}(\pi, \tau, f')$, i.e., $\mathcal{C} = (x_a^l, x_a^r, P_1, x_i^l, x_i^r, x_b^l, x_b^r, P_2, x_j^l, x_j^r, x_a^l)$. (As shown in Fig. 5-(d).) Assume that $\mathcal{C}$ is an s-cycle, $\mathcal{C}'$ is an s'-cycle and $\mathcal{C}''$ is an s''-cycle, then $s' + s'' = s$. Similarly, we can verify that $c_e^{f'}(u) - c_e^{f}(u) \in \{-1, 1\}$.

A similar proof can show that $c_e^{f'}(v) - c_e^{f}(v) \in \{-1, 1\}$.

Now we construct the multi-common adjacency graph $MCA(\pi)$ of π. Construct a vertex for each element, and there is an edge between a pair of vertices u and v if and only if the adjacency $\langle u^r, v^l \rangle$ or $\langle v^r, u^l \rangle$ appears more than once in $\mathbb{A}[\pi]$.

Theorem 4. *Let π and τ be two related genomes, and f be an arbitrary bijection between $\mathbb{A}[\pi]$ and $\mathbb{A}[\tau]$. π can be transformed into τ by a series of flanked transpositions if and only if $\mathbb{A}[\pi] = \mathbb{A}[\tau]$ and the number of unsafe elements in each connected component of $MCA(\pi)$ is even.*

The pseudo-code of the algorithm for the general case is shown in **Algorithm 2**.

Analysis of the Running Time of Algorithm 2: Step 1 to 11 of Algorithm 2 is the same as Algorithm 1. Thus the expected running time of these steps is $O(n)$. Since the length of input genomes is $n + 1$, the number of elements and adjacencies are less than $n + 1$, so the number of common adjacencies also less than $n + 1$. Thus, in the multi-common adjacency graph $MCA(\pi)$ of π, both

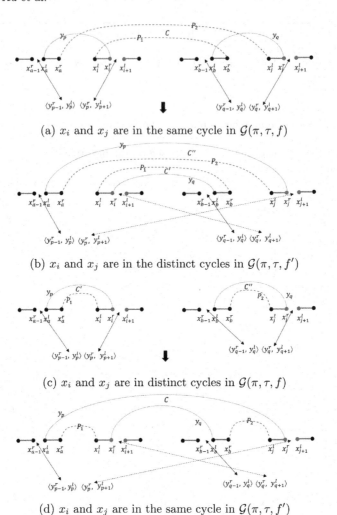

(a) x_i and x_j are in the same cycle in $\mathcal{G}(\pi, \tau, f)$

(b) x_i and x_j are in the distinct cycles in $\mathcal{G}(\pi, \tau, f')$

(c) x_i and x_j are in distinct cycles in $\mathcal{G}(\pi, \tau, f)$

(d) x_i and x_j are in the same cycle in $\mathcal{G}(\pi, \tau, f')$

Fig. 5. Two cases while changing the matching of two common adjacencies. (Color figure online)

the number of vertices (represent elements) and the number of edges(represent common adjacencies) are $O(n)$. The breadth-first search of $MCA(\pi)$ will find out all the connected components of $MCA(\pi)$, which takes $O(n)$ time. Finally, it takes $O(n)$ time to count the number of unsafe vertices in each connected component of $MCA(\pi)$. Therefore, the time complexity of Algorithm 2 is $O(n)$ expected time.

Finally, in the next section we show that sorting by the minimum flanked transpositions is NP-hard.

5 Hardness Result for the Optimization Version

In this section, we show that the optimization version of the sorting by flanked transpositions problem *SMFT* is NP-hard, even if the two input genomes are simple and both of them have a duplicate number of three. Firstly, we recall the traditional problem of sorting by transpositions.

Definition 3. *Sorting by transpositions, abbreviated as **SBT**.*
   Instance: *A permutation* $\chi = [\chi_0, \chi_1, \ldots, \chi_n]$, *where* $\chi_0 = 0$, $\chi_n = n$, $\chi_i \in \{1, 2, \ldots, n-1\}$ *for* $1 \leq i \leq n-1$, *and* $\chi_i \neq \chi_j$ *if* $i \neq j$, *an identity permutation* $\iota = [0, 1, \ldots, n]$.
   Question: *A sequence of transpositions* $\rho_1, \rho_2, \ldots, \rho_K$ *that transform* χ *into* ι, *such that* K *is minimized.*

As introduced in [16], we can construct a breakpoint graph $BP(\chi)$ as follows. Construct two nodes, named χ_i^l and χ_i^r, for each χ_i $(0 \leq i \leq n)$. Connect χ_i^r and χ_{i+1}^l with a red edge, and connect i^r and $(i+1)^l$ with a blue edge. The breakpoint graph $BP(\chi)$ is composed of disjoint cycles, on which red edges and blue edges appear alternatively. The breakpoint graph also fulfills the proposition that, all blue and green edges form an path while connecting χ_i^r and χ_{i+1}^l with a green edge, for $1 \leq i \leq n$. An s-cycle is a cycle that contains s red edges and s blue edges. A permutation is called a *3-permutation* if all the cycles in its breakpoint graph are 3-cycles [30]. In [20], whether a 3-permutation of length $n+1$ can be sorted by $n/3$ transpositions is shown to be NP-hard.

Theorem 5. *It is NP-hard to find a minimum number of flanked transpositions to transform a genome* π *into another genome* τ *when the transformation is possible, even if* π *and* τ *are simple and* $dp[\pi] = dp[\tau] = 3$.

6 Discussion and Conclusion

It has been observed that transpositions are associated with 3 identical repeats at the ends of the two swapped segments in many cases. Thus, the proposed model (flanked transposition) is biological meaningful compared with the traditional model. However, it is worth to point out that some repeats may disappear due to various reasons. Thus, it is perhaps only of theoretical interest, if we merely consider flanked transpositions. A more interesting open problem is to add minimum number of occurrences of elements to the two input genomes so that the two new genomes can be transformed by flanked transpositions from one to the other.

Acknowledgments. This research is supported by NSF of China under grant 62272279 and 62272272. We also thank the anonymous reviewers for their valuable revision advice.

References

1. Sturtevant, A.: A crossover reducer in Drosophila melanogaster due to inversion of a section of the third chromosome. Biol. Zent. Bl. **46**, 697–702 (1926)
2. Sturtevant, A., Dobzhansky, T.: Inversions in the third chromosome of wild races of drosophila pseudoobscura, and their use in the study of the history of the species. Proc. Nat. Acad. Sci. USA **22**, 448–450 (1936)
3. Nadeau, J.H., Taylor, B.A.: Lengths of chromosomal segments conserved since divergence of man and mouse. Proc. Nat. Acad. Sci. USA **81**, 814–818 (1984)
4. Makaroff, C., Palmer, J.: Mitochondrial DNA rearrangements and transcriptional alternatives in the male sterile cytoplasm of Ogura radish. Mol. Cell. Biol. **8**, 1474–1480 (1988)
5. Palmer, J., Herbon, L.: Plant mitochondrial DNA evolves rapidly in structure, but slowly in sequence. J. Mol. Evolut. **27**, 87–97 (1988)
6. Sankoff, D., Leduc, G., Antoine, N., Paquin, B., Lang, B.F., Cedergran, R.: Gene order comparisons for phylogenetic interferce: evolution of the mitochondrial genome. Proc. Nat. Acad. Sci. USA **89**, 6575–6579 (1992)
7. Watterson, G.A., Ewens, W.J., Hall, T.E., Morgan, A.: The chromosome inversion problem. J. Theor. Biol. **99**(1), 1–7 (1982)
8. Hannenhalli, S., Pevzner, P.: Transforming cabbage into turnip: polynomial algorithm for sorting signed permutations by reversals. J. ACM **46**(1), 1–27 (1999)
9. Caprara, A.: Sorting by reversals is difficult. In: Proceedings of the first annual international Conference on Computational Molecular Biology, pp. 75–83 (1997)
10. Berman, P., Karpinski, M.: On some tighter inapproximability results (extended abstract). Languages, and Programming. In: International Colloquium on Automata (1999)
11. Kececioglu, J., Sankoff, D.: Exact and approximation algorithms for sorting by reversals, with application to genome rearrangement. Algorithmica **13**(1), 180–210 (1995)
12. Christie, D.A.: A 3/2-approximation algorithm for sorting by reversals. In Proceedings of the Ninth Annual ACM-SIAM Symposium on Discrete Algorithms, pp. 244–252 (1998)
13. Berman, P., Hannenhalli, S., Karpinski, M.: 1.375-approximation algorithm for sorting by reversals. In: Möhring, R., Raman, R. (eds.) ESA 2002. LNCS, vol. 2461, pp. 200–210. Springer, Heidelberg (2002). https://doi.org/10.1007/3-540-45749-6_21
14. Kaplan, H., Shamir, R., Tarjan, R.E.: A faster and simpler algorithm for sorting signed permutations by reversals. SIAM J. Comput. **29**(3), 880–892 (2000)
15. Tannier, E., Bergeron, A., Sagot, M.-F.: Advances on sorting by reversals. Dis. Appli. Math. **155**(6–7), 881–892 (2007)
16. Bafna, V., Pevzner, P.A.: Sorting by transpositions. SIAM J. Discrete Math. **11**(2), 224–240 (1998)
17. Hartman, T., Shamir, R.: A simpler and faster 1.5-approximation algorithm for sorting by transpositions. Inform. Comput. **204**, 275–290 (2006)
18. Feng, J., Zhu, D.: Faster algorithms for sorting by transpositions and sorting by block-interchanges. ACM Trans. Algorithms **3**(3), 25 (2007)
19. Elias, I., Hartman, T.: A 1.375-approximation algorithm for sorting by transpositions. IEEE/ACM Trans. Comput. Biol. Bioinform. **3**(4), 369–379 (2006)
20. Bulteau, L., Fertin, G., Rusu, I.: Sorting by transpositions is difficult. SIAM J. Discret. Math. **26**(3), 1148–1180 (2012)

21. Eriksson, H., Eriksson, K., Karlander, J., Svensson, L., Wastlund, J.: Sorting a bridge hand. Discret. Appl. Math. **241**(1–3), 289–300 (2001)
22. Longo, M.S., Carone, D.M., Green, E.D., O'Neill, M.J., O'Neill, R.J., et al.: Distinct retroelement classes define evolutionary breakpoints demarcating sites of evolutionary novelty. BMC Genomics **10**(1), 334 (2009)
23. Sankoff, D.: The where and wherefore of evolutionary breakpoints. J. Biol. **8**, 66 (2009)
24. Thomas, A., Varré, J.S., Ouangraoua, A.: Genome dedoubling by dcj and reversal. BMC Bioinform. **12**(9), S20 (2011)
25. Bailey, J.A., Baertsch, R., Kent, W.J., Haussler, D., Eichler, E.E.: Hotspots of mammalian chromosomal evolution. Genome Biol. **5**(4), R23 (2004)
26. Armengol, L., Pujana, M.A., Cheung, J., Scherer, S.W., Estivill, X.: Enrichment of segmental duplications in regions of breaks of synteny between the human and mouse genomes suggest their involvement in evolutionary rearrangements. Hum. Mol. Genet. **12**(17), 2201–2208 (2003)
27. Small, K., Iber, J., Warren, S.T.: Emerin deletion reveals a common X-chromosome inversion mediated by inverted repeats. Nat. Genet. **16**, 96–99 (1997)
28. Wang, D., Wang, L.: Grsr: a tool for deriving genome rearrangement scenarios from multiple unichromosomal genome sequences. BMC Bioinform. **19**(9), 11–19 (2018)
29. Wang, D., Li, S., Guo, F., Wang, L.: Core genome scaffold comparison reveals the prevalence that inversion events are associated with pairs of inverted repeats. BMC Genomics **18**, 268 (2017)
30. Vineet, B., Pavel P.A.: Sorting permutations by tanspositions. In: Proceedings of the sixth annual ACM-SIAM Symposium on Discrete Algorithms (SODA 1995), pp. 614–623 (1995)
31. Han, Y.: Improving the efficiency of sorting by reversals. In: Proceedings of the 2006 International Conference Bioinformatics & Computational Biology (BIOCOMP 2006), pp. 406–409 (2006)
32. Tong, X.: Men Can't Always be Transformed into Mice: Decision Algorithms and Complexity for Sorting by Symmetric Reversals. arXiv:2302.03797 (2023)

Integrative Analysis of Gene Expression and Alternative Polyadenylation from Single-Cell RNA-seq Data

Shuo Xu[1,2], Liping Kang[1], Xingyu Bi[1], and Xiaohui Wu[1(✉)]

[1] Pasteurien College, Suzhou Medical College of Soochow University, Soochow University, Suzhou 215000, China
xhwu@suda.edu.cn
[2] Department of Automation, Xiamen University, Xiamen 361005, China

Abstract. Single-cell RNA-seq (scRNA-seq) is a powerful technique for assaying transcriptional profile of individual cells. However, high dropout rate and overdispersion inherent in scRNA-seq hinders the reliable quantification of genes. Recent bioinformatic studies switched the conventional gene-level analysis to APA (alternative polyadenylation) isoform level, and revealed cell-to-cell heterogeneity in APA usages and APA dynamics in different cell types. The additional layer of APA isoforms creates immense potential to develop cost-efficient approaches for dissecting cell types by integrating multiple modalities derived from existing scRNA-seq experiments. Here we proposed a pipeline called scAPAfuse for enhancing cell type clustering and identifying of novel/rare cell types by combing gene expression and APA profiles from the same scRNA-seq data. scAPAfuse first maps gene expression and APA profiles to a shared low-dimensional space using partial least squares. Then anchors (i.e., similar cells) between gene and APA profiles were identified by constructing the nearest neighbors of cells in the low-dimensional space, using algorithms like hyperplane local sensitive hash and shared nearest neighbor. Finally, gene and APA profiles were integrated to a fused matrix, using the Gaussian kernel function. Applying scAPAfuse on four public scRNA-seq datasets including human peripheral blood mononuclear cells (PBMCs) and Arabidopsis roots, new subpopulations of cells that were undetectable using the gene expression or APA profile alone were found. scAPAfuse provides a unique strategy to mitigate the high sparsity of scRNA-seq by fusing gene expression and APA profiles to improve cell type clustering, which can be included in many other routine scRNA-seq pipelines.

Keywords: single-cell RNA-seq · cell type clustering · alternative polyadenylation

1 Introduction

Single-cell RNA-seq (scRNA-seq) is a powerful technique for assaying transcriptional profile of individual cells, which can be utilized for discovering cell types, reconstructing developmental trajectories, and spatially modeling complex tissues [1]. Unsupervised

X. Guo et al. (Eds.): ISBRA 2023, LNBI 14248, pp. 312–324, 2023.
https://doi.org/10.1007/978-981-99-7074-2_24

clustering based on the transcriptome similarity is probably a central component of most scRNA-seq analytical workflows to identify putative cell types. However, high dropout rate and overdispersion inherent in scRNA-seq hinders the reliable quantification of genes, especially the lowly and/or moderately expressed ones [2, 3], resulting in an extremely sparse and noisy gene expression profile. Consequently, little satisfactory overlap of expressed genes can be observed among cells, affecting the performance of computational methods that are solely based on the gene expression profile.

Recent bioinformatic analysis [4–6] has been extended to extract information on transcript levels at single-cell resolution from diverse scRNA-seq protocols, such as 10x Genomics [7], Drop-seq [8], and CEL-seq [9]. Approaches, such as scAPAtrap [10], Sierra [11], and scAPA [12], began to emerge for identifying and quantifying polyadenylation [poly(A)] sites in single cells [13]. These pioneering studies switched the conventional gene-level analysis to APA (alternative polyadenylation) isoform level, and revealed cell-to-cell heterogeneity in APA usages and APA dynamics in different cell types. Although both genes and poly(A) sites were identified from the same scRNA-seq data, latest tools, like scAPAtrap [10], can efficiently capture poly(A) sites at the whole genome level, including those poorly expressed or minor isoforms caused by lack of coverage in low-expression regions. In contrast, most routine scRNA-seq pipelines for gene expression analysis normally require fairly good read coverage to detect genes and tend to discard lowly expressed genes in the quality control step. A considerable number poly(A) sites are present in genes that may be discarded or unrecognized in traditional scRNA-seq pipelines [13, 14], providing valuable information about cell-cell associations that are missing in the standard gene-cell expression matrix. We anticipate that this additional layer of APA isoforms may encapsulate complementary information about cell-cell associations that is not manifested by the conventional gene expression profile from the same scRNA-seq data, which creates immense potential to develop cost-efficient approaches for dissecting cell types by integrating multiple modalities derived from existing scRNA-seq experiments.

Here we proposed a pipeline called scAPAfuse for enhancing cell type clustering and identifying of novel/rare cell types by combing gene expression profiles with an additional layer of APA knowledge derived from the same scRNA-seq data. By employing partial least squares for dimensionality reduction, scAPAfuse maps gene expression and APA profiles to a shared low-dimensional space. Then anchors (i.e., similar cells) between gene and APA profiles were identified by constructing the nearest neighbors of cells in the low-dimensional space, using algorithms like hyperplane local sensitive hash and shared nearest neighbor. Finally, gene and APA profiles were integrated to a fused matrix, using the Gaussian kernel function. We applied scAPAfuse on four public scRNA-seq datasets from human and Arabidopsis. Results showed that scAPAfuse found new subpopulations of cells in peripheral blood mononuclear cells (PBMCs) and plant roots that were undetectable using the gene expression or APA profile alone. scAPAfuse provides a unique strategy to mitigate the high sparsity of scRNA-seq by fusing gene expression and APA profiles to improve cell type clustering, which can be included in many other routine scRNA-seq pipelines.

2   Materials and Methods

2.1   Data Preprocessing

We used four public scRNA-seq datasets from human peripheral blood mononuclear cells (PBMCs) [7] and Arabidopsis roots [15]. The corresponding gene-cell expression matrices (hereinafter referred to as GE-matrix) were obtained from the respective public sources (10xgenomics.com for PBMCs; Accession Nos. GSM4212550 and GSM4212551 for roots). Poly(A) sites of each cell from these scRNA-seq datasets were identified by scAPAtrap [10]. Expression levels of poly(A) sites of the same gene in each cell were summed to form a gene-level matrix (hereinafter referred to as PA-matrix) also by scAPAtrap. Finally, for each paired GE and PA-matrix, common genes were retained to make the two matrices of the same dimension (m genes and n cells). The $L2$ standardization (Eq. 1) was applied for each matrix.

$$z = ||x||_2 = \sqrt{\sum_{i=1}^{n} x_i^2} \tag{1}$$

2.2   Joint Dimensionality Reduction Based on Partial Least Squares

We used partial least squares (PLS) for dimensionality reduction, which can map GE- and PA-matrix to a shared low-dimensional space. Let G denote the GE-matrix and P the PA-matrix (Eq. 2).

$$G = \begin{bmatrix} x_{11} & \cdots & x_{1m} \\ \vdots & \ddots & \vdots \\ x_{n1} & \cdots & x_{nm} \end{bmatrix} \quad P = \begin{bmatrix} y_{11} & \cdots & y_{1m} \\ \vdots & \ddots & \vdots \\ y_{n1} & \cdots & y_{nm} \end{bmatrix} \tag{2}$$

The first principal component (PC) can be extracted (Eq. 3).

$$\begin{cases} t_1 = G \times w_1 \\ u_1 = P \times v_1 \end{cases} \tag{3}$$

Here, the variance of t_1 and u_1, and the correlation coefficient of t_1 and u_1 should be as large as possible. The maximum covariance of t_1 and u_1 is then calculated (Eq. 4).

$$\text{Cov}(t_1, u_1) = w_1^T \times G^T \times P \times v_1 \tag{4}$$

Then we established a regression model of G and P versus t_1. Here p and r are parameter vectors, and G_1 and P_1 are residual matrices.

$$\begin{cases} G = t_1 p_1^T + G_1 \\ P = t_1 r_1^T + P_1 \end{cases} \tag{5}$$

Above steps were repeated by replacing the residual matrix G_1 and P_1 with G and P to extract the second PC until the absolute value of all elements of the residual matrix P_1 approached 0.

$$
\begin{cases}
G = t_1 p_1^T + t_2 p_2^T + G_2 \\
P = t_1 r_1^T + t_2 r_2^T + P_2
\end{cases}
\tag{6}
$$

We implemented the above process using the *PLSRegression* function of the *sklearn.cross_decomposition* library in python.

2.3 Identification of Anchor Correspondences Between GE-Matrix and PA-Matrix

After dimensionality reduction, we corrected the expression matrix at different levels by constructing the nearest neighbors of cells between GE-matrix and PA-matrix in a low-dimensional space. The mutual nearest neighbors (called anchors) are cells with a high degree of similarity. In order to quickly identify nearest neighbors, we referred to a method of calculating K nearest neighbors [16] called *annoy* in Annoy python package, which conducted an approximate search based on locality sensitive hashing (LSH), where multiple trees of random hyperplanes, used as hash functions, divided the search space of the points in the query set, for accomplishing more efficient nearest neighbor identification. Moreover, in order to enhance the effect of data integration, we used cell label or cell type information to filter anchors to ensure that each anchor comes from the same cell group. If the cell type is given, the anchors can be filtered and calculated directly according to the cell type. Otherwise, cell labels were obtained by clustering with *Seurat v3* process. Moreover, to filter more representative anchors between datasets, we further implemented a graph-based anchor scoring algorithm, shared nearest neighbor (SNN) [17]. That is, for anchors from GE and PAG-matrix, find their K nearest neighbors ($K = 30$) in GE and PAG-matrix respectively, so we got 4 neighborhood matrices, which we combined to form a neighborhood graph. For anchors, we calculate their shared neighborhood, call this value "anchor score", and use 0.01 and 0.9 quantiles to adjust the anchor score to between 0–1. Deleted the anchors with an anchor score of 0 to make our anchors accurate enough. Anchor scores of SNN adds an extra level of robustness to edge recognition by detecting the consistency of the edges between cells in the same local neighborhood.

2.4 Data Integration

After determining anchors between GE-matrix and PA-matrix, the two matrices can be integrated to a fused matrix. First, we used the Gaussian kernel function [16, 18] to construct the weight matrix. Considering GE-matrix, PA-matrix, and a set of matching vectors M_{ij}, we denoted gene expression values and poly(A) expression levels as E_i^{match} and E_j^{match}, with each row of E_i^{match} and E_j^{match} corresponding to a pair of anchors in M_{ij}. The matching vectors were therefore the rows of $E_j^{match} - E_i^{match}$. To integrate the

GE-matrix or PA-matrix using either matrix as the reference, we calculated the weight between GE-matrix and matching vectors M_{ij} by Gaussian kernel function, which as well as following procedures was also appropriate for PA-matrix. We computed weights between the cells in GE-matrix and the matched cells in GE-matrix (Eq. 7), letting E_i be the vectors of GE-matrix.

$$\gamma_i = exp(-\frac{\sigma}{2} \| E_i - E_i^{match} \|_2^2) \tag{7}$$

Here σ is set to 15 by default. Then we constructed the bias according to the average value of the matching vector of the Gaussian smoothing matrix (Eq. 8).

$$bias = \frac{\gamma_i (E_j^{match} - E_i^{match})}{\sum \gamma i} \tag{8}$$

Then, the corrected matrix of GE-matrix was obtained (Eq. 9).

$$\widehat{E_i} = E_i + bias \tag{9}$$

Finally, after the same process of PA-matrix, Fused-matrix was averaged over two corrected matrices.

2.5 Single-Cell Clustering

In this article, we referred to the *Seurat v3* [19] process to perform cluster analysis on GE-matrix, PA-matrix and the integrated Fused-matrix with scAPAfuse. We used the *Louvain* method in the *FindClusters* function for cell clustering which contained a parameter *resolution* that set the resolution of the downstream cluster, resulting in an appropriate quantity of clusters with referring to number of cell types in PBMCs or Arabidopsis roots. In actual operation, set the resolution of PBMC4K, TAIR-WT1 and TAIR-WT2 to 1, and set the resolution of PBMC8K to 1.1. Cell types in PBMCs was provided by *CellMarker* website in which we obtained their marker genes either, while with regard to Arabidopsis roots, we consulted former research to determine.

3 Results

3.1 Overview of the Integrative Framework Pipeline

See Fig. 1.

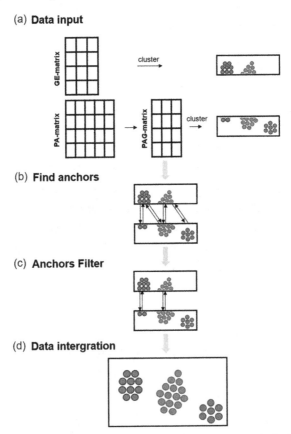

Fig. 1. Schema of scAPAfuse. scAPAfuse consists of four modules: (**a**) The input module. (**b**) Find anchors. (**c**) Anchors Filter. (**d**) Data integration.

3.2 Single-Cell APA Profile Distinguishes Cells

It is a routine step to perform clustering on the GE-matrix, while it would be interesting to examine whether the PA-matrix can also be used to distinguish cell types and/or discover new cell populations. Here we compared clustering results obtained from PA-matrix and GE-matrix of the PBMC4K data. In total, 15 clusters corresponding to 12 cell types (Fig. 2a) and 14 clusters corresponding to 12 cell types (Fig. 2b) were obtained from the GE-matrix and PA-matrix, respectively. Eleven cell types were common in the two clustering results. Particularly, using the PA-matrix, some T cell subtypes were detected, including CD4+ Memory T cell, Naïve CD8 T cell, and Naïve CD4 T cell. Two B cell subtypes were also detected, Naïve B cell and Memory B cell. Next, we calculated differentially expressed poly(A) sites (DEPAs) from the PA-matrix and differentially expressed genes (DEGs) from the GE-matrix by DEseq2 [20] (*adjusted P value* < 0.05; *log2FC* > 1). A total of 318 genes were common in DEPAs and DEGs between Monocytes and B cells, while 44 DEPAs were exclusively found by the PA-matrix (Fig. 2c). For example, *TNFSF13* is rarely expressed in GE-matrix. In contrast, according

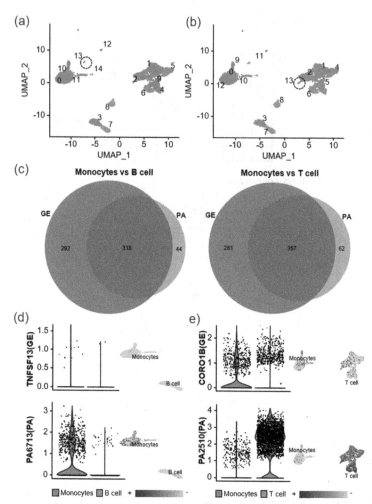

Fig. 2. Clustering and DE analysis using the PA-matrix and GE-matrix of PBMC4K data. (**a**) Clustering based on the GE-matrix. A total of 15 clusters were obtained, and 14 clusters were annotated according to known marker genes. Naïve CD4+ T (1), Naïve CD8+ T (5, 9), CD4+ Memory T (2), CD8+ T (4, 6), NK (8), Naïve B (3), Memory B (7), CD14+ Monocytes (0), CD16+ Monocytes (10), Monocyte Derived Dendritic (11), Megakaryocyte Progenitors (13), Plasmacytorid Dendritic (12). The dashed circle is the cluster not identified by the PA-matrix. (**b**) Clustering based on the PA-matrix. A total of 14 clusters were obtained. Naïve CD4+ T (1), Naïve CD8+ T (4), CD4+ Memory T (2), CD8+ T (5, 6), Regulatory T (13), NK (8), Naïve B (3), Memory B (7), CD14+ Monocytes(0, 12), CD16+ Monocytes (9), Monocyte Derived Dendritic (10), Plasmacytorid Dendritic (11). The dashed circle is the cluster not identified by the GE-matrix. (**c**) DEGs between Monocytes and T cells or Monocytes and B cells. (**d**) *TNFSF13* is not a DEG between Monocytes and B cells, but it has at least one DEPA. (**e**) *CORO1B* is not a DEG between Monocytes and T cells, but it has at least one DEPA.

to the PA-matrix, this gene has a poly(A) site *PA6713* (coordinate: 7561617), which is much higher expressed in Monocytes than in B cells (Fig. 2d). Similarly, *CORO1B* has higher expression level in Monocytes based on the GE-matrix. However, it has a poly(A) site *PA2510* (coordinate: 67435508) which is much higher expressed in T cells than in Monocytes (Fig. 2e).

3.3   scAPAfuse Identifies Subtypes in PBMC

Next, we applied scAPAfuse to integrate GE-matrix and PA-matrix from the PBMC4K data. A total of 16 clusters were obtained (Fig. 3a). Notably, a small cell sub-type was found, Regulatory T cells (cluster 13 with the marker gene *FOXP3*). We also observed two marker genes of Megakaryocyte Progenitors (cluster 15), *PPBP* and *PF4*.

In order to further verify the effectiveness of scAPAfuse in integrating different datasets, we conducted a similar analysis on PBMC8K. And we further applied scAPA-fuse to integrate the four matrices – PBMC4K's GE-matrix, PA-matrix and PBMC8K's GE-matrix and PA-matrix. Clustering using this integrated data generated all cell types identified in individual PBMC4K and PBMC8K experiments (Fig. 4a). According to the expression of *CCR10* and *PPBP*, cluster 13 is determined as Regulatory T cell (Fig. 4b), and cluster 17 is Megakaryocyte Progenitors (Fig. 4b).

3.4   scAPAfuse Identifies Subtypes in Arabidopsis Root Cells

We obtained 6049 single-cell transcriptomes from two replicates (WT1 and WT2) of wild-type Arabidopsis root tip protoplasts. Using the fused GE- and PA-matrix of WT1, a total of 28 clusters were obtained (Fig. 5.a). Referring to the marker genes given in the published literature [21, 22], we assigned cell type labels to these clusters –Tri-choblasts/hair (clusters 3, 4, 8, 15, 25), Atricholblasts/non-hair (clusters 2, 13, 16, 19), LRC (cluster 7), Columella (cluster 27), Cortex (clusters 6, 21, 24), Endodermis (clusters 0, 1, 7), Phloem (cluster 9), Phericycle (clusters 5, 10, 12, 22), Xylem (18, 23, 26), Procambium (clusters 11, 14). Particularly, cluster 20 is the junction of Trichoblasts/hair and Atricholblasts/non-hair, indicating that it may be the origin of the common development of the two epidermal cell types, i.e., the root tip meristem. For comparison, a cluster analysis was also performed on the GE-matrix of WT1, and 24 clusters were obtained. We found that cluster 27 was exclusively identified by scAPAfuse, which was not recognized only using the GE-matrix. The expression profile of the marker gene *AT4G34970* confirmed that the cluster is a sub-type of Columella in Root Cap (Fig. 5b).

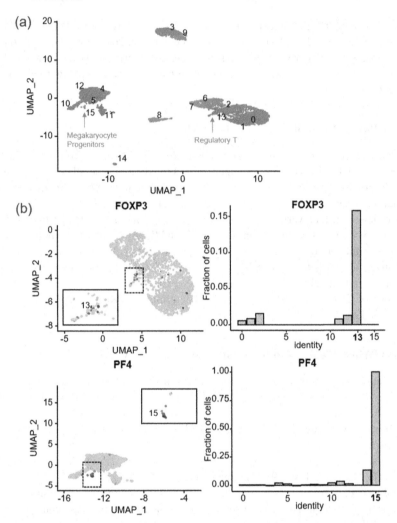

Fig. 3. scAPAfuse identifies hidden subpopulations of cells from PBMC4K. (a) UMAP represents the clustering result of the fused matrix of GE- and PA-matrix, and a total of 16 clusters were obtained. Naïve CD4+ T (1), Naïve CD8+ T (0), CD4+ Memory T (2), CD8+ T (6, 7), Regulatory T (13), NK (8), Naïve B (3), Memory B (9), CD14+ Monocytes (4, 5, 12), CD16+ Monocytes (10), Monocyte Derived Dendritic (11), Plasmacytorid Dendritic (14), Megakaryocyte Progenitors (15). The arrow points to the cell type exclusively identified by scAPAfuse. (b) The gene expression of *FOXP3* distinguishes Regulatory T cell from other T cell types. The gene expression of *PF4* distinguishes Megakaryocyte Progenitors from other cell types. The bar charts show the expression ratio of *FOXP3* and *PF4* in each cluster.

Next, we applied scAPAfuse on WT2 and obtained 19 clusters (Fig. 5c) – Trichoblasts/hair (clusters 3, 9), Atricholblasts/non-hair (clusters 7, 10), LRC (clusters 0, 2), Columella (cluster 18), Cortex (clusters 11, 13, 16), Endodermis (clusters 4, 5, 6),

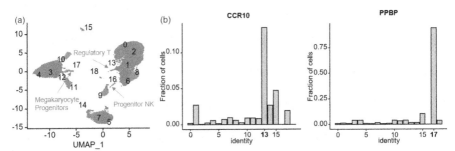

Fig. 4. scAPAfuse identifies hidden cell subtypes from integrated data of PBMC4K and PBMC-8K. (**a**) After integrating the four matrices in PBMC4K and PBMC8K, a total of 19 clusters were obtained. Naïve CD4+ T (0), Naïve CD8+ T (2), CD4+ Memory T (1), CD8+ T (6, 8), Regulatory T (13), NK (9), Progenitor NK (16), Naïve B(5), Memory B (7, 14), CD14+ Monocytes (3, 4, 12), CD16+ Monocytes (10), Monocyte Derived Dendritic (11), Plasmacytorid Dendritic (15), Megakaryocyte Progenitors (17). (**b**) Fraction of cells shows the expression ratio of *CCR10* and *PPBP* in each cluster, distinguishing Regulatory T cell and Megakaryocyte Progenitors from other cell subtypes.

Phloem (cluster 12), Phericycle (cluster 1), Xylem (clusters 14, 17), Procambium (cluster 8). By observing the expression profile of the marker genes of QC and SCN in RAM (e.g., *AT3G15357*) (Fig. 5d), it is confirmed that this cluster 25 is RAM. These results demonstrate that integration of GE- and PA-matrix by scAPAfuse can better distinguish cell types that are not easily distinguishable using only gene expression profiles.

3.5 scAPAfuse Identifies Rare QC Cells in Arabidopsis Root Cells

In order to evaluate the ability of scAPAfuse in identifying cell populations of small size, we re-clustered RAM cells which are composed of QC and SCN cells of WT1 and WT2, respectively (Figs. 6a and 6b, top). Based on the 52 QC marker genes provided in Ryu [22] (Figs. 6a and 6b, bottom), we found that QC cells are more dominant in cluster 0 of WT1 and cluster 1 of WT2. UMAP visualization of meristem and epidermal tissues in both WT1 and WT2 showed that QC was located in the lower part of RAM (Figs. 6c and 6d), providing additional evidence for the cell type clustering results by scAPAfuse.

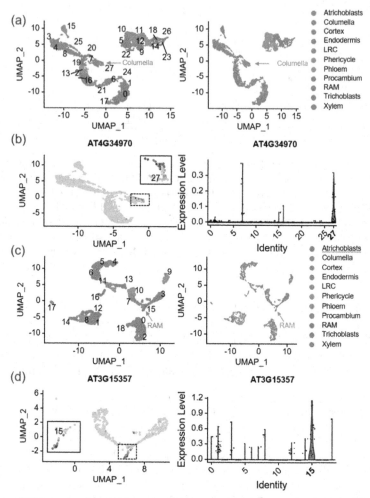

Fig. 5. scAPAfuse recognizes hidden sub-types in Arabidopsis root tips. **(a)** Clusters using the fused GE- and PA-matrix of WT1 (left) and cell type annotations according to the collected marker gens (right). The arrow points to the new sub-types identified by scAPAfuse compared to using only the GE-matrix. **(b)** According to the UMAP visualization of the gene expression profile, the gene expression of *AT4G34970* distinguishes Columella from other cell types. The detailed information in the dashed box is displayed in the solid box (left). The bar chart shows the expression of *AT4G349970* in each cluster (right). **(c)** Applying scAPAfuse on WT2 obtained 19 clusters (left), and 11 cell types are annotated (right). The arrow points to cell types that are only recognized by scAPAfuse but not by the GE-matrix. **(d)** The gene expression profile of *AT3G15357* distinguishes RAM from other cell types. The detailed information in the dashed box is displayed in the solid box (left). The bar chart shows the expression of *AT3G15357* in each cluster (right).

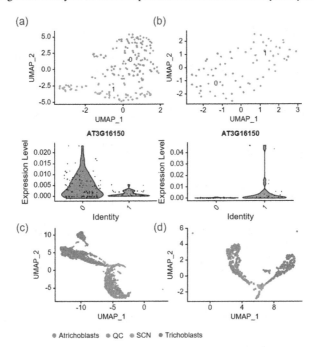

Fig. 6. scAPAfuse recognizes a rare cell type in the root tip of Arabidopsis, the quiescent center cell (QC). **(a)** Re-clustering the RAM cells identified by scAPAfuse in WT1 generated two sub-clusters (top). Expression profiles of a QC marker gene (*AT3G16150*) in the two clusters (bottom). **(b)** As in (a) except that WT2 was analyzed. **(c)** UMAP visualization of meristem and epidermal tissues of WT1 showed that QC was located in the lower part of RAM. **(d)** As in (c) except that WT2 was analyzed.

4   Conclusion

The layer of APA isoforms identified from scRNA-seq encapsulates complementary information about cell-cell associations that is not manifested by the conventional gene expression profile, which creates potential to improve cell type dissection by integrating multiple modalities. We proposed a pipeline called scAPAfuse for enhancing cell type clustering and identifying of novel/rare cell types by combing gene expression and APA profiles from the same scRNA-seq data. scAPAfuse combines several algorithms, including PLS, LSH, SNN, and Gaussian kernel function, to fuse gene and APA profiles to the same low-dimensional space. Results by applying scAPAfuse on several scRNA-seq datasets showed that scAPAfuse can effectively identify new subpopulations of cells that were undetectable using the gene expression or APA profile alone. scAPAfuse provides a unique strategy to improve cell type clustering, which can be included in many other routine scRNA-seq pipelines.

Acknowledgements. This work was supported by the National Natural Science Foundation of China (Grant No. T2222007 to XW).

References

1. Butler, A., et al.: Integrating single-cell transcriptomic data across different conditions, technologies, and species. Nat. Biotechnol. **36**, 411 (2018)
2. Kharchenko, P.V., et al.: Bayesian approach to single-cell differential expression analysis. Nat. Methods **11**, 740 (2014)
3. Grun, D., et al.: Validation of noise models for single-cell transcriptomics. Nat. Methods **11**, 637–640 (2014)
4. Saliba, A.-E., et al.: Single-cell RNA-seq: advances and future challenges. Nucleic Acids Res. **42**, 8845–8860 (2014)
5. Chen, W., et al.: Alternative polyadenylation: methods, findings, and impacts. Genomics Proteomics Bioinf. **15**, 287–300 (2017)
6. Ye, C., et al.: Discovery of alternative polyadenylation dynamics from single cell types. Comput. Struct. Biotechnol. J. **18**, 1012–1019 (2020)
7. Zheng, G.X., et al.: Massively parallel digital transcriptional profiling of single cells. Nat. Commun. **8**, 14049 (2017)
8. Macosko, E.Z., et al.: Highly parallel genome-wide expression profiling of individual cells using nanoliter droplets. Cell **161**, 1202–1214 (2015)
9. Hashimshony, T., et al.: CEL-seq: single-cell RNA-seq by multiplexed linear amplification. Cell Rep. **2**, 666–673 (2012)
10. Wu, X., et al.: scAPAtrap: identification and quantification of alternative polyadenylation sites from single-cell RNA-seq data. Brief. Bioinform. **22** (2021)
11. Patrick, R., et al.: Sierra: discovery of differential transcript usage from polyA-captured single-cell RNA-seq data. Genome Biol. **21**, 167 (2020)
12. Shulman, E.D., Elkon, R.: Cell-type-specific analysis of alternative polyadenylation using single-cell transcriptomics data. Nucleic Acids Res. **47**, 10027–10039 (2019)
13. Ye, W., et al.: A survey on methods for predicting polyadenylation sites from DNA sequences, bulk RNA-seq, and single-cell RNA-seq. Genomic Proteomics Bioinf. **21**, 63–79 (2023)
14. Ji, G., et al.: stAPAminer: mining spatial patterns of alternative polyadenylation for spatially resolved transcriptomic studies. Genomic Proteomics Bioinf. (2023)
15. Wendrich, J.R., et al.: Vascular transcription factors guide plant epidermal responses to limiting phosphate conditions. Science **370** (2020)
16. Hie, B., et al.: Efficient integration of heterogeneous single-cell transcriptomes using Scanorama. Nat. Biotechnol. **37**, 685–691 (2019)
17. Levine, J.H., et al.: Data-driven phenotypic dissection of AML reveals progenitor-like cells that correlate with prognosis. Cell **162**, 184–197 (2015)
18. Haghverdi, L., et al.: Batch effects in single-cell RNA-sequencing data are corrected by matching mutual nearest neighbors. Nat. Biotechnol. **36**, 421 (2018)
19. Stuart, T., et al.: Comprehensive integration of single-cell data. Cell **177**, 1888–1902 (2019). e1821
20. Love, M.I., et al.: Moderated estimation of fold change and dispersion for RNA-seq data with DESeq2. Genome Biol. **15**, 1–21 (2014)
21. Shahan, R., et al.: A single cell Arabidopsis root atlas reveals developmental trajectories in wild type and cell identity mutants. Front. Genet. **370** (2020)
22. Ryu, K.H., et al.: Single-cell RNA sequencing resolves molecular relationships among individual plant cells. Plant Physiol. **179**, 1444–1456 (2019)

SaID: Simulation-Aware Image Denoising Pre-trained Model for Cryo-EM Micrographs

Zhidong Yang[1,3], Hongjia Li[1,3], Dawei Zang[1], Renmin Han[4(⊠)], and Fa Zhang[2(⊠)]

[1] High Performance Computer Research Center, Institute of Computing Technology, Chinese Academy of Sciences, Beijing, China
[2] School of Medical Technology, Beijing Institute of Technology, Beijing, China
zhangfa@bit.edu.cn
[3] University of Chinese Academy of Sciences, Beijing, China
[4] Research Center for Mathematics and Interdisciplinary Sciences, Shandong University, Qingdao, China
hanrenmin@sdu.edu.cn

Abstract. Cryo-Electron Microscopy (cryo-EM) is a revolutionary technique for determining the structures of proteins and macromolecules. Physical limitations of the imaging conditions cause a very low Signal-to-Noise Ratio (SNR) in cryo-EM micrographs, resulting in difficulties in downstream analysis and accurate ultrastructure determination. Hence, the effective denoising algorithm for cryo-EM micrographs is in demand to facilitate the quality of analysis in macromolecules. However, lacking rich and well-defined dataset with ground truth images, supervised image denoising methods generalize poorly to experimental micrographs.

To address this issue, we present a Simulation-aware Image Denoising (SaID) pre-trained model for improving the SNR of cryo-EM micrographs by only training with the accurately simulated dataset. Firstly, we devise a calibration algorithm for the simulation parameters of cryo-EM micrographs to fit the experimental micrographs. Secondly, with the accurately simulated dataset, we propose to train a deep general denoising model which can well generalize to real experimental cryo-EM micrographs. Extensive experimental results demonstrate that our pre-trained denoising model can perform outstandingly on experimental cryo-EM micrographs and simplify the downstream analysis. This indicates that a network only trained with accurately simulated noise patterns can reach the capability as if it had been trained with rich real data. Code and data are available at https://github.com/ZhidongYang/SaID.

Keywords: Cryo-EM · Image Denoising · Noise Simulation · Deep Learning

---

Z. Yang and H. Li—Two authors contribute equally to this work.

© The Author(s), under exclusive license to Springer Nature Singapore Pte Ltd. 2023
X. Guo et al. (Eds.): ISBRA 2023, LNBI 14248, pp. 325–336, 2023.
https://doi.org/10.1007/978-981-99-7074-2_25

1 Introduction

Cryo-Electron Microscopy (cryo-EM) is a prominent imaging technique providing convincing proof of determining structures of proteins and macro-molecules at near-atomic resolution. However, visualization of biological specimens captured by cryo-EM is affected by the low electron dose conditions which are required to overcome the radiation damage to proteins. Consequently, the low Signal-to-Noise Ratio (SNR) in cryo-EM micrographs usually exists in experimental cryo-EM data. The typical SNR of cryo-EM is estimated to be only as high as 0.1 [2], which may have side-effect on downstream tasks like particle picking and alignment. To address this issue, denoising will be a reasonable solution to improving the quality of cryo-EM micrographs. With decades of research, image denoising has been an essential and fundamental task in the field of low-level computer vision and signal processing. The research on image denoising started by removing the additive white Gaussian noise (AWGN) [3]. Numerous iterative traditional methods are proposed to solve this problem with outstanding performance [3,6,14]. However, noise in the real imaging system emerges from several sources (like shot noise and dark current noise). Moreover, the noise in cryo-EM micrographs will be additionally affected by the interactions between electrons and biological specimens, which may be more sophisticated than the situations in the imaging system of natural images.

Various methods have been proposed to deal with the complex noise in cryo-EM micrographs denoising. The methods start from traditional filtering algorithms, such as median filter, low-pass filter [17] and Wiener filter [19]. Nonlinear anisotropic diffusion (NAD) [24] is a procedure based on nonlinear evolution partial differential equations that can be applied to denoising. Non-local mean [3] and BM3D [6] are two methods that can reduce the noise via averaging the non-local patches. Most of the traditional methods depend on the thorough formulation of the noise model in an image. The performance of such methods will be limited by the complexity of the noise model in noisy images. As the noise model in cryo-EM micrographs varies in different biological specimens with corresponding configurations, conventional methods using the pre-defined prior knowledge can not correctly distinguish the signal from complex noise in cryo-EM micrographs. Consequently, conventional denoising methods may not be well generalized to cryo-EM micrographs in some special configurations.

With the advances in deep convolutional neural networks (CNNs), the performance of image denoising has been significantly improved [15,20,28,29], such deep CNNs-based denoisers mainly advance the performance of removing AWGN. However, when removing the non-AWGN or non-additive noise, such deep denoisers can not well generalize to the images with complex noise and tend to be over-fitted to Gaussian noise. To tackle this issue, several improved deep denoisers introduced a real-world noise modeling module to improve the generalization to sophisticated noise. [5] firstly proposed a GAN-CNN deep denoiser to model the noise in real noisy images using a patches-based strategy. [7] generalized this issue to blind real-world image denoising. The proposed denoiser incorporates a sub-network to directly estimate a feature map with spatial-invariant noise which will be additional input for the training of the denoising network. Due to the unavailable ground truth in experimental cryo-EM imaging, most of the deep denoiser based on supervised learning are not suitable. A straightforward solution is to learn the signal from only noisy dataset. Consequently, self-supervised denoisers are proposed. [12] firstly propose a deep learning framework called Noise2Noise to recover the signal from noisy image pairs with the same signal. [11] goes a step further by introducing blind-spot inference to implement constructing noisy pairs from the single noisy image, such a method is applied to cryo-ET denoising with limited performance. Recently, several improvements are tried to enhance the performance of self-supervised denoisers [1, 8, 23].

With the advances in Noise2Noise framework, [2] proposed a pre-trained general deep learning model called Topaz-Denoise for cryo-EM and cryo-ET denoising. Especially, [27] proposed a self-supervised sparsity constrained network for the restoration of cryo-ET volume. However, the hypothesis of Noise2Noise is quite ideal in most cases, that is, the noise in the noisy dataset is zero-mean, identically distributed, and independent from the signal. Consequently, the general model in Topaz-Denoise based on this hypothesis only probably removes the detector noise. To circumvent this problem, [13] developed a novel protocol called NoiseTransfer2Clean (NT2C) by introducing Generative Adversarial Network (GAN) to directly learn the noise pattern in cryo-EM micrographs and transferring the noise pattern to simulated clean signal with re-weighting technique. The NT2C protocol solved the problem of introducing supervised learning to deep denoiser for cryo-EM, but such a solution is not suitable to be integrated into the pipeline of single particle analysis (SPA) because of its complicated procedures.

In this work, we summarize the currently existing problems in the design of deep denoiser for cryo-EM micrographs into two folds: (i) Firstly, it is confirmed that supervised learning can achieve better performance for a deep denoiser, but the pre-trained supervised denoiser for cryo-EM micrographs is not available due to the lack of well-defined dataset; (ii) Secondly, if the noise characteristics in simulated micrographs are homogeneous to real micrographs, the supervised denoiser will be available. But the current simulation of cryo-EM micrographs has a bias between real experimental and simulated cryo-EM micrographs.

To address these issues, we propose a supervised pre-trained denoiser with a large-scale simulated cryo-EM micrographs dataset, called Simulated-aware Image Denoising (SaID) framework for experimental cryo-EM micrographs. The proposed SaID framework includes two essential contributions: (i) A simulation strategy to generate dataset homogeneous to experimental cryo-EM datasets by calibrating the simulation parameters to fit the distribution of experimental micrographs is proposed; (ii) A simulation-aware pre-trained denoising network for experimental cryo-EM micrographs is proposed with an outstanding performance.

2   Method

2.1   Process Overview

The complete workflow of the proposed SaID method can be summarized as following three steps: (i) calibration of the simulation parameters to fit the noise characteristics in experimental cryo-EM micrographs; (ii) simulating the cryo-EM micrographs with calibrated parameters; (iii) training and inference of the denoising network. Figure 1 illustrates the overall processing pipeline of the proposed SaID framework.

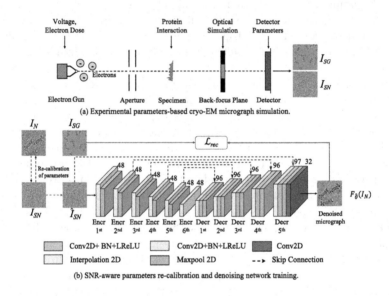

(a) Experimental parameters-based cryo-EM micrograph simulation.

(b) SNR-aware parameters re-calibration and denoising network training.

Fig. 1. Overall processing pipeline of the proposed SaID framework.

2.2   Simulation of cryo-EM Micrographs

In order to correctly simulate ground truth and noisy cryo-EM micrographs according to experimental parameters, we utilize the software InsilicoTEM [22] to synthesize micrographs according to physical imaging principles. Figure 1. (a) shows the main pipeline of simulation. The simulation is based on the parameters matching the experimental conditions during data collection, which includes voltage, electron dose, defocus values, type of detector, pixel size and etc. With well-defined ground truth, the adoption of supervised learning will be possible for cryo-EM micrographs.

2.3   Calibration of the Parameters for Simulation

With our observation, we notice that the naive simulation with InsilicoTEM under the experimental parameters has the issue that the simulated micrographs have a bias on contrast compared with the real experimental micrographs. This bias mainly depends on the defocus values and may hinder the training of the denoising network. To address this issue, we propose a calibration algorithm on defocus values to reduce this bias. Firstly, the defocus values is initialized with the experimental parameters reported by each utilized PDB structure. Secondly, we will evaluate the SNR [13] of simulated and experimental micrographs. Then, the defocus values will be updated with step length δ_{step} in a loop. The loop will be interrupted when the SNR of simulated micrographs has reached the SNR of experimental micrographs. Algorithm 1 presents the complete procedure.

Algorithm 1: Procedure of calibrating the parameters for simulation

---

Input: Initial experimental defocus values d^i_{min}, d^i_{max};
            Simulated noisy micrograph with initial experimental parameters I^{IS};
            Real experimental micrograph I^N
Output: Calibrated defocus values d^c_{min}, d^c_{max}.

1   **Procedure** Calibration(d^i_{min}, d^i_{max}, I^{IS}):
2     $SNR_s \leftarrow \mathrm{SNR}(I^{IS})$;
3     $SNR_r \leftarrow \mathrm{SNR}(I^N)$;
4     $d^t_{min} \leftarrow d^i_{min}$;
5     $d^t_{max} \leftarrow d^i_{max}$;
6   **if** $SNR_s \leq SNR_r$:
7       $\delta_{step} \leftarrow -0.1 \times d^i_{min}$;
8   **else:**
9       $\delta_{step} \leftarrow 0.1 \times d^i_{min}$;
10   **while** **True:**
11      $d^t_{min} \leftarrow d^i_{min} + \delta_{step}$;
12      $d^t_{max} \leftarrow d^i_{max} + \delta_{step}$;
13   Simulated a noisy micrograph I^S with updated d^t_{min}, d^t_{max};
14      $SNR_s \leftarrow \mathrm{SNR}(I^S)$;
15   **if** $|SNR_s - SNR_r| \to 0$:
16        $d^c_{min} \leftarrow d^t_{min}$;
17        $d^c_{max} \leftarrow d^t_{max}$;
18   break;
19   **return** d^c_{min}, d^c_{max};

2.4   Detailed Architecture of the Denoising Network

In the SaID framework, the backbone network is improved from a UNet-based structure [18], as shown in Fig. 1. (b). The denoising network is trained in a supervised learning manner only with our calibrated simulated cryo-EM datasets.

During the inference phase, the pre-trained model will be directly fed with experimental cryo-EM micrographs to denoise.

The noisy micrographs will be first fed into the encoders. Six levels of encoders are adopted in the denoising network. Each encoder consists of a 3×3 bias-free convolutional layer together with an affined BatchNorm and a LeakyReLU whose slope $k = 0.1$. Especially, the sixth encoder plays the role of bottleneck layer with abundant features, the BatchNorm and LeakyReLU layers are removed. Five levels of decoders are adopted in the denoising network, the decoder in each level consists of the nearest interpolation to execute up-sampling and then followed by two CONV blocks with a 3×3 convolutional layer together with an affined BatchNorm and a LeakyReLU whose slope $k = 0.1$. To promisingly recover structural information in the images, a skip-connection operation is used between encoder and decoder blocks at corresponding spatial resolution.

2.5 Loss Function

SaID is an end-to-end deep-learning-based framework for cryo-EM micrograph denoising. The denoising network in SaID is completely trained in a supervised manner, utilizing simulated noisy and clean cryo-EM micrographs generated with experimental parameters. The mathematical formulation of the loss function is shown as follows:

$$\hat{\theta} = \arg\min_{\theta} \frac{1}{N} \sum_{i=1}^{N} ||F_\theta(\hat{I}_i^{SN}) - I_i^{SG}||_2^2 + \mathcal{R}(\theta), \tag{1}$$

where $\hat{\theta}$ is the optimal parameter set of the denoising network in SaID. $F(\cdot)$ denotes the denoising network. $\mathcal{R}(\theta)$ denotes the regularization term to avoid over-fitting. I^{SN} and I^{SG} denote the simulated noisy (SN) micrograph and simulated ground truth (SG) micrograph separately. With the trained network $F(\cdot)$, real experimental cryo-EM micrographs I^N will be denoised with $F_{\hat{\theta}}(\cdot)$. That is:

$$\hat{I} = F_{\hat{\theta}}(I^N). \tag{2}$$

3 Experimental Results

3.1 Implementation Details

The denoising network in SaID was implemented by PyTorch. For all the experiments, the model was trained on two NVIDIA Tesla A100 GPUs in a patches-based manner. The batch size was set to 32 with a 640×640 patch size during the training phase [13]. The model was trained by 100 epochs with simulated micrograph pairs requiring 2 h. The optimizer of denoising network is Adam [10] parameterized by $\beta_1 = 0.5$ and $\beta_2 = 0.999$. The learning rate was set to 0.001. 450 pairs of simulated cryo-EM micrographs sized by 4096×4096 are synthesized for the pre-training of denoising network. 80% of the simulated micrographs were randomly selected as the training set, and 20% of them were randomly selected as the validation set. The denoising network is trained in a patches-based manner, each micrograph is cropped into a patch sized by 640 × 640. During the test phase, the real experimental cryo-EM micrographs will be denoised with the pre-trained denoising network.

3.2   Dataset

Training Simulated Datasets. The denoising network in the proposed SaID framework utilizes a large number of simulated cryo-EM datasets. The training dataset consists of 450 pairs of simulated cryo-EM micrographs with the size of 4096 × 4096. The micrographs are generated with the following PDB structures: 5LZF.pdb, 7ABI.pdb, 1RYP.pdb. The parameters of the simulation are summarized in Table 1 as follow.

Table 1. Parameters for the simulated training dataset.

Protein	Dose $(e^-/\mathring{A}^2)$	Voltage (kV)	Calibrated Defocus (μm)	Original Defocus (μm)	Pixel Size $(\mathring{A})$	Detector	Size
5LZF	20	300	2.1–3.8	0.9–2.6	1.16	K2	4096 × 4096
7ABI	45	300	1.4–3.8	0.9–3.3	1.16	FalconII	4096 × 4096
1RYP	53	300	1.0–1.5	0.9–2.4	0.66	K2	4096 × 4096

Experimental Datasets. To evaluate the performance of SaID, five experimental cryo-EM datasets captured in real experimental environments are evaluated and compared in our experiments. Table 2 describes the biological structure information of each cryo-EM dataset.

Table 2. Real experimental cryo-EM datasets utilized in the experiments.

Dataset	Biological structure
EMPIAR [9]-10025 [4]	Thermoplasma acidophilum 20S proteasome
EMPIAR-10028 [25]	Plasmodium falciparum 80S ribosome
EMPIAR-10077 [16]	Elongation factor SelB on the 70s ribosome complexes
EMPIAR-10090 [26]	Activated human 26S proteasome
EMPIAR-10616 [21]	Human pre-Bact spliceosome

3.3   Results

Proposed SaID Can Improve the SNR and Contrast of Micrographs. In this experiment, BM3D [6], Lowpass Filter, Gaussian Filter, Topaz-Denoise [2] and NT2C [13] are selected as methods for comparisons. The selections of the parameters of BM3D, Lowpass Filter, and Gaussian Filter follow the selections in previous work [2,13]. The evaluation metric is SNR, detailed definition

Table 3. Quantitative results of the real experimental cryo-EM datasets for comparison methods estimated by SNR (in dB, the larger means the better).

Methods	Experimental cryo-EM dataset		
	EM25	EM28	EM77
Noisy	–0.16	–0.34	0.14
BM3D	0.07	–0.27	0.30
Lowpass Filter	–0.04	–0.24	0.33
Gaussian Filter	–0.14	–0.18	0.53
Topaz-Denoise	0.21	1.04	0.55
NT2C	4.88	**7.65**	6.29
Ours	**5.58**	7.57	**6.93**

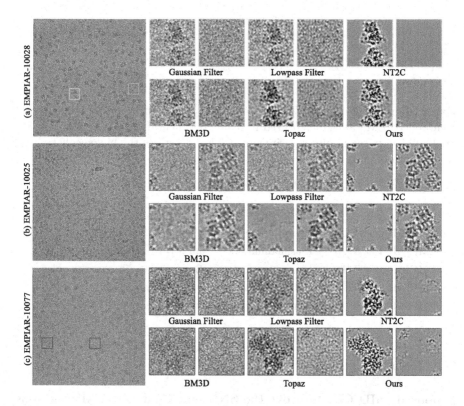

Fig. 2. The benchmark results of the proposed SaID and other comparison methods. Comparisons between each method are evaluated on EM28, EM25, and EM77 datasets.

of SNR can be found in [13]. The datasets EM28 [25], EM25 [4] and EM77 [16] are tested in our experiment. Figure 2 shows visualized results of all benchmark datasets. Table 3 gives the SNR results of all experimental results in our experiments. Judging from the visualized results in Fig. 2, we can conclude that the background noise is nearly completely removed by our proposed SaID compared with Topaz-Denoiser and the other non-learning-based methods. Although the Topaz-Denoise method is a general model trained with enormous paired noisy micrographs, the background noise is not completely removed due to the lacking of prior knowledge of noise. The SaID method has solved this problem. Compared with NT2C, the SaID method achieves a better performance in recovering the particle without missing the structure, as shown in the results of EM77. Judging from the quantitative results assessed by SNR, we can conclude that the SaID method achieved almost the best performance in most cases. Because the signal in EM28 is easy to be recovered, NT2C and SaID both perform well on this dataset, the SaID can recover high-frequency information better than NT2C (shown in Fig. 1).

Proposed SaID Can Improve the Interpretability of Microgrpahs. In this part, we select the experimental dataset EM90 [26] for the illustration. Although the noise within this dataset is relatively weak, some of the projections of specific orientations remain blurred. As shown in Fig. 3, we can find that the region pointed out by the red arrow is blurred. The results recovered by Topaz lose too much signal because of the blur. The signal recovered by our proposed SaID is more clear.

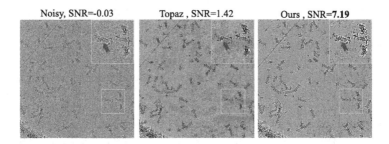

Fig. 3. Denoising with SaID improves the interpretability of real experimental cryo-EM micrographs. Comparisons between each method are evaluated on the EM90 dataset.

Proposed SaID Can Simplify the Particle Picking. The particle picking is affected by the high noise in micrographs thus benefiting from the denoising. Figure 4 shows the results of the particle picking in the micrographs denoised by our SaID method. We can find that, with a cleaner micrograph denoised by our method, the wrongly picked particles are reduced, which simplifies the particle picking. Although the NT2C [13] achieved a similar goal in this task, SaID provides a set of pre-trained models which will be more flexible to be integrated into a SPA pipeline and simplifies the particle picking.

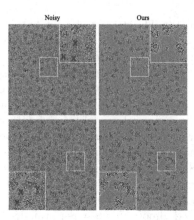

Fig. 4. The particle picking results for original micrograph and SaID denoised. The algorithm of particle picking is Auto-Picking in Relion [30]. The results show that, with better interpretability in micrographs, the wrongly picked particles are reduced.

4   Discussion and Conclusion

In this article, we proposed a simulation-aware image denoising (SaID) framework for cryo-EM micrographs. The contributions of this work can be summarized in two folds: (i) A calibration algorithm for the simulation parameters is proposed to provide a reliable simulated training dataset; (ii) A novel supervised training framework for the denoising of cryo-EM micrographs is proposed, resulting in pre-trained denoising model for the recovery of signal in real micrographs. This work proved that with accurate simulation, the deep denoiser only trained with simulated dataset can well generalize to the experimental dataset. Comprehensive experimental results have demonstrated that the pre-trained model provided by the SaID framework can promisingly enhance the signal and suppress the noise in experimental cryo-EM micrographs. The current SaID framework performs well on protein particles with relatively stable structures, in future work, we will improve our method from two perspectives: (i) The extension of biological structures for the simulation of the training dataset, in order to ensure that our method can be applied to more various dataset; (ii) The improvement of the calibration algorithm. Defocus values are currently regarded as the main factor for the simulation. We will take more imaging parameters into consideration in our algorithm, in order to improve the quality of simulation.

Acknowledgements. This research was supported by the National Key Research and Development Program of China [2021YFF0704300], the National Natural Science Foundation of China [61932018, 32241027, 62072280 and 62072441], and the Natural Science Foundation of Shandong Province [ZR2023YQ057], the Youth Innovation Promotion Association CAS, the Young Scientists Fund of the National Natural Science Foundation of China [Grant No. 61902373], the Foundation of the Chinese Academy of Sciences, China [Grant No. JCPYJJ 22013], the Strategic Priority Research Program of the Chinese Academy of Sciences, China [Grant No.XDB24050300, XDB44030300].

References

1. Batson, J., Royer, L.: Noise2Self: blind denoising by self-supervision. In: Proceedings of the 36th International Conference on Machine Learning (ICML), pp. 524–533 (2019)
2. Bepler, T., Kelley, K., Noble, A.J., Berger, B.: Topaz-denoise: general deep denoising models for cryoem and cryoet. Nat. Commun. **11**(1), 1–12 (2020)
3. Buades, A., Coll, B., Morel, J.M.: A non-local algorithm for image denoising. In: 2005 IEEE Computer Society Conference on Computer Vision and Pattern Recognition (CVPR 2005), vol. 2, pp. 60–65. IEEE (2005)
4. Campbell, M.G., Veesler, D., Cheng, A., Potter, C.S., Carragher, B.: 2.8 Å resolution reconstruction of the Thermoplasma acidophilum 20s proteasome using cryo-electron microscopy. eLife **4**, e06380 (2015)
5. Chen, J., Chen, J., Chao, H., Yang, M.: Image blind denoising with generative adversarial network based noise modeling. In: 2018 IEEE/CVF Conference on Computer Vision and Pattern Recognition, pp. 3155–3164 (2018). https://doi.org/10.1109/CVPR.2018.00333
6. Dabov, K., Foi, A., Katkovnik, V., Egiazarian, K.: Image denoising by sparse 3-d transform-domain collaborative filtering. IEEE Trans. Image Process. **16**(8), 2080–2095 (2007)
7. Guo, S., Yan, Z., Zhang, K., Zuo, W., Zhang, L.: Toward convolutional blind denoising of real photographs. In: Proceedings of the IEEE/CVF Conference on Computer Vision and Pattern Recognition (CVPR) (June 2019)
8. Huang, T., Li, S., Jia, X., Lu, H., Liu, J.: Neighbor2neighbor: self-supervised denoising from single noisy images. In: Proceedings of the IEEE/CVF Conference on Computer Vision and Pattern Recognition (CVPR), pp. 14781–14790 (June 2021)
9. Iudin, A., et al.: Empiar: the electron microscopy public image archive. Nucleic Acids Res. **51**(D1), D1503–D1511 (2022). https://doi.org/10.1093/nar/gkac1062
10. Kingma, D.P., Ba, J.: Adam: a method for stochastic optimization. In: 3rd International Conference on Learning Representations, ICLR (2015)
11. Krull, A., Buchholz, T.O., Jug, F.: Noise2void - learning denoising from single noisy images. In: CVPR (June 2019)
12. Lehtinen, J., et al.: Noise2noise: learning image restoration without clean data. In: ICML, pp. 2965–2974 (2018)
13. Li, H., et al.: Noise-Transfer2Clean: denoising cryo-EM images based on noise modeling and transfer. Bioinformatics **38**(7), 2022–2029 (2022)
14. Maggioni, M., Katkovnik, V., Egiazarian, K., Foi, A.: Nonlocal transform-domain filter for volumetric data denoising and reconstruction. IEEE Trans. Image Process. **22**(1), 119–133 (2013). https://doi.org/10.1109/TIP.2012.2210725
15. Mao, X.J., Shen, C., Yang, Y.: Image restoration using very deep convolutional encoder-decoder networks with symmetric skip connections. In: NIPS (2016)
16. Niels, F., et al.: The pathway to gtpase activation of elongation factor selb on the ribosome. Nature **540**, 80–85 (2016)
17. Penczek, P.A.: Chapter two - image restoration in cryo-electron microscopy. In: Jensen, G.J. (ed.) Cryo-EM, Part B: 3-D Reconstruction, Methods in Enzymology, vol. 482, pp. 35–72 (2010)
18. Ronneberger, O., Fischer, P., Brox, T.: U-Net: convolutional networks for biomedical image segmentation. In: Navab, N., Hornegger, J., Wells, W.M., Frangi, A.F. (eds.) MICCAI 2015. LNCS, vol. 9351, pp. 234–241. Springer, Cham (2015). https://doi.org/10.1007/978-3-319-24574-4_28

19. Sindelar, C.V., Grigorieff, N.: An adaptation of the wiener filter suitable for analyzing images of isolated single particles. J. Struct. Biol. **176**(1), 60–74 (2011)
20. Tai, Y., Yang, J., Liu, X., Xu, C.: Memnet: a persistent memory network for image restoration. In: 2017 IEEE International Conference on Computer Vision (ICCV), pp. 4549–4557 (2017). https://doi.org/10.1109/ICCV.2017.486
21. Townsend, C., et al.: Mechanism of protein-guided folding of the active site u2/u6 rna during spliceosome activation. Science **370**(6523), eabc3753 (2020)
22. Vulović, M., et al.: Image formation modeling in cryo-electron microscopy. J. Struct. Biol. **183**(1), 19–32 (2013)
23. Wang, Z., Liu, J., Li, G., Han, H.: Blind2unblind: Self-supervised image denoising with visible blind spots. In: Proceedings of the IEEE/CVF Conference on Computer Vision and Pattern Recognition (CVPR), pp. 2027–2036 (June 2022)
24. Weickert, J.: Coherence-enhancing diffusion filtering. Int. J. Comput. Vision **31**(1), 111–127 (1999)
25. Wong, W., et al.: Cryo-em structure of the plasmodium falciparum 80s ribosome bound to the anti-protozoan drug emetine. eLife **3**, e03080 (2014)
26. Yanan, Z., Weili, W., Daqi, Y., Qi, O., Ying, L., Youdong, M.: Structural mechanism for nucleotide-driven remodeling of the aaa-atpase unfoldase in the activated human 26s proteasome. Nat. commun. **9**(1360) (2018)
27. Yang, Z., Zhang, F., Han, R.: Self-supervised cryo-electron tomography volumetric image restoration from single noisy volume with sparsity constraint. In: Proceedings of the IEEE/CVF International Conference on Computer Vision (ICCV), pp. 4056–4065 (October 2021)
28. Zhang, K., Zuo, W., Chen, Y., Meng, D., Zhang, L.: Beyond a gaussian denoiser: Residual learning of deep CNN for image denoising. IEEE Trans. Image Process. **26**(7), 3142–3155 (2017)
29. Zhang, K., Zuo, W., Zhang, L.: Ffdnet: toward a fast and flexible solution for CNN-based image denoising. IEEE Trans. Image Process. **27**(9), 4608–4622 (2018). https://doi.org/10.1109/TIP.2018.2839891
30. Zivanov, J., Otón, J., et al.: A bayesian approach to single-particle electron cryo-tomography in relion-4.0. eLife **11**, e83724 (2022). https://doi.org/10.7554/eLife.83724

Reducing the Impact of Domain Rearrangement on Sequence Alignment and Phylogeny Reconstruction

Sumaira Zaman[1] and Mukul S. Bansal[1,2(✉)]

[1] Department of Computer Science and Engineering, University of Connecticut, Storrs, CT 06269, USA
sumaira.zaman@uconn.edu
[2] Institute for Systems Genomics, University of Connecticut, Storrs, CT 06269, USA
mukul.bansal@uconn.edu

Abstract. Existing computational approaches for studying gene family evolution generally do not account for domain rearrangement within gene families. However, it is well known that protein domain architectures often differ between genes belonging to the same gene family. In particular, domain shuffling can lead to out-of-order domains which, unless explicitly accounted for, can significantly impact even the most fundamental of tasks such as multiple sequence alignment and phylogeny inference.

In this work, we make progress towards addressing this important but often overlooked problem. Specifically, we (i) demonstrate the impact of protein domain shuffling and rearrangement on multiple sequence alignment and gene tree reconstruction accuracy, (ii) propose two new computational methods for *correcting* gene sequences and alignments for improved gene tree reconstruction accuracy and evaluate them using realistically simulated datasets, and (iii) assess the potential impact of our new methods and of two existing approaches, MDAT and ProDA, in practice by applying them to biological gene families. We find that the methods work very well on simulated data but that performance of all methods is mixed, and often complementary, on real biological data, with different methods helping improve different subsets of gene families.

1 Introduction

Protein domains, or just *domains* for short, are independently folding structural and/or functional units that recur across multiple protein coding gene families [4]. Domains can be viewed as recurrent building blocks of proteins and are known to play an important role in the function and evolution of many gene families [20,28,29]. In fact, it is estimated that the majority of protein coding genes in eukaryotes and almost half of protein coding genes in prokaryotes contain at least one domain [10,12]. Known domain sequences can be clustered into different domain families and many thousands of distinct domain families have already been identified [5].

As a gene evolves, one or more of its domains can get duplicated or be lost, and new domains can be acquired from other genes. The resulting gain and loss of domains during gene family evolution can lead to genes from the same gene family having different domain contents and architectures (i.e., sequential orderings). This is illustrated

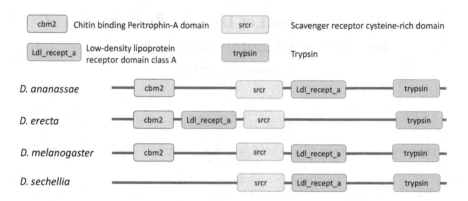

Fig. 1. Different domain architectures within a gene family. The four depicted fly proteins belong the same gene family but show different domain architectures (orderings). In particular, the order of "srcr" and "ldl_recept_a" domains appears to be inverted between *D. erecta* and *D. ananassae*. Also observe that the gene from *D. sechellia* does not have the "cbm2" domain. Note that the figure does not depict the exact location or length of any domain and only shows domain orderings.

in Fig. 1. Such changes in domain content and architecture through domain shuffling are believed to be key drivers of protein evolution and proteome complexity [6]. As a result, mechanisms of domain shuffling and domain architecture evolution have been extensively studied in the literature [2,7,8,11,17,25].

A frequent outcome of domain content and architecture changes within gene families is that genes belonging to the same gene family can have incompatible domain orders. For example, a gene in some gene family may have two domains A and B (from different domain families) in the order $\langle A, B \rangle$, while a different gene from the same gene family may have those domains in the order $\langle B, A \rangle$. This could occur, for example, if there is a tandem duplication of domains A and B, resulting in domain order $\langle A, B, A, B \rangle$, followed by losses of the first and last domains, resulting in the order $\langle B, A \rangle$. Such domain rearrangements, unless explicitly accounted for, can significantly impact even the most fundamental of tasks such as multiple sequence alignment and phylogeny inference. Yet, traditional approaches for computing multiple sequence alignments (MSAs) and reconstructing gene trees do not account for domain rearrangement within gene families. This is because traditional MSA algorithms perform a linear alignment of the given sequences, assuming that any variation in gene sequences is a result of point mutations or indels [1]. Domain rearrangements can violate this assumption, directly affecting the quality of the resulting MSA and of any gene trees inferred using that MSA.

Previous Work. To the best of our knowledge, only three multiple sequence alignment methods, ABA [23], ProDA [22], and MDAT [13], currently exist that explicitly take domain contents and architectures into account. ABA represents a sequence alignment as a directed (possibly cyclic) graph [23], which allows for domain architecture changes and rearrangements to be detected and taken into account when analyzing evolutionary relationships between the aligned sequences. However, to our knowledge, ABA does not compute a global multiple sequence alignment, as needed for gene tree reconstruction, and the ABA software is no longer available. ABA was also shown to have poor residue level accuracy when applied to gene families with rearranged, out-of-order domains [22]. ProDA [22] takes as input a set of unaligned sequences, uses local alignment and clustering to identify all homologous regions appearing in one or more sequences, and outputs a collection of local multiple alignments for the identified homologous regions. ProDA was shown to work well at detecting conserved domain boundaries and clustering domain segments, and at recovering known domain organizations [22]. ProDA can detect local protein homology and construct local multiple alignments, but it cannot be directly used to obtain a global alignment when the input gene sequences contain multiple domain copies from any domain family. The more recent method MDAT [13] seeks to compute more accurate MSAs by computing multiple domain alignments and restricting the global alignment such that domains from different families cannot align to each other. A limitation of MDAT is that it respects the linear arrangement of domains within each input sequence and cannot correct for rearranged, out-of-order domains. Importantly, despite the development of these previous methods, the impact of domain rearrangement on MSAs and subsequent gene tree reconstruction has not been systematically evaluated and remains largely unknown.

Our Contribution. In this work, we propose two new, easy-to-apply computational methods to mitigate the impact of rearranged, out-of-order domains on gene tree reconstruction. We also carefully assess the impact of the new and previous methods on real biological data. Specifically, we first use simulated gene families, modeled after real fly gene families, to assess the impact of domain shuffling and rearrangement on MSA and gene tree reconstruction accuracy. Second, we propose two new computational approaches, referred to as *Door-S* and *Door-A* (where *Door* is short for "domain organizer"), for *correcting* gene sequences and alignments for improved gene tree reconstruction accuracy. The key idea behind our two methods is to identify known domains within the input gene sequences and then reorganize the domains to remove any domain ordering incompatibilities between the different gene sequences. This allows for an improved MSA inference for that gene family, leading to improved gene tree reconstruction. Essentially, our methods leverage the fact that standard phylogeny inference algorithms assume that sites evolve independently of each other and treat each column (site) of an MSA independently. Thus, homologous sites within gene sequences can be rearranged (together) without affecting phylogeny inference. Third, we demonstrate the impact of applying *Door-S* and *Door-A* on realistically simulated gene families. And finally, we carefully evaluate the applicability and impact of *Door-S*, *Door-A*, and the previous methods MDAT and ProDA, on biological gene families from 12 fly species. We find that the new methods result in an almost 70% average reduction in gene tree reconstruction error for the simulated gene families. However, we find that the performance of all methods is mixed when applied to the biological gene families, with the best performing methods resulting in significantly improved gene tree reconstruction for about a quarter of the gene families but showing either comparable or worse reconstruction accuracy for the other gene families. Interestingly, the performance of the different methods on biological data is often complementary, with different methods helping improve different subsets of gene families. Scripts implementing *Door-S* and *Door-A* are freely available from https://github.com/suz11001/Door/tree/main.

2 Description of Methods

2.1 Proposed Methods: *Door-S* and *Door-A*

Both *Door-S* and *Door-A* seek to identify and reorganize domains within each input gene sequence to enable and improve the alignment of homologous regions in the final global MSA. The main steps in the *Door-S* and *Door-A* methods are as follows:

1. Identification of domain families present within the gene family.
2. Identification of domain sequence boundaries and non-domain regions within each gene sequence.
3. Ordering of non-domain regions and domain families for each gene.
4. Ordering of domains copies from same domain family within each gene.
5. Computation of final global MSA.

Door-S and *Door-A* differ only in their implementation of Step 5 above. Specifically, *Door-S* uses a traditional multiple sequence aligner, such as MUSCLE [9], to globally align the reorganised gene sequences, while *Door-A* separately aligns the different domain families and non-domain regions and concatenates these alignments to create a global concatenated alignment for the gene family. Figure 2 illustrates the shared and individual steps of *Door-S* and *Door-A*. We elaborate on these steps below.

1. Identification of domain families present: Domain families present within gene sequences can be identified using protein domain databases or tools such as Pfam [19], SMART [26], PANTHER [18] or InterPro [21]. For our biological dataset from 12 fly species, we used UniProt gene IDs to determine their protein domain constituents from the Pfam A database.

2. Identification of domain sequence boundaries and non-domain regions: Domain annotations are imperfect and the domain sequences found in PFAM or any other domain database may not be an exact match to the domain sequence present in the gene. We therefore align each annotated domain sequence back to the gene and extract the precise genic region where the annotated domain aligns. In case multiple annotated domains from different domain families overlap in the genic space, we duplicate the regions of alignment where the domains overlap. Once all the domain regions of the gene have been identified, these domain regions are removed from the gene sequences and are placed as domain sequences as part of their respective domain families.

3. Ordering of non-domain regions and domain families: The domain and non-domain (genic) regions within each gene sequence of a gene family are ordered such that the genic regions appear first, followed by the domain family sequences in a fixed order. This ensures that the ordering of domain family sequences remains consistent between all genes belonging to the same gene family. This is illustrated in the top half of Fig. 2.

4. Ordering of domain copies from same domain family: If a gene sequence contains multiple domain copies from the same domain family then we place these copies

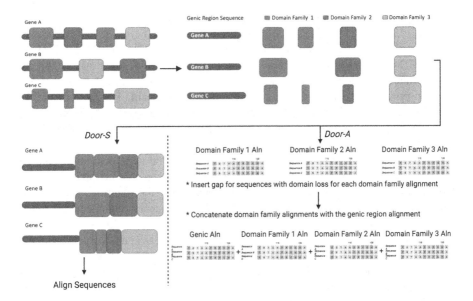

Fig. 2. Overview of *Door-S* and *Door-A*. The key difference between the two methods is that *Door-S* first concatenates all genic (non-domain) and domain sequences in a consistent order across all gene sequences and then performs global sequence alignment of the resulting, reordered gene sequences as the final step (see bottom left of figure). In contrast, *Door-A* first separately aligns the genic sequences and sequences from each domain family and concatenates the resulting alignments, introducing gaps for sequences with domain losses, to obtain the final global alignment (see bottom right of figure).

contiguously in the same order as in the original gene sequence. A different approach was used for the simulated gene families since we did not explicitly simulate domain orderings in the gene sequences; specifically, for each domain family, we choose a reference gene g_{ref} with the most number of domain copies of that domain family and greedily align the domain copies in the other genes from that gene family to the most similar domain copy in g_{ref}.

5. *Computation of final global MSA.* *Door-S* and *Door-A* take different approaches for this final step, as illustrated in the bottom half of Fig. 2. *Door-S* concatenates the reordered genic and domain region sequences within each gene to create a reordered version of each original gene sequence. These reordered gene sequences, all from the same gene family, are then globally aligned using a standard global aligner. In this work, we used MUSCLE v. 3.8.31 [9] with default parameters to compute all alignments. Instead of first concatenating the reordered genic and domain regions and then aligning the resulting concatenated sequences, *Door-A* first aligns the genic regions and each domain family separately, and then concatenates the resulting alignments to obtain a final global alignment of the reordered gene sequences. As part of this process, to ensure a well-formed final global alignment, gaps are artificially introduced if a domain family is completely absent from a gene sequence.

2.2   Existing Methods: MDAT and ProDA

We also evaluate the two related previous methods ProDA [22] and MDAT [13]. Even though MDAT cannot correct for rearranged, out-of-order domains, it can be directly used to compute a global sequence alignment for multi-domain gene families. MDAT relies on protein domain annotations generated using a specific version (version 27) of the Pfam domain database and uses this annotation to restrict the global alignment, ensuring that domains belonging to different domain families cannot be aligned together.

ProDA takes as input a set of unaligned sequences and uses local alignment and clustering to identify all homologous regions appearing in one or more input sequences. It outputs a collection of local multiple alignments for the identified homologous regions. However, ProDA does not compute a global sequence alignment and cannot be directly used to compute one based on the output alignment blocks. For instance, some sequence segments, or even entire genes, do not appear in any output alignment, and each alignment block can contain multiple homologous regions from the same gene sequence. Nonetheless, ProDA's effectiveness at identifying regions of local homology can be leveraged to identify and correct for out-of-order domains or other regions. Accordingly, to apply ProDA to this problem, we use a scheme similar to that used for *Door-A* to compute global sequence alignments from ProDA's output: First, we modify each block of homologous sequences by identifying domain copies from the same gene sequence and arrange them linearly according to their ordering in the gene sequence. This step is similar to Step 3 of *Door-S* and *Door-A*. Second, we compute a sequence alignment (using MUSCLE) for each modified block of homologous sequences (similar to Step 5 of *Door-A*). Third, we add back the genes not represented in the resulting alignment by introducing gaps in the alignment for that gene. Finally, we concatenate the alignment for each blocks of homologous sequences to obtain an overall global alignment for the gene sequences of that gene family.

ProDA also has an input parameter which controls for the minimum size of a homologous sequence block. In our evaluation of ProDA, we used two settings for this parameter; one in which the minimum block size is set to 50 amino acids (aa), and another in which the parameter value is set to the length of the shortest Pfam domain sequence found in that gene family. We refer to these two executions as ProDA_50 and ProDA, respectively.

3   Dataset Description and Experimental Setup

3.1   Simulated Dataset

We first used simulated gene family sequences, with known ground truth, to assess the impact of domain rearrangements on gene tree reconstruction and demonstrate the impact of *Door-S* and *Door-A*. To enhance the biological realism of this simulated dataset, we selected key parameter values, such as for gene length, average number of domain families per gene famiy, average domain length, and average number of domain rearrangements, based on a real datset from 12 fly species (described later in this section). Specifically, starting with a biological dataset of 2307 multi-domain gene families from 12 fly species (see Sect. 3.2), we first identified 198 gene families with plausible out-of-order domains using the simple procedure described in Sect. 3.2. Essentially, this procedure identifies those gene families which contain at least one pair of genes whose domain orderings are incompatible with each other. For these 198 gene families, we find a median genic (not counting domain sequences) length of 452 aa, median domain sequence length of 78 aa, median of 3.6 *unique* domain families per gene family, and median of 1 for the number of unique out-of-order domain-family pairs present. We also estimated the probability of any given gene sequence having out-of-order domains. This probability depended on gene family size y, and was estimated to be 0.45, 0.27, 0.24, 0.15, 0.34, and 0.22 for $y \leq 10$, $10 < y \leq 25$, $25 < y \leq 50$, $50 < y \leq 75$, $75 < y \leq 100$, and $y > 100$, respectively.

Simulating Gene Trees and Domain Trees. Based on these parameter estimates, we used the phylogenetic simulation framework SaGePhy [14] to generate 100 gene families and their corresponding domain families. First, we simulated 100 species trees with SaGePhy using a birth-death model with birth and death rate of 5 and 2, respectively, and height 1. A gene tree was then evolved inside each species tree under a duplication-loss model with gene duplication and gene loss rates of 0.3 each. Finally, we evolved 3 domain trees inside each gene tree with domain duplication and domain loss rates of 0.3 each. This yielded gene families with similar domain characteristics as the biological dataset.

Simulating Sequence Data. We then used SaGePhy to simulate protein sequences along both the gene and the three domain trees under the LG amino acid substitution model [15] and appended together (in a predetermined order) the genic and domain sequences belonging to the same gene. Hence, each gene consists of a genic (non-domain) sequence and a variable number of domain sequences from one or more domain families. Each genic sequence is 450 aa long and each domain sequence is 100 aa long, so that each full gene sequence has length 450 aa or more depending on the number of domain sequences present in it.

Introducing Rearrangements. After creating these baseline sequences for each gene in the gene family, we introduce domain rearrangement in a randomly chosen subset of the gene sequences based on the probabilities previously estimated from the biological dataset. We follow a conservative procedure for introducing domain rearrangements where we only make one rearrangement (exchange the positions of a single pair of domains) in each selected gene sequence. In most cases, we only exchange two neighboring domains. For example, if the simulated gene sequence shows the domain ordering $([A1, A2, A3], [B1, B2], [C1])$, where A, B, and C represent the three domain families, then, in most cases, we only exchange either $A3$ with $B1$ or $B2$ with $C1$, thereby creating exactly one pair of out-of-order domain sequences in that gene sequence. Based on observations in the biological dataset, we also sometimes perform rearrangements so as not to disrupt the tandem ordering of domain copies. For example, if the simulated gene sequence shows domain ordering $([A1, A2], [B1], [C1])$, then we rearrange the sequence to $([B1], [A1, A2], [C1])$ with a small probability based on biological data.

3.2 Biological Dataset

As our real biological dataset, we used the 12-flies dataset assembled by Li et al. [16] in their study of protein domain evolution. This dataset consists of 7165 gene families in which at least one gene has at least one Pfam A domain. Of these 7165 gene families, 2307 gene families contain domains from at least two domain families. Among these 2307 gene families, we identified 198 as having plausible out-of-order domains and our experimental results are based on these 198 gene families.

The 198 gene families with plausible out-of-order domains were identified as follows: We first represent each gene sequence by its ordering of domains. For example, a gene sequence consisting of 8 distinct domains from 4 different domain families A, B, C and D would be represented as follows, based on the specific ordering of the 8 domain sequences: [(A),(A),(B),(B),(C),(D),(C),(B)]. For simplicity and to avoid possible overcounting of out-of-order domains, we then condense the above representation by merging together contiguous domains from the same domain family. Thus, the representation for the above gene would be condensed to [(A),(B),(C),(D),(C),(B)]. We then consider the condensed representations of each pair of gene sequences from the gene family and check if that pair of genes has incompatible domain orders. More precisely, we check if there exists a domain family pair $\{X,Y\}$ such that this pair occurs only in the order $\langle X, Y \rangle$ in one of the gene sequences and only in the order $\langle Y, X \rangle$ in the other gene sequence. If we find any pair of gene sequences to have incompatible domain orders then we flag that gene family as plausibly having out of order domains.

3.3 Evaluation of Results

The most commonly used accuracy metric for multiple sequence alignments is the sum-of-pairs (SP) score. SP scores are computed by comparing every pair of amino acids in an aligned column to assign an alignment quality score to that column, and then summing up these scores across all columns in the alignment. The higher the total score, the better the quality of the alignment. However, this scoring scheme is only appropriate when the sequences being aligned are actually alignable. For sequences with out-of-order domains, the SP score can yield misleading results and need not be correlated with gene tree reconstruction accuracy. We will see a clear example of this in the next section. Consequently, we assess the impact of out-of-order domains and of the different correction methods based on reconstructed gene tree accuracy. We measure gene tree accuracy by comparing each reconstructed gene tree against the corresponding ground truth gene tree using the standard Robinson-Fould's metric [24]. Specifically, we count the number of splits present in only one of the two trees being compared (the reconstructed vs the true gene tree). We refer to the resulting number as the RF-score, with a lower RF-score implying greater gene tree reconstruction accuracy. Note that the reported RF-scores count unique splits of both trees (i.e., we do not divide the computed score by 2).

Since ground truth gene trees are only available for the simulated dataset, gene tree accuracy cannot be directly measured for the biological dataset. To overcome this challenge, we use the reconciliation cost (specifically the duplication-loss reconciliation cost) of each reconstructed gene tree against the known 12-flies species tree as a proxy for gene tree accuracy. We compute this reconciliation cost under a parsimony framework [3] using a loss cost of 1 and a duplication cost of 2. We refer to the resulting cost as the DL-score. In Sect. 4.1, using the simulated datasets, we show that the DL-score generally increases or decreases in line with the RF-score (i.e., greater gene tree error results in a higher DL-score), thereby justifying its use as a proxy for the RF-score.

3.4 Gene Tree Reconstruction

For each simulated gene family, we reconstruct four gene trees based on the following four gene family alignments: (i) an alignment (using MUSCLE [9]) of the simulated baseline sequences with no domain rearrangement, (ii) an alignment (using MUSCLE) of the rearranged gene sequences, (iii) the alignment produced by applying *Door-S* to the rearranged sequences, (iv) and the alignment produced by applying *Door-A* to the rearranged sequences. Thus, the first tree represents the baseline scenario when there is no domain rearrangement in the gene sequences and captures baseline alignment and gene tree reconstruction error. The second tree represents the scenario when domain rearrangement is present but is not accounted for in the gene family alignment. The third and fourth trees represent the scenarios when domain rearrangement is present and has been corrected for using *Door-S* and *Door-A*, respectively. All simulated dataset gene trees were reconstructed using RAxML v8.2.11 [27] with thorough search settings (-f a -N 100) and under the same model (PROTGAMMAILG) used for the simulation.

For the biological dataset, we reconstruct six gene trees for each of the 198 gene families. These six gene trees correspond to the original (uncorrected) MUSCLE alignment and the corrected alignments obtained by applying *Door-S*, *Door-A*, MDAT, ProDA, and ProDA_50. All biological dataset gene trees were reconstructed using RAxML v8.2.11 [27] with thorough search settings under the PROTGAMMAAUTO model.

4 Results

4.1 Simulated Dataset Results

Table 1 summarises our results for the simulated dataset. As the table shows, introducing domain rearrangements in the gene sequences leads to a dramatic worsening of gene tree reconstruction accuracy, with the average RF-score increasing from 9 for the baseline sequences without rearrangement to 43 for the aligned rearranged sequences. The table also shows the drastic improvement in gene tree accuracy obtained after correcting the rearranged sequences using *Door-S* and *Door-A*. Specifically, the RF-score decreases from 43 to only 14 and 13, respectively, after *Door-S* and *Door-A* are applied. Overall, among the 100 simulated gene families in this dataset, *Door-S* resulted in an improved RF-score for 95 gene families and *Door-A* for 97 gene families. These results show that both *Door-S* and *Door-A* are highly effective at correcting MSAs for improved gene tree reconstruction, with *Door-A* slightly outperforming *Door-S*.

Relationship between RF-Score and DL-Score. Table 1 also shows average DL-scores for the gene trees reconstructed using the four alignment types. As the table shows,

Table 1. Gene tree reconstruction accuracy using different alignment types for the simulated gene families. Accuracy is shown in terms of RF-scores, averaged across the 100 gene families in the simulated dataset. Corresponding average DL-scores and SP-scores are also shown. Lower values are better for RF-score and DL-score, while higher values are better for SP-score. Observe that DL-scores are well-aligned with RF-scores, but that SP-scores are not, with the *Door-A* corrected alignment showing the worst (lowest) SP-score among all four alignment types.

Alignment type	**RF-score**	DL-score	SP-score
Baseline sequences alignment (no rearrangement)	9	40	626
Rearranged sequences alignment	43	117	576
Door-S corrected alignment	14	51.25	635
Door-A corrected alignment	13	48	550

these DL-scores are highly correlated with corresponding RF-scores, increasing and decreasing by similar degrees as the RF-scores. Overall, we observed that application of *Door-S* and *Door-A* resulted in improved (decreased) DL-score in 95 of the 100 gene families. These observations justify the use of DL-score as a proxy for gene tree reconstruction error for the biological dataset where true gene trees are unknown.

Inapplicability of SP-Score. We also computed SP-scores for the *Door-S* and *Door-A* alignments and compared them to SP-scores for the rearranged sequence alignments (Table 1). Based on the drastic improvement in gene tree accuracy enabled by *Door-S* and *Door-A*, one would expect the *Door-S* and *Door-A* alignments to show much better (higher) SP-scores. While all *Door-S* alignments do show an improvement, we found that only 11 of the 100 *Door-A* alignments had an improved SP-score compared to rearranged sequence alignments. In other words, 89% of the *Door-A* alignments actually had worse SP-scores than the rearranged sequence alignments. The finding that *Door-A* alignments have worse SP-scores than *Door-S* alignments is not is not surprising; specifically, *Door-A* alignments are composed of concatenated alignments of smaller sequence blocks and are therefore more "restricted" compared to the *Door-S* alignments where the aligner has greater opportunity to improve the SP-score by aligning matching nucleotides (or amino acids) across domain boundaries. These results demonstrate how SP-scores need not be correlated with alignment quality or gene tree reconstruction accuracy in the presence of domain rearrangement.

Note that we did not apply MDAT and ProDA to the simulated dataset. ProDA (or ProDA_50) could not offer any improvement over *Door-A* since exact domain families and domain sequence boundaries are already known for the simulated dataset. MDAT could not be applied since it requires specifically formatted PFam annotations which are unavailable for the simulated data.

4.2   Biological Dataset Results

We applied all five methods, *Door-S*, *Door-A*, MDAT, ProDA, and ProDA_50 to the 198 biological gene families and compared the accuracies of the resulting gene trees against the gene trees constructed using the original (uncorrected) gene sequence alignments.[1] Since true gene trees are unavailable for the biological dataset, relative gene tree accuracies were estimated based on DL-scores, as described previously. Table 2 shows the results of this analysis. In contrast with the results on simulated datasets, we observed that none of the methods could consistently improve all gene families and that the majority of gene families showed worse accuracy after the methods were applied. The best performing methods on this dataset were MDAT and *Door-A*, which both improved approximately 25% of the gene families and worsened 57.6% of the gene families.

Table 2. Number of gene families improving or worsening, per the DL-score, when applying MDAT, ProDA, ProDA_50, *Door-S*, and *Door-A* to the 198 biological gene families.

Method	No. of Families Improved	Avg. Percent Improvement	No. of Families Worsened	Avg. Percent Worsening
MDAT	50	16.4	114	37.4
ProDA	39	19.1	120	26.3
ProDA_50	45	17	119	31.6
Door-S	42	15.4	125	36
Door-A	49	16	114	34.6

We also observed that the different methods tended to improve different subsets of gene families; see Fig. 3. As expected, the greatest overlap in improved gene families occurs for *Door-S* and *Door-A* and for ProDA and ProDA_50. When considering only the best performing method of each type, MDAT, ProDA_50, and *Door-A*, we find that they all show an improvement for 15 shared gene families (Fig. 3(a)). This level of overlap is highly unlikely to occur by chance (p value < 0.0001). In fact, based on 10,000 randomization experiments, we observed an average overlap of only 2.8 gene families for the three methods. This suggests that it may be possible to predict which gene families would benefit from the application of such methods.

These results also highlight the difficulty of dealing with domain rearrangement in real biological gene families. In particular, error-prone identification of domains and domain boundaries, and inability to identify all homologous regions affected by rearrangement can all greatly impact *Door-S* and *Door-A*, as well as the other methods. The competitive performance of MDAT on these gene families also suggests that, for several of the gene families, the gene sequences may actually be linearly alignable. E.g., the seemingly incompatible domain orders $\langle A, B \rangle$ and $\langle B, A \rangle$ become linearly alignable in the presence of a third sequence with domain order $\langle A, B, A \rangle$.

5 Discussion and Conclusion

In this work, we considered the problem of out-of-order domains within gene families. We used carefully simulated gene families to demonstrate the impact of protein domain

---

[1] ProDA and ProDA_50 could only be run successfully on 183 and 191 gene families, resp.

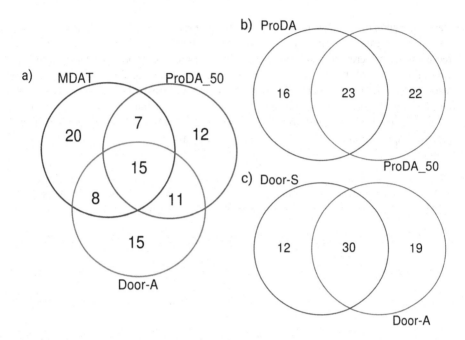

Fig. 3. Venn diagrams for gene families improved by different methods. (a): Venn diagram showing intersections of improving gene families for the three primary methods, MDAT, ProDA_50, and Door-A. (b) and (c): Venn diagrams showing intersections of improving gene families for similar methods ProDA & ProDA_50 (b) and *Door-S* & *Door-A* (c).

shuffling and rearrangement on multiple sequence alignment and gene tree reconstruction accuracy, proposed two new computational methods, *Door-S* and *Door-A*, for correcting gene sequences and alignments for improved gene tree reconstruction accuracy, demonstrated their drastic impact on gene tree reconstruction accuracy on the simulated dataset, and assessed the potential real-world impact of the new methods and MDAT and ProDA by applying them to biological gene families. Our findings demonstrate the significant impact that proper handling of domain rearrangements can have on gene tree reconstruction accuracy, and identify the substantial challenges that such methods must overcome to become widely applicable in practice. Notably, none of the evaluated methods could consistently improve the accuracy of reconstructed gene trees for the biological gene families.

Between *Door-S* and *Door-A*, our experimental results on both simulated and biological gene families indicate that the concatenated alignment approach implemented in *Door-A* may be slightly superior, overall, to the simpler approach implemented in *Door-S*. However, we found that there were several biological gene families that were improved by *Door-S* but not by *Door-A* (Fig. 3), and further work is needed to better understand the scenarios in which one or the other method works better. Our results on the biological dataset suggest that both *Door-S* and *Door-A* could be further improved by combining protein domain annotations with the local alignment approach of ProDA to better identify out-of-order domains and other homologous regions and their boundaries. Our results also suggest that first constructing an order-preserving alignment, as done by MDAT, may help to better identify gene families with true out-of-order domains which could benefit from the reordering-based approach of *Door-S* and *Door-A*. Finally, our results are based on using MUSCLE as the underlying sequence aligner and it could be instructive to assess the impact of using other sequence aligners to compute baseline alignments and *Door-S* and *Door-A* corrected alignments.

Funding. This work was supported in part by NSF award IIS 1553421 to MSB.

References

1. Krogh, A., Eddy, S., Durbin, R.M.: Biological Sequence Analysis. Cambridge University Press, Probabilistic Models of Proteins and Nucleic Acids (1998)
2. Baker, E.P., et al.: Evolution of host-microbe cell adherence by receptor domain shuffling. Elife **11** (2022)
3. Bansal, M.S., Kellis, M., Kordi, M., Kundu, S.: RANGER-DTL 2.0: rigorous reconstruction of gene-family evolution by duplication, transfer and loss. Bioinformatics **34**(18), 3214–3216 (2018)
4. Björklund, A.K., Ekman, D., Light, S., Frey-Skött, J., Elofsson, A.: Domain rearrangements in protein evolution. J. Mol. Biol. **353**(4), 911–923 (2005)
5. Blum, M., Chang, H.Y., Chuguransky, S., et al.: The InterPro protein families and domains database: 20 years on. Nucleic Acids Res. **49**(D1), D344–D354 (2020)
6. Choudhuri, S.: Chapter 2 - fundamentals of molecular evolution. In: Choudhuri, S. (ed.) Bioinformatics for Beginners, pp. 27–53. Academic Press, Oxford (2014)
7. Cohen-Gihon, I., Sharan, R., Nussinov, R.: Processes of fungal proteome evolution and gain of function: gene duplication and domain rearrangement. Phys. Biol. **8**(3), 035009 (2011)
8. Dohmen, E., Klasberg, S., Bornberg-Bauer, E., Perrey, S., Kemena, C.: The modular nature of protein evolution: domain rearrangement rates across eukaryotic life. BMC Evol. Biol. **20**(1), 30 (2020)
9. Edgar, R.C.: MUSCLE: a multiple sequence alignment method with reduced time and space complexity. BMC Bioinformatics **5**, 113 (2004)
10. Ekman, D., Björklund, Å.K., Frey-Skött, J., Elofsson, A.: Multi-domain proteins in the three kingdoms of life: orphan domains and other unassigned regions. J. Mole. Biol. **348**(1), 231–243 (2005)
11. Forslund, K., Sonnhammer, E.L.L.: Evolution of protein domain architectures. In: Anisimova, M. (ed.) Evolutionary Genomics: Statistical and Computational Methods, vol. 2, pp. 187–216. Humana Press, Totowa, NJ (2012)
12. Han, J.H., Batey, S., Nickson, A.A., Teichmann, S.A., Clarke, J.: The folding and evolution of multidomain proteins. Nat. Rev. Mol. Cell Biol. **8**, 319–330 (2007)
13. Kemena, C., Bitard-Feildel, T., Bornberg-Bauer, E.: MDAT- aligning multiple domain arrangements. BMC Bioinform. **16**, 19 (2015)
14. Kundu, S., Bansal, M.S.: SaGePhy: an improved phylogenetic simulation framework for gene and subgene evolution. Bioinformatics **35**(18), 3496–3498 (2019)
15. Le, S.Q., Gascuel, O.: An improved general amino acid replacement matrix. Mol. Biol. Evol. **25**(7), 1307–1320 (2008)
16. Li, L., Bansal, M.S.: An integrated reconciliation framework for domain, gene, and species level evolution. IEEE/ACM Trans. Comput. Biol. Bioinform. **16**(1), 63–76 (2019)

17. Marsh, J.A., Teichmann, S.A.: How do proteins gain new domains? Genome Biol. **11**(7), 126 (2010)
18. Mi, H., Thomas, P.: PANTHER pathway: an ontology-based pathway database coupled with data analysis tools. Methods Mol. Biol. **563**, 123–140 (2009)
19. Mistry, J., et al.: Pfam: the protein families database in 2021. Nucleic Acids Res. **49**(D1), D412–D419 (2021)
20. Miyata, T., Suga, H.: Divergence pattern of animal gene families and relationship with the cambrian explosion. BioEssays **23**(11), 1018–1027 (2001)
21. Paysan-Lafosse, T., Blum, M., Chuguransky, S., et al.: Interpro in 2022. Nucleic Acids Res. **51**(D1), D418–D427 (2023)
22. Phuong, T.M., Do, C.B., Edgar, R.C., Batzoglou, S.: Multiple alignment of protein sequences with repeats and rearrangements. Nucleic Acids Res. **34**(20), 5932–5942 (2006)
23. Raphael, B., Zhi, D., Tang, H., Pevzner, P.: A novel method for multiple alignment of sequences with repeated and shuffled elements. Genome Res. **14**(11), 2336–2346 (2004)
24. Robinson, D.F., Foulds, L.R.: Comparison of phylogenetic trees. Math. Biosci. **53**(1), 131–147 (1981)
25. Sato, P.M., Yoganathan, K., Jung, J.H., Peisajovich, S.G.: The robustness of a signaling complex to domain rearrangements facilitates network evolution. PLoS Biol. **12**(12), e1002012 (2014)
26. Schultz, J., Copley, R.R., Doerks, T., Ponting, C.P., Bork, P.: SMART: a web-based tool for the study of genetically mobile domains. Nucleic Acids Res. **28**(1), 231–234 (2000)
27. Stamatakis, A.: RAxML version 8: a tool for phylogenetic analysis and post-analysis of large phylogenies. Bioinformatics **30**(9), 1312–1313 (2014)
28. Tordai, H., Nagy, A., Farkas, K., Banyai, L., Patthy, L.: Modules, multidomain proteins and organismic complexity. FEBS J. **272**(19), 5064–5078 (2005)
29. Vogel, C., Bashton, M., Kerrison, N.D., Chothia, C., Teichmann, S.A.: Structure, function and evolution of multidomain proteins. Curr. Opin. Struct. Biol. **14**(2), 208–216 (2004)

Identification and Functional Annotation of circRNAs in Neuroblastoma Based on Bioinformatics

Jingjing Zhang[1], Md. Tofazzal Hossain[2], Zhen Ju[1], Wenhui Xi[1(✉)], and Yanjie Wei[1(✉)]

[1] Shenzhen Key Laboratory of Intelligent Bioinformatics and Center for High Performance Computing, Shenzhen Institute of Advanced Technology, Chinese Academy of Sciences, Shenzhen 518055, China
{wh.xi,yj.wei}@siat.ac.cn
[2] Bangabandhu Sheikh Mujibur Rahman Science and Technology University, Gopalganj 8100, Bangladesh

Abstract. Neuroblastoma is a prevalent solid tumor affecting children, with a low 5-year survival rate in high-risk patients. Previous studies have shed light on the involvement of specific circRNAs in neuroblastoma development. However, there is still a pressing need to identify novel therapeutic targets associated with circRNAs. In this study, we performed an integrated analysis of two circRNA sequencing datasets, the results revealed dysregulation of 36 circRNAs in neuroblastoma tissues, with their parental genes likely implicated in tumor development. In addition, we identified three specific circRNAs, namely hsa_circ_0001079, hsa_circ_0099504, and hsa_circ_0003171, that exhibit interaction with miRNAs, modulating the expression of genes associated with neuroblastoma. Additionally, by analyzing the translational potential of differentially expressed circRNAs, we uncovered seven circRNAs with the potential capacity for polypeptide translation. Notably, structural predictions suggest that the protein product derived from hsa_circ_0001073 belongs to the TGF-beta receptor protein family, indicating its potential involvement in promoting neuroblastoma occurrence.

Keywords: circular RNA · neuroblastoma · miRNA · RBP · translation

1 Introduction

Neuroblastoma (NB) is the most prevalent extracranial solid tumor among children, originating from primitive neural crest cells during embryonic development [1]. NB accounts for about 7% of malignant tumors in children and about 15% of cancer-related deaths [2]. Metastases are present at diagnosis in around 50% of patients, with bone marrow, bone, and lymph nodes being the major sites of involvement4. Currently, the overall survival rate for low-risk and

X. Guo et al. (Eds.): ISBRA 2023, LNBI 14248, pp. 351–363, 2023.
https://doi.org/10.1007/978-981-99-7074-2_27

intermediate-risk neuroblastoma patients exceeds 90% [3]. However, the long-term survival rate for high-risk neuroblastoma patients remains below 50% [4]. Thus, there is an urgent need to identify novel biomarkers and treatment targets for high-risk neuroblastoma, intending to improve the cure rate and long-term survival of these patients.

Circular RNA (circRNA) is an emerging class of RNA molecules that possess a unique circular structure formed through reverse splicing [5]. Research has found that circRNAs play critical roles in regulating fundamental biological processes such as cell proliferation, differentiation, and apoptosis, essential for maintaining normal cellular function and tissue homeostasis. Notably, dysregulation of circRNA expression has been associated with the onset and progression of various diseases, including cancer [6], cardiovascular disorders [7], and neurological conditions [8]. Consequently, circRNAs have emerged as potential biomarkers or therapeutic targets for disease diagnosis and treatment [9].

In the quest for more convenient and effective biomarkers and therapeutic targets for the diagnosis and treatment of NB, researchers have investigated the role of circRNA in this disease [10]. Zhang et al. conducted a study in which they observed the upregulation of seven circRNAs within the MYCN amplification region in cell lines exhibiting high MYC signaling activity and MYCN amplification [11]. Another study by Lin et al. identified the relevance of circRNA-TBC1D4, circRNA-NAALAD2, and circRNA-TGFBR3 to the clinical manifestations of NB [12]. Previous investigations have also suggested that circRNA expression levels can serve as diagnostic biomarkers for NB. However, the precise mechanisms underlying the specific regulation of NB by circRNAs remain unclear, and there are numerous circRNAs whose functions in NB are yet to be explored.

In this study, we conducted a comprehensive analysis of two NB circRNA sequencing datasets, PRJNA721263 and PRJNA554935, obtained from the SRA database. By comparing the expression levels of circRNAs between normal tissues and NB tissues, we identified differentially expressed circRNAs (DECs) and performed functional annotation and pathway analysis of their parent genes. To gain insight into the underlying mechanisms of circRNA action, we constructed a circRNA-miRNA-mRNA regulatory network by integrating differentially expressed genes (DEGs) and miRNAs (DEMs) related to NB from Chen's research findings. Furthermore, we investigated the relationship between circRNAs and RBPs and examined the translation potential of the differentially expressed circRNAs. Notably, the translation product of hsa_circ_0001073 has potential relevance to the occurrence and development of NB.

2   Methodology

2.1   Datasets Collection

The circRNA sequencing datasets of NB and normal tissue were downloaded from SRA (https://www.ncbi.nlm.nih.gov/sra/), the PRJNA721263 dataset is composed of 5 NB and 5 control tissues and the PRJNA554935 dataset includes

3 NB and 3 control tissues. In addition, based on the research results of Chen [13,14] and Shao [15], differentially expressed miRNAs and differentially expressed genes related to NB were collected. The reference genome of human (GRCh38/hg38) was downloaded from UCSC.

2.2 Identification of CircRNAs and Differentially Expressed CircRNAs

Previous studies have shown that the combination of multiple circRNA prediction methods yields more reliable results [16]. Here, we combined three circRNA prediction methods, including CIRI combined with BWA [17], CIRCexplorer combined with BWA [18], and DCC combined with STAR [19], and circRNAs detected by three methods at the same time are used further analysis. Next, we used R package DEseq to analyze the differential expression of circRNAs in two datasets, and the circRNAs detected in less than half of the samples were considered as lowly expressed and were filtered. CircRNAs with p<0.05 and |log2FC|>1 were considered significantly different and were used for further analysis.

2.3 GO and KEGG Enrichment Analysis

Gene Ontology (GO) and the Kyoto Encyclopedia of Genes and Genomes (KEGG) functional enrichment analyses were carried out with clusterProfiler R package to predict the function of parental genes that DECs.

2.4 Construction of a CircRNA-miRNA-mRNA Regulatory Network

The potential target miRNAs of the circRNAs were predicted using the circbank [20]. At the same time, we selected the DEMs through the research results of Chen. Next, we conducted a Venn diagram analysis; overlapping miRNAs were chosen for further investigation and research. Furthermore, we used miRTarBase [21] to predicate target mRNAs of these selected miRNAs. The DEGs were gained from the research results of Chen and Shao. Similarly, overlapping mRNAs were selected for the following bioinformatics analysis. Cytoscape (3.10.0) was used to create a circRNA-miRNA-mRNA network [22].

2.5 RNA Binding Protein (RBP) Prediction of CircRNAs

As a protein sponge, circRNAs can bind to RNA-associated proteins to form RNA-protein complexes that regulate gene transcription. The RBPs that bind the differential circRNAs were predicted by the CircInteractome web tool [23], which were visualized by Cytoscape.

2.6   Computational Tools Used for Prediction of Protein Structures, and Domain Analysis

The list of human circRNAs potentially translating into proteins was downloaded from the riboCIRC database [24]. The structure of predicted polypeptides encoded by circRNA splice variants was determined using Alpha Fold 2 with default parameters [25]. InterProScan was used to identify functional protein domains reported by various prediction tools such as Pfam, Phobius, and PANTHER in the circRNA-derived polypeptides [26].

3   Results

3.1   Identification of DECs in Neuroblastoma

In this study, we employed three distinct identification methods to identify circRNAs in two datasets. CircRNAs that were consistently identified by all three methods were considered highly reliable. The results demonstrated that dataset PRJNA721263 detected a total of 38,438 unique circRNAs, and dataset PRJNA554935 revealed 11,368 unique circRNAs. And, there were 10,060 circR-NAs shared between the two datasets.

Subsequently, using a statistical threshold of p<0.05 and |log2FC|>1, we identified 2,234 DECs between normal tissue and NB tissue from two datasets. In the PRJNA721263 dataset, we detected 2,221 DECs, with 846 circRNAs upregulated and 1,375 circRNAs downregulated (Fig. 1A). The PRJNA554935 dataset exhibited 50 DECs, consisting of 36 upregulated and 14 downregulated circRNAs (Fig. 1B). Additionally, through Venn diagram analysis, we identified 36 DECs that overlapped between the two circRNA sequencing datasets and displayed the same regulation trend (Fig. 1C). Among these, 12 circRNAs upregulated, while 24 circRNAs downregulated.

3.2   Reconstruction of DECs Full-Length Sequences

The functionality of circRNAs, including their interactions with miRNA or RNA-binding proteins (RBPs) and their potential for translation, is primarily determined by their unique sequences. To gain deeper insights into the functions of the identified DECs, we employed CIRI-full with default parameters to assemble the full-length sequences of the 36 DECs. This assembly process successfully generated 82 distinct sequences, among which 21 circRNAs exhibited variable splicing isoforms. Upon analyzing the full-length circRNA sequences, we observed that 60 out of the 82 sequences originated exclusively from the exon regions of the genome. Then, we compared assembled sequences with sequences available in the circbank, CircInteractome, and riboCIRC databases to establish a robust foundation for conducting functional analyses of circRNAs using these databases. At least one full-length sequence of each differentially expressed circRNA has been confirmed across three databases and will be used for further analysis.

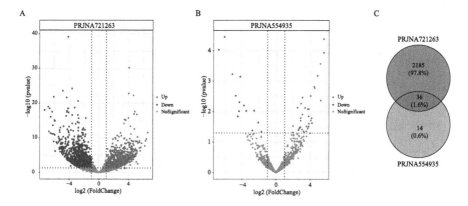

Fig. 1. Volcano plot of (A) PRJNA721263 and (B) PRJNA554935. The blue points indicate the screened down-regulated DECs, the red dots indicate the screened up-regulated DECs, and the gray dots indicate genes with no significant differences. (C) Venn diagram of 36 overlapped dysregulated DECs between PRJNA721263 and PRJNA554935. (Color figure online)

3.3   Function Enrichment Analysis of CircRNA Parental Genes

To gain further insights into the potential biological functions of the dysregulated circRNAs in NB, we conducted GO and KEGG pathway enrichment analyses using the parental genes associated with the 36 DECs.

The GO analysis revealed that a majority of the genes were significantly enriched in various biological processes (Fig. 2A). These processes encompassed cognition, substrate adhesion-dependent cell spreading, cellular response to epinephrine stimulus, negative regulation of neuroinflammatory response, maintenance of synapse structure, and macrophage proliferation, among others. These enrichment results indicate that the parental genes of the DECs likely play pivotal roles in neural cell differentiation and inflammatory responses.

In the KEGG pathway enrichment analysis (Fig. 2B), we observed prominent involvement of the parental genes in neuroblastoma-related signaling pathways. The enriched pathways included the Neurotrophin signaling pathway, Pathways of neurodegeneration - multiple diseases, Transcriptional misregulation in cancer, Hedgehog signaling pathway, TNF signaling pathway, Spinocerebellar ataxia, Rap1 signaling pathway, ErbB signaling pathway, and others. These enriched pathways further suggest that these DECs may contribute to the processes underlying neuroblastoma occurrence and development.

3.4   CircRNA-miRNA-mRNA Network

To explore the regulatory relationships among circRNAs, miRNAs, and mRNAs, we constructed a circRNA-miRNA-mRNA network using 36 candidate DECs (12 upregulated and 24 downregulated), 481 candidate DEGs (177 upregulated and 304 downregulated genes), and 81 candidate DEMs (30 upregulated and

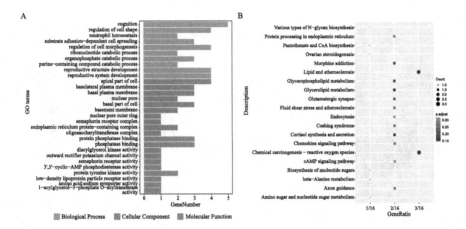

Fig. 2. Functional enrichment analysis of parental gene of 36 overlapped DECs. (A) The top ten GO items of parental genes of overlapped DECs. (B) KEGG enrichment analysis of parental genes of overlapped DECs.

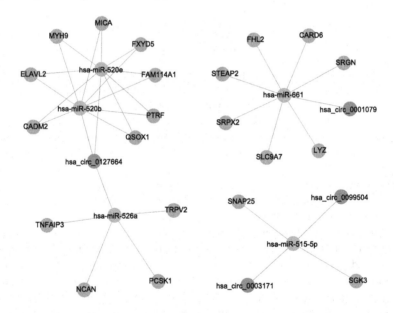

Fig. 3. circRNA-miRNA-mRNA regulatory Network. The red dot represents circRNA, the green dot represents mRNA, the yellow dot represents miRNA. (Color figure online)

51 downregulated DEMs). After filtering, the resulting network, visualized in Cytoscape, consisted of 30 nodes and 31 regulatory axes (Fig. 3). Notably, 4 DECs (hsa_circ_0127664, hsa_circ_0001079, hsa_circ_0099504, hsa_circ_0003171) were identified within the network, connecting with 21 DEGs through 5 DEMs.

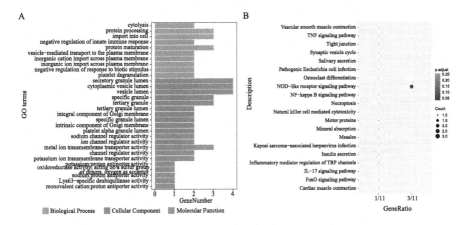

Fig. 4. (A) GO enrichment analysis of DEGs in the regulatory network. (B). KEGG enrichment analysis of DEGs in the regulatory network.

To gain insights into the functions of these DEGs, we conducted a comprehensive functional enrichment analysis. GO-based BP analysis showed that DEGs were significantly enriched in protein processing, import into the cell, and protein maturation (Fig. 4A). Additionally, the CC of GO analysis demonstrated significant enrichment in secretory granule organization, cytoplasmic vesicle lumen, vesicle lumen, specific granule, and tertiary granule. Regarding MF, three DEGs were enriched in metal ion transmembrane transporter activity.

Furthermore, the KEGG pathway analysis identified significant enrichment of DEGs in several pathways (Fig. 4B), such as the NOD-like receptor signaling pathway, NF-κB signaling pathway, natural killer cell-mediated cytotoxicity, and PI3K-Akt signaling pathway.

3.5 Construction of a CircRNA-RBP Network in NB

Apart from their role as miRNA sponges, circRNAs can also serve as protein sponges. To investigate potential interactions between DECs and RBPs, we employed the CircInteractome database to construct a circRNA-RBP network. This network consists of 32 DECs and 22 RBPs. By calculating the Degree value of each node, we identified the top three hub RBPs, namely EIF4A3, AGO2, and HuR (Fig. 5A). EIF4A3 plays a pivotal role in various aspects of RNA metabolism and translation initiation. AGO2 is centrally involved in the RNA interference (RNAi) pathway and post-transcriptional gene silencing. HuR has been implicated in cancer development and progression. Previous studies have reported the relevance of these three RBPs to the occurrence and development of NB.

Furthermore, KEGG enrichment analysis of all RBPs in the network suggests their potential involvement in essential metabolic pathways, including transcriptional regulation in cancer, pathways of neurogenesis with multiple releases, mRNA surveillance pathways, and others(Fig. 5B).

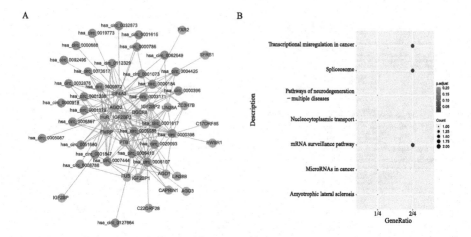

Fig. 5. (A) Construction of the RBP-circRNA regulatory network in NB. (B) KEGG enrichment analysis of RBPs in the regulatory network.

3.6   Analysis of CircRNA Translation Potential

To identify potentially translatable DECs, we utilized the riboCIRC database, which employs multiple lines of evidence to predict the translation potential of circRNAs. From this database, we extracted a set of human circRNAs predicted to have translation ability. Some circRNA sequences show the potential to generate multiple peptide chain subtypes, while others are short peptides with a length of less than 50 amino acids (aa). Domain prediction using InterProScan was performed to identify functionally relevant sites within these circ-proteins. The analysis results indicated that peptide chains with fewer than 50aa do not possess specific functions, whereas longer peptide chains contain distinct functional domains. Finally, we retained 7 long circRNA peptide chains, which are summarized in Table 1, and we employed AlphaFold2 analysis to predict their three-dimensional structures, as illustrated in Fig. 6.

Table 1. DECs with translation potential in riboCIRC database

circRNA_ID	location	length	evidences	riboCIRL_ID
hsa_circ_0001073	chr2:147896300-147899898|+	570nt	RPF, cORF	hsa_circACVR2A_001
hsa_circ_0001238	chr22:41808874-41810291|+	326nt	RPF, cORF, MS, m6A	hsa_circCCDC134_001
hsa_circ_0005087	chr1:26942659-26943065|+	406nt	RPF, cORF	hsa_circNUDC_001
hsa_circ_0006988	chr11:36227084-36227430|+	346nt	RPF, cORF	hsa_circLDLRAD3_001
hsa_circ_0007444	chr5:95755395-95763620|+	479nt	RPF, cORF, m6A	hsa_circRHOBTB3_001
hsa_circ_0051680	chr19:47362475-47362693|+	218nt	RPF, cORF, m6A	hsa_circDHX34_002
hsa_circ_0073517	chr5:108186116-108348530|-	371nt	RPF, cORF, MS	hsa_circFBXL17_001

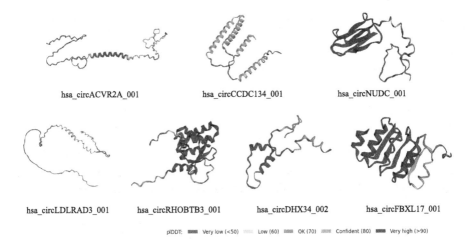

Fig. 6. AlphaFold2 generated potential longest circRNA peptide structure coming from 7 DECs.

After conducting the comparison, we did not find any effective structural domains in hsa_circLDLRAD3_001. However, we have performed a functional analysis of the other six circRNA translation proteins:

hsa_circACVR2A_001: This protein is likely a member of the SerThr protein kinase, TGFβ receptor protein family, with a TMhelix formed at 58-80aa. Gene Ontology (GO) analysis suggests its involvement in the transmembrane receptor protein serinethreonine kinase signaling pathway (GO:0007178).

hsa_circCCDC134_001: This protein is likely a member of the Coiled-coil domain-containing protein 134 (CCDC134) protein family, with a TMhelix formed at 6-28aa. CCDC134 is a secretory protein that regulates the mitogen-activated protein kinase (MAPK) pathway [27].

hsa_circNUDC_001: This protein is likely a member of the NudC family, with a CS domain formed at 1-88aa. Previous studies have shown that the CacyBPSIP protein, which contains a CS p23-like domain, is redundant in neurons and neuroblastoma NB2a cells.

hsa_circRHOBTB3_001: This protein sequence, 26-137aa forms a BTBPOZ domain. Many BTB proteins are transcriptional regulators that control chromatin structure.

hsa_circDHX34_002: This protein contains the P-loop_NTPase (P-loop containing nucleoside triphosphate hydrolase) domain superfamily, which is the most prevalent domain among several distinct nucleotide-binding protein folds. It exhibits a significant substrate preference for either ATP or GTP [28].

hsa_circFBXL17_001: The peptide chain contains the LRR_dom_sf (Leucine-rich repeat domain superfamily). LRRs are present in proteins ranging from viruses to eukaryotes and are involved in various biological processes, including signal transduction, cell adhesion, DNA repair, recombination, transcription,

RNA processing, disease resistance, apoptosis, and the immune response. Proteins containing LRRs include tyrosine kinase receptors, cell adhesion molecules, viral factors, and extracellular matrix-binding glycoproteins [29].

4 Discussion

Neuroblastoma is the most prevalent extracranial malignant solid tumor in children, with a long-term survival rate of less than 50% for high-risk patients. Previous studies have primarily focused on investigating the role of epigenetic factors, genes, and ncRNA molecules in the development and progression of NB. Recently, the involvement of circRNAs in NB has gained attention, as they have been found to contribute to tumor cell proliferation, invasion, and migration. A comprehensive study of circRNAs can significantly enhance our understanding of their functions in NB and pave the way for novel approaches in NB diagnosis and treatment.

In this study, we conducted an integrated analysis of circRNA sequencing data obtained from NB tissues and normal tissues. Our main objective was to identify differentially expressed genes (DEGs) between NB and normal tissues, predict their targeted miRNAs, RNA-binding proteins (RBPs), and translated proteins using bioinformatics methods, and gain further insights into the underlying mechanisms of NB pathogenesis.

In dataset PRJNA721263, we identified a total of 38,438 unique circRNAs. This dataset comprises 10 samples, each with a data size of approximately 10 GB. However, in dataset PRJNA554935, we detected 11,368 unique circRNAs, consisting of 6 samples, each with a data size of roughly 5 GB. The significant difference in the number of detected circRNAs in the two datasets may be attributed to two primary reasons. Firstly, the variance can be attributed to differences in sequencing depth. Larger data sizes are more likely to capture low-expression circRNAs. Secondly, circRNAs are known to exhibit tissue-specific and development stage-specific expression patterns. The dissimilarity in the samples collected between the two datasets could also have a significant impact on the number of detected circRNAs.

The circRNA-RBP regulatory network consisted of 32 DECs and 22 RBPs, Based on their Degree values (Fig. 5A), EIF4A3, AGO2, and HuR emerged as the top hub RBPs in the network. Previous studies have implicated these RBPs in tumor development and progression. In addition to hub RBPs, previous studies have highlighted the involvement of IGF2BP family members in NB development [30], all family members are reported to be expressed in up to 100% of neuroblastoma tumors, with high expression of IGF2BP1 associated with poor prognosis and MYCN amplification in neuroblastoma tumors [31].

Many solid tumors, including neuroblastoma (NB), evade immune control by generating a suppressive tumor microenvironment, dominated largely by the cytokine transforming growth factor beta [32]. Through Sequence alignment, it is found that hsa_circ ACVR2A_001 is likely a member of the SerThr protein kinase, TGFβ receptor protein family, with a TMhelix formed at 58-80aa. Therefore, we speculate that circRNA hsa_circ_0001073 (hsa_circACVR2A_001) is upregulated in NB tissues, and its protein product may promote the development of NB. Through our comprehensive analysis of DECs associated with NB, including their interactions with miRNAs, RBPs, and translation into proteins, we identified potential biomarkers for NB diagnosis and treatment. hsa_circ_0001079 may influence the expression of NB-related genes through its interaction with hsa-miR-661, while hsa_circ_0099504 and hsa_circ_0003171 may affect the expression of the SGK3 gene through their interaction with hsa-miR-515-5p. Additionally, hsa_circ_0001073 (hsa_circACVR2A_001) may contribute to NB development through protein translation. However, the specific pathways involving these DECs in NB and their potential as new therapeutic targets require further validation through biological experiments.

In summary, our study provides a more comprehensive investigation of circRNAs would enhance our understanding of their function in NB, and provide broader research avenues for NB diagnosis and treatment.

Acknowledgment. This work was partly supported by the Key Research and Development Project of Guangdong Province under grant No. 2021B0101310002, National Key Research and Development Program of China Grant No. 2021YFF1200100 Strategic Priority CAS Project XDB38050100, National Science Foundation of China under grant No. 62272449, the Shenzhen Basic Research Fund under grant, No. RCY X20200714114734194, KQTD20200820113106007 and ZDSYS20220422103800001. We would also like to thank the funding support by the Youth Innovation Promotion Association (Y2021101), CAS to Yanjie Wei.

References

1. Qiu, B., Matthay, K.K.: Advancing therapy for neuroblastoma. Nat. Rev. Clin. Oncol. **19**(8), 515–533 (2022)
2. Steliarova-Foucher, E., et al.: International incidence of childhood cancer, 2001–10: a population-based registry study. Lancet Oncol. **18**(6), 719–731 (2017)
3. Strother, D.R., et al.: Outcome after surgery alone or with restricted use of chemotherapy for patients with low-risk neuroblastoma: results of Children's Oncology Group study P9641. J. Clin. Oncol. **30**(15), 1842–1848 (2012)
4. Holmes, K., et al.: Influence of surgical excision on the survival of patients with stage 4 high-risk neuroblastoma: a report from the HR-NBL1/SIOPEN study. J. Clin. Oncol. **38**(25), 2902–2915 (2020)
5. Liu, C.X., Chen, L.L.: Circular RNAs: characterization, cellular roles, and applications. Cell **185**(12), 2016–2034 (2022)
6. Wang, C.C., Han, C.D., Zhao, Q., Chen, X.: Circular RNAs and complex diseases: from experimental results to computational models. Brief. Bioinform. **22**, bbab286 (2021)
7. Ju, J., Song, Y.N., Chen, X.Z., Wang, T., Liu, C.Y., Wang, K.: circRNA is a potential target for cardiovascular diseases treatment. Mol. Cell. Biochem. **477**, 417–430 (2022)
8. Floris, G., Zhang, L., Follesa, P., Sun, T.: Regulatory Role of Circular RNAs and Neurological Disorders. Mol. Neurobiol. **54**, 5156–5165 (2017)

9. Liu, X., Zhang, Y., Zhou, S., Dain, L., Mei, L., Zhu, G.: Circular RNA: An emerging frontier in RNA therapeutic targets, RNA therapeutics, and mRNA vaccines. J. Control. Release **348**, 84–94 (2022)

10. Wu, K., Tan, J., Yang, C.: Recent advances and application value of circRNA in neuroblastoma. Front. Oncol. **13**, 1180300 (2023)

11. Zhang, L., et al.: Comprehensive characterization of circular RNAs in neuroblastoma cell lines. Technol. Can. Res. Treatment **19**, 1533033820957622 (2020)

12. Lin, W., et al.: circRNA-TBC1D4, circRNA-NAALAD2 and circRNA-TGFBR3: selected key circRNAs in neuroblastoma and their associations with clinical features. Cancer Manage. Res. 4271–4281 (2021)

13. Chen, B., Ding, P., Hua, Z., Qin, X., Li, Z.: Analysis and identification of novel biomarkers involved in neuroblastoma via integrated bioinformatics. Invest. New Drugs **39**, 52–65 (2021)

14. Chen, B., Hua, Z., Qin, X., Li, Z.: Integrated Microarray to Identify the Hub miRNAs and Constructed miRNA-mRNA Network in Neuroblastoma Via Bioinformatics Analysis. Neurochem. Res. **46**, 197–212 (2021)

15. Shao, F.-L., Liu, Q., Wang, S.: Identify potential miRNA-mRNA regulatory networks contributing to high-risk neuroblastoma. Invest. New Drugs **39**(4), 901–913 (2021). https://doi.org/10.1007/s10637-021-01064-y

16. Rebolledo, C., Silva, J.P., Saavedra, N., Maracaja-Coutinho, V.: Computational approaches for circRNAs prediction and in silico characterization. Brief. Bioinform. **24**(3), bbad154 (2023)

17. Gao, Y., Zhang, J., Zhao, F.: Circular RNA identification based on multiple seed matching. Brief. Bioinform. **19**(5), 803–810 (2018)

18. Zhang, X.O., Wang, H.B., Zhang, Y., Lu, X., Chen, L.L., Yang, L.: Complementary sequence-mediated exon circularization. Cell **159**(1), 134–147 (2014)

19. Cheng, J., Metge, F., Dieterich, C.: Specific identification and quantification of circular RNAs from sequencing data. Bioinformatics **32**(7), 1094–1096 (2016)

20. Liu, M., Wang, Q., Shen, J., Yang, B.B., Ding, X.: Circbank: a comprehensive database for circRNA with standard nomenclature. RNA Biol. **16**(7), 899–905 (2019)

21. Huang, H.Y., et al.: miRTarBase update 2022: an informative resource for experimentally validated miRNA-target interactions. Nucleic Acids Res. **50**(D1), D222–D230 (2022)

22. Shannon, P., et al.: Cytoscape: a software environment for integrated models of biomolecular interaction networks. Genome Res. **13**(11), 2498–2504 (2003)

23. Dudekula, D.B., Panda, A.C., Grammatikakis, I., De, S., Abdelmohsen, K., Gorospe, M.: Circinteractome: a web tool for exploring circular RNAs and their interacting proteins and microRNAs. RNA Biol. **13**(1), 34–42 (2016)

24. Li, H., Xie, M., Wang, Y., Yang, L., Xie, Z., Wang, H.: riboCIRC: a comprehensive database of translatable circRNAs. Genome Biol. **22**, 1–11 (2021)

25. Jumper, J., et al.: Highly accurate protein structure prediction with AlphaFold. Nature **596**(7873), 583–589 (2021)

26. Quevillon, E., et al.: InterProScan: protein domains identifier. Nucleic Acids Res. **33**, W116–W120 (2005)

27. Huang, J., et al.: CCDC134, a novel secretory protein, inhibits activation of ERK and JNK, but not p38 MAPK. Cell. Mol. Life Sci. **65**, 338–349 (2008)

28. Leipe, D.D., Koonin, E.V., Aravind, L.: STAND, a class of P-loop NTPases including animal and plant regulators of programmed cell death: multiple, complex domain architectures, unusual phyletic patterns, and evolution by horizontal gene transfer. J. Mol. Biol. **343**(1), 1–28 (2004)

29. Rothberg, J.M., Jacobs, J.R., Goodman, C.S., Artavanis-Tsakonas, S.: slit: an extracellular protein necessary for development of midline glia and commissural axon pathways contains both EGF and LRR domains. Genes Dev. **4**(12A), 2169–2187 (1990)

30. Chen, S.T., et al.: Insulin-like growth factor II mRNA-binding protein 3 expression predicts unfavorable prognosis in patients with neuroblastoma. Cancer Sci. **102**(12), 2191–2198 (2011)

31. Bell, J.L., Turlapati, R., Liu, T., Schulte, J.H., Hüttelmaier, S.: IGF2BP1 harbors prognostic significance by gene gain and diverse expression in neuroblastoma. J. Clin. Oncol. **33**(11), 1285–1293 (2015)

32. Bottino, C., et al.: Natural killer cells and neuroblastoma: tumor recognition, escape mechanisms, and possible novel immunotherapeutic approaches. Front. Immunol. **5**, 56 (2014)

SGMDD: Subgraph Neural Network-Based Model for Analyzing Functional Connectivity Signatures of Major Depressive Disorder

Yan Zhang, Xin Liu, Panrui Tang, and Zuping Zhang[(✉)]

School of Computer Science and Engineering, Central South University, Changsha 410083, China
zpzhang@csu.edu.cn

Abstract. Biomarkers extracted from brain functional connectivity (FC) can assist in diagnosing various psychiatric disorders. Recently, several deep learning-based methods are proposed to facilitate the development of biomarkers for auxiliary diagnosis of depression and promote automated depression identification. Although they achieved promising results, there are still existing deficiencies. Current methods overlook the subgraph of braingraph and have a rudimentary network framework, resulting in poor accuracy. Conducting FC analysis with poor accuracy model can render the results unreliable. In light of the current deficiencies, this paper designed a subgraph neural network-based model named SGMDD for analyzing FC signatures of depression and depression identification. Our model surpassed many state-of-the-art depression diagnosis methods with an accuracy of 73.95%. To the best of our knowledge, this study is the first attempt to apply subgraph neural network to the field of FC analysis in depression and depression identification, we visualize and analyze the FC networks of depression on the node, edge, motif, and functional brain region levels and discovered several novel FC feature on multi-level. The most prominent one shows that the hyperconnectivity of postcentral gyrus and thalamus could be the most crucial neurophysiological feature associated with depression, which may guide the development of biomarkers used for the clinical diagnosis of depression.

Keywords: Major depressive disorder · Subgraph neural network · Analysis of brain functional connectivity · resting-state fMRI

1 Introduction

Major depressive disorder (MDD) is a severe mental disorder that leads to a substantial economic burden on public systems [1]. Thus, it is essential to develop precise and reliable biomarkers for clinical diagnosis. Although there exist several hypotheses that explain the pathogenesis of depression [2], there is still no biomarker that is widely accepted and employed for the clinical diagnosis of MDD, leading to overdiagnosis or misdiagnosis [3].

© The Author(s), under exclusive license to Springer Nature Singapore Pte Ltd. 2023
X. Guo et al. (Eds.): ISBRA 2023, LNBI 14248, pp. 364–375, 2023.
https://doi.org/10.1007/978-981-99-7074-2_28

The analysis of functional connectivity (FC) networks in the brain derived from fMRI data has emerged as an effective way to develop highly accurate biomarkers for neurological or psychiatric disorders [4]. Recently, several deep learning-based methods have been proposed to extract universal features from depression FC network and automatically identify depression. In 2022, Noman et al. [5] utilize a GCN-based graph autoencoder to diagnose depression and analyze the FC network of depression with an accuracy of 72.5%, although the accuracy is relatively high, the dataset they used only contains a few dozen MDD patients, resulting in the analysis of FC unreliable. In 2023, Gallo et al. [4] and Yuqi Fang et al. [6] utilized the rest-metamdd dataset for model training and brain FC analysis. The size of dataset is improved, but their deep learning models encode the entire braingraph directly, have overlooked the subgraph structure of braingraph. Additionally, the network framework they used is rudimentary, resulting in insufficient feature extraction capabilities. The accuracy of their models for depression detection is only 61.47% and 59.73%, resulting in a lack of precision when analyzing brain FC.

Although there has been significant progress in deep learning-based analysis of FC signatures on depression and depression identification over the past few years, there are three remaining issues that need to be addressed:

1) Existing deep learning-based research has disregarded many functionally and structurally oriented FC features. The brain regions can be classified according to their anatomical location and function. Each of these sub-structures is responsible for distinct functions. Current research [4, 5, 7] encoding the entire braingraph directly, has overlooked the subgraph structure of braingraph.
2) Existing deep learning-based research lacks precision when analyzing FC. The deep learning models they used have poor accuracy due to the rudimentary network framework [3–5, 7].
3) Existing deep learning-based research [4–6] related to analyzing the functional connectivity of depression is mainly focusing on the edge level, which means they all focus on finding the most discriminative edge, the analysis on the motif, node, and functional brain region levels is insufficient.

In light of the three deficiencies mentioned above, this paper proposes a subgraph neural network-based framework for analyzing FC of depression and automated depression identification. The primary contributions of our research are briefly outlined as follows:

1) Our network framework preserves more functionally and structurally oriented FC features by using S-BFS to generate sub-braingraphs in nine different functional brain regions and applying subgraph neural network to extract FC features from discriminative sub-braingraphs.
2) Our network framework extracts FC features more precisely. The accuracy of our model surpasses many state-of-the-art depression diagnosis models with an accuracy of 73.95%, which offers the potential for automated diagnosis of depression in clinical settings.

3) We visualize and analyze the FC network on the motif, edge, node, and functional brain region levels. Our findings have discovered the hyperconnectivity of postcentral gyrus and thalamus could be the most crucial neurophysiological feature associated with MDD, which may guide the development of biomarkers used for the clinical diagnosis of MDD. Also, we have discovered some new FC signatures on multi-level, such as the hyperconnectivity of sensorimotor network (SMN) and default mode network (DMN) and the hyperconnectivity between SMN and DMN, which can promote the analysis of fMRI for MDD patients in the clinical scenario.

2   Related Work

2.1   Construction of Braingraph

The process of constructing a braingraph is illustrated in Fig. 1. In our study, the Ledoit-Wolf estimator is used to evaluate the functional connectivities between ROIs. The connectivity pattern of braingraph is derived through proportional thresholding. The Automated Anatomical Labeling (AAL) atlas with N = 116 regions-of-interest (ROIs) is used in our method. The partition method of functional brain region is based on standard (7+1) system template defined in literature [3]. In addition, Cerebellum and Vermis were categorized as the ninth functional brain regions named the CV, these 116 ROIs are divided into nine functional brain regions as shown in Fig. 1 (Visualization of braingraph); the nodes with different colors correspond to nine different functional brain regions.

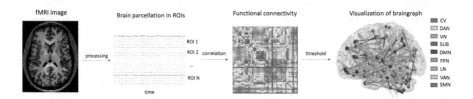

Fig. 1. An illustration of constructing a braingraph.

3   Method

As shown in Fig. 2. The overall framework of SGMDD involves three fundamental steps: (1) Sub-braingraph Sampling And Encoding; (2) Sub-braingraph Selection And Sub-braingraph's Node Selection; (3) Sub-braingraph Sketching And Classification. In Sects. 3.1, 3.2, and 3.3, we provide a detailed introduction to each of these three steps.

The braingraph in our study for every subject is denoted as $G = (V, X, A)$. $V = \{v_1, v_2, \cdots, v_N\}$ represents a set of N nodes (ROIs), The adjacency matrix is denoted as $\mathbf{A} = [a_{ij}] \in \{0, 1\}^{N \times N}$, the node feature for G is denoted as $\mathbf{X} = [\mathbf{x}_1, \ldots, \mathbf{x}_N]^T \in \mathbb{R}^{N \times d}$.

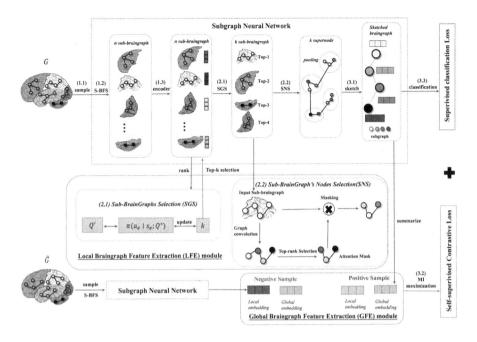

Fig. 2. An illustration of the SGMDD architecture.

3.1 Sub-braingraph Sampling and Encoding

The process of subbraingraph sampling and encoding can be divided into three steps as shown in Fig. 2 (1.1, 1.2, 1.3):

First (1.1), nodes in nine functional brain regions are sorted in descending order based on their degree. The top N nodes are selected in nine functional brain regions to serve as centers of each sub-braingraph.

Second (1.2), a sub-braingraph is generated for each central node using the Selective-Breadth First Search (S-BFS) algorithm. The S-BFS algorithm prioritizes the expansion of central node in the same functional region defined in Fig. 1, and each subgraph is limited to a maximum of s nodes. Ultimately, the set of sub-braingraphs is obtained as $\{g_1, g_2, \cdots, g_n\}$.

Third (1.3), a GNN-based encoder, denoted by $\varepsilon : \mathbb{R}^{s \times d} \times \mathbb{R}^{s \times s} \to \mathbb{R}^{s \times d_1}$, is learned to get node representation of sub-braingraphs, where d_1 represents the dimension of node representation. The generalized equation for obtaining the node representations $\mathbf{H}(g_i) \in \mathbb{R}^{s \times d_1}$ for nodes within sub-braingraph g_i is:

$$H(g_i) = \mathcal{E}(g_i) = \{h_j \mid v_j \in V(g_i)\} \tag{1}$$

By incorporating $\mathcal{E}$ into a message-passing framework, the unified formulation is defined as follows:

$$\mathbf{h}_i^{(l+1)} = U^{(l+1)}\left(\mathbf{h}_i^{(l)}, \text{AGG}\left(M^{(l+1)}\left(\mathbf{h}_i^{(l)}, \mathbf{h}_j^{(l)}\right) \mid v_j \in N(v_i)\right)\right) \tag{2}$$

where $U(\cdot)$ denotes the state updating function, $AGG(\cdot)$ denotes the aggregation function and $M(\cdot)$ denotes the message generation function. An intra-subgraph attention mechanism is utilized for encoding node representations, this is given by the following equation:

$$c_j^{(i)} = \sigma \left(a_{\text{intra}}^{\text{T}} \mathbf{W}_{\text{intra}} \mathbf{h}_j^{(i)} \right) \tag{3}$$

where $\mathbf{W}_{\text{intra}} \in \mathbb{R}^{d_1 \times d_1}$ and $a_{\text{intra}} \in \mathbb{R}^{d_1}$ represent weight matrix and weight vector. the representations $\mathbf{z}_i$ of g_i is given by:

$$z_i = \sum_{v_j \in V(g_i)} c_j^{(i)} \mathbf{h}_j^{(i)}. \tag{4}$$

3.2 Sub-braingraph Selection and Sub-braingraph's Node Selection

SGMDD utilizes a Local braingraph Feature Extraction (LFE) module to select important Sub-braingraphs and Sub-braingraph' Nodes. The LFE module can be divided into two sections (2.1, 2.2) as shown in Fig. 2.

Sub-BrainGraph Selection (SGS) section (2.1): In order to select significant sub-braingraphs, we utilize the technique of top-k sampling involves adjusting the adaptive pooling ratio k; we employ a trainable vector p to project all sub-braingraphs features to 1D footprints $\{val_i \mid g_i \in G\}$, we select the top $n' = \lceil k \cdot n \rceil$ sub-braingraphs, val_i of sub-braingraph g_i with respect to p is determined as follows:

$$val_i = \frac{\mathbf{z}_i \mathbf{p}}{\|\mathbf{p}\|}, idx = \text{rank}\left(\{val_i\}, n'\right) \tag{5}$$

The function $rank\left(\{val_i\}, n'\right)$ is applied to rank the importance of subgraphs and returns the indices of the n'-largest values in $\{val_i\}$. In order to identify the most significant sub-braingraphs, a reinforcement learning (RL) algorithm is used to dynamically update the pooling ratio $k \in (0, 1]$. We model updating k as a finite horizon Markov decision process, the specific details of the RL algorithm can refer to the literature [8].

Once the significant sub-braingraphs are determined, we use self-attention pooling [9] to perform Sub-braingraph's Node Selection section (SNS) (2.2): SNS uses GNN layer to obtain self-attention scores. Specifically, the self-attention score $Z \in \mathbb{R}^{N \times 1}$ is computed according to the following equation:

$$Z = \sigma \left(\dot{D}^{-\frac{1}{2}} \dot{A} \dot{D}^{-\frac{1}{2}} X \Theta_{att} \right) \tag{6}$$

In our model, σ denotes the activation function. The adjacency matrix with self-connections is denoted by $\dot{A}$. Furthermore, we use $\dot{D} \in \mathbb{R}^{N \times N}$ as the degree matrix of $\dot{A}$. Input features are denoted by $X \in \mathbb{R}^{N \times F}$, where N represent the number of nodes and F represents the dimensional of feature.

To retain a proportion of nodes from the input graph, we use Gao et al. [10]'s node selection method. The top $\lceil kN \rceil$ nodes are chosen based on the corresponding value of Z.

$$idx = \text{top} - \text{rank}(Z, \lceil kN \rceil), Z_{\text{mask}} = Z_{\text{idx}} \tag{7}$$

The function top-rank yields the indices of the top $\lceil kN \rceil$ values. Z_{mask} indicates the feature attention mask, and idx denotes an indexing operation. In Fig. 2 (Sub-BrainGraph's Node Selection part), the 'masking' operation is employed to process the input sub-braingraph.

$$X' = X_{\text{idx},:}, X_{\text{out}} = X' \odot Z_{\text{mask}}, A_{\text{out}} = A_{\text{idx,idx}} \tag{8}$$

where $X_{\text{idx},:}$ represents the feature matrix with row-wise indexing (i.e., node-wise), $\odot$ is the elementwise product that has been broadcasted, and $A_{\text{idx,idx}}$ is the adjacency matrix with both row-wise and column-wise indexing. X_{out} and A_{out} correspond to the resulting feature matrix and adjacency matrix.

3.3 Sub-braingraph Sketching and Classification

The sub-braingraph sketching and classification can be divided into three steps:

First (3.1), as shown in Fig. 2 (3.1), our method considers selected sub-braingraphs as super nodes, by doing so, the original braingraph is reduced to a sketched graph. Edge (g_i, g_j) donates the number of edges between different sub-braingraphs. The sketched graph is donated as $G^{\text{sk}} = \left(V^{\text{sk}}, E^{\text{sk}}\right)$. We set a predefined threshold of b_{th}. If the number of edges between subgraphs g_i and g_j surpasses this threshold, an edge $e(i,j)$ will be added to the sketched graph

$$V^{\text{sk}} = \{g_i\}, \forall i \in idx;$$
$$E^{\text{sk}} = \{e_{i,j}\}, \forall \text{Edge} e(g_i, g_j) > b_{\text{th}} \tag{9}$$

We adopt an inter-subgraph attention mechanism defined in [11] to model the mutual influence among sub-braingraphs. With these attention coefficients, we obtain the subgraph embeddings as follows:

$$z_i' = \frac{1}{M} \sum_{m=1}^{M} \sum_{e_{ij} \in E^{\text{ske}}} \alpha_{ij}^m \mathbf{W}_{\text{inter}}^m \mathbf{z}_i \tag{10}$$

where M is the number of independent attention. α_{ij} is the attention coefficient, and $\mathbf{W}_{\text{inter}} \in \mathbb{R}^{d_2 \times d_1}$ is weight matrix.

Second (3.2), we improve the quality of subgraph embeddings by maximizing mutual information (MI) between local and global braingraph representation. A READOUT function summarizes the obtained global graph representation $\mathbf{r}$:

$$\mathbf{r} = \text{READOUT}\left(\{\mathbf{z}_i'\}_{i=1}^{n'}\right) \tag{11}$$

In this case, we utilize averaging strategy as the READOUT function. To maximize the estimated MI, we use the Jensen-Shannon (JS) MI estimator [12] on the local/global pairs. A discriminator function $\mathcal{D} : \mathbb{R}^{d_2} \times \mathbb{R}^{d_2} \to \mathbb{R}$ is introduced, as $\mathbf{W}_{MI}$ is a scoring matrix and $\sigma(\cdot)$ donates the sigmoid function. A bilinear score function is applied as the discriminator:

$$\mathcal{D}\left(\mathbf{z}_i', \mathbf{r}\right) = \sigma\left(\mathbf{z}_i'^T \mathbf{W}_{MI} \mathbf{r}\right) \tag{12}$$

Our self-supervised mutual information (MI) approach is based on contrastive learning, we use the same way in literature [8] to generate negative samples. To optimize the self-supervised MI objective and maximize the mutual information between z_i' and $\mathbf{r}$, we use a standard binary cross-entropy (BCE) loss $\mathcal{L}_{MI}^G$, where n_{neg} denotes the number of negative samples.

$$\mathcal{L}_{MI}^G = \frac{1}{n' + n_{neg}} \left(\sum_{g_i \in G}^{n'} \mathbb{E}_{pos} \left[\log \left(\mathcal{D} \left(\mathbf{z}_i', \mathbf{r} \right) \right) \right] + \sum_{g_j \in \hat{G}}^{n_{neg}} \mathbb{E}_{neg} \left[\log \left(1 - \mathcal{D} \left(\dot{\mathbf{z}}_j', \mathbf{r} \right) \right) \right] \right) \tag{13}$$

Third (3.3), to predict the labels of the sub-braingraphs, we apply a softmax function to the sub-braingraph embeddings. The graph classification results are determined by voting among the sub-braingraphs. We merge the supervised classification loss $\mathcal{L}_{Classify}$ and the self-supervised MI loss $\mathcal{L}_{MI}^G$ in Eq. 13 to serve as a form of regularization. The graph classification loss function $\mathcal{L}_{Classify}$ is established on cross-entropy. The loss $\mathcal{L}$ for SGMDD is defined as:

$$\mathcal{L} = \mathcal{L}_{Classify} + \beta \sum_{G \in \mathcal{G}} \mathcal{L}_{MI}^G + \lambda \|\Theta\|^2 \tag{14}$$

SGMDD preserves more functionally and structurally oriented FC features by using S-BFS to generate sub-braingraphs in nine different functional brain regions. While LFE module and GFE module is used to improve the feature extraction capabilities in both local and global levels. By combining S-BFS, LFE module, and GFE module, the SGMDD can extracts FC features more precisely.

4   Experiments

4.1   Data Acquisition and Parameter Settings

The rest-metamdd dataset is the most comprehensive resting-state fMRI database of individuals with depression available at present [13]. It was utilized to verify the efficacy of our proposed SGMDD and analyze FC. The detail of fMRI data preprocessing can refer to literature [5]. 5-fold cross-validation was used to evaluate these competing models and the proposed SGMDD. Common parameters for SGMDD training were set as follows: $Momentum = 0.8$, $Dropout = 0.4$, and $L2$ Norm Regularization weight $decay = 0.01$. The parameter settings of competing models can find in literature [3,5,6,14].

4.2   Overall Evaluation

As shown in Table 1, the performance of the SGMDD is evaluated by benchmarking them against BrainNetCNN [15], Wck-CNN [16], XGBoost [17] and five GCN-based methods [5], which are cutting-edge connectome-based models. The proposed SGMDD surpasses many state of art methods by a large margin show that our model can extract the features from FC network more precisely and offer the potential for automated diagnosis of depression in clinical settings. In 2023, Gallo et al. [4] and Fang et al. [6] used the same rest-metamdd dataset for model training and brain FC analysis, achieving an accuracy of 61.47% and 59.73%. This result indicates that our analysis of FC features for depression in Sects. 4.3, and 4.4 is more accurate than the latest research.

Table 1. Comparing performance (average accuracy±standard deviation) with other state-of-the-art methods.

Feature Extraction Model	Acc	Sen	Spe	Pre	F1
BrainNetCNN (NeuroImage, 2017)	56.47 ± 6.01	51.43 ± 2.31	49.24 ± 1.56	53.24 ± 6.13	55.28 ± 7.24
Wck-CNN (Medical image analysis, 2020)	58.27 ± 5.24	51.43 ± 4.37	48.65 ± 4.46	47.08 ± 6.38	52.46 ± 1.37
XGBoost (ACM SIGKDD, 2016)	60.13 ± 5.27	57.26 ± 6.43	60.74 ± 8.15	48.36 ± 5.43	54.23 ± 3.18
Hi-GCN (CIBM, 2020)	55.16 ± 4.17	52.15 ± 4.15	47.21 ± 2.18	44.24 ± 2.39	48.43 ± 3.37
E-Hi-GCN (Neuroinformatics, 2021)	58.45 ± 1.89	59.37 ± 5.37	57.37 ± 5.46	55.25 ± 4.39	57.43 ± 5.16
Population-based GCN (BSPC, 2023)	61.46 ± 4.76	58.37 ± 4.76	57.38 ± 5.46	60.46 ± 5.27	61.37 ± 2.27
GAE-FCNN (ArXiv:2107.12838, 2022)	59.55 ± 7.45	55.86 ± 4.37	56.47 ± 8.65	57.17 ± 5.95	54.16 ± 4.23
GroupINN (ACM SIGKDD, 2019)	58.17 ± 4.27	53.46 ± 2.24	57.15 ± 5.34	56.27 ± 5.39	56.43 ± 4.79
SGMDD	**73.95 ± 4.63**	**75.37 ± 3.59**	**72.38 ± 5.56**	**69.58 ± 7.42**	**71.27 ± 3.21**

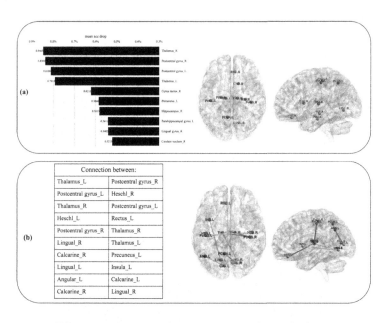

Fig. 3. Top-10 most discriminative nodes and edges.

4.3 Discriminative Nodes, Edges, and Motifs Analysis

Discriminative Nodes: In order to determine which nodes (ROIs) have higher discriminative ability in diagnosing depression, we conducted ablation experiments on different nodes (ROIs) in Fig. 3 (a, left) and visualize it in Fig. 3 (a, right): We found that the postcentral gyrus and thalamus had the most significant impact on accuracy compared to other nodes, indicating that the postcentral gyrus and thalamus are the most discriminative nodes (ROIs).

Discriminative Edges: We used the same method defined in literature [4,6] to determine which edge has higher discriminative ability. We summarize the top 10 discriminative edges in Fig. 3 (b, left) and visualize it in Fig. 3 (b, right): we find that there are many top 10 discriminative edges are connected to the postcentral gyrus and thalamus.

In 2023, Gallo et al. [4] and Fang et al. [6] find that a lot of discriminative edges are symmetric between left hemisphere and right hemisphere. However, this conclusion is not reflected in our study; this could be due to the differences in the deep learning frameworks, their framework encode the entire braingraphs directly, while the human brain is not entirely symmetrical both in structure and function; by selecting asymmetric discriminative sub-braingraphs from nine different functional brain regions, the Sub-GNN network framework provides a unique perspective when analyzing the FC of depression. In the meantime, the high accuracy of our model may be attributed to the asymmetric discriminative sub-braingraphs selected by the Sub-GNN component.

Discriminative Motif: In order to identify motifs with higher discrimination ability in the FC network of depression, we sorted the accuracy of sub-

Table 2. Top-10 most discriminative FC motifs in MDD patient.

Top-10 most discriminative motifs
IFGoperc.L-IFGtriang.R-SFGdor.L-CUN.L-ANG.R-THA.R-CRBL9.L-ROL.R
SFGdor.L-SMA.L-REC.L-ORBsupmed.L-PCUN.L-PCG.L-CRBL6.L-CRBL7b.L-R-ORBsupmed
MTG.L-ITG.L-ROL.R-THA.R-ORBinf.L-CUN.L-PCG.R-CAU.L
SFGdor.R-ORBmid.L-CUN.R-PCG.L-PCG.R-CAU.R-ANG.R-PCG.R-CUN.L
ORBinf.R-REC.L-ACG.R-CRBL3.L-MTG.L-MTG.R-CRBL3.R-REC.R
IPL.L-FFG.R-SFGdor.R-ORBinf.R-ITG.L-CRBL7b.L-CRBL45.L
PUT.L-ORBinf.R-SPG.R-CUN.R-ORBsupmed.R-PCG.R-THA.L-CRBL9.R
ORBinf.R-MFG.L-ACG.L-CUN.L-PCUN.R-MTG.R-SFGmed.R-CAU.L-ROL.R
ORBinf.R-FFG.R-ROL.R-PreCG.L-SFGdor.L-MFG.L-ORBinf.L-ANG.L
PCG.R-THA.R-ROL.R-CUN.R-REC.R-CUN.R-HIP.L

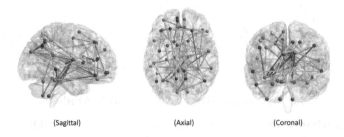

(Sagittal)   (Axial)   (Coronal)

Fig. 4. Visualization of top-10 most discriminative FC motifs in different planes.

braingraphs to filter out motifs with high repetition rates from sub-braingraphs with high accuracy. The result and topological representations of it are shown in Table 2 and Fig. 4. By analyzing these motifs, we found that discriminative motifs are mainly located between the DMN and frontoparietal network (FPN). To the best of our knowledge, this is a newly discovered FC feature.

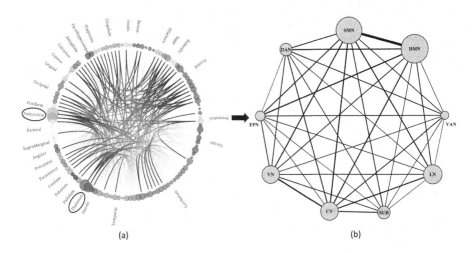

(a)   (b)

Fig. 5. Visualization of top-200 discriminative FCs.

4.4   Overall Functional Connectivity Network Analysis

We visualize the top-200 discriminative edges in Fig. 5 (a). The visualization of the discriminative edges between nine functional brain regions is shown in Fig. 5 (b), the size of node is related to how many discriminative edges connect to it.

Edge level analysis: we find many discriminative FCs located between the thalamus and several ROIs associated with emotional processing and regulation, such as temporal regions and cerebellum. Previous research [4], has indicated the hyperactivity of thalamus both during rest and cognitive and emotional processing in patients with depression. The connectivity profile observed in the previous study is consistent with the connectivity pattern found in our study.

Node (ROIs) level analysis: We analyze the discriminative FCs at the node level, as shown in Fig. 5 (a). We find that the number of discriminative FCs connected to the postcentral gyrus and thalamus is higher compared to other ROIs, indicating the significant importance of the postcentral gyrus and thalamus in diagnosing depression.

Functional brain region level analysis: We analyze the overall discriminative FCs between nine functional brain regions. As shown in Fig. 5 (b). The width of line is related to how many discriminative FCs exist between two functional brain regions. We find that the number of discriminative FCs between the SMN and DMN is significantly higher than other functional brain regions. Also, we found that there are more discriminative FCs connected to SMN and DMN compared to other functional brain regions, which indicate the hyperconnectivity of SMN and DMN as well as hyperconnectivity between SMN and DMN could be crucial neurophysiological features associated with MDD.

5 Conclusion

This paper design a subgraph neural network-based model named SGMDD for analyzing FC signatures in depression and depression identification. The performance of SGMDD is superior than many latest depression diagnostic models, which offers potential for automated diagnosis of depression in clinical settings.

Compared to other studies [4–6] related to analyzing the FC of depression, this study has advantages mainly in the following three aspects:

1) our network framework preserves more functionally and structurally oriented FC features by using S-BFS to generate sub-braingraphs in nine different functional brain region and applying subgraph neural network to extract FC features from discriminative sub-braingraphs;
2) Our network framework extracts FC features more precisely, the accuracy of our model surpasses many state-of-the-art depression diagnosis models by a large margin.
3) We visualize and analyze the FC on the motif, edge, node, and functional brain region levels and discovered several new FC features on mutil-level. Additionally, the asymmetric discriminative sub-braingraphs selected by the Sub-GNN component offer a unique perspective for analyzing the FC of depression.

Multi-angle analysis of FC network has revealed the importance of postcentral gyrus and thalamus in the diagnosis of depression, it suggests that the hyperconnectivity of postcentral gyrus and thalamus could be the most crucial neurophysiological feature associated with MDD, which may guide the development of biomarkers used for the clinical diagnosis of MDD. Also, our findings have discovered some new brain FC signatures related to depression on multi-level, such as the hyperconnectivity of SMN and DMN and the hyperconnectivity between SMN and DMN, which can promote the analysis of fMRI for MDD patients in the clinical scenario.

References

1. Abdoli, N., et al.: The global prevalence of major depressive disorder (MDD) among the elderly: a systematic review and meta-analysis. Neurosci. Biobehav. Rev. **132**, 1067–1073 (2022)
2. Kishi, T., et al.: Antidepressants for the treatment of adults with major depressive disorder in the maintenance phase: a systematic review and network meta-analysis. Mol. Psychiatry **28**(1), 402–409 (2023)

3. Shi, Y., et al.: Multivariate machine learning analyses in identification of major depressive disorder using resting-state functional connectivity: a multicentral study. ACS Chem. Neurosci. **12**(15), 2878–2886 (2021)
4. Gallo, S., et al.: Functional connectivity signatures of major depressive disorder: machine learning analysis of two multicenter neuroimaging studies. Mol. Psych. 1–10 (2023)
5. Noman, F., et al.: Graph autoencoders for embedding learning in brain networks and major depressive disorder identification. arXiv preprint arXiv:2107.12838 (2021)
6. Fang, Y., Wang, M., Potter, G.G., Liu, M.: Unsupervised cross-domain functional MRI adaptation for automated major depressive disorder identification. Med. Image Anal. **84**, 102707 (2023)
7. Venkatapathy, S., et al.: Ensemble graph neural network model for classification of major depressive disorder using whole-brain functional connectivity. Front. Psych. **14**, 1125339 (2023)
8. Sun, Q., et al.: Sugar: Subgraph neural network with reinforcement pooling and self-supervised mutual information mechanism. In: Proceedings of the Web Conference 2021, pp. 2081–2091 (2021)
9. Lee, J., Lee, I., Kang, J.: Self-attention graph pooling. In: International Conference on Machine Learning, pp. 3734–3743. PMLR (2019)
10. Cangea, C., Veličković, P., Jovanović, N., Kipf, T., Liò, P.: Towards sparse hierarchical graph classifiers. arXiv preprint arXiv:1811.01287 (2018)
11. Veličković, P., Cucurull, G., Casanova, A., Romero, A., Lio, P., Bengio, Y.: Graph attention networks. arXiv preprint arXiv:1710.10903 (2017)
12. Nowozin, S., Cseke, B., Tomioka, R.: f-GAN: training generative neural samplers using variational divergence minimization. In: Advances in Neural Information Processing Systems, vol. 29 (2016)
13. Yan, C.G., et al.: Reduced default mode network functional connectivity in patients with recurrent major depressive disorder. Proc. Natl. Acad. Sci. **116**(18), 9078–9083 (2019)
14. Zhang, Y., Qing, L., He, X., Zhang, L., Liu, Y., Teng, Q.: Population-based GCN method for diagnosis of Alzheimer's disease using brain metabolic or volumetric features. Biomed. Signal Process. Control **86**, 105162 (2023). https://doi.org/10.1016/j.bspc.2023.105162, https://www.sciencedirect.com/science/article/pii/S1746809423005955
15. Kawahara, J., et al.: BrainnetCNN: convolutional neural networks for brain networks; towards predicting neurodevelopment. Neuroimage **146**, 1038–1049 (2017)
16. Jie, B., Liu, M., Lian, C., Shi, F., Shen, D.: Designing weighted correlation kernels in convolutional neural networks for functional connectivity based brain disease diagnosis. Med. Image Anal. **63**, 101709 (2020)
17. Chen, T., Guestrin, C.: Xgboost: a scalable tree boosting system. In: Proceedings of the 22nd ACM SIGKDD International Conference on Knowledge Discovery and Data Mining, pp. 785–794 (2016)

PDB2Vec: Using 3D Structural Information for Improved Protein Analysis

Sarwan Ali[✉], Prakash Chourasia, and Murray Patterson

Georgia State University, Atlanta, GA, USA
{sali85,pchourasia1,mpatterson30}@gsu.edu

Abstract. In recent years, machine learning methods have shown remarkable results in various protein analysis tasks, including protein classification, folding prediction, and protein-to-protein interaction prediction. However, most studies focus only on the 3D structures or sequences for the downstream classification task. Hence analyzing the combination of both 3D structures and sequences remains comparatively unexplored. This study investigates how incorporating protein sequence and 3D structure information influences protein classification performance. We use two well-known datasets, STCRDAB and PDB Bind, for classification tasks to accomplish this. To this end, we propose an embedding method called PDB2Vec to encode both the 3D structure and protein sequence data to improve the predictive performance of the downstream classification task. We performed protein classification using three different experimental settings: only 3D structural embedding (called PDB2Vec), sequence embeddings using alignment-free methods from the biology domain including on k-mers, position weight matrix, minimizers and spaced k-mers, and the combination of both structural and sequence-based embeddings. Our experiments demonstrate the importance of incorporating both three-dimensional structural information and amino acid sequence information for improving the performance of protein classification and show that the combination of structural and sequence information leads to the best performance. We show that both types of information are complementary and essential for classification tasks.

Keywords: Protein structures · Sequences · Structural Information · Embeddings · PDB2Vec · Classification · Representation Learning

1 Introduction

Proteins are crucial in various biological processes such as metabolic reactions, cell signaling, and structural support [8,9,18]. Their three-dimensional (3D) structures are essential in understanding their functions and interactions with

---

A. Ali and P. Chourasia—Equal Contribution.

X. Guo et al. (Eds.): ISBRA 2023, LNBI 14248, pp. 376–386, 2023.
https://doi.org/10.1007/978-981-99-7074-2_29

other proteins or small molecules [20, 22, 30]. The availability of structural data for proteins through the Protein Data Bank (PDB) has opened up new avenues for researchers to delve deeper into the analysis of proteins. By analyzing the structural information of proteins, researchers can develop new approaches for drug discovery [6], design new enzymes [26], and understand protein-protein interactions better [32]. Therefore, investigating the relationship between protein sequence and structure can significantly improve our understanding and their roles in biological processes. One result of the ongoing coronavirus pandemic is the creation of an abundance of protein data, thanks to the rapid testing process facilitated the availability of the data through popular databases like GISAID[1]. This has, in turn, resulted in the recent appearance of many methods for protein sequence classification, both alignment-based [27] and alignment-free [5, 38].

In this study, we aim to perform sequence-based classification by investigating the use of the protein 3D structure information for classifying proteins. To this end, we use the PDB structure of proteins and extract their sequences, which we then use to generate numerical embeddings through different methods. In addition, we also generate embeddings directly from the 3D coordinates of the proteins using an autoencoder. Finally, we combine the embeddings from both the sequence and 3D structural representations to perform protein classification. We evaluate the performance of our methods on two benchmark datasets: STCRDAB and PDB bind. Our results show that the classification performance is the lowest when only 3D structural information is used. However, the performance improves significantly when only the protein sequences are used. We then observe that combining the embeddings from both the 3D structures and sequences leads to further improvement in the classification performance. These findings highlight the importance of incorporating 3D structural information and sequences in protein classification.

Our contributions to this paper are the following:

- We investigate the performance of protein classification using three different settings: (i) directly from 3D structural coordinates, (ii) from protein sequences, and (iii) from a combination of the two.
- We evaluate the results on two benchmark datasets: STCRDAB and PDB bind, and show that our proposed PDB2Vec outperforms the recently proposed sequence-only-based baselines in terms of predictive performance.
- We demonstrate that combining structural and sequence information can improve performance compared to using only one source of information.
- We provide insight into the value of incorporating structural and sequence information for protein classification tasks.

The rest of the paper is organized as follows. In Sect. 2, we provide related work. Section 3 describes the proposed methodology used in this study. Section 4 describes the datasets used and classification models. In Sect. 5, we present and discuss the results of our experiments. In Sect. 6, we summarize our findings and discuss future directions for research in this area.

---

[1] https://gisaid.org/.

2   Literature Review

In the field of computational biology, much research has been done in the areas of protein structure prediction [1], function prediction [31], and protein-protein interaction prediction [32]. One of the most commonly used methods for protein structure prediction is the homology-based method [17,40]. Another commonly used method is the ab initio method [23,36], in which the protein structure is predicted without prior knowledge of similar proteins. In recent years, there has been growing interest in using deep learning methods for protein analysis [13,19,39]. Deep learning has been used in various applications such as protein folding prediction [42], protein-protein interaction prediction [43], and protein classification [37]. One of the key challenges in using deep learning methods for protein analysis is the representation of protein information [25,34,39]. For the analysis of protein sequences, different domains offer a range of methods for the classification task. A set of pre-trained models to perform protein classification like Protein Bert [10], Seqvec [24], and UDSMProt [37] are proposed in the literature. Another domain for sequence analysis involves using kernel function [4,16]. Another domain of research for protein classification/clustering involves designing low dimensional embeddings directly from the sequences using the ideas of k-mers [3,5,38], minimizers [11,12,33], position weight matrix [2] and gapped k-mers [35]. Although these methods showed higher predictive performance, they do not incorporate structural information into the embeddings.

3   Methodology

This section describes the approach we used to design the numerical representation from 3D coordinates of protein sequences. Along with the process of extracting amino acids to generate the protein sequences. Finally, we describe how we combined the 3D coordinates-based representation with sequence-based representation to get the embedding.

3.1   PDB2Vec

Given a PDB file as input containing 3D (x, y, and z-axis) coordinates for amino acids within a protein sequence, our first goal is to design a fixed-length numerical representation in the output. For this purpose, we use an autoencoder-based architecture. For reference, we call this method as PDB2Vec. The process involves the following steps:

Flatten the 3D Coordinates. In the first step, we concatenate all the 3D coordinates within a single PDB file of a protein into a vector since the autoencoder takes a vector-based representation as input. This process is repeated for all PDB files to generate a set of vectors containing all 3D coordinates. We use data padding to get fixed-length vectors for all PDB files. The resultant set of vectors is represented by X.

Autoencoder Architecture. Autoencoder is designed to learn a compressed input data representation while preserving the most critical information. In the context of protein structure analysis, the main goal is to extract a low-dimensional representation of the 3D coordinates, which can capture the essential structural features of the protein. Autoencoder can be a good fit for this task because it can learn non-linear mapping from the high-dimensional input space to the low-dimensional feature space. Our autoencoder architecture takes X as input and gives a low dimensional representation X' as output. It is composed of an encoder, a bottleneck, and a decoder. The encoder consists of 4 fully connected dense layers with increasing units, followed by Batch Normalization and Leaky ReLU activation functions. The bottleneck layer is a fully connected dense layer with ReLU activation. The decoder consists of 4 fully connected dense layers with decreasing units, each followed by Batch Normalization and Leaky ReLU activation functions. The output layer is a fully connected dense layer with linear activation. We use the Adam optimizer and Mean Squared Error (MSE) loss function for optimization. The number of epochs is set to 100.

Remark 1. As a result of applying autoencoder to get low dimensional representation, we get X', which we call as PDB2Vec.

Sequence Extraction from PDB Files. Protein Data Bank (PDB) files contain detailed information about the structure of proteins, including the coordinates of individual atoms. To extract the amino acid sequence of a protein from a PDB file, we follow the following steps:

1. Parse the PDB file using a PDB parser.
2. Use the coordinates of the alpha-carbon atom in each amino acid residue to define the protein backbone. The alpha-carbon atom represents each amino acid residue in a protein structure, commonly used to define the protein backbone.
3. Map the amino acid residues to their corresponding one-letter codes using a dictionary that associates each residue name with its corresponding code.

The amino acid residues and their mapped corresponding one-letter code is a single letter that represents the amino acid. For example, the code for alanine is "A", and the code for tryptophan is "W". The one-letter code is usually used in molecular biology to represent a sequence of amino acids. The code is more convenient to use than the full name of the amino acid, which can be pretty long. Using the one-letter code, a sequence s of amino acids is represented in a compact and readable format. We repeat this sequence extraction process for all PDB files to get a set S of all sequences.

Sequence Embeddings. After extracting the set of sequences S from PDB files, the next step is to generate fixed-length numerical embeddings from the sequences to use these embeddings as input for supervised analysis. For this purpose, we use different alignment-free embedding methods such as a k-mers based embedding (called Spike2Vec [3]), a position weight matrix based embedding (called PWM2Vec [2]), a minimizer based embedding, and a spaced k-mers based embedding.

Remark 2. Sequence alignment is computationally costly. It also requires a reference genome [14,15]. Moreover, it could introduce bias into the result [21].

Spike2Vec: It is a k-mers-based representation learning approach [3], which utilizes fixed-length substrings, called mers or n-grams, to create embeddings while maintaining the order of characters in protein sequences. The frequency vector's length is $|\Sigma|^k$, where Σ is the set of unique amino acid characters and k is a user-defined parameter.

PWM2Vec: This method is designed to produce fixed-length numerical embeddings using the concept of the position-weight matrix (PWM) [2]. Given a protein sequence, it first generates a $|\Sigma| \times k$ dimensional PWM matrix, which contains the count of each amino acid within k-mers of the sequence. Based on the counts, each k-mer is assigned a numerical weight. In the end, all weights are concatenated to get the final representation.

Minimizer: The approach is based on the concept of minimizers [33], which provides a compact representation of a biological sequence. The minimizer is a substring of consecutive amino acids with a fixed length m, obtained from a k-mer ($m < k$) by selecting the lexicographically smallest substring in forward and backward order.

Spaced k-mers: A method to generate spaced g-mers [35] from the k-mers, where $g < k$, to decrease sparsity and size. We generate our fixed-length numerical embeddings using the idea of spaced k-mers from [35]. We set $k = 4$ and $g = 9$ using the validation set approach in our experiments.

Combining Sequences and Structures. We are also interested in combining the structure-based embedding (i.e., PDB2Vec) and the sequence-based embeddings (i.e., Spike2Vec, PWM2Vec, Minimizer, and Spaced k-mers) to improve the predictive performance further. For this purpose, we concat the PDB2Vec-based embedding with those sequence-based embeddings to generate new embeddings containing richer structural and sequence-based information. Therefore, given a PDB file, we generate PDB2Vec-SP (PDB2Vec + Spike2Vec), PDB2Vec-P (PDB2Vec + PWM2Vec), PDB2Vec-M (PDB2Vec + Minimizer), and PDB2Vec-Sk (PDB2Vec + Spaced k-mers), which basically contains the concatenated information from both PDB2Vec and sequence-based embeddings.

4   Experimental Setup

This section describes the datasets and machine learning (ML) classifiers and evaluation metrics. The experiments were conducted on an Intel(R) Xeon(R) CPU E7-4850 v4 at 2.10GHz processor with a 3023 GB of memory, using the

Ubuntu 64-bit OS (16.04.7 LTS Xenial Xerus). To ensure that the models are highly accurate, we used a train-test split of 70–30%, with a 10% validation set from the training data used to fine-tune the hyperparameters for all the methods. To guarantee reliable and consistent results, we repeated the experiments five times and reported average results.

4.1 Baseline Model

biLSTM [7]. Authors in [7] proposed a representation learning method for protein sequences that uses a bidirectional long short-term memory (LSTM) model with a two-part feedback mechanism. It incorporates global structural similarity information between proteins and pairwise residue contact maps information for individual proteins. In this paper, we use this method to design the embeddings and use those as input for supervised analysis.

Unsupervised Protein Embeddings (UPE) [41]. Authors in [41] proposed a deep learning-based unsupervised solution to generate embeddings for protein sequences that considers sequences and structural information while designing the feature vector representation. Authors used a method proposed in [24] to generate the initial embeddings from sequences. The structural features are based on the one-hot encoding of the secondary structure angles computed from the 3D structure of proteins. In the end, sequence and structural features are combined to get the final representation of the proteins.

Dataset Statistics: In this paper, we use the following two datasets.

STCRDAB. A dataset called Structural T-Cell Receptor (STCRDab) [28], which is an automated, curated set of T-Cell Receptor structural data from the Protein Data Bank (PDB), comprised of 512 protein structures in total, which were downloaded on 5/27/21. The class/target labels for this dataset contain two species, namely "Humans" and "Mouse" (hence the binary class classification problem). After preprocessing, we selected 480 PDB files for our experimentation, which contain 325 Human and 155 Mouse species.

PDB Bind. The PDB bind dataset with version 2020 [29]. We use the total core set comprised of 14127 protein structures for this dataset. After preprocessing, we selected 3792 PDB files for experimentation. Protein names are the class/target labels, and their counts are as follows: Serine/threonine-protein: 404, Tyrosine-protein: 381, Mitogen-activated: 325, Beta-secretase: 299, Beta-lactamase: 220, Bromodomain-containing: 174, HIV-1: 164, Carbonic: 159, Cell: 157, Glycogen: 144, Protein: 138, E3: 138, Cyclin-dependent: 128, Glutamate: 112, Dual: 111, Heat: 110, Proteasome: 110, Tankyrase-2: 108, Lysine-specific: 105, DNA: 104, Coagulation: 101, Phosphatidylinositol-45-bisphosphate: 100.

The Minimum, Maximum, and Average sequence lengths extracted from the PDB files for the STCRDAB database are 109, 5415, and 1074.38, respectively. Whereas for PDB Bind database is 33, 3292, and 403.60. We can observe that the protein structures are comparatively bigger in the STCRDAB dataset compared to the PDB Bind dataset.

Classifiers and Evaluation Metrics: To classify the protein structures, we employed several ML models, including Support Vector Machine (SVM), Naive Bayes (NB), Multi-Layer Perceptron (MLP), K-Nearest Neighbors (KNN), Random Forest (RF), Logistic Regression (LR), and Decision Tree (DT). We evaluated classification performance using average accuracy, precision, recall, F1 (weighted), F1 (macro), Receiver Operator Characteristic Curve (ROC), Area Under the Curve (AUC), and training runtime.

5   Results and Discussion

The classification results are reported in Table 1. For the Structure Only category, we can observe that the random forest classifier outperforms all other classifiers for the STCRDAB dataset. In general, it is well known that the PDB-bind dataset is a challenging benchmark for structure-based prediction methods. Therefore, we observed a low predictive performance for this data while using structure-based embedding.

For sequence-only embedding methods, we can observe > 90% predictive performance in most cases for both datasets compared to the structure-only results. This is because the functional regions of a protein sequence are often more conserved across different proteins than the 3D structure, making them easier to identify and predict. This means that sequence-based models can be more effective at predicting protein function or classifying proteins based on their function. Moreover, the sequence-based models are more straightforward than the 3D structure-based autoencoder model, as they do not need to consider the complexities of protein folding and interactions. This makes them easier to train and interpret, leading to better results. Overall, we can observe that spaced k-mers with Naive Bayes classifier in the case of the STCRDAB dataset showed the highest performance among sequence-only embedding methods, hence beating the baselines Spike2Vec and PWM2Vec. Spike2Vec with Logistic Regression shows the highest performance for the PDB Bind dataset.

When we combine sequences and structure-based embeddings (i.e., PDB2Vec-SP, PDB2Vec-P, PDB2Vec-M, and PDB2Vec-Sk), we can observe that spaced k-mers with Naive Bayes shows almost a perfect performance. This is because when combining structure and sequence-based embeddings, we are incorporating more information about the protein and its environment, which can help improve the accuracy of the classification task. The sequence-based embeddings capture the amino acid composition and ordering in the protein sequence. In contrast, the structure-based embeddings capture the 3D spatial arrangement of the atoms in the protein structure. By combining these two sources of information, we can leverage the strengths of both methods and obtain a more comprehensive representation of the protein. Moreover, the proposed sequence + structural PDB2Vec outperforms the baselines UPE [41] and biLSTM [7] for all evaluation metrics and both datasets.

Table 1. Average Classification results (of 5 runs) for different methods and datasets using different evaluation metrics. The best values are shown in bold.

Category	Embedding	Algo	STCRDAB Acc.↑	Prec.↑	Recall↑	F1 (Weig.)↑	F1 (Macro)↑	ROC-AUC↑	Train Time (Sec.)↓	PDB Bind Acc.↑	Prec.↑	Recall↑	F1 (Weig.)↑	F1 (Macro)↑	ROC-AUC↑	Train Time (Sec.)↓
Structure Only	biLSTM [7]	SVM	0.838	0.860	0.838	0.822	0.785	0.758	0.017	0.808	0.874	0.808	0.815	0.830	0.888	4.765
		NB	0.654	0.693	0.654	0.664	0.634	0.655	0.008	0.475	0.630	0.475	0.487	0.495	0.737	0.170
		MLP	0.774	0.774	0.774	0.773	0.740	0.741	1.068	0.705	0.729	0.705	0.705	0.697	0.845	12.493
		KNN	0.840	0.839	0.840	0.836	0.807	0.796	0.016	0.854	0.876	0.854	0.853	0.845	0.920	0.216
		RF	0.829	0.866	0.829	0.808	0.768	0.742	0.369	0.886	0.892	0.886	0.886	0.887	0.935	5.497
		LR	0.885	0.899	0.885	0.887	0.874	0.898	0.023	0.853	0.930	0.853	0.877	0.871	0.926	128.075
		DT	0.831	0.830	0.831	0.829	0.802	0.799	0.103	0.832	0.835	0.832	0.832	0.830	0.911	3.265
Sequence + Structure	UPE [41]	SVM	0.916	0.989	0.916	0.988	0.909	0.907	0.961	0.891	0.912	0.891	0.942	0.929	0.899	6.581
		NB	0.897	0.908	0.897	0.895	0.896	0.911	0.975	0.922	0.941	0.922	0.918	0.919	0.896	1.675
		MLP	0.915	0.929	0.915	0.928	0.983	0.971	1.097	0.963	0.932	0.963	0.921	0.905	0.896	4.254
		KNN	0.921	0.928	0.921	0.929	0.981	0.979	0.482	0.959	0.923	0.959	0.949	0.938	0.893	0.234
		RF	0.894	0.885	0.894	0.892	0.881	0.893	0.813	0.921	0.944	0.921	0.932	0.928	0.948	4.563
		LR	0.957	0.942	0.957	0.954	0.975	0.963	0.128	0.954	0.925	0.954	0.930	0.929	0.965	9.753
		DT	0.901	0.899	0.901	0.900	0.921	0.943	0.042	0.939	0.928	0.939	0.935	0.912	0.945	0.973
Structure Only	PDB2Vec	SVM	0.710	0.693	0.710	0.688	0.613	0.609	0.321	0.100	0.067	0.100	0.054	0.031	0.502	4.166
		NB	0.504	0.657	0.504	0.507	0.504	0.577	0.001	0.036	0.047	0.036	0.020	0.021	0.500	0.053
		MLP	0.728	0.735	0.728	0.730	0.686	0.680	1.130	0.094	0.053	0.094	0.053	0.027	0.500	3.624
		KNN	0.742	0.737	0.742	0.738	0.680	0.686	0.014	0.064	0.059	0.064	0.054	0.041	0.502	0.055
		RF	0.803	0.800	0.803	0.790	0.742	0.722	0.267	0.059	0.053	0.059	0.054	0.039	0.497	3.420
		LR	0.688	0.668	0.688	0.671	0.597	0.595	0.008	0.066	0.056	0.066	0.057	0.040	0.499	10.847
		DT	0.717	0.724	0.717	0.719	0.673	0.676	0.007	0.056	0.056	0.056	0.056	0.043	0.499	0.597
Sequence Only	Spike2Vec [3]	SVM	0.976	0.977	0.976	0.976	0.972	0.967	1.824	0.960	0.963	0.960	0.961	0.954	0.975	263.112
		NB	0.978	0.978	0.978	0.978	0.974	0.967	0.189	0.943	0.956	0.943	0.944	0.931	0.964	8.230
		MLP	0.983	0.984	0.983	0.983	0.981	0.982	5.145	0.934	0.939	0.934	0.934	0.919	0.958	85.427
		KNN	0.963	0.963	0.963	0.962	0.956	0.948	0.087	0.896	0.954	0.896	0.896	0.910	0.897	1.961
		RF	0.975	0.975	0.975	0.975	0.971	0.967	0.482	0.960	0.966	0.960	0.961	0.954	0.975	6.888
		LR	0.986	0.986	0.986	0.986	0.984	0.982	0.119	0.966	0.967	0.966	0.966	0.959	0.978	8.471
		DT	0.957	0.957	0.957	0.957	0.950	0.948	0.204	0.939	0.942	0.939	0.939	0.929	0.962	4.682
	Minimizer	SVM	0.979	0.980	0.979	0.979	0.976	0.975	1.687	0.947	0.949	0.947	0.947	0.938	0.967	232.581
		NB	0.986	0.986	0.986	0.986	0.984	0.980	0.178	0.933	0.946	0.933	0.935	0.925	0.960	7.663
		MLP	0.974	0.973	0.974	0.974	0.970	0.975	5.018	0.929	0.932	0.929	0.929	0.917	0.957	107.245
		KNN	0.964	0.965	0.964	0.964	0.959	0.962	0.088	0.898	0.940	0.898	0.907	0.894	0.943	1.531
		RF	0.974	0.974	0.974	0.974	0.970	0.973	0.449	0.950	0.955	0.950	0.950	0.943	0.968	6.508
		LR	0.986	0.987	0.986	0.986	0.984	0.984	0.084	0.951	0.953	0.951	0.951	0.943	0.969	3.185
		DT	0.953	0.955	0.953	0.953	0.946	0.948	0.105	0.933	0.934	0.933	0.932	0.924	0.960	1.981
	Spaced k-mers	SVM	0.978	0.979	0.978	0.978	0.974	0.974	6.828	0.942	0.965	0.942	0.944	0.936	0.946	890.622
		NB	0.994	0.995	0.994	0.994	0.993	0.994	2.757	0.965	0.961	0.965	0.965	0.946	0.973	93.051
		MLP	0.983	0.984	0.983	0.983	0.980	0.985	76.321	0.898	0.919	0.898	0.902	0.888	0.940	1098.180
		KNN	0.960	0.960	0.960	0.959	0.949	0.942	0.993	0.885	0.897	0.885	0.905	0.892	0.937	23.986
		RF	0.971	0.972	0.971	0.971	0.965	0.962	2.279	0.951	0.966	0.951	0.955	0.947	0.971	77.044
		LR	0.983	0.984	0.983	0.983	0.980	0.981	1.117	0.961	0.965	0.961	0.962	0.955	0.977	21.214
		DT	0.965	0.967	0.965	0.965	0.958	0.955	2.475	0.943	0.950	0.943	0.944	0.935	0.966	89.690
Sequence + Structure (ours)	PDB2Vec-SP	SVM	0.981	0.981	0.981	0.980	0.978	0.976	1.735	0.944	0.948	0.944	0.945	0.937	0.966	294.827
		NB	0.988	0.988	0.988	0.987	0.986	0.982	0.203	0.942	0.958	0.942	0.943	0.931	0.964	6.449
		MLP	0.985	0.986	0.985	0.985	0.983	0.987	4.255	0.927	0.934	0.927	0.926	0.912	0.954	89.376
		KNN	0.924	0.924	0.924	0.923	0.912	0.909	0.263	0.083	0.038	0.083	0.070	0.052	0.508	1.263
		RF	0.974	0.974	0.974	0.974	0.970	0.968	0.581	0.947	0.953	0.947	0.948	0.937	0.966	3.546
		LR	0.982	0.982	0.982	0.982	0.979	0.975	0.131	0.927	0.930	0.927	0.928	0.912	0.953	45.593
		DT	0.949	0.949	0.949	0.948	0.940	0.934	0.214	0.928	0.930	0.928	0.927	0.911	0.954	4.273
	PDB2Vec-P	SVM	0.971	0.972	0.971	0.971	0.967	0.960	0.414	0.943	0.948	0.943	0.942	0.939	0.964	31.976
		NB	0.983	0.984	0.983	0.983	0.981	0.977	0.113	0.939	0.952	0.939	0.941	0.918	0.961	34.941
		MLP	0.975	0.975	0.975	0.975	0.973	0.975	2.607	0.921	0.919	0.921	0.919	0.904	0.950	34.941
		KNN	0.892	0.892	0.892	0.890	0.878	0.865	0.057	0.084	0.088	0.084	0.072	0.055	0.509	0.505
		RF	0.964	0.964	0.964	0.964	0.960	0.959	0.518	0.953	0.958	0.953	0.954	0.948	0.972	5.991
		LR	0.979	0.979	0.979	0.979	0.977	0.973	0.095	0.921	0.924	0.921	0.921	0.912	0.952	34.746
		DT	0.947	0.948	0.947	0.947	0.942	0.941	0.213	0.918	0.921	0.918	0.918	0.905	0.950	5.692
	PDB2Vec-M	SVM	0.981	0.981	0.981	0.980	0.977	0.971	2.118	0.941	0.942	0.941	0.940	0.931	0.963	255.117
		NB	0.983	0.984	0.983	0.983	0.981	0.979	0.219	0.936	0.948	0.936	0.938	0.926	0.961	5.839
		MLP	0.990	0.990	0.990	0.990	0.989	0.992	5.812	0.923	0.926	0.923	0.923	0.908	0.952	113.741
		KNN	0.942	0.942	0.942	0.941	0.932	0.931	0.074	0.113	0.141	0.113	0.100	0.081	0.521	1.041
		RF	0.974	0.974	0.974	0.974	0.969	0.967	0.515	0.941	0.944	0.941	0.940	0.929	0.961	5.481
		LR	0.988	0.988	0.988	0.987	0.985	0.981	0.067	0.929	0.932	0.929	0.929	0.915	0.954	12.721
		DT	0.950	0.952	0.950	0.950	0.944	0.951	0.121	0.920	0.920	0.920	0.918	0.907	0.951	2.303
	PDB2Vec-Sk	SVM	0.974	0.974	0.974	0.973	0.970	0.966	6.953	0.940	0.951	0.940	0.941	0.934	0.962	874.034
		NB	**0.999**	**0.999**	**0.999**	**0.999**	**0.998**	**0.999**	0.990	**0.967**	**0.968**	**0.967**	**0.967**	**0.966**	**0.979**	93.332
		MLP	0.989	0.989	0.989	0.989	0.988	0.991	48.372	0.909	0.920	0.909	0.911	0.892	0.945	925.806
		KNN	0.910	0.910	0.910	0.909	0.898	0.887	8.820	0.076	0.085	0.076	0.066	0.050	0.506	17.964
		RF	0.974	0.974	0.974	0.974	0.970	0.969	2.394	0.940	0.959	0.940	0.944	0.930	0.963	54.800
		LR	0.986	0.987	0.986	0.986	0.984	0.983	0.767	0.928	0.936	0.928	0.930	0.917	0.956	94.031
		DT	0.967	0.967	0.967	0.966	0.961	0.953	2.845	0.935	0.935	0.935	0.937	0.924	0.961	76.458

Discussion: Overall, we can observe that if we only use structure information, the proposed PDB2Vec can achieve reasonable predictive performance using STCRDAB dataset while showing poor performance for a more challenging PDB Bind dataset. Using the sequence information only shows higher predictive performance. Similarly, combining structure and sequence information shows almost a perfect predictive performance. The structure-based embeddings can help account for the effect of structural features, such as solvent accessibility, hydrogen bonding, and electrostatic interactions, which can influence the protein function or interactions. The sequence-based embeddings can help account for the effect of sequence variations, such as point mutations or indels, which can also affect protein function or interactions. Overall, the combination of structure and sequence-based embeddings provides a more comprehensive representation of the protein, which can lead to improved performance in classification tasks.

6 Conclusion

In this study, we have investigated the effect of incorporating both protein sequence and 3D structure information on protein classification performance. Our experiments show that the combination of structural and sequence information leads to the best performance, indicating that both types of information are complementary and essential for protein classification tasks. We have demonstrated this by performing protein classification using three different experimental settings: only 3D structural embeddings, sequence embeddings, and combining both embeddings. Our results show that the classification performance is the lowest when only 3D structural information is used, but the performance improves significantly when only the protein sequences are used. Finally, we have evaluated our methods on two benchmark datasets: STCRDAB and PDB bind. Overall, our findings suggest that it is important to consider both structural and sequence information in protein analysis tasks and that combining these sources of information can lead to improved performance. In the future, we will develop a deep learning-based model to combine sequences and structural information more effectively. We will also explore graph-based models for efficiently embedding 3D structural information. Using more datasets to test the proposed model's scalability, robustness, and interpretability is also exciting future work.

Author contributions. Sarwan Ali and Prakash Chourasia–Equal Contribution

References

1. Al-Lazikani, B., Jung, J., Xiang, Z., Honig, B.: Protein structure prediction. Curr. Opin. Chem. Biol. **5**(1), 51–56 (2001)
2. Ali, S., Bello, B., Chourasia, P., Punathil, R.T., Zhou, Y., Patterson, M.: Pwm2vec: An efficient embedding approach for viral host specification from coronavirus spike sequences. MDPI Biology (2022)
3. Ali, S., Patterson, M.: Spike2vec: an efficient and scalable embedding approach for covid-19 spike sequences. In: IEEE International Conference on Big Data (Big Data), pp. 1533–1540 (2021)
4. Ali, S., Sahoo, B., Khan, M.A., Zelikovsky, A., Khan, I.U., Patterson, M.: Efficient approximate kernel based spike sequence classification. IEEE/ACM Transactions on Computational Biology and Bioinformatics (2022)
5. Ali, S., Sahoo, B., Ullah, N., Zelikovskiy, A., Patterson, M., Khan, I.: A k-mer based approach for sars-cov-2 variant identification. In: International Symposium on Bioinformatics Research and Applications, pp. 153–164 (2021)
6. Batool, M., Ahmad, B., Choi, S.: A structure-based drug discovery paradigm. Int. J. Mol. Sci. **20**(11), 2783 (2019)

7. Bepler, T., Berger, B.: Learning protein sequence embeddings using information from structure. In: International Conference on Learning Representations (2019)
8. Bigelow, D.J., Squier, T.C.: Redox modulation of cellular signaling and metabolism through reversible oxidation of methionine sensors in calcium regulatory proteins. Biochimica et Biophysica Acta (BBA)-Proteins and Proteomics **1703**(2), 121–134 (2005)
9. Boscher, C., Dennis, J.W., Nabi, I.R.: Glycosylation, galectins and cellular signaling. Curr. Opin. Cell Biol. **23**(4), 383–392 (2011)
10. Brandes, N., Ofer, D., Peleg, Y., Rappoport, N., Linial, M.: ProteinBERT: a universal deep-learning model of protein sequence and function. Bioinformatics **38**(8), 2102–2110 (2022)
11. Chourasia, P., Ali, S., Ciccolella, S., Della Vedova, G., Patterson, M.: Clustering sars-cov-2 variants from raw high-throughput sequencing reads data. In: International Conference on Computational Advances in Bio and Medical Sciences, pp. 133–148. Springer (2021)
12. Chourasia, P., Ali, S., Ciccolella, S., Vedova, G.D., Patterson, M.: Reads2vec: Efficient embedding of raw high-throughput sequencing reads data. J. Comput. Biol. **30**(4), 469–491 (2023)
13. Chourasia, P., Tayebi, Z., Ali, S., Patterson, M.: Empowering pandemic response with federated learning for protein sequence data analysis. In: 2023 International Joint Conference on Neural Networks (IJCNN), pp. 01–08. IEEE (2023)
14. Chowdhury, B., Garai, G.: A review on multiple sequence alignment from the perspective of genetic algorithm. Genomics **109**(5–6), 419–431 (2017)
15. Denti, L., Pirola, Y., Previtali, M., Ceccato, T., Della Vedova, G., Rizzi, R., Bonizzoni, P.: Shark: fishing relevant reads in an rna-seq sample. Bioinformatics **37**(4), 464–472 (2021)
16. Farhan, M., Tariq, J., Zaman, A., Shabbir, M., Khan, I.: Efficient approximation algorithms for strings kernel based sequence classification. In: Advances in neural information processing systems (NeurIPS), pp. 6935–6945 (2017)
17. Fiser, A., Šali, A.: Modeller: generation and refinement of homology-based protein structure models. In: Methods in Enzymology, vol. 374, pp. 461–491 (2003)
18. Freeman, B.A., O'Donnell, V.B., Schopfer, F.J.: The discovery of nitro-fatty acids as products of metabolic and inflammatory reactions and mediators of adaptive cell signaling. Nitric Oxide **77**, 106–111 (2018)
19. Gao, W., Mahajan, S.P., Sulam, J., Gray, J.J.: Deep learning in protein structural modeling and design. Patterns **1**(9), 100142 (2020)
20. Gohlke, H., Klebe, G.: Approaches to the description and prediction of the binding affinity of small-molecule ligands to macromolecular receptors. Angew. Chem. Int. Ed. **41**(15), 2644–2676 (2002)
21. Golubchik, T., Wise, M.J., Easteal, S., Jermiin, L.S.: Mind the gaps: evidence of bias in estimates of multiple sequence alignments. Molecular Biol. Evol. **24**(11), 2433–2442 (2007). https://doi.org/10.1093/molbev/msm176
22. Groom, C.R., Allen, F.H.: The cambridge structural database: experimental three-dimensional information on small molecules is a vital resource for interdisciplinary research and learning. Wiley Interdisciplinary Rev. Comput. Molecular Sci. **1**(3), 368–376 (2011)
23. Hardin, C., Pogorelov, T.V., Luthey-Schulten, Z.: Ab initio protein structure prediction. Curr. Opin. Struct. Biol. **12**(2), 176–181 (2002)
24. Heinzinger, M., Elnaggar, A., Wang, Y., Dallago, C., Nechaev, D., Matthes, F., Rost, B.: Modeling aspects of the language of life through transfer-learning protein sequences. BMC Bioinform. **20**(1), 1–17 (2019)

25. Jisna, V., Jayaraj, P.: Protein structure prediction: conventional and deep learning perspectives. Protein J. **40**(4), 522–544 (2021)
26. Kubinyi, H.: Structure-based design of enzyme inhibitors and receptor ligands. Curr. Opin. Drug Discov. Devel. **1**(1), 4–15 (1998)
27. Kuzmin, K., et al.: Machine learning methods accurately predict host specificity of coronaviruses based on spike sequences alone. Biochem. Biophys. Res. Commun. **533**(3), 553–558 (2020)
28. Leem, J., de Oliveira, S.H.P., Krawczyk, K., Deane, C.M.: Stcrdab: the structural t-cell receptor database. Nucleic Acids Res. **46**(D1), D406–D412 (2018)
29. Liu, Z., Li, Y., Han, L., Li, J., Liu, J., Zhao, Z., Nie, W., Liu, Y., Wang, R.: Pdb-wide collection of binding data: current status of the pdbbind database. Bioinformatics **31**(3), 405–412 (2015)
30. Oshima, A., Tani, K., Hiroaki, Y., Fujiyoshi, Y., Sosinsky, G.E.: Three-dimensional structure of a human connexin26 gap junction channel reveals a plug in the vestibule. Proc. Natl. Acad. Sci. **104**(24), 10034–10039 (2007)
31. Radivojac, P., Clark, W.T., Oron, T.R., Schnoes, A.M., Wittkop, T., Sokolov, A., Graim, K., Funk, C., Verspoor, K., Ben-Hur, A., et al.: A large-scale evaluation of computational protein function prediction. Nat. Methods **10**(3), 221–227 (2013)
32. Reynolds, C., Damerell, D., Jones, S.: Protorp: a protein-protein interaction analysis server. Bioinformatics **25**(3), 413–414 (2009)
33. Roberts, M., Haynes, W., Hunt, B., Mount, S., Yorke, J.: Reducing storage requirements for biological sequence comparison. Bioinformatics **20**, 3363–9 (2004)
34. Sapoval, N., et al.: Current progress and open challenges for applying deep learning across the biosciences. Nat. Commun. **13**(1), 1728 (2022)
35. Singh, R., Sekhon, A., Kowsari, K., Lanchantin, J., Wang, B., Qi, Y.: Gakco: a fast gapped k-mer string kernel using counting. In: Joint European Conference on Machine Learning and Knowledge Discovery in Databases, pp. 356–373 (2017)
36. Spencer, M., Eickholt, J., Cheng, J.: A deep learning network approach to ab initio protein secondary structure prediction. IEEE/ACM Trans. Comput. Biol. Bioinf. **12**(1), 103–112 (2014)
37. Strodthoff, N., Wagner, P., Wenzel, M., Samek, W.: Udsmprot: universal deep sequence models for protein classification. Bioinformatics **36**(8), 2401–2409 (2020)
38. Tayebi, Z., Ali, S., Patterson, M.: Robust representation and efficient feature selection allows for effective clustering of sars-cov-2 variants. Algorithms **14**(12), 348 (2021)
39. Torrisi, M., Pollastri, G., Le, Q.: Deep learning methods in protein structure prediction. Comput. Struct. Biotechnol. J. **18**, 1301–1310 (2020)
40. Tramontano, A., Morea, V.: Assessment of homology-based predictions in casp5. Proteins: Struct. Function Bioinform. **53**(S6), 352–368 (2003)
41. Villegas-Morcillo, A., Makrodimitris, S., van Ham, R.C., Gomez, A.M., Sanchez, V., Reinders, M.J.: Unsupervised protein embeddings outperform hand-crafted sequence and structure features at predicting molecular function. Bioinformatics **37**(2), 162–170 (2021)
42. Xu, J.: Distance-based protein folding powered by deep learning. Proc. Natl. Acad. Sci. **116**(34), 16856–16865 (2019)
43. Yao, Y., Du, X., Diao, Y., Zhu, H.: An integration of deep learning with feature embedding for protein-protein interaction prediction. PeerJ **7**, e7126 (2019)

Hist2Vec: Kernel-Based Embeddings for Biological Sequence Classification

Sarwan Ali[1][(✉)], Haris Mansoor[2], Prakash Chourasia[1], and Murray Patterson[1]

[1] Georgia State University, Atlanta, GA, USA
{sali85,pchourasia1,mpatterson30}@gsu.edu
[2] Lahore University of Management Sciences, Lahore, Pakistan
16060061@lums.edu.pk

Abstract. Biological sequence classification is vital in various fields, such as genomics and bioinformatics. The advancement and reduced cost of genomic sequencing have brought the attention of researchers for protein and nucleotide sequence classification. Traditional approaches face limitations in capturing the intricate relationships and hierarchical structures inherent in genomic sequences, while numerous machine-learning models have been proposed to tackle this challenge. In this work, we propose Hist2Vec, a novel kernel-based embedding generation approach for capturing sequence similarities. Hist2Vec combines the concept of histogram-based kernel matrices and Gaussian kernel functions. It constructs histogram-based representations using the unique k-mers present in the sequences. By leveraging the power of Gaussian kernels, Hist2Vec transforms these representations into high-dimensional feature spaces, preserving important sequence information. Hist2Vec aims to address the limitations of existing methods by capturing sequence similarities in a high-dimensional feature space while providing a robust and efficient framework for classification. We employ kernel Principal Component Analysis (PCA) using standard machine-learning algorithms to generate embedding for efficient classification. Experimental evaluations on protein and nucleotide datasets demonstrate the efficacy of Hist2Vec in achieving high classification accuracy compared to state-of-the-art methods. It outperforms state-of-the-art methods by achieving $> 76\%$ and $> 83\%$ accuracies for DNA and Protein datasets, respectively. Hist2Vec provides a robust framework for biological sequence classification, enabling better classification and promising avenues for further analysis of biological data.

Keywords: Kernel Function · Embeddings · Histogram · PCA · Classification · Representation Learning · Sequence Analysis · COVID-19

S. Ali, H. Mansoor and P. Chourasia—Equal Contribution

1 Introduction

The rapid advancement of sequencing technology has led to an increase in the quantity of sequence data [16], presenting new opportunities and difficulties for biological sequence analysis. Biological sequence classification, particularly in the domains of protein and nucleotide sequences, is of significant importance in genomics, drug discovery [41], bioinformatics [10,15], and various other research areas [8,26]. Classification of these sequences is crucial in understanding their functions, identifying potential disease-causing variants, and predicting protein structures [5,11,38,40,42]. Traditional classification approaches often rely on feature engineering [1,3,13,23] or sequence alignment methods, having limitations in capturing complex sequence similarities and high-dimensional representations [17,23]. In recent years, kernel methods emerged as powerful techniques for extracting meaningful features and facilitating classification tasks [3]. Motivated by the need for improved biological sequence classification [27,32], this work proposes Hist2Vec, which combines the concept of histogram-based kernel matrices with the effectiveness of Gaussian kernel functions.

Hist2Vec aims to address the limitations of existing methods by capturing sequence similarities in a high-dimensional feature space while providing a robust and efficient framework for classification. By utilizing the unique k-mers present in sequences and leveraging the power of kernel methods, Hist2Vec offers a promising avenue for enhancing sequence classification accuracy and enabling further analysis of biological data. We introduce Hist2Vec as a novel kernel-based approach for biological sequence classification. We demonstrate the theoretical properties and advantages of Hist2Vec, including its ability to satisfy Mercer's condition, ensure continuity, and exhibit the universal approximation property. Additionally, we investigate the practical aspects of Hist2Vec, such as generating kernel matrices, converting these matrices into low-dimensional embeddings using kernel PCA, and applying ML algorithms for sequence classification.

In summary, this paper introduces Hist2Vec as a powerful and effective approach for biological sequence classification. Improved classification accuracy and practical implementation guidelines collectively advance sequence analysis and facilitate a deeper understanding of biological data. Our contributions to this paper are listed below:

1. Introducing Hist2Vec: We propose Hist2Vec, a novel kernel-based approach that combines histogram-based kernel matrices with Gaussian kernels for efficient and accurate biological sequence classification.
2. High-dimensional Feature Representation: Hist2Vec captures sequence similarities by constructing histogram-based representations using the unique k-mers present in the sequences. By leveraging the power of Gaussian kernels, Hist2Vec transforms these representations into high-dimensional feature spaces, preserving important sequence information.
3. Improved Classification Accuracy: Through extensive evaluations of protein and nucleotide datasets, we demonstrate that Hist2Vec outperforms state-of-the-art methods in terms of classification accuracy.

4. Advancing Sequence Analysis: Hist2Vec contributes to the advancement of sequence analysis by providing a robust framework for capturing and exploiting sequence similarities. The method offers new insights, fostering further research in genomics, bioinformatics, and related fields.

The remainder of this paper is organized as follows. Section 2 provides a detailed overview of the related work. Section 3 presents the methodology of Hist2Vec, including the computation of histogram-based kernel matrices and the application of Gaussian kernels. Section 4 describes the experimental setup and dataset. Section 5 presents the results of extensively evaluating protein and nucleotide datasets. Finally, Sect. 6 concludes the paper, summarizes the contributions of Hist2Vec, and discusses future research directions.

2   Related Work

Several methods have been proposed for biological sequence classification to capture sequence similarities and facilitate classification [12]. These methods can be broadly categorized into alignment-based methods, feature engineering approaches, and kernel-based methods.

Alignment-based methods, such as BLAST (Basic Local Alignment Search Tool [6]) and Smith-Waterman algorithm [21], rely on sequence alignment techniques to identify similarities between sequences. While these methods effectively detect homologous sequences, they may struggle to capture more complex relationships and handle large-scale datasets efficiently.

Feature engineering approaches involve extracting informative features from sequences and using them for classification [2,22,37]. These features include amino acid composition, dipeptide composition, and physicochemical properties. While these methods are computationally efficient, they often rely on handcrafted features that may not capture all relevant sequence characteristics.

Kernel-based methods [4,29] can capture complex relationships between sequences. Methods such as the spectrum kernel and string kernel [24] have been proposed for sequence classification, leveraging the concept of k-mer frequencies to construct similarity measures. However, these methods may suffer from high computational complexity and memory requirements, limiting their scalability [34]. Kernel methods have been successfully applied to various bioinformatics tasks, including protein fold recognition [31], protein-protein interaction prediction [39], and protein function prediction [7]. These methods leverage the power of kernel functions to map sequences into high-dimensional feature spaces, where the relationships between sequences can be effectively captured.

Mercer's Condition is a fundamental property of kernel methods, ensuring the kernel matrix is positive semidefinite [28,33]. Many kernel functions, such as the Gaussian kernel, satisfy Mercer's Condition [35] and have been widely used in bioinformatics applications [19,25]. Hist2Vec addresses the computational complexity and memory requirements by utilizing histogram-based representations of sequences. By counting the frequencies of k-mers and constructing histograms, Hist2Vec reduces the dimensionality of the kernel matrix, making it more tractable for large-scale datasets. Furthermore, using the Gaussian kernel in Hist2Vec allows for preserving important sequence information in a high-dimensional feature space.

3 Proposed Approach

In this section, we present the proposed approach Hist2Vec. It consists of two main steps: computing histogram-based representations and generating embeddings using kernel PCA.

3.1 Histogram-Based Representations

A histogram serves as an approximate visual depiction of the distribution pattern of numerical data. Histogram creation involves partitioning the range of values into discrete intervals, commonly called bins, and subsequently tallying the frequency of data points falling within each interval. These bins are defined as consecutive intervals that do not overlap, typically possessing uniform size. By ensuring that adjacent bins are contiguous, histograms effectively eliminate gaps between the rectangular bars, resulting in mutual contact.

Histograms provide an approximate indication of the concentration or density of the underlying data distribution, thereby facilitating the estimation of the probability density function associated with the respective variable. Visually representing the data distribution, histograms depict the frequency or count of observations falling within individual bins. This analytical tool proves valuable in discerning patterns and trends within the data and facilitating comparisons between disparate datasets.

Suppose a dataset $D = \{x_1, x_2, ..., x_n\} \in \mathbb{R}^d$. The first step of Hist2Vec involves computing histogram-based representations of the input sequences. This process captures the frequencies of specific k-mers within the sequences, providing a compact representation that preserves important sequence characteristics.

Given a protein or nucleotide sequence, we compute the k-mer spectrum by extracting all possible substrings of length k. Each unique k-mer represents a distinct feature. We then construct a histogram, where each bin corresponds to a specific k-mer and captures its frequency within the sequence. The histogram-based representation provides a compact and informative sequence summary, facilitating efficient computations and capturing sequence similarities.

The data used to construct a histogram are generated via a function m_i that counts the number of observations that fall into each disjoint category (bins). Thus, if we let z be the total number of observations and k be the total number of bins, the histogram data m_i meet the following conditions:

$$z = \sum_{i=1}^{k} m_i \tag{1}$$

where z is the total number of $k-$mers in a sequence, and m_i represents the number of $k-$mers belonging to the i_{th} bin.

3.2   Gaussian Kernel Transformation

Once the histogram-based representations are obtained, Hist2Vec applies the Gaussian kernel transformation to map the histogram values into a high-dimensional feature space. The Gaussian kernel is a popular choice for capturing complex relationships between data points and is well-suited for capturing sequence similarities. The Gaussian kernel function is defined as:

$$k(x_i', x_j') = exp(-\frac{||x_i' - x_j'||^2}{2\sigma^2}) \tag{2}$$

where x_i' and x_j' represent the histogram-based representations of sequence x_i and x_j respectively, and σ controls the width of the kernel. The kernel function measures the similarity between sequences in the high-dimensional feature space, with higher values indicating greater similarity.

Hist2Vec converts the histogram-based representations into a feature space where sequence similarities are kept by performing the Gaussian kernel transformation. This further helps apply standard machine learning (ML) algorithms to the modified feature space, leading to better classification results.

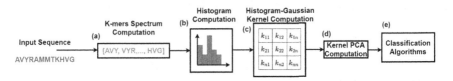

Fig. 1: Workflow of Hist2Vec. The input represents a sequence vector computed using k-mers spectrum (a). The sequence is converted into the histogram (b) and kernel (c) matrix using His2Vec-Gaussian kernel, and then the kernel matrix is processed through Kernel-PCA (d) and used in the classification algorithms (e).

3.3   Kernel PCA for Embeddings

To further reduce the dimensionality and extract essential features, Hist2Vec employs kernel Principal Component Analysis (KPCA) to generate low-dimensional embeddings from the kernel matrix. Kernel PCA is a nonlinear extension of traditional PCA that operates in the feature space defined by the kernel function. The kernel matrix, denoted as K, is computed by applying the Gaussian kernel function to all pairs of histogram-based representations. Kernel PCA then performs eigendecomposition on the kernel matrix to extract the principal components, which capture the most crucial information. By selecting a subset of the principal components, Hist2Vec generates low-dimensional embeddings, which capture the similarities and variations between sequences in a compact representation, enabling efficient classification.

The algorithmic pseudocode for Hist2Vec is presented in Algorithm 1 and 2, while Fig. 1 provides flowchart. The Algorithm 1 takes biological sequences and the total bins as input and calculates histogram embeddings based on the k-mers spectrum. First, the sequence is converted into K-mers (lines 2–5), then K-mers is converted into histogram embedding according to the number of bins (lines 6-9). While Algorithm 2 applies the Gaussian kernel to obtain the final kernel value (line 6). The kernel matrix is further processed for kernel-PCA and subsequently utilized in classification algorithms.

Algorithm 1. Histogram computation based on k-mers

1: **function** COMPUTEKMERHISTOGRAM(sequence, k, totBins)
2:    kmers{}   ▷ Initialize an empty dictionary
3:    **for** i = 0 to n - k + 1 **do**
4:       kmer = sequence[i:i+k]   ▷ Extract k-mer from sequence starting at index i
5:       kmers[kmer] += 1   ▷ Increment count of kmer in kmers
6:    histogram = np.zeros(totBins)   ▷ Initialize an array histogram of size totBins with all elements as 0
7:    **for** kmer, count **in** kmers.items() **do**
8:       bin_index = hash(kmer) ▷ Compute bin index by hashing kmer and taking the modulus of totBins
9:       histogram[bin_index] + = count   ▷ Increment histogram
10:    **return** histogram

Algorithm 2. Hist2Vec-based kernel matrix generation

1: **function** COMPUTEKERNELVALUE(sequence1, sequence2, k, totBins)
2:    Hist1 ← COMPUTEKMERHISTOGRAM(sequence1)
3:    Hist2 ← COMPUTEKMERHISTOGRAM(sequence2)
4:    Hist1 ← NORMALIZE(Hist1)
5:    Hist2 ← NORMALIZE(Hist2)
6:    K ← GAUSSIANKERNEL(Hist1,Hist2)
7:    **return** K

---

4 Experimental Setup

In this section, we detail the spike sequence dataset used for experimentation. We also discuss the baselines used for classification. In the end, we talk about the evaluation metrics used to test the performance of the models.

All experiments use an Intel(R) Core i5 system @ 2.10 GHz having Windows 10 64 bit OS with 32 GB memory. For the classification algorithms, we use 70% of the data for training and 30% for testing. The 10% data from the training set is used as a validation set for hyperparameter tuning. We use two datasets to evaluate the performance of the proposed embedding method, which are explained below.

Human DNA Dataset: This dataset uses the unaligned human DNA nucleotide sequences from [20]. It includes sequences and details on the relevant gene family for each sequence. It encodes data for seven distinct gene families (class labels). The count of sequences for each class label are G Protein Coupled (531), Tyrosine Kinase (534), Tyrosine Phosphatase (349), Synthetase (672), Synthase (711), Ion Channel (240), and Transcription Factor (1343) with the total count of 4380 sequences. The classification task classifies the gene family using the DNA sequence as input. Due to the variable length of sequences, since they are unaligned sequences, we have the maximum, minimum, and average lengths of 18921, 5, and 1263.59, respectively in our dataset.

Coronavirus Host Dataset: The spike sequences of the clades of the Coronaviridae family are extracted from ViPR [1,30] and GISAID[1], along with their metadata (genus/ subgenus, infected host, etc.), and the spike sequence and its corresponding host information are used to create our Coronavirus Host dataset. The number of hosts in our dataset is as follows: Bats (153), Bovines (88), Cats (123), Cattle (1), Equine (5), Fish (2), Humans (1813), Pangolins (21), Rats (26), Turtle (1), Weasel (994), Birds (374), Camels (297), Canis (40), Dolphins (7), Environment (1034), Hedgehog (15), Monkey (2), Python (2), Swines (558), and Unknown (2). We used 5558 spike sequences, which contain 21 unique hosts. Our classification jobs for this dataset use the hostname as the class label and sequences as input.

Baseline Methods: To establish baselines, we choose newly suggested methods from several embedding generation categories, including feature engineering, conventional kernel matrix creation (including kernel PCA), neural networks, pretrained language models, and pre-trained transformers for protein sequences.

PWM2Vec: Feature Engineering method takes a biological sequence as input and designs fixed-length numerical embeddings [1].

String Kernel: Kernel Matrix-based method designs $n \times n$ kernel matrix that can be used with kernel classifiers or kernel PCA to get feature vector based on principal components [4,14].

WDGRL: A neural network (NN) based method takes the one-hot representation of biological sequence as input and designs an NN-based embedding method by minimizing loss [36].

AutoEncoder: This method uses a neural network (NN) to teach itself how to encode data as features. To iteratively optimize the objective, it applies the non-linear mapping approach. We used a 2 multilayer network with an ADAM optimizer and MSE loss function for our experiments [43].

SeqVec: It is a pre-trained Language Model which takes biological sequences as input and fine-tunes the weights based on a pre-trained model to get final embedding [18].

---

[1] https://www.gisaid.org/.

ProteinBert: It is a pre-trained Transformer, a protein sequence model to classify the given biological sequence using Transformer/Bert [9].

4.1   Evaluation Metrics and Classification Algorithms

We use average accuracy, precision, recall, F1 (weighted), F1 (macro), Receiver Operator Characteristic Curve (ROC), Area Under the Curve (AUC), and training runtime to compare the performance of various models. When using metrics created for the binary classification problem, we employ the one-vs-rest method for multi-class classification. Support Vector Machine (SVM), Naive Bayes (NB), Multi-Layer Perceptron (MLP), K-Nearest Neighbours (KNN), Random Forest (RF), Logistic Regression (LR), and Decision Tree (DT) are just a few of the linear and non-linear classifiers used in supervised analysis.

5   Results and Discussion

In Table 1, we present the classification results for the suggested and baseline methods for the **Human DNA** dataset. In terms of average accuracy, precision, recall, and F1 (weighted) and F1 (macro) scores, the Random Forest with Hist2Vec feature vectors outperformed all other baseline embedding approaches, while KNN using Hist2Vec-based embedding has the highest ROC-AUC score when compared to other embedding techniques. This suggests that the Hist2Vec embedding method is the method that performs the best for classifying sequences using ML models. Despite minimal train time, WDGRL's classification performance is quite subpar. Compared to other embedding techniques, the WDGRL fails to preserve the overall data structure in its embeddings, which is one reason for this behavior. We can see that, except for one assessment criterion (training duration), Hist2Vec outperforms all baselines.

The results are presented in Table 2 for the **Coronavirus Host** dataset. Here also observe the performance of Hist2Vec on the host data sequences as compared to other baseline approaches. Because PWM2Vec is used to classify hosts (see [1]), we can compare Hist2Vec to PWM2Vec and other baselines to better understand its effectiveness. According to the results from the experiments, the RF classifier with Hist2Vec embedding works better than even PWM2Vec and different baselines in terms of accuracy, precision, recall, F1 weighted score, and AUC ROC score. Similarly, WDGRL has a low train time with the NB classifier in this instance, but the classifier performance is not even comparable. These findings suggest that the Hist2Vec method outperforms all other baseline techniques, including PWM2Vec.

Table 1: Classification results (averaged over 5 runs) on **Human DNA** dataset. The best classifier for respective embeddings is shown with the underline. Overall best values are shown in bold.

Embeddings	Algo	Acc. ↑	Prec. ↑	Recall ↑	F1 (Weig.) ↑	F1 (Macro) ↑	ROC AUC ↑	Train Time (sec.) ↓
PWM2Vec	SVM	0.302	0.241	0.302	0.168	0.091	0.505	10011.3
	NB	0.084	0.442	0.084	0.063	0.066	0.511	4.565
	MLP	0.310	0.380	0.310	0.175	0.107	0.510	320.555
	KNN	0.121	0.337	0.121	0.093	0.077	0.509	2.193
	RF	0.309	0.332	0.309	0.181	0.110	0.510	65.250
	LR	0.304	0.257	0.304	0.167	0.094	0.506	23.651
	DT	0.306	0.284	0.306	0.181	0.111	0.509	1.861
String Kernel	SVM	0.618	0.617	0.618	0.613	0.588	0.753	39.791
	NB	0.338	0.482	0.338	0.347	0.333	0.617	0.276
	MLP	0.597	0.595	0.597	0.593	0.549	0.737	331.068
	KNN	0.645	0.657	0.645	0.646	0.612	0.774	1.274
	RF	0.731	0.776	0.731	0.729	0.723	0.808	12.673
	LR	0.571	0.570	0.571	0.558	0.532	0.716	2.995
	DT	0.630	0.631	0.630	0.630	0.598	0.767	2.682
WDGRL	SVM	0.318	0.101	0.318	0.154	0.069	0.500	0.751
	NB	0.282	0.214	0.282	0.196	0.138	0.517	0.004
	MLP	0.326	0.286	0.326	0.263	0.186	0.535	8.613
	KNN	0.317	0.317	0.317	0.315	0.286	0.574	0.092
	RF	0.453	0.501	0.453	0.430	0.389	0.625	1.124
	LR	0.323	0.279	0.323	0.177	0.095	0.507	0.041
	DT	0.368	0.372	0.368	0.369	0.328	0.610	0.047
Auto-Encoder	SVM	0.621	0.638	0.621	0.624	0.593	0.769	22.230
	NB	0.260	0.426	0.260	0.247	0.268	0.583	0.287
	MLP	0.621	0.620	0.621	0.620	0.578	0.756	111.809
	KNN	0.565	0.577	0.565	0.568	0.547	0.732	1.208
	RF	0.689	0.738	0.689	0.683	0.668	0.774	20.131
	LR	0.692	0.700	0.692	0.693	0.672	0.799	58.369
	DT	0.543	0.546	0.543	0.543	0.515	0.718	10.816
SeqVec	SVM	0.656	0.661	0.656	0.652	0.611	0.791	0.891
	NB	0.324	0.445	0.312	0.295	0.282	0.624	0.036
	MLP	0.657	0.633	0.653	0.646	0.616	0.783	12.432
	KNN	0.592	0.606	0.592	0.591	0.552	0.717	0.571
	RF	0.713	0.712	0.701	0.702	0.698	0.752	1.209
	LR	0.723	0.715	0.726	0.725	0.685	0.784	1.209
	DT	0.580	0.553	0.585	0.577	0.557	0.736	0.24
Protein Bert	_	0.542	0.580	0.542	0.514	0.447	0.675	58681.57
Hist2Vec (ours)	SVM	0.306	0.094	0.306	0.143	0.067	0.500	2.294
	NB	0.240	0.403	0.240	0.238	0.219	0.554	0.081
	MLP	0.707	0.712	0.707	0.707	0.683	0.818	5.879
	KNN	0.729	0.741	0.729	0.730	0.697	0.833	0.190
	RF	0.760	0.804	0.760	0.760	0.759	0.828	3.174
	LR	0.306	0.094	0.306	0.143	0.067	0.500	0.856
	DT	0.609	0.613	0.609	0.610	0.577	0.756	1.070

Table 2: Classification results (averaged over 5 runs) on **Coronavirus** dataset. The best classifier for respective embeddings is shown with the underline. Overall best values are shown in bold.

Embeddings	Algo	Acc. ↑	Prec. ↑	Recall ↑	F1 (Weig.) ↑	F1 (Macro) ↑	ROC AUC ↑	Train Time (sec.) ↓
PWM2Vec	SVM	0.799	0.806	0.799	0.801	0.648	0.859	44.793
	NB	0.381	0.584	0.381	0.358	0.400	0.683	2.494
	MLP	0.782	0.792	0.782	0.778	0.693	0.848	21.191
	KNN	0.786	0.782	0.786	0.779	0.679	0.838	12.933
	RF	0.839	0.839	0.836	0.828	0.739	0.862	7.690
	LR	0.809	0.815	0.809	0.800	0.728	0.852	274.91
	DT	0.801	0.802	0.801	0.797	0.633	0.829	4.537
String Kernel	SVM	0.601	0.673	0.601	0.602	0.325	0.624	5.198
	NB	0.230	0.665	0.230	0.295	0.162	0.625	0.131
	MLP	0.647	0.696	0.647	0.641	0.302	0.629	42.322
	KNN	0.613	0.623	0.613	0.612	0.310	0.629	0.434
	RF	0.668	0.692	0.668	0.663	0.360	0.658	4.541
	LR	0.554	0.724	0.554	0.505	0.193	0.568	5.096
	DT	0.646	0.674	0.646	0.643	0.345	0.653	1.561
WDGRL	SVM	0.329	0.108	0.329	0.163	0.029	0.500	2.859
	NB	0.230	0.098	0.004	0.007	0.002	0.496	0.008
	MLP	0.328	0.136	0.328	0.170	0.082	0.499	5.905
	KNN	0.235	0.198	0.235	0.211	0.058	0.499	0.081
	RF	0.261	0.196	0.261	0.216	0.051	0.499	1.288
	LR	0.332	0.149	0.332	0.177	0.034	0.500	0.365
	DT	0.237	0.202	0.237	0.211	0.054	0.498	0.026
Auto-Encoder	SVM	0.602	0.588	0.602	0.590	0.519	0.759	2575.9
	NB	0.261	0.520	0.261	0.303	0.294	0.673	21.74
	MLP	0.486	0.459	0.486	0.458	0.216	0.594	29.93
	KNN	0.763	0.764	0.763	0.735	0.547	0.784	18.51
	RF	0.800	0.796	0.800	0.791	0.648	0.815	57.90
	LR	0.717	0.750	0.717	0.702	0.564	0.812	11072.6
	DT	0.772	0.767	0.772	0.765	0.571	0.808	121.36
SeqVec	SVM	0.711	0.745	0.711	0.698	0.497	0.747	0.751
	NB	0.503	0.636	0.503	0.554	0.413	0.648	0.012
	MLP	0.718	0.748	0.718	0.708	0.407	0.706	10.191
	KNN	0.802	0.806	0.815	0.809	0.588	0.800	0.418
	RF	0.833	0.824	0.833	0.825	0.678	0.839	1.753
	LR	0.673	0.683	0.673	0.654	0.332	0.660	1.177
	DT	0.786	0.786	0.778	0.781	0.618	0.825	0.160
Protein Bert	_	0.799	0.806	0.799	0.789	0.715	0.841	15742.9
Hist2Vec (ours)	SVM	0.320	0.102	0.320	0.155	0.027	0.500	5.189
	NB	0.543	0.627	0.543	0.537	0.425	0.722	0.108
	MLP	0.741	0.747	0.741	0.737	0.470	0.730	3.548
	KNN	0.802	0.793	0.802	0.796	0.612	0.805	0.237
	RF	0.837	0.847	0.837	0.830	0.740	0.863	3.954
	LR	0.320	0.102	0.320	0.155	0.027	0.500	2.504
	DT	0.786	0.791	0.786	0.784	0.545	0.788	1.216

6 Conclusion

We propose the Hist2Vec approach, an effective and alignment-free embedding method that outperforms state-of-the-art methods. Combining histogram-based embedding with a Gaussian kernel provides an efficient and effective framework for capturing sequence similarities and generating high-quality feature representations. Hist2Vec achieves the greatest accuracy of 76% and ROC AUC score of 83.3% for the classification of human DNA data and the highest accuracy of 83.7% and ROC AUC score of 86.3% for the classification of coronavirus hosts. Future studies involve assessing the Hist2Vec on other viruses, such as Zika.

References

1. Ali, S., Bello, B., Chourasia, P., et al.: Pwm2vec: an efficient embedding approach for viral host specification from coronavirus spike sequences. MDPI Biology (2022)
2. Ali, S., Bello, B., Chourasia, P., et al.: Virus2vec: Viral sequence classification using machine learning. arXiv preprint arXiv:2304.12328 (2023)
3. Ali, S., Patterson, M.: Spike2vec: An efficient and scalable embedding approach for covid-19 spike sequences. CoRR arXiv:2109.05019 (2021)

4. Ali, S., Sahoo, B., Khan, M.A., Zelikovsky, A., Khan, I.U., Patterson, M.: Efficient approximate kernel based spike sequence classification. IEEE/ACM Transactions on Computational Biology and Bioinformatics (2022)
5. Ali, S., Tamkanat-E-Ali, Khan, M.A., Khan, I., Patterson, M., et al.: Effective and scalable clustering of sars-cov-2 sequences. Accepted for publication at "International Conference on Big Data Research (ICBDR)" (2021)
6. Altschul, S.F., Gish, W., Miller, W., Myers, E.W., Lipman, D.J.: Basic local alignment search tool. J. Mol. Biol. **215**(3), 403–410 (1990)
7. Bokharaeian, B., et al.: Automatic extraction of ranked snp-phenotype associations from text using a bert-lstm-based method. BMC Bioinform. **24**(1), 144 (2023)
8. Bonidia, R.P., Sampaio, L.D., et al.: Feature extraction approaches for biological sequences: a comparative study of mathematical features. Briefings in Bioinform. **22**(5), bbab011 (2021)
9. Brandes, N., Ofer, D., Peleg, Y., Rappoport, N., Linial, M.: Proteinbert: a universal deep-learning model of protein sequence and func. Bioinformatics **38**(8) (2022)
10. Chen, J., Li, K., et al.: A survey on applications of artificial intelligence in fighting against covid-19. ACM Comput. Surv. (CSUR) **54**(8), 1–32 (2021)
11. Chourasia, P., Ali, S., Ciccolella, S., Vedova, G.D., Patterson, M.: Reads2vec: Efficient embedding of raw high-throughput sequencing reads data. J. Comput. Biol. **30**(4), 469–491 (2023)
12. Chourasia, P., Ali, S., et al.: Clustering sars-cov-2 variants from raw high-throughput sequencing reads data. In: International Conference on Computational Advances in Bio and Medical Sciences, pp. 133–148. Springer (2021)
13. Corso, G., et al.: Neural distance embeddings for biological sequences. In: Advances in Neural Information Processing Systems, vol. 34, pp. 18539–18551 (2021)
14. Farhan, M., Tariq, J., Zaman, A., Shabbir, M., Khan, I.: Efficient approximation algorithms for strings kernel based sequence classification. In: Advances in neural information processing systems (NeurIPS), pp. 6935–6945 (2017)
15. Gabler, F., Nam, S.Z., et al.: Protein sequence analysis using the mpi bioinformatics toolkit. Curr. Protoc. Bioinformatics **72**(1), e108 (2020)
16. Golestan Hashemi, F.S., et al.: Intelligent mining of large-scale bio-data: bioinformatics applications. Biotech Biotechnol. Equipment **32**(1), 10–29 (2018)
17. Guan, M., Zhao, L., Yau, S.S.T.: Classification of protein sequences by a novel alignment-free method on bacterial and virus families. Genes **13**(10), 1744 (2022)
18. Heinzinger, M., et al.: Modeling aspects of the language of life through transfer-learning protein sequences. BMC Bioinform. **20**(1), 1–17 (2019)
19. Hsu, C.W., et al.: A practical guide to support vector classification (2003)
20. Human DNA: https://www.kaggle.com/code/nageshsingh/demystify-dna-sequencing-with-machine-learning/data (2022). Accessed 10 Oct 2022
21. Khajeh-Saeed, A., Poole, S., Perot, J.B.: Acceleration of the smith-waterman algorithm using single and multiple graphics processors. J. Comput. Phys. **229**(11), 4247–4258 (2010)
22. Khandelwal, M., Kumar Rout, R., Umer, S., Mallik, S., Li, A.: Multifactorial feature extraction and site prognosis model for protein methylation data. Brief. Funct. Genomics **22**(1), 20–30 (2023)
23. Kuzmin, K., et al.: Machine learning methods accurately predict host specificity of coronaviruses based on spike sequences alone. Biochem. Biophys. Res. Commun. **533**, 553–558 (2020)
24. Leslie, C., Eskin, E., Noble, W.S.: The spectrum kernel: A string kernel for svm protein classification. In: Biocomputing, pp. 564–575 (2001)

25. Lin, S.W., Ying, K.C., Chen, S.C., Lee, Z.J.: Particle swarm optimization for parameter determination and feature selection of support vector machines. Expert Syst. Appl. **35**(4), 1817–1824 (2008)
26. Lou, H., Schwartz, M., Bruck, J., Farnoud, F.: Evolution of k-mer frequencies and entropy in duplication and substitution mutation systems. IEEE Trans. Inf. Theory **66**(5), 3171–3186 (2019)
27. Mitchell, A.L., Attwood, T.K., Babbitt, P.C., Blum, M., Bork, P., Bridge, A., Brown, S.D., Chang, H.Y., El-Gebali, S., Fraser, M.I., et al.: Interpro in 2019: improving coverage, classification and access to protein sequence annotations. Nucleic Acids Res. **47**(D1), D351–D360 (2019)
28. Otto, M.P.: Scalable and interpretable kernel methods based on random fourier features (2023)
29. P. Kuksa, P., Khan, I., Pavlovic, V.: Generalized similarity kernels for efficient sequence classification. In: Proceedings of the 2012 SIAM International Conference on Data Mining, pp. 873–882. SIAM (2012)
30. Pickett, B.E., Sadat, E.L., Zhang, Y., Noronha, J.M., Squires, R.B., et al.: Vipr: an open bioinformatics database and analysis resource for virology research. Nucleic acids research, pp. D593–D598 (2012)
31. Qi, R., Guo, F., Zou, Q.: String kernels construction and fusion: a survey with bioinformatics application. Front. Comp. Sci. **16**(6), 166904 (2022)
32. Rao, R., Bhattacharya, N., et al.: Evaluating protein transfer learning with tape. Advances in neural information processing systems 32 (2019)
33. Roman, I., Santana, R., et al.: In-depth analysis of svm kernel learning and its components. Neural Comput. Appl. **33**(12), 6575–6594 (2021)
34. Saifuddin, K.M., et al.: Seq-hygan: Sequence classification via hypergraph attention network. arXiv preprint arXiv:2303.02393 (2023)
35. Scholkopf, B., Sung, K.K., et al.: Comparing support vector machines with gaussian kernels to radial basis function classifiers. IEEE Trans. Signal Process. **45**(11), 2758–2765 (1997)
36. Shen, J., Qu, et al.: Wasserstein distance guided representation learning for domain adaptation. In: AAAI Conference on Artificial Intelligence (2018)
37. Sikander, R., Ghulam, A., Ali, F.: Xgb-drugpred: computational prediction of drug-gable proteins using extreme gradient boosting and optimized features set. Sci. Rep. **12**(1), 5505 (2022)
38. Solis-Reyes, S., Avino, M., Poon, A., Kari, L.: An open-source k-mer based machine learning tool for fast and accurate subtyping of hiv-1 genomes. Plos One (2018)
39. Sun, C., Ai, X., Zhang, Z., Hancock, E.R.: Labeled subgraph entropy kernel. arXiv preprint arXiv:2303.13543 (2023)
40. Taslim, M., Prakash, C., et al.: Hashing2vec: Fast embedding generation for sars-cov-2 spike sequence classification. In: ACML, pp. 754–769. PMLR (2023)
41. Vamathevan, J., Clark, et al.: Applications of machine learning in drug discovery and development. Nature Rev. Drug Discovery **18**(6), 463–477 (2019)
42. Wood, D., Salzberg, S.: Kraken: ultrafast metagenomic sequence classification using exact alignments. Genome Biol. **15** (2014)
43. Xie, J., Girshick, R., Farhadi, A.: Unsupervised deep embedding for clustering analysis. In: International Conference on Machine Learning, pp. 478–487 (2016)

DCNN: Dual-Level Collaborative Neural Network for Imbalanced Heart Anomaly Detection

Ying An[1], Anxuan Xiong[2], and Lin Guo[1(✉)]

[1] Big Data Institute, Central South University, Changsha, China
guolincsu@csu.edu.cn
[2] School of Computer Science and Engineering, Central South University, Changsha, China

Abstract. The electrocardiogram (ECG) plays an important role in assisting clinical diagnosis such as arrhythmia detection. However, traditional techniques for ECG analysis are time-consuming and laborious. Recently, deep neural networks have become a popular technique for automatically tracking ECG signals, which has demonstrated that they are more competitive than human experts. However, the minority class of life-threatening arrhythmias causes the model training to skew towards the majority class. To address the problem, we propose a dual-level collaborative neural network (DCNN), which includes data-level and cost-sensitive level modules. In the Data Level module, we utilize the generative adversarial network with Unet as the generator to synthesize ECG signals. Next, the Cost-sensitive Level module employs focal loss to increase the cost of incorrect prediction of the minority class. Empirical results show that the Data Level module generates highly accurate ECG signals with fewer parameters. Furthermore, DCNN has been shown to significantly improve the classification of the ECG.

Keywords: ECG · Arrhythmia classification · Generative adversarial networks · Data imbalance

1 Introduction

Cardiovascular diseases (CVDs) kill nearly one million people per year in the US under a report by the American Heart Association [1]. In the therapy of CVDs and detection of arrhythmias, ECG is commonly used for auxiliary diagnosis. It is a non-invasive, inexpensive tool to obtain a patient's ECG signals during a period. The cardiologist can observe the amplitude of the signal as well as the cardiac cycle to identify whether the patient's heart electrical activity is normal.

However, the manual diagnosis of ECG is time-consuming and laborious. Thus, some artificial intelligence technologies have been applied to ECG classification and recognition [2]. For ECG classification, convolutional neural networks (CNN) and recurrent neural networks (RNN) focus on capturing the spatial and temporal features of ECG, respectively [3,4]. For the detection of atrial fibrillation, Unet has the characteristic of compensation for feature information [5]. Although there are many methods for ECG classification and recognition, they mainly focus on the feature learning and classifier of ECG signals, lacking the consideration of data imbalance and feature extraction for rare heartbeats [6].

For data imbalance issue, the minority class has difficulty learning valuable features, and the majority class will be overtrained. They are extremely adverse for the identification of arrhythmias [7]. Methods to address the ECG data imbalance have three classes: algorithm level, cost-sensitive level, and data level. For the algorithm level, the SMOTE algorithm cluster is taken to generate fake samples or to dynamically move the boundaries of the classification while training [8]. For the cost-sensitive level, a loss function was designed to increase the weight of the minority class samples to alleviate the imbalance of ECG data [9]. For ECG data augmentation, the dominant method utilized is Generative Adversarial Networks (GANs). In summary, the SMOTE algorithm is not available for multi-classification, and a large number of studies reveal that the data augmentation by synthesis of heartbeats is most effective to alleviate data imbalance of ECG in the three levels of methods [10].

In this paper, we propose an Unet-based Generative Adversarial Network to synthesize ECG signals for balancing the number of different heartbeats. The architecture of the Data Level module consists of two components, a discriminator based on a convolutional neural network and a generator constructed by Unet. We further incorporate focal loss in downstream tasks to mitigate the side effects of data imbalance on a model as much as possible [11]. The empirical results on the gold-standard dataset show that DCNN can significantly alleviate the training effects of data imbalance for improving classification accuracy. Our contributions are summarized as follows:

1. To the best of our knowledge, we are the first to propose a Dual-Level Collaborative Neural Network (DCNN) to alleviate ECG data imbalance at both data and cost-sensitive levels. The results verified that the data-level module has fewer model parameters and generates high-quality ECG signals.
2. Experimental results on different datasets verified that identification of the minority class of heartbeat samples, which DCNN is effective. F1 scores for the minority class improved by an average of 4.9% and 19.3% on the MIT-BIH and INCART datasets, respectively.

2   Related Work

The imbalance for medical signals. Most effective deep neural networks are been employed in earlier stages of ECG signal data imbalance. Petmezas et al. (2021) proposed a CNN-LSTM architecture to alleviate data imbalance for ECG of atrial fibrillation (AF), where the features got by CNN are fed into LSTM for the sake of capturing dynamic temporal features [12]. Pandey &Janghel (2019) use a mixture of CNN and SMOTE techniques to first enlarge the sample size of the minority classes with SMOTE and then classify ECG using an end-to-end structure [13]. Gao et al. (2019) and Romdhane &Pr (2020) select LSTM and CNN respectively for feature extraction of ECG, except that the training loss function is converted to focal loss (FL) to improve the precision of ECG identification. The prevalence of GAN as a data augmentation technique has inspired much effort to trial GAN in its varying manners for ECG data imbalance [14,15]. Lima et al. (2019) proposed a combination of 2D CNN and InfoGAN to synthesize a few abnormal heartbeats. InfoGAN involves an extra code being added to the initial input vector for controlling synthetic fake signals. Xia, Xu et al. (2023) treated the transformer as the generator of GAN to synthetic fake signal, which is composed of the transformer, upscaling, and fully connected layer. Classification accuracy of different classes of heartbeat will be elevated after incorporating fake signals into the training process [16,17]. In contrast, our work focuses on mitigating ECG data imbalances at both the data and loss levels.

Generative Adversarial Networks. With the remarkable success of GAN in such a wide range of fields, it recently has been implemented into the synthesis of ECG signals. Golany et al. (2020a) exploited convolution and deconvolution as components of GAN architecture to synthetic ECG signals with improvement in ECG classification. Also in the same year, Golany et al. (2020b) explored the optimization of the training process of synthesizing fake signals by modeling the dynamic processes of ECG through ordinary differential equations. Apart from traditional random noise vector inputs, other forms of inputs are present in GAN [18,19]. For example, Sarkar &Etemad (2021) take PPG as input of GAN to synthetic fake signals, moreover, the dual discriminator preserves the salient features in time and frequency domains to generate an ECG signal of integrity [20]. Besides, patients could detect multi-lead (e.g., 6, 8, 12-lead) ECG signals, that is, ECG signals of different views. Chen et al. (2022) proposed ME-GAN, which the idea is using angular information of leads to obtain the features of multi-view ECG signals, which these features are subsequently leveraged to synthetic multi-view ECG signals. However, in terms of specific cardiac abnormal signals, generation using GAN is still rather scarce [21]. Li et al. (2022) thereby attempted SLC-GAN to synthetic myocardial infarction (MI) ECG signals for better detection accuracy of single-lead MI. In contrast, we employed the Unet-generator to synthesize ECG signals, which not only ensures stable training but also results in the synthesis of high-quality signals. [22].

3   Methods

In this section, we will introduce the proposed DCNN (Dual-Level Collaborative neural network). The architecture consists of a Data Level module for data augmentation, a Cost-sensitive Level module for increasing the weight of the minority class samples, and a residual module for feature extraction. The overall architecture of the model is shown in Fig. 1. Next, we will elaborate on the details of the proposed model.

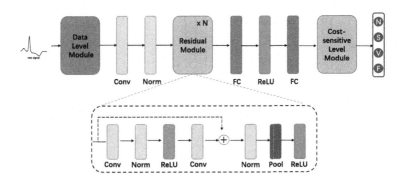

Fig. 1. The proposed Dual-Level Collaborative neural network for alleviating ECG data imbalance. the raw ECG signal as input first enters the Data Level module, followed by the residual module, and finally the Cost-sensitive module.

Data Level Module

Unet-based Generator: Inspired by Unet, we propose an Unet generator to synthesize fake signals. As shown in Fig. 2, the generator consists of a contracting path(downsampling) and an expensive path(upsampling). To capture the global feature information, the contracting path is based on convolutional network architecture. It consists of two 1D convolutions layers, each followed by a rectified linear unit (ReLU) and a max pooling operation with stride 2 for downsampling. At each step of downsampling, we double the number of feature channels. Repeated contraction occurs a total of three times during the process of heartbeat synthesis using the Unet generator. In order to facilitate the subsequent feature upsampling, feature maps by the contracting path are additionally followed by convolution and ReLU. Every step in the expensive path includes upsampling followed by convolution, concatenation of feature map with feature map in the contracting path, and multiple convolutions of concatenated features. The referred feature incorporation is to trade off feature information loss in downsampling.

Discriminator: The discriminator takes the real and generated signals as input and eventually outputs a binary decision score. It judges whether the signal is from the training set or synthesized by the generator. The objective of the discriminator is that the score of the real sample is close to 1, while the score of the fake sample is close to 0. The discriminator was built with six convolutional layers, each of them followed by batch normalization together with a LeakyReLU activation function that ended with a sigmoid activation function.

More formally, we denote the generator network as $G(x; \theta_G) : X \rightarrow X_{\text{fake}}$, and the discriminator network as $D(x; \theta_D) : X \rightarrow [0, 1]$, where X denotes raw signal while X_{fake} is generated fake signal. Both θ_G and θ_D represent the parameters of the generator and discriminator networks, respectively.

Using the above notation, the following specific objective:

$$V(\theta_G, \theta_D) = E_{x \sim p_{\text{data}}}[log D(x; \theta_D)] + E_{z \sim p_{\text{fake}}}[1 - log D(G(z; \theta_G); \theta_D)] \quad (1)$$

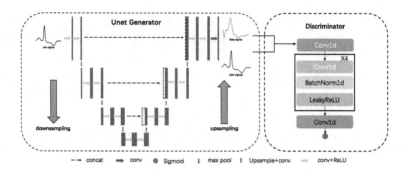

Fig. 2. Data Level module includes a generator and discriminator. Real signal (black solid line) as input and fake signal (red dashed line) as output. The generator contains the contracting path and the expansive path, and the discriminator scoring signals to identify their source. (Color figure online)

Where p_{data} denotes the distribution of training data, p_{fake} denotes the distribution of false signals. $E[\cdot]$ denotes expected value. The objective of the discriminator is to maximize this expression with respect to parameter θ_D. The objective of the generator is to minimize this objective with respect to parameter θ_G.

Residual Module

As for the classifier layout, we introduced a residual network as an architectural framework to construct the ECG classification model, as shown in Fig. 1. Signals first go through a convolution layer with a normalization layer, followed by a residual module, which ultimately makes use of a fully connected layer in combination with the ReLU activation function to output a multi-classification score for signals. The number of modules employed in residual modules is varying to fit signals in different datasets. For example, the number of residual modules used for signal feature extraction in the MIT-BIH dataset is 5, while the corresponding number of residual modules in the INCART dataset is 6. Moreover, in contrast to the primitive Resnet, the batch normalization layer is attached behind convolution to enhance the robustness of data for more stable training.

Cost-Sensitive Level Module

For addressing data imbalance at different levels, we further relieve imbalance issues at the loss level in multi-classification leveraging focal loss. Focal loss takes two key factors to alleviate the training skewness of data imbalance in binary classification for Object Detection. Under this premise, Focal loss can be extended to multi-classification as well. The formulas are following:

$$F_f = -\alpha_t(1 - p_t)^\gamma \cdot log(p_t + b) \tag{2}$$

F_f represents the objective function of focal loss, the key to loss is two factors α_t and γ. p_t is the predicted value of signals via softmax, in which training raises the penalty for incorrect prediction of the minority class if the p_t is smaller the bigger the previous term in the equation. α_t is class weight, inversely related to the number of classes. The b in the latter term is acted as a bias to prevent overflow.

4 Experiments

Compared Methods

LSTM + Attention: The characteristics of LSTM for capturing temporal features and Attention for calculating feature weights have been verified in a variety of fields, thus we use a combination of LSTM and Attention to extract feature information from ECG as a basic comparison.

AlexNet: The typical pattern in the domain of convolutional neural networks uses a composition of five convolutional layers with fully connected layers to learn feature information [23].

VGG: VGG is designed to explore the effect of deep neural network depth on feature extraction. Extensive practice can conclude that the increase in network depth could enhance the capability of feature learning within a certain number of layers [24].

CBM20: We treat this methodology as a beginner competitor in handling ECG data imbalance. To improve the recognition of abnormal heartbeats, Romdhane &Pr (2020) focus on the loss level to tackle the issue and employ DNN to work together with the focal loss [15].

We carried out the consistency data augmentation using the Data Level module for different datasets (different types of samples were synthesized into the same number of ECG signals and then put into the training set). Sufficient samples existed of the N class in both datasets to meet the expected recognition outcome, thus based on the original MIT-BIH, we enlarged the samples of the S, V, and F classes to 5000, 10000, and 2000, whereas in the data augmentation of INCART dataset, we increased S, V, and F to 30,000, 3800, and 400.

Implementation

Our model and other experiments were implemented with PyTorch 1.4 and Python 3.6 on an RTX2080Ti GPU.

We evaluate the model on the MIT-BIH [25] and INCART datasets [26]. Firstly, the datasets are divided into training and testing sets, where the training set is 80% and the testing set is 20%. Besides, the batch size in the model training is 128, and the epoch is set to 2000 and 100 for DCNN training, respectively. Further, we picked the Adam optimizer to update the network parameters, where the learning rate for training the MIT-BIH and INCART datasets was fixed at 0.08 and 0.0005. It is worth noting that in case of data imbalance, Accuracy always shows a high score under the impact of the majority class when evaluating the model. Hence, we use precision, recall, and F1 as the evaluation metrics for the model.

5 Results

Quantitative Results

Results from experiments of different methods on two datasets are reported in Table 3. It can be seen that our proposed DCNN outperforms the existing abnormal heartbeat identification methods in the recognition of the minority classes from the experimental results.

The results of the experiments performed on the MIT-BIH dataset from Table 1, reveal that the recognition of the minority class samples (S, F) of ECG signals gets improvement in different metrics with the utilization of the DCNN. Compared with the method of CBM20, the S class samples gained 13.09%, 15.46%, and 14.51% on Precision, Recall, and F1, respectively, and the F class samples gained 5.79%, 0.62%, and 2.42%. The volume of N and V class ECG signals supports the models to learn the majority class features easily and the different models all have high identification rates, therefore the improvement on the majority class samples is not noticeable.

Meanwhile, DCNN has distinguished performance over other methods on the INCART dataset as well. The bottom part of Table 3, indicates that the metrics of the minority classes show obvious enhancement. The Precision, Recall, and F1 of the F class signals have an average improvement of 21.6%, 26.13%, and 28.18% compared with the other methods. The Precision of the S class signals is approximately equal to the results of the other models, while the Recall and F1 have an elevation of 16.24%, and 10.51%, respectively.

Table 1. Comparison of DCNN with other methods on the MIT-BIH dataset and the INCART dataset. Pre, Re, and F1 denote Precision, Recall, and F1. The best results are shown in bold.

Models	MIT-BIH											
	N			S			V			F		
	Pre(%)	Re(%)	F1(%)	Pre(%)	Re(%)	F1(%)	Pre(%)	Re(%)	F1(%)	Pre(%)	Re(%)	F1(%)
LSTM + Attention	98.94	99.65	99.29	92.77	78.42	84.99	95.89	96.43	96.16	95.37	64.38	76.87
AlexNet	99.08	99.49	98.99	72.14	80.58	76.13	96.71	96.36	96.53	92.79	64.38	76.01
VGG	99.03	99.49	99.26	92.44	79.14	85.27	94.11	96.79	95.43	90.52	65.62	76.09
CBM20	98.57	99.26	98.91	80.08	67.99	73.54	95.30	95.44	95.37	89.66	65.00	75.36
DCNN	**99.15**	**99.72**	**99.43**	**93.17**	**83.45**	**88.05**	**96.94**	**97.29**	**97.12**	**95.45**	**65.62**	**77.78**
Models	INCART											
	N			S			V			F		
	Pre(%)	Re(%)	F1(%)	Pre(%)	Re(%)	F1(%)	Pre(%)	Re(%)	F1(%)	Pre(%)	Re(%)	F1(%)
LSTM + Attention	98.83	98.90	98.87	82.16	71.68	76.57	92.54	93.85	93.19	20.00	6.82	10.17
AlexNet	98.51	**99.49**	98.99	85.90	68.47	76.14	**96.32**	91.65	93.93	50	13.64	21.43
VGG	98.69	99.19	98.94	**87.31**	71.94	78.88	93.94	92.58	93.25	28.57	9.09	13.79
CBM20	96.61	98.95	97.77	76.16	54.59	63.60	91.64	77.79	84.15	35.71	11.36	17.24
DCNN	**99.05**	99.37	**99.21**	85.75	**82.91**	**84.31**	96.21	**94.48**	**95.34**	**55.17**	**36.36**	**43.84**

Qualitative Results

We performed a visual qualitative analysis of the S, V, and F class signals synthesized by different models. As shown in Fig. 3, the gap between the signal synthesized by the Data Level module (UnetGAN) and the original signal is rather small, on the contrary, the gap between the signal synthesized by DCGAN and the original signal is greater. Among them, it is observed from the upper and lower half figures that the signal synthesized using MIT-BIH as the source data is closer to the original signal than the signal synthesized using INCART as the source data. The cause is that the data of MIT-BIH is neater than the data of INCART, and the inconsistent signal length in INCART leads to the poor training effect of DCGAN. Besides, the signals synthesized by DCGAN contain more noise than those synthesized by the Data Level module. In other words, data augmentation using UnetGAN will enhance the detector training and then improve the recognition of ECG.

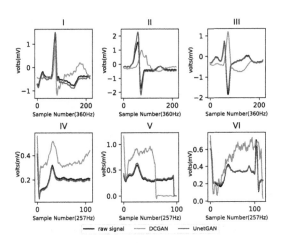

Fig. 3. Comparison of the morphology of heartbeats synthesized using UnetGAN, DCGAN, and the raw heartbeats. There are six subplots, where subplots I, II, and III are S, V, and F heartbeats synthesized from MIT-BIH, and subplots IV, V, and VI are S, V, and F heartbeats synthesized from INCART.

In terms of the complexity of the model parameters, as shown in Table 2, UnetGAN is more lightweight than DCGAN. The parameter size of the model is 139.15 MB when UnetGAN synthesizes the ECG signal on MIT-BIH, while DCGAN is 165.25 MB. And DCGAN synthesizes the heartbeat on INCART, not only is the parameter size higher than UnetGAN 19.94 MB, as well as DCGAN is prone to mode collapse during training.

Ablation Study

We perform an ablation study on both datasets. Numerous experiments have

Table 2. Comparison of different model parameter complexities in the MIT-BIH and INCART.

Method	Parm(MITBIH)	Parm(INCART)
DCGAN	165.25	185.04
UnetGAN	139.15	165.10

demonstrated high accuracy for the majority class (N, V), and we are therefore concerned about the performance of the minority class (S, F). DCNN w/o FL is a variant of DCNN without taking Focal loss into account, while DCNN w/o FS is a variant of Unet not utilizing the fake signal synthesized by the generator for data augmentation. we can see in Table 3 that the experimental results of the proposed DCNN are superior to the different variants.

Table 3. Ablation study using the MIT-BIH and INCART datasets.

Models	MIT-BIH						INCART					
	S			F			S			F		
	Pre(%)	Re(%)	F1(%)	Pre(%)	Re(%)	F1(%)	Pre(%)	Re(%)	F1(%)	Pre(%)	Re(%)	F1(%)
DCNN w/o FS	78.65	60.97	68.69	91.51	60.62	72.93	84.73	75.00	79.57	26.32	11.36	15.87
DCNN w/o FL	67.17	71.76	69.39	93.14	59.38	72.52	85.39	76.02	80.43	35.00	31.82	33.33
DCNN	**93.17**	**83.45**	**88.05**	**95.45**	**65.62**	**77.78**	**85.75**	**82.91**	**84.31**	**55.17**	**36.36**	**43.84**

6   Conclusions

In this work, we try to solve the issue of data imbalance of ECG signals at different levels to boost the identification of abnormal heartbeats. At the data level, we propose a Data Level Module to synthesize the high-quality heartbeats, which are then fed into the training set to enhance the model training and ultimately improve the model representation learning capability. At the Cost-sensitive level, the focal loss is applied to further upgrade the recognition of anomalous heartbeats. Experimental results on the MIT-BIH and INCART datasets demonstrate the effectiveness of our approach. In particular, DCNN also has prominent performance in the case where ECG signals in the dataset are not aligned. Abnormal heartbeats due to rare diseases are potential for information extraction. In the future, we will investigate how to synthesize the minority class of heartbeats with insufficient volume of samples to better identify the rare heartbeats further.

Acknowledgement. This work was supported in part by the National Key Research and Development Program of China (2021YFF1201300); in part by the Changsha Municipal Natural Science Foundation (kq2202106); and in part by the National Natural Science Foundation of China (62102456).

References

1. Tsao, C.W., et al.: Heart disease and stroke statistics-2023 update: a report from the American heart association. Circulation **147**(8), e93–e621 (2023)
2. Essa, E., Xie, X.: Multi-model deep learning ensemble for ECG heartbeat arrhythmia classification. In 2020 28th European Signal Processing Conference (EUSIPCO), pp. 1085–1089. IEEE (2021)
3. Li, Y., Pang, Y., Wang, J., Li, X.: Patient-specific ECG classification by deeper CNN from generic to dedicated. Neurocomputing **314**, 336–346 (2018)
4. Singh, S., Pandey, S.K., Pawar, U., Janghel, R.R.: Classification of ECG arrhythmia using recurrent neural networks. Procedia Comput. Sci. **132**, 1290–1297 (2018)
5. Ronneberger, O., Fischer, P., Brox, T.: U-Net: convolutional networks for biomedical image segmentation. In: Navab, N., Hornegger, J., Wells, W.M., Frangi, A.F. (eds.) MICCAI 2015. LNCS, vol. 9351, pp. 234–241. Springer, Cham (2015). https://doi.org/10.1007/978-3-319-24574-4_28
6. Wei, L., Hou, H., Chu, J.: Feature fusion for imbalanced ECG data analysis. Biomed. Signal Process. Control **41**, 152–160 (2018)
7. Japkowicz, N., Stephen, S.: The class imbalance problem: a systematic study. Intell. Data Anal. **6**(5), 429–449 (2002)
8. Kandala NVPS Rajesh and Ravindra Dhuli: Classification of imbalanced ECG beats using re-sampling techniques and AdaBoost ensemble classifier. Biomed. Signal Process. Control **41**, 242–254 (2018)
9. Zhou, Z.-H., Liu, X.-Y.: Training cost-sensitive neural networks with methods addressing the class imbalance problem. IEEE Trans. Knowl. Data Eng. **18**(1), 63–77 (2005)
10. Goodfellow, I., et al.: Generative adversarial networks. Commun. ACM **63**(11), 139–144 (2020)
11. Lin, T.-Y., Goyal, P., Girshick, R., He, K., Dollár, P.: Focal loss for dense object detection. In: Proceedings of the IEEE International Conference on Computer Vision, pp. 2980–2988 (2017)
12. Petmezas, G., et al.: Automated atrial fibrillation detection using a hybrid CNN-LSTM network on imbalanced ECG datasets. Biomed. Signal Process. Control **63**, 102194 (2021)
13. Saroj Kumar Pandey and Rekh Ram Janghel: Automatic detection of arrhythmia from imbalanced ECG database using CNN model with smote. Australas. Phys. Eng. Sci. Med. **42**(4), 1129–1139 (2019)
14. Gao, J., Zhang, H., Lu, P., Wang, Z., et al.: An effective lstm recurrent network to detect arrhythmia on imbalanced ecg dataset. J. Healthc. Eng. **2019** (2019)
15. Taissir Fekih Romdhane and Mohamed Atri Pr: Electrocardiogram heartbeat classification based on a deep convolutional neural network and focal loss. Comput. Biol. Med. **123**, 103866 (2020)
16. Lima, J.L.P., Macêdo, D., Zanchettin, C.: Heartbeat anomaly detection using adversarial oversampling. In: 2019 International Joint Conference on Neural Networks (IJCNN), pp. 1–7. IEEE (2019)
17. Xia, Y., Yangyang, X., Chen, P., Zhang, J., Zhang, Y.: Generative adversarial network with transformer generator for boosting ECG classification. Biomed. Signal Process. Control **80**, 104276 (2023)
18. Golany, T., Lavee, G., Yarden, S.T., Radinsky, K.: Improving ECG classification using generative adversarial networks. In: Proceedings of the AAAI Conference on Artificial Intelligence, vol. 34, pp. 13280–13285 (2020)

19. Golany, T., Radinsky, K., Freedman, D.: SimGANs: simulator-based generative adversarial networks for ECG synthesis to improve deep ECG classification. In: International Conference on Machine Learning, pp. 3597–3606. PMLR (2020)
20. Sarkar, P., Etemad, A.: CardioGAN: attentive generative adversarial network with dual discriminators for synthesis of ECG from PPG. In: Proceedings of the AAAI Conference on Artificial Intelligence, vol. 35, pp. 488–496 (2021)
21. Chen, J., Liao, K., Wei, K., Ying, H., Chen, D.Z., Wu, J.: ME-GAN: Learning panoptic electrocardio representations for multi-view ECG synthesis conditioned on heart diseases. In: International Conference on Machine Learning, pp. 3360–3370. PMLR (2022)
22. Li, W., Tang, Y.M., Yu, K.M., To, S.: SLC-GAN: an automated myocardial infarction detection model based on generative adversarial networks and convolutional neural networks with single-lead electrocardiogram synthesis. Inf. Sci. **589**, 738–750 (2022)
23. Krizhevsky, A., Sutskever, I., Hinton, G.E.: ImageNet classification with deep convolutional neural networks. In: Advances in Neural Information Processing Systems, 25 (2012)
24. Simonyan, K., Zisserman, A.: Very deep convolutional networks for large-scale image recognition. arXiv preprint arXiv:1409.1556 (2014)
25. Moody, G.B., Mark, R.G.: The impact of the MIT-BIH arrhythmia database. IEEE Eng. Med. Biol. Mag. **20**(3), 45–50 (2001)
26. Tihonenko, V., Khaustov, A., Ivanov, S., Rivin, A., Yakushenko, E.: St petersburg incart 12-lead arrhythmia database. PhysioBank PhysioToolkit and PhysioNet (2008)

On the Realisability of Chemical Pathways

Jakob L. Andersen[1], Sissel Banke[1(✉)], Rolf Fagerberg[1], Christoph Flamm[2,8],
Daniel Merkle[1], and Peter F. Stadler[2,3,4,5,6,7]

[1] Department of Mathematics and Computer Science, University of Southern
Denmark, Odense DK-5230, Denmark
{jlandersen,banke,rolf,daniel}@imada.sdu.dk
[2] Institute for Theoretical Chemistry, University of Vienna, Wien 1090, Austria
xtof@tbi.univie.ac.at
[3] Bioinformatics Group, Department of Computer Science, and Interdisciplinary
Center for Bioinformatics, University of Leipzig, Leipzig 04107, Germany
stadler@bioinf.uni-leipzig.de
[4] Max Planck Institute for Mathematics in the Sciences, Leipzig 04103, Germany
[5] Fraunhofer Institute for Cell Therapy and Immunology, Leipzig 04103, Germany
[6] Center for non-coding RNA in Technology and Health, University of Copenhagen,
Frederiksberg DK-1870, Denmark
[7] Santa Fe Institute, 1399 Hyde Park Rd, Santa Fe NM 87501, USA
[8] Research Network Chemistry Meets Microbiology, University of Vienna, Wien
1090, Austria

Abstract. The exploration of pathways and alternative pathways that
have a specific function is of interest in numerous chemical contexts. A
framework for specifying and searching for pathways has previously been
developed, but a focus on which of the many pathway solutions are realisable, or can be made realisable, is missing. Realisable here means that
there actually exists some sequencing of the reactions of the pathway
that will execute the pathway. We present a method for analysing the
realisability of pathways based on the reachability question in Petri nets.
For realisable pathways, our method also provides a certificate encoding
an order of the reactions which realises the pathway. We present two
extended notions of realisability of pathways, one of which is related
to the concept of network catalysts. We exemplify our findings on the
pentose phosphate pathway. Lastly, we discuss the relevance of our concepts for elucidating the choices often implicitly made when depicting
pathways.

1 Introduction

Large *Chemical Reaction Networks* (CRN) lie at the heart of many questions and
challenges in research, industry, and society. Examples include understanding
metabolic networks and their regulation in health and biotechnology; planning
and optimising chemical synthesis in industry and research labs; modelling the
fragmentation of molecular ions inside mass spectrometers; probing hypotheses
on the origins of life; and monitoring environmental pollution in air, water and

soil. Subnetworks with desirable properties, often called *pathways*, such as a synthesis plan for a given target molecule, or a metabolic subsystem, are of particular interest. The ability to specify and search for pathways in a given CRN thus is a core objective in chemical modelling, exploration, and design.

CRNs can be modelled as directed hypergraphs [3,4,19,25], where each molecule is represented by a vertex and each reaction is modelled by a directed hyperedge. Viewing pathways in CRNs as sets of reactions with integer multiplicities, [3] formally defined pathways as integer hyperflows in hypergraphs. In contrast to real-valued hyperflows, integer hyperflows account for molecules as indivisible entities and allows a more direct mechanistic interpretation. In [3], also the concept of a *chemical transformation motif* in a CRN was introduced, providing a versatile framework for querying reaction networks for pathways. A chemical transformation motif is the specification of a pathway by prescribing the input and output compounds (intermediate products may appear but must be used up again). Finding and enumerating pathways fulfilling a chemical transformation motif can be treated computationally via Integer Linear Programming (ILP) [3]. ILP is NP-hard, both in the general case and in the restricted setting of finding integer hyperflows in CRNs [1]. However, for many networks and pathways of practical interest, current ILP solvers perform well [3].

The starting point of this paper is the following observation: While integer hyperflows specify reactions and their multiplicities, they do not determine in which *order* the individual reactions take place to perform the specified overall chemical transformation. In fact, there may be *no* sequencing possible. Figure 4 gives an example of this. Specifically, no sequencing of the reactions e_1 and e_2 in the hyperflow of Fig. 4 will make it executable—in essence, C or D must be present *before* they can be produced. We introduce the term *realisable* for hyperflows where the corresponding chemical transformation is executable by some sequence of the constituent reactions of the hyperflows. We develop a framework that converts integer hyperflows into corresponding Petri nets, which then allow us to use Petri net methodology to express and decide whether integer hyperflows are realisable. Petri nets have already been used extensively to model metabolic networks [5].

For realisable flows we introduce the concept of a *realisability certificate* that specifies an order in which the reactions can occur along the pathway. Finding an explicit order both facilitates a mechanistic understanding of pathways and is a necessity for investigations where the identity of individual atoms matter, such as computing atom traces [2]. We also study ways in which non-realisable integer hyperflows can be extended to realisable ones. One option is a scaling of the flow itself, another is borrowing additional molecules which are then returned. The latter construction is closely related to the concept of a "network catalyst" (see e.g. [8,18]). An algorithmic approach to deciding realisability by borrowing thus forms an important basis for a future formal computational treatment of higher-level chemical motifs such as autocatalysis and even hypercycles [10,11,23,24]. Finally, we utilise the non-oxidative phase of the pentose phosphate pathway (PPP) to demonstrate our approach and explore how to find potential catalysts within the network. PPP is a well-known example which highlights the importance of simplicity in finding solutions [17,20].

The main thrust of our paper lies in formally defining and exploring the concept of realisability of pathways. However, we would like to point out that commonly used representations of pathways in the life science literature often fall in between the two extremes of integer hyperflows and realisability certificates. We believe that our formalisation of these concepts may help raise awareness of the choices one often subconsciously makes when creating illustrations of pathways. We elaborate on this viewpoint in Sect. 5.

2 Preliminaries

2.1 Chemical Reaction Networks and Pathways

A CRN can be modelled by a directed hypergraph $\mathcal{H} = (V, E)$, where V is the set of vertices representing the molecules. Reactions are represented as directed hyperedges E, where each edge $e = (e^+, e^-)$ is an ordered pair of multisets of vertices, i.e., $e^+, e^- \subseteq V$.[1] We call e^+ the *tail* of the edge e, and e^- the *head*. In the interest of conciseness we will refer to directed hypergraphs simply as hypergraphs, directed hyperedges simply as edges, and CRNs as networks. For a multiset Q and an element q we use $m_q(Q)$ to denote its multiplicity, i.e., the number of occurrences of q in Q. When denoting multisets we use the notation $\{\!\!\{ \ldots \}\!\!\}$, e.g., $Q = \{\!\!\{ a, a, b \}\!\!\}$ is a multiset with $m_a(Q) = 2$ and $m_b(Q) = 1$. For a vertex $v \in V$ and a set of edges A we use $\delta_A^+(v)$ and $\delta_A^-(v)$ to denote respectively the set of out-edges and in-edges of v contained in A, i.e., the edges in A that have v in their tail and v in their head, respectively.

In order to later define pathways we first introduce an extension of the network for representing input and output of compounds. Given a hypergraph $\mathcal{H} = (V, E)$ we define the *extended hypergraph* $\overline{\mathcal{H}} = (V, \overline{E})$ with $\overline{E} = E \cup E^- \cup E^+$, where

$$E^- = \{e_v^- = (\emptyset, \{\!\!\{ v \}\!\!\}) \mid v \in V\} \qquad E^+ = \{e_v^+ = (\{\!\!\{ v \}\!\!\}, \emptyset) \mid v \in V\} \qquad (1)$$

The hypergraph $\overline{\mathcal{H}}$ has additional "half-edges" e_v^- and e_v^+, for each $v \in V$. These explicitly represent potential input and output channels to and from $\mathcal{H}$, i.e., what is called exchange reactions in metabolic networks. An example of an extended hypergraph is shown in Fig. 1.

In [3] it was proposed to model a pathway in a network $\mathcal{H} = (V, E)$ as an *integer hyperflow*. This is an integer-valued function f on the extended network, $f \colon \overline{E} \to \mathbb{N}_0$, which satisfies the following *flow conservation constraint* on each vertex $v \in V$:

$$\sum_{e \in \delta_{\overline{E}}^+(v)} m_v(e^+) f(e) - \sum_{e \in \delta_{\overline{E}}^-(v)} m_v(e^-) f(e) = 0 \qquad (2)$$

---

[1] When comparing a multiset M and a set S, we view M as a set. I.e., $M \subseteq S$ holds if every element in M is an element of S.

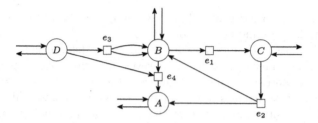

Fig. 1. Example of an extended hypergraph. It has vertices $\{A, B, C, D\}$, edges $\{e_1, e_2, e_3, e_4\}$, and a half-edge to and from each vertex. An edge e is represented by a box with arrows to (from) each element in e^- (e^+).

Note in particular that $f(e_v^-)$ is the input flow for vertex v and $f(e_v^+)$ is its output flow. An example of an integer hyperflow is shown in Fig. 2.

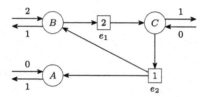

Fig. 2. Example integer hyperflow f on the extended hypergraph from Fig. 1. Vertex D has been omitted as it has no in- or out-flow. Edges leaving or entering D have also been omitted as they have no flow. The flow on an edge is represented by an integer. For example, the half edge into B has flow $f(e_B^-) = 2$, the half edge leaving B has flow $f(e_B^+) = 1$, and edge e_1 has flow $f(e_1) = 2$.

2.2 Petri Nets

Petri nets are an alternative method to analyse CRNs. Each molecule in the network forms a *place* in the Petri net and each reaction corresponds to a transition [16,21,22]. The stoichiometric matrix commonly used in chemistry has an equivalent in Petri net terminology, called the incidence matrix [16]. In Sect. 3 we will describe a transformation of a hyperflow to a Petri net. The following notation for Petri nets (with the exception of arc weights) follows [12].

A *net* is a triple (P, T, W) with a set of places P, a set of transitions T, and an arc weight function $W: (P \times T) \cup (T \times P) \rightarrow \mathbb{N}_0$. A *marking* on a net is a function $M: P \rightarrow \mathbb{N}_0$ assigning a number of tokens to each place. With $M_\emptyset$ we denote the empty marking, i.e., $M_\emptyset(p) = 0$, $\forall p \in P$. A *Petri net* is a pair (N, M_0) of a net N and an initial marking M_0. For all $x \in P \cup T$, we define the *pre-set* as $^\bullet x = \{y \in P \cup T \mid W(y, x) > 0\}$ and the *post-set* as $x^\bullet = \{y \in P \cup T \mid W(x, y) > 0\}$. We say that a transition t is enabled by the marking M if $W(p,t) \leq M(p), \forall p \in P$. When a transition t is enabled it can *fire*, resulting in a marking M' where $M'(p) = M(p) - W(p,t) + W(t,p)$, $\forall p \in P$. Such a firing is denoted by $M \xrightarrow{t} M'$. A *firing sequence* σ is a sequence of firing transitions $\sigma = t_1 t_2 \ldots t_n$. Such a firing sequence gives rise to a sequence of markings $M_0 \xrightarrow{t_1} M_1 \xrightarrow{t_2} M_2 \xrightarrow{t_3} \ldots \xrightarrow{t_n} M_n$ which is denoted by $M_0 \xrightarrow{\sigma} M_n$.

3   Realisability of Integer Hyperflows

The paper [3] gave a method (summarized in Sect. 2.1) for specifying pathways in CRNs and then proceeded to use ILP to enumerate pathway solutions fulfilling the specification.

In this paper, we focus on assessing the realisability of such a pathway solution and on determining a specific order of reactions that proves its realisability. To this end, we map integer hyperflows into Petri nets and rephrase the question of realisability as a particular reachability question in the resulting Petri net.

3.1   Flows as Petri Nets

We convert a hypergraph $\mathcal{H} = (V, E)$ to a net $N = (P, T, W)$ by using the vertices V as the places P and the edges E as the transitions T, and by defining the weight function from the incidence information as follows: for each vertex/place $v \in V$ and edge/transition $e = (e^+, e^-) \in E$ let $W(v, e) = m_v(e^+)$ and $W(e, v) = m_v(e^-)$. This conversion also works for extended hypergraphs, where the half-edges result in transitions with either an empty pre-set or post-set. The transitions corresponding to input reactions are thus always enabled. Denote by $M_\emptyset$ the empty marking on N. Every firing sequence σ starting and ending in $M_\emptyset$, i.e., $M_\emptyset \xrightarrow{\sigma} M_\emptyset$, therefore implies a flow $f \colon \overline{E} \to \mathbb{N}_0$ simply by setting $f(e)$ to be the number of occurrences of the transition e in the sequence σ. The flow conservation constraint at $v \in V$ is satisfied as a consequence of the execution semantics of Petri nets.

Given a flow, we would like to constrain the Petri net to only yield firing sequences for that particular flow. We therefore further convert the extended hypergraph $\overline{\mathcal{H}}$ into an extended net $(V \cup V_E, \overline{E}, W \cup W_E)$ by adding for each edge $e \in \overline{E}$ an "external place" $v_e \in V_E$ with connectivity $W(v_e, e) = 1$. In the following, we will denote the extended Petri net again by N. We then proceed by translating the given flow f of $\overline{\mathcal{H}}$ into an initial marking M_0 on the extended net. To this end, we set $M_0(v) = 0$ for $v \in V$ and $M_0(v_e) = f(e)$ for places $v_e \in V_E$. Transitions in (N, M_0) therefore can fire at most the number of times specified by the flow. Furthermore, any firing sequence $M_0 \xrightarrow{\sigma} M_\emptyset$ ending in the empty marking must use each transition exactly the number of times specified by the flow. As an example, the hyperflow in Fig. 2 is converted to the Petri net in Fig. 3.

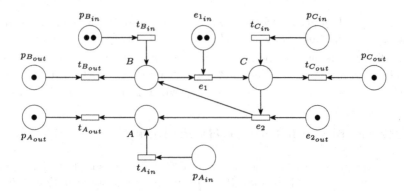

Fig. 3. The integer hyperflow from Fig. 2 converted to a Petri net. Places are circles, transitions are squares, and tokens are black dots. Arrows indicate pairs of places and transitions for which the weight function W is non-zero (in this example, all non-zero weights are equal to one). We have omitted the part of the net that corresponds to the omitted part of Fig. 2.

3.2   Realisability of Integer Hyperflows

We are interested in whether a given pathway, represented by a flow f on an extended hypergraph $\overline{\mathcal{H}} = (V, \overline{E})$, is realisable in the following sense: Given the input molecules specified by the input flow, is there a sequence of reactions that respects the flow, which in the end produces the specified output flow? In the light of the construction of (N, M_0) from $(\overline{\mathcal{H}}, f)$, this question translates into a reachability problem on a Petri net.

Definition 1. *A flow f on $\overline{\mathcal{H}}$ is realisable if there is a firing sequence $M_0 \overset{*}{\to} M_\emptyset$ on the Petri net (N, M_0) constructed from $(\overline{\mathcal{H}}, f)$.*

Figure 4 shows that not all flows f on $\overline{\mathcal{H}}$ are realisable. In this example it is impossible to realise the flow as long as there is no flow entering either C or D. For the flow in Fig. 2, on the other hand, such a firing sequence exists. The firing sequences corresponding to a realisable flow are not unique in general. For instance, the Petri net constructed from the integer hyperflow presented in Fig. 3 can reach the empty marking $M_\emptyset$, in essentially two different manners. Modulo the firing of input/output transitions, those two firing subsequences are $e_1 e_1 e_2$ and $e_1 e_2 e_1$.

Showing the existence of a firing sequence as specified in Definition 1 is one way of proving the realisability of an integer hyperflow. Making use of occurrence nets [6,13,15] and processes [15], a *realisability certificate* can be defined which constitute an ordered sequence of reactions together with an individual token interpretation [14]. Thus it contains the exact dependencies between reactions in the realisation of the integer hyperflow and explicitly expresses which individual molecule is used when and for which reaction. Due to space constraints, we defer a formal description of realisability certificates to the forthcoming full version of this paper. However, we do note here that a realisability certificate uniquely

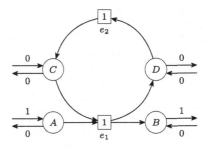

Fig. 4. Example of a flow which is not realisable. Observe that the flow is indeed viable as it fulfils the flow conservation constraint. Furthermore, notice that there is no input flow to neither C nor D, and therefore in the corresponding Petri net it will not be possible to fire either of e_1 or e_2 which is necessary for it to be realised. However, if C or D was borrowed the related flow with this borrowing would be realisable.

determines a corresponding integer hyperflow while integer hyperflows, on the other hand, do not specify the order of the reactions or which one of multiple copies of a molecules is used in which reaction. An integer hyperflow therefore may correspond to multiple different realisability certificates, each representing a different mechanism. For an example of a realisability certificate see Fig. 6.

4   Extended Realisability

Although we have seen above that some integer hyperflows are not realisable, they can be turned into realisable hyperflows by means of certain modifications.

Definition 2 (Scaled-Realisable). *An integer hyperflow f on an extended hypergraph $\overline{\mathcal{H}} = (V, \overline{E})$ is scaled-realisable, if there exists an integer $k \geq 1$ such that the resulting integer hyperflow $k \cdot f$ is realisable.*

Asking if an integer hyperflow f is scaled-realisable corresponds to asking if k copies of f can be realised concurrently. This is of interest as in the real world, a pathway is often not just happening once, but multiple times. Therefore, even if the integer hyperflow is not realisable, it still has value to consider if the scaled integer hyperflow is. However, not all integer hyperflows are scaled-realisable. An example is the integer hyperflow presented in Fig. 4: no integer scaling can alleviate the fact that firing requires that C and D is present at the outset.

Definition 3 (Borrow-Realisable). *Let f be a flow on an extended hypergraph, $\overline{\mathcal{H}} = (V, \overline{E})$ and $b: V \rightarrow \mathbb{N}$. Set $f'(e_v^-) = b(v) + f(e_v^-)$ and $f'(e_v^+) = b(v) + f(e_v^+)$ for all $v \in V$, and $f'(e) = f(e)$ for all $e \in E$. Then f is borrow-realisable if there exists a borrowing function b such that f' is realisable.*

We say that f' is the flow f where $v \in V$ has been borrowed $b(v)$ times. This allows intermediary molecules required for reactions in the pathway to be available in the environment. Formally, this is modelled by having an additional input

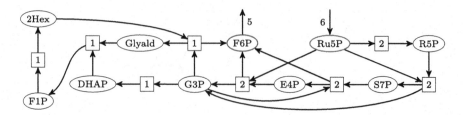

Fig. 5. Example of a hyperflow for the pentose phosphate pathway that is not scaled-realisable. The hyperflow is borrow-realisable. The input compound is marked with green and the output compound is marked with blue. The edges without flow have been omitted. (Color figure online)

and output flow $b(v)$ for species v. Furthermore, for a borrowing function b we define $|b| = \sum_{v \in V} b(v)$, i.e., the total count of molecules borrowed. The idea of borrowing tokens in the corresponding Petri net setting has been proposed in [9, Proposition 10] together with a proof that f' is realisable for some b with sufficiently large $|b|$. That is, every integer hyperflow is borrow-realisable.

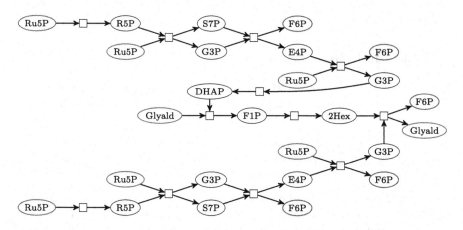

Fig. 6. A realisability certificate for the hyperflow in Fig. 5 where the molecule Glyald is borrowed in order to make it borrow-realisable. The input compounds are marked with green, the output compounds are marked with blue and the borrowed compound is marked with purple.

The combinatorics underlying the non-oxidative phase of the PPP has been analysed less formally in a series of studies focusing, e.g., on simplifying principles that explain the structure of metabolic networks, see e.g. [17,20]. An example of a simple integer hyperflow that is not scaled-realisable is shown in Fig. 5. Here, the production of glyceraldehyde (Glyald) is dependent of the presence of Hex-2-ulose (2Hex), which depends on fructose-1-phosphate (F1P), which in turn depends on Glyald. This cycle of dependencies implies that firing is impossible unless one of the molecules in this cycle is present at the outset, which cannot be achieved by scaling. As illustrated in Fig. 6 and proven by the existence of the realisability certificate, the flow is borrow-realisable with just one borrowing, namely of the compound Glyald. Thus Glyald can be seen as a network catalyst for this pathway.

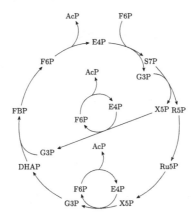

Fig. 7. Example of a pathway drawing for the cyclic non-oxidative glycolysis (NOG) pathway. Recreated from [7, Fig. 2a].

5 Representations of Pathways

We have described two ways of modelling pathways: integer hyperflows and realisability certificates. Here we want to point out that commonly used representations of pathways in the life science literature fall in between these two extremes, see Fig. 7 for an example. In this example, the order of reactions is not fully resolved—for instance, is F6P produced before E4P or after? Indeed, some unspecified choice of borrowing is needed to set the pathway in motion. Additionally, the semantics of a molecule identifier appearing in several places is unclear—for instance, are the three appearances of G3P interchangeable in the associated reactions or do they signify different individual instances of the same type of molecule? In the former case, the figure corresponds to a much larger number of different realisability certificates than in the latter case. The answers to these questions have important consequences for investigations where the identity of individual atoms matter, such as atom tracing.

Furthermore, when there is a choice between different pathway suggestions, avoiding borrow-realisable pathways often gives simpler depictions. However, this introduces a bias among the possible pathways, which may be unwanted, as borrow-realisable solutions are usually equally simple in chemical terms. We note that the need for borrowing in pathways is usually not discussed in the literature.

Additionally, there has been a lack of computational methods to systematically look for borrow-realisable pathways, even if they could equally likely form part of what happens in nature. We believe that our focus on the realisability of pathways may help raise awareness of the choices one often subconsciously makes when creating pathway illustrations.

Acknowledgements. This work is supported by the Novo Nordisk Foundation grant NNF19OC0057834 and by the Independent Research Fund Denmark, Natural Sciences, grant DFF-0135-00420B.

References

1. Andersen, J.L., Flamm, C., Merkle, D., Stadler, P.F.: Maximizing output and recognizing autocatalysis in chemical reaction networks is NP-complete. J. Syst. Chem. **3**(1) (2012)
2. Andersen, J.L., Flamm, C., Merkle, D., Stadler, P.F.: 50 Shades of rule composition. In: Fages, F., Piazza, C. (eds.) FMMB 2014. LNCS, vol. 8738, pp. 117–135. Springer, Cham (2014). https://doi.org/10.1007/978-3-319-10398-3_9
3. Andersen, J.L., Flamm, C., Merkle, D., Stadler, P.F.: Chemical transformation motifs – modelling pathways as integer hyperflows. IEEE/ACM Trans. Comput. Biol. Bioinf. **16**(2), 510–523 (2019). https://doi.org/10.1109/TCBB.2017.2781724
4. Andersen, J.L., Flamm, C., Merkle, D., Stadler, P.F.: Defining autocatalysis in chemical reaction networks. Tech. rep., Cornell University Library, arXiv.org (2021)
5. Baldan, P., Cocco, N., Marin, A., Simeoni, M.: Petri nets for modelling metabolic pathways: a survey. Nat. Comput. **9**, 955–989 (2010). https://doi.org/10.1007/s11047-010-9180-6
6. Best, E., Merceron, A.: Discreteness, K-density and D-continuity of occurrence nets. In: Cremers, A.B., Kriegel, H.P. (eds.) Theoretical Computer Science, pp. 73–83. Springer, Berlin, Heidelberg (1982). https://doi.org/10.1007/BFb0036470
7. Bogorad, I.W., Lin, T.S., Liao, J.C.: Synthetic non-oxidative glycolysis enables complete carbon conservation. Nature (London) **502**(7473), 693–697 (2013)
8. Braakman, R., Smith, E.: The compositional and evolutionary logic of metabolism. Phys. Biol. **10**(1) (2013)
9. Desel, J.: Basic linear algebraic techniques for place/transition nets. In: Reisig, W., Rozenberg, G. (eds.) ACPN 1996. LNCS, vol. 1491, pp. 257–308. Springer, Heidelberg (1998). https://doi.org/10.1007/3-540-65306-6_18
10. Eigen, M.: Selforganization of matter and the evolution of biological macromolecules. Naturwissenschaften **58**(10), 465–523 (1971)
11. Eigen, M., Schuster, P.: The hypercycle: a principle of natural self-organization. Die Naturwissenschaften (1977)
12. Esparza, J.: Decidability and complexity of petri net problems-an introduction. Lectures on Petri Nets I: Basic models, pp. 374–428 (1998)
13. Genrich, H.J., Stankiewicz-Wiechno, E.: A dictionary of some basic notions of net theory. In: Brauer, W. (ed.) Net Theory and Applications. LNCS, vol. 84, pp. 519–531. Springer, Heidelberg (1980). https://doi.org/10.1007/3-540-10001-6_39
14. Glabbeek, R.J.: The individual and collective token interpretations of petri nets. In: Abadi, M., de Alfaro, L. (eds.) CONCUR 2005. LNCS, vol. 3653, pp. 323–337. Springer, Heidelberg (2005). https://doi.org/10.1007/11539452_26

15. Goltz, U., Reisig, W.: The non-sequential behaviour of petri nets. Inf. Control **57**(2), 125–147 (1983). https://doi.org/10.1016/S0019-9958(83)80040-0

16. Koch, I.: Petri nets - a mathematical formalism to analyze chemical reaction networks. Mol. Inf. **29**(12), 838–843 (2010). https://doi.org/10.1002/minf.201000086

17. Meléndez-Hevia, E., Isidoro, A.: The game of the pentose phosphate cycle. J. Theor. Biol. **117**(2), 251–263 (1985). https://doi.org/10.1016/S0022-5193(85)80220-4

18. Morowitz, H.J., Copley, S.D., Smith, E.: Core Metabolism as a Self-Organized System, chap. 20. Protocells, The MIT Press (2008)

19. Müller, S., Flamm, C., Stadler, P.F.: What makes a reaction network "chemical"? J. Cheminformat. **14**(1), 63–63 (2022)

20. Noor, E., Eden, E., Milo, R., Alon, U.: Central carbon metabolism as a minimal biochemical walk between precursors for biomass and energy. Mol. Cell **39**(5), 809–820 (2010). https://doi.org/10.1016/j.molcel.2010.08.031

21. Reddy, V.N., Liebman, M.N., Mavrovouniotis, M.L.: Qualitative analysis of biochemical reaction systems. Comput. Biol. Med. **26**(1), 9–24 (1996)

22. Reddy, V.N., Mavrovouniotis, M.L., Liebman, M.N.: Petri net representations in metabolic pathways. In: Proceedings of the International Conference on Intelligent Systems for Molecular Biology, pp. 328–36 (1993)

23. Szathmáry, E.: A hypercyclic illusion. J. Theor. Biol. **134**(4), 561–563 (1988)

24. Szathmáry, E.: On the propagation of a conceptual error concerning hypercycles and cooperation. J. Syst. Chem. **4**, 1 (2013)

25. Zeigarnik, A.V.: On hypercycles and hypercircuits in hypergraphs. Discrete Math. Chem. **51**, 377–383 (2000). https://doi.org/10.1090/dimacs/051/28

A Brief Study of Gene Co-expression Thresholding Algorithms

Carissa Bleker[1]([✉]) [iD], Stephen K. Grady[2] [iD], and Michael A. Langston[2] [iD]

[1] Department of Biotechnology and Systems Biology,
National Institute of Biology, Ljubljana, Slovenia
carissa.bleker@nib.si
[2] Department of Electrical Engineering and Computer Science,
University of Tennessee, Knoxville, USA

Abstract. The thresholding problem is considered in the context of high-throughput biological data. Several approaches are reviewed, implemented, and tested over an assortment of transcriptomic data.

Keywords: biological network analysis · thresholding methodology

1 Introduction

Graphs are frequently employed in the analysis of high throughput biological data, with thresholding used to eliminate weak or irrelevant edges. Numerous methods to determine a most appropriate threshold have been proposed, but no consensus has been reached [9]. The main goals of this brief study are to review, implement, and compare these methods systematically over a testbed of representative data.

In the next section, we briefly review and classify several state-of-the-art thresholding algorithms. In Sect. 3, we describe a modification to an existing technique. In Sect. 4, we present the data, procedures, implementations, and methodology we employed for testing. In Sect. 5, we discuss our results. And in a final section, we draw a few conclusions from this work.

2 Popular Thresholding Approaches

Let G denote a finite, simple, undirected graph, each of whose edges is weighted by a measure of similarity between its endpoints. Correlation provides a standard measure of this similarity, with Pearson's r the metric most frequently employed. Thresholding algorithms retain an edge only if its weight is at least some pre-computed value. Relevant techniques can generally be divided into those based on graph structure versus those that perform statistical modeling on underlying data distributions. As a general rule, too low a threshold runs the risk of false positives, while too high a threshold may produce false negatives.

X. Guo et al. (Eds.): ISBRA 2023, LNBI 14248, pp. 420–430, 2023.
https://doi.org/10.1007/978-981-99-7074-2_33

2.1   Thresholding Based on Graph Structure

Knowledge or assumptions about graph structure can herald contextual meaning, in which case thresholding can be viewed as an attempt to make use of that structure. Most thresholding methods can be interpreted in this context.

Scale Free Models. In what's termed a "scale free" graph, vertex degrees follow a power law distribution, $p(k) \approx k^{-\gamma}$, which simply means that the number of vertices with degree k is approximately $k^{-\gamma}$. It turns out that γ usually lies in the interval $[2, 3]$, and thus only a few vertices will have high degree, while the degree of most vertices is very small. It is sometimes argued that biological networks generally follow an empirical scale free topology, and that therefore a biologically relevant threshold can be found by fitting a scale free model to the graph [35]. To accomplish this, the linear model $\log p(k) \sim \log k$ may be fitted to the graph for each threshold under consideration, and the R^2 value of the model then used to measure goodness of fit. In [35], for example, a plateau is sought in the relationship between the threshold and R^2. An R package is available that can assist in threshold selection using this approach [21]. More recently, however, chi-squared goodness of fit [20] and likelihood ratio tests [6] have found that biological networks are not overwhelmingly scale free, or are only weakly scale free. Therefore, any such assumption should probably be called into question, and adopted only once a scale free model has truly been shown to fit the data under analysis.

Spectral Graph Theory. Spectral graph theory studies the eigenvalues of adjacency or Laplacian matrices of graphs [7], and makes it possible to discover information about graph connectivity. For example, the number of zero-valued eigenvalues of a graph's Laplacian corresponds to the number of its connected components. Moreover, if the first non-zero eigenvalue is small enough, ordering its corresponding eigenvector results in a step-like function, with each step corresponding to a transition between nearly-disconnected-components (dense subgraphs that are cross-connected by only a few bridge edges) [13]. Motivated in part by the well-known "guilt-by-association" principle [34], the work described in [29] argues for the threshold that maximizes the number of nearly-disconnected-components, equivalently minimizing the number of bridges (edges connecting dissimilar parts).

Random Matrix Theory (RMT). RMT was developed to analyze nuclear spectra, but has also been applied to biological networks [16,24,25]. RMT thresholding is based on the correlation of the eigenvalues of the graph, measured by their Nearest Neighbor Spacing Distribution (NNSD). NNSD is defined as the distribution of the differences between subsequent (ordered) values. A Gaussian orthogonal ensemble (GOE) is a random, symmetric matrix with independent, normally distributed entries. According to RMT for real, symmetric matrices, the eigenvalue NNSD of a GOE matrix has a Wigner-Dyson distribution, while the eigenvalue NNSD of a non-random matrix follows a Poisson distribution

[25]. This approach thus presupposes that an initial correlation matrix is GOE distributed, and that the deletion of low-weighted edges will result in a non-random matrix with biological signal. To find an appropriate threshold, such edges are iteratively removed until the NNSD converts from GOE to Poisson, with Chi-square tests performed at each iteration [24].

Maximal Clique. A subgraph is a clique if it contains all possible edges. A clique is maximal if it cannot be enlarged by the addition of more vertices (and remain a clique). Thresholding may be performed by incrementally lowering the threshold, and searching for inflection points at which the number of maximal cliques increases dramatically. In [5], both clique doubling and tripling were considered in a top-down fashion, with doubling especially shown to perform well in the analysis of mRNA microarray co-expression data. A superficially similar idea can be seen in work on k-clique communities [28], now renamed clique percolation [12]. Unfortunately, this method requires pre-selecting the clique size (k) and requiring it to be unrealistically small (usually between 3 and 6). Moreover, as an exhaustive bottom-up approach, it has been found to be relatively inefficient by design [19].

Graph Density. A thresholding objective common to a variety of application domains seeks to maximize the number of edges within each cluster, while at the same time minimizing the number of edges between disjoint clusters. This simple goal is generally ill suited to problems in network biology, however, because it overlooks important principles such as gene pleiotropy and cluster overlap. Nevertheless, attempts have been made to pursue it [2, 26]. Oddly enough, they proceed not by finding a threshold that largely maximizes component density (as most other methods described here actually do), but instead they focus solely on inter-component edges and operate by iteratively lowering the threshold and ignoring isolated vertices until the overall graph density is minimized.

Edge Weight Distribution. A somewhat similar approach to graph density is described in [3]. There an algorithm is proposed that sets the threshold to the smallest value that produces a non-negative derivative when the number of edges is divided by the number of non-isolated vertices. It is argued that this methodology is appropriate for protein sequence similarity graphs, because it tends to preserve edges within protein families while eliminating edges between different families as well as those to outliers. For this technique to work, however, the edge weight distribution must be left skewed, so that the number of edges rapidly decreases as the threshold is increased. While this condition held for the specific protein similarity graphs studied, the extent to which it holds across a wider spectrum of biological network graphs is unclear. ·

Global-Local Measures. A method that combines both local and global measures was introduced in [18], where it was applied to semantic similarity graphs. Pendant vertices are first removed, and the local threshold assigned to each remaining vertex is a function of its incident edge weights and a global threshold parameter. Only if an edge's weight exceeds the threshold of either of its endpoints is it is included in the graph. If the graph contains nearly-disconnected-components (as defined by spectral methods), the process stops, otherwise the global parameter is adjusted, and the process is repeated.

Clustering Coefficient. A clustering coefficient can be viewed as a local measure of graph density, and is generally assigned separately to each vertex. One definition is the number of edges within the vertex's open neighborhood divided by the maximum number possible. The clustering coefficient of the graph, C, is the average across the graph [33]. For a complete graph $C = 1$. A method to employ this measure is proposed in [17]. Initially, as the threshold is increased, edges and isolated vertices are discarded, and thus C decreases. It is argued that at higher thresholds, signal in the network manifests in highly connected sub-networks, which results in C increasing. A threshold is chosen that best corresponds to an inflection upwards of C. When compared to predefined thresholds, and thresholds based on significance on a number of micro-array datasets, the work presented in [15] found that this method resulted in graphs with better ontological enrichment. Following on the work of [17], the discussion of [15] introduces C^r, calculated from a random counterpart to the original graph which preserves the original degree distribution. It is proposed that, as the threshold is increased, edges eliminated are due to noise as long as the difference between C and C^r is monotonically increasing, and so the smallest threshold is selected such that the difference between C and C^r is larger than the same difference at the next threshold. In contrast to the method of [17], this can be calculated analytically.

3   Algorithmic Advances

We implemented several of these thresholding techniques and found utility in a handful of algorithmic enhancements. In the process, we also devised a previously-unreported thresholding method suitable for comparison.

Scale Free Models. To implement a scale free model, we applied maximum likelihood to estimate γ and a Kolmogorov-Smirnov test to compare the fitted distribution to the input degree distribution [10]. This is in contrast to the ad-hoc linear model method used in [35], which has a number of issues that can lead to unreliability in identifying scale free graphs [8, 20].

Maximal Clique. We implemented what we term "maximal clique ratio" as a generalization of maximal clique doubling and tripling. To accomplish this, we simply stepped through threshold increments and divided the number of maximal cliques at the current value by that at the previous value, setting the threshold where this ratio was maximized. We retained a step size of 0.01 as in [5], although other increments can of course be used. Note that the increment should be kept constant, so that the ratios are comparable.

Global-Local Measures. We generalized the global-local thresholding method of [18] so that it would handle edge-weighted graphs. This required that we increase the default limits of the global threshold parameter, employ absolute correlation values, and refrain from removing pendant vertices.

4 Implementation and Testing

4.1 Computational Milieu

Each of the thresholding techniques we have discussed is reasonably fast, and thus we were not preoccupied with relative speed or competitive timings. Accordingly, we employed a variety of hardware architectures and operating systems based largely on availability. These include a SUSE Linux Enterprise Server 15 operating system running on a Cray XC40 with two 2.3 GHz 16-core Haswell processors and 128 GB DDR4 2133 MHz memory per node, a CentOS Linux release 7.6.1810 operating system running on a multi-platform cluster with Intel® Xeon® CPUs and 4682MB memory per core, and a Ubuntu 18.0.10 LTS operating system running on a Dell Precision 5510 laptop with eight 2.70GHz Intel® Core™ i7-6820HQ processors and 15.5 GB of memory. For software support, we employed the igraph C library [10] for graph theoretical analysis and the ALGLIB library [4] for statistical distributions.

4.2 Methods Studied

We chose eight of the aforementioned thresholding algorithms for testing. Selections for testing were based on the following exclusionary criteria: (1) methods must be entirely data dependent, which eliminated relevance networks, and (2) they must be scalable, which eliminated the local-global method (we found iterative eigenvalue calculations of very large matrices unfeasible during parameterization). One of these techniques we newly devised (seen as shaded rows), and seven we implemented as described and/or enhanced in the preceding discussion. See Table 1.

4.3 Benchmarking

We created a testbed of 39 finite, simple, undirected graphs from a variety of real-world data repositories. Using Pearson correlation coefficients to weight their edges, these graphs were produced from the following sources:

- EntropyExplorer [32], which contains transcriptomic disease case/control microarray data (19 graphs)
- ManyMicroarray [22,23], which contains pre-normalized human microarray data (20 graphs)

Table 1. Thresholding algorithms studied.

Algorithmic Approach	Name	Description
Clustering Coefficient	Coefficient-Random	Chooses the smallest threshold at which the difference between the clustering coefficient and that of a random counterpart graph is larger than at the next threshold [15].
Edge Weight Distribution	Cluster-Separation	Employs the heuristic of [3] to estimate a threshold separating inter-cluster from intracluster edges.
Maximal Clique	MCR-2	Determines a threshold at which the change in the ratio in the number of maximal cliques doubles [5].
	MCR-Max	Generalizes MCR-2 by choosing the threshold at which the ratio of change in maximal clique size is maximized.
Random Matrix Theory	RMT	Chooses a threshold based on eigenvalue changes as in [16,24,25].
Scale Free Models	Scale-Free	Uses maximum likelihood and a Kolmogorov-Smirnov test to determine goodness-of-fit of the degree distribution to a power law.
Significance and Power	Power (β)	Selects the minimum threshold that limits the Type II error rate to $1 - \beta$. (Pearson only)
Spectral Graph Theory	Spectral-Methods	Exploits the second-smallest eigenvector as described in [29].

A preliminary threshold of 0.10 was used for each graph. To meet memory and file transfer limitations, graphs were downsized if necessary by starting at this preliminary threshold and increasing it in increments of 0.10 until the resultant file size was no larger than 3GB. The eight aforementioned thresholding methods were then applied to this testbed. In some cases, methods did not converge or computational limits were exceeded. This resulted in the identification of 303 thresholds. Of these, 170 were at least 0.60 and considered for further study. The bound of 0.60 was based on multiple criteria. Not only are clustering times prohibitive at lower values, but it has been shown that transcriptomic data can require a threshold as high as 0.84 to achieve even a 50% true positive rate of associations [1]. Moreover, we found that the median consensus threshold across all methods tested was 0.81, while the methods purportedly developed with transcriptomic data in mind all produced thresholds in the range [0.80,0.96]. Clustering was next performed on these 170 thresholded graphs. For this we used paraclique [14,30], a state-of-the-art algorithm known to produce clusters of superior quality [19]. Paraclique calls were limited to no more than 300 clusters, with 4 calls that terminated after 60 h, resulting a total of 211 clustered graphs. In total, we examined 8,392 clusters distributed over these 211 graphs. Vertices within a cluster were annotated with gene symbols so that enrichment via the Gene Ontology project [31] could be employed to gauge biological fidelity. For this we used the Panther web API [27] and a custom Python script. A cluster was deemed significant if it contained at least one significantly over-represented GO term.

5   Empirical Results and Discussion

Threshold ranges computed by each technique are displayed in Fig. 1. A few may be noteworthy. The Cluster-Separation algorithm found no thresholds at all for most of the EntropyExplorer graphs, and only extremely high thresholds otherwise. Presumably it was hobbled by its reliance on skewed edge weight distributions. The MCR and Scale-Free techniques too returned only excessively high thresholds, perhaps due to MCR's inherent dependence on changes in graph density and Scale-Free's assumptions about graph structure. At the other end of the spectrum, the Power approach computed thresholds so low that they were

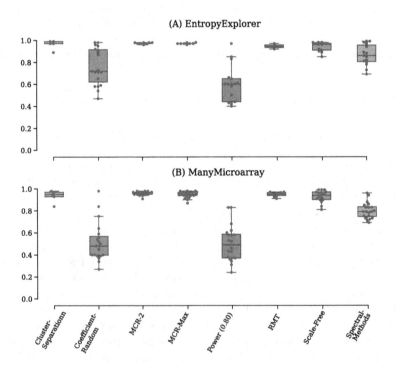

Fig. 1. Threshold ranges across data sources. (A) EntropyExplorer and (B) ManyMicroarray.

almost always below our 0.60 cut-off. This may be due to their reliance on a large number of samples. And finally, MCR-2 seems to have struggled to find suitable thresholds, perhaps because of their stringent doubling requirements.

Results of GO cluster annotation are shown in Table 2. Four values are reported for each method: the number of graphs on which a threshold of 0.60 or greater was found, the number of gene clusters produced, the number of significant gene clusters produced, and most importantly a metric we call the Total Significant Clusters Ratio (TSCR). We calculated the TSCR by summing individual significance ratios over all clusters isolated within each graph. Its intent is to prevent any one graph from overwhelming the metric and to provide a measure of cluster quality independent of the number of clusters found. The Cluster-Separation method, for example, produced clusters from five graphs. In one graph, only a single cluster was found, and it was enriched. In a second graph, two clusters were found of which only one was enriched. In a third, seven out of ten clusters were enriched, and so on. The Cluster-Separation TSCR calculation was therefore: $\frac{1}{1} + \frac{1}{2} + \frac{7}{10} + \frac{1}{1} + \frac{4}{5} = 4$.

Table 2. Enrichment analysis. Thresholding algorithms ranked by Total Significant Clusters Ratio (TSCR).

	Thresholds Found	Gene Clusters	Significant Clusters	TSCR
MCR-Max	28	488	94	12.38
Scale-Free	29	206	91	11.84
Spectral-Methods	35	3,098	809	10.89
MCR-2	26	438	76	10.28
Coefficient-Random	18	1,221	348	8.84
RMT	12	34	25	7.73
Cluster-Separation	10	19	14	4.00
Power (0.80)	12	2,907	509	2.04

Unsurprisingly, methods designed with transcriptomic data in mind performed quite well. Moreover, three of the four highest-scoring algorithms (the two MCR-based techniques plus Spectral-Methods) are all focused in one way or another on the largest change in graph density. In contrast, Scale-Free seems to rely on graph sparsity [11], which allowed it to perform well at higher thresholds.

6   Concluding Remarks

Thresholding is a foundational technique for creating unweighted graphs from correlational data. This makes it a crucial first step in many analytical toolchains, with numerous algorithms proposed for this purpose. In this paper, we described, implemented, and systematically tested an assortment of these methods in the context of gene co-expression. We also devised a new technique, built a test-bed of relevant data, and created a simple metric to evaluate algorithmic performance.

Acknowledgements. This research was supported in part by the National Institutes of Health under grant R01HD092653 and by the Environmental Protection Agency under grant G17D112354237. It also employed resources of the National Energy Research Scientific Computing Center, a U.S. Department of Energy Office of Science User Facility located at Lawrence Berkeley National Laboratory and operated under Contract No. DE-AC02-05CH11231.

Author contributions. Conceptualization: CB, ML; Methodology: CB, ML; Formal analysis: CB, SG, ML; Software: CB, SG; Writing: CB, SG, ML; Supervision: ML; Funding acquisition: ML.

References

1. Allocco, D.J., Kohane, I.S., Butte, A.J.: Quantifying the relationship between co-expression, co-regulation and gene function. BMC Bioinform. **5**(18) (2004)
2. Aoki, K., Ogata, Y., Shibata, D.: Approaches for extracting practical information from gene co-expression networks in plant biology. Plant Cell Physiol. **48**(3), 381–390 (2007). https://doi.org/10.1093/pcp/pcm013
3. Apeltsin, L., Morris, J.H., Babbitt, P.C., Ferrin, T.E.: Improving the quality of protein similarity network clustering algorithms using the network edge weight distribution. Bioinform. **27**(3), 326–333 (2011). https://doi.org/10.1093/bioinformatics/btq655
4. Bochkanov, S.: Alglib. https://www.alglib.net/ (2019)
5. Borate, B.R., Chesler, E.J., Langston, M.A., Saxton, A.M., Voy, B.H.: Comparison of threshold selection methods for microarray gene co-expression matrices. BMC Res. Notes **2**, 240 (2009). https://doi.org/10.1186/1756-0500-2-240
6. Broido, A.D., Clauset, A.: Scale-free networks are rare. Nat. Commun. **10**(1), 1017 (2019)
7. Chung, F.R.: Spectral graph theory. American Mathematical Soc, Providence, RI (1997)
8. Clauset, A., Shalizi, C.R., Newman, M.E.: Power-law distributions in empirical data. SIAM Rev. **51**(4), 661–703 (2009)
9. CR, B.: Data-Driven analytics for high-throughput biological applications. Ph.D. thesis, University of Tennessee (2020)
10. Csardi, G., Nepusz, T., et al.: The igraph software package for complex network research. Int. J. Complex Syst. **1695**(5), 1–9 (2006)
11. Del Genio, C.I., Gross, T., Bassler, K.E.: All scale-free networks are sparse. Phys. Rev. Lett. **107**, 178701 (2011). https://doi.org/10.1103/PhysRevLett.107.178701
12. Derényi, I., Palla, G., Vicsek, T.: Clique percolation in random networks (2005). https://doi.org/10.1103/PhysRevLett.94.160202
13. Ding, C.H.Q., He, X., Zha, H.: A spectral method to separate disconnected and nearly-disconnected Web graph components. In: Proceedings of the Seventh ACM SIGKDD International Conference on Knowledge Discovery and Data Mining - KDD '01, pp. 275–280. ACM Press, New York, USA (2004). https://doi.org/10.1145/502512.502551

14. Ej, C., Ma, L.: Combinatorial genetic regulatory network analysis tools for high throughput transcriptomic data. Lect. Notes Comput. Sci. **4023**, 150–165 (2005)
15. Elo, L.L., Järvenpää, H., Orešič, M., Lahesmaa, R., Aittokallio, T.: Systematic construction of gene coexpression networks with applications to human T helper cell differentiation process. Bioinformatics **23**(16), 2096–2103 (2007). https://doi.org/10.1093/bioinformatics/btm309
16. Gibson, S.M., Ficklin, S.P., Isaacson, S., Luo, F., Feltus, F.A., Smith, M.C.: Massive-scale gene co-expression network construction and robustness testing using random matrix theory. PLoS ONE **8**(2), e55871 (2013). https://doi.org/10.1371/journal.pone.0055871
17. Gupta, A., Maranas, C.D., Albert, R.: Elucidation of directionality for co-expressed genes: predicting intra-operon termination sites. Bioinformatics **22**(2), 209–214 (2006). https://doi.org/10.1093/bioinformatics/bti780
18. Guzzi, P.H., Veltri, P., Cannataro, M.: Thresholding of semantic similarity networks using a spectral graph-based technique. In: Appice, A., Ceci, M., Loglisci, C., Manco, G., Masciari, E., Ras, Z.W. (eds.) New Frontiers in Mining Complex Patterns, pp. 201–213. Springer International Publishing, Cham (2014). https://doi.org/10.1007/978-3-319-08407-7_13
19. Jay, J.J., et al.: A systematic comparison of genome scale clustering algorithms. BMC Bioinformatics **13**(10) (2012). https://doi.org/10.1186/1471-2105-13-S10-S7
20. Khanin, R., Wit, E.: How scale-free are biological networks. J. Comput. Biol. **13**(3), 810–818 (2006)
21. Langfelder, P., Horvath, S.: WGCNA: an R package for weighted correlation network analysis. BMC Bioinform. **9**(1), 559 (2008)
22. Lee, H.K., Hsu, A.K., Sajdak, J., Qin, J., Pavlidis, P.: Coexpresion analysis of human genes across many microarray data sets. Genome Res. **14**(6), 1085–1094 (2004). https://doi.org/10.1101/gr.1910904
23. Lee, H.K., Hsu, A.K., Sajdak, J., Qin, J., Pavlidis, P.: Coexpression analysis of human genes across many microarray data sets (2019). https://doi.org/10.5683/SP2/JOJYOP
24. Luo, F., et al.: Constructing gene co-expression networks and predicting functions of unknown genes by random matrix theory. BMC Bioinform. **8**, 1–17 (2007). https://doi.org/10.1186/1471-2105-8-299
25. Luo, F., Zhong, J., Yang, Y., Scheuermann, R.H., Zhou, J.: Application of random matrix theory to biological networks. Phys. Lett. A **357**(6), 420–423 (2006). https://doi.org/10.1016/j.physleta.2006.04.076
26. Mao, L., Van Hemert, J.L., Dash, S., Dickerson, J.A.: Arabidopsis gene co-expression network and its functional modules. BMC Bioinformatics **10**(1), 346 (2009). https://doi.org/10.1186/1471-2105-10-346
27. Mi, H., et al.: Panther version 11: expanded annotation data from Gene Ontology and Reactome pathways, and data analysis tool enhancements. Nucleic Acids Res. **45**(D1), D183–D189 (2017). https://doi.org/10.1093/nar/gkw1138
28. Palla, G., Derényi, I., Farkas, I., Vicsek, T.: Uncovering the overlapping community structure of complex networks in nature and society. Nature **435**(7043), 814–818 (2005). https://doi.org/10.1038/nature03607
29. Perkins, A.D., Langston, M.A.: Threshold selection in gene co-expression networks using spectral graph theory techniques. BMC Bioinform. **10**(Suppl 11), S4 (2009). https://doi.org/10.1186/1471-2105-10-S11-S4
30. Hagan, R.D., Langston, M.A., Wang, K.: Lower bounds on paraclique density. Discrete Appl. Math. **204**, 208–212 (2016)

31. The Gene Ontology, C., et al.: Gene ontology: tool for the unification of biology. Nat. Gen. **25**(1), 25–29 (2000). https://doi.org/10.1038/75556

32. Wang, K., Phillips, C.A., Saxton, A.M., Langston, M.A.: EntropyExplorer: an R package for computing and comparing differential Shannon entropy, differential coefficient of variation and differential expression. BMC. Res. Notes **8**, 832 (2015). https://doi.org/10.1186/s13104-015-1786-4

33. Watts, D.J., Strogatz, S.H.: Collective dynamics of small-world networks. Nature **393**(6684), 440–442 (1998). https://doi.org/10.1038/30918

34. Wolfe, C.J., Kohane, I.S., Butte, A.J.: Systematic survey reveals general applicability of guilt-by-association within gene coexpression networks. BMC Bioinformatics **6**(227) (2005). https://doi.org/10.1186/1471-2105-6-227

35. Zhang, B., Horvath, S.: A general framework for weighted gene co-expression network analysis. Statistical applications in genetics and molecular biology **4**(1) (2005). https://doi.org/10.2202/1544-6115.1128

Inferring Boolean Networks from Single-Cell Human Embryo Datasets

Mathieu Bolteau[1](✉)[iD], Jérémie Bourdon[1][iD], Laurent David[2][iD],
and Carito Guziolowski[1][iD]

[1] Nantes Université, Ecole Centrale Nantes, CNRS, LS2N, UMR 6004, 44000 Nantes,
France
{mathieu.bolteau, jeremie.bourdon, carito.guziolowski}@ls2n.fr
[2] Nantes Université, CHU Nantes, Inserm, CR2TI, 44000 Nantes, France
laurent.david@univ-nantes.fr

Abstract. This study aims to understand human embryonic development and cell fate determination, specifically in relation to trophectoderm (TE) maturation. We utilize single-cell transcriptomics (scR-
NAseq) data to develop a framework for inferring computational models that distinguish between two developmental stages. Our method
selects pseudo-perturbations from scRNAseq data since actual perturbations are impractical due to ethical and legal constraints. These pseudo-
perturbations consist of input-output discretized expressions, for a limited set of genes and cells. By combining these pseudo-perturbations with
prior-regulatory networks, we can infer Boolean networks that accurately
align with scRNAseq data for each developmental stage. Our publicly
available method was tested with several benchmarks, proving the feasibility of our approach. Applied to the real dataset, we infer Boolean
network families, corresponding to the medium and late TE developmental stages. Their structures reveal contrasting regulatory pathways,
offering valuable biological insights and hypotheses within this domain.

Keywords: Boolean networks · Answer Set Programming · Human
preimplantation development · scRNAseq modeling

1 Introduction

One of the outstanding questions of the field of in vitro fertilization is to understand the chain of events regulating human preimplantation development leading to an implantation-competent embryo. To address this question, in [9],
we analyzed single-cell transcriptomic data (scRNAseq) from preimplantation
human embryos. Our analysis proposed some hierarchy of transcription factors
in epiblast, trophectoderm and primitive endoderm lineages. Individual cell fate
within heterogeneous samples, such as human embryos, can be followed from
scRNAseq data but presents multiple computational challenges with normalization and "zero-inflation", complicating network models [7]. The state-of-the-art

X. Guo et al. (Eds.): ISBRA 2023, LNBI 14248, pp. 431–441, 2023.
https://doi.org/10.1007/978-981-99-7074-2_34

tools used to propose a temporal distribution of such data are based on statistical approaches, such as manifolds (UMAP [8]) or graphs theory (pseudo-time [11]). In [3], the authors used such pseudo-time distributions along with scRNAseq expression data to infer Boolean networks for modeling gene regulation in cancer progression. They focus, however, on the hypothesis of an averaged cell expression at each stage defined by pseudo-time analysis, allowing to model the dynamics of the cell fate decision. In the context of embryo development, Dunn et al. [4] proposed computational models on transcriptional networks from knockout data on mouse stem cells. This data type is ideal since the proposed perturbations add crucial information to the inferring process.

In this work, we propose a framework to discover a family of Boolean networks (BNs) of human preimplantation development that capture the progression from one developmental stage to the next. This framework uses prior-knowledge networks (PKN) as a base on which the scRNAseq data is mapped. Then, it identifies *pseudo-perturbations* specific for two developmental stages. These pseudo-perturbations are used in the last step to infer stage-specific BNs models. Since perturbation data is rarely available due to practical and legal concerns, our main contribution was to extract pseudo-perturbation data from scRNAseq data, considering its high redundancy and sparsity. We used the Pathway Commons database [12] to build a PKN and discovered 20 pseudo-perturbations (across 10 genes) characterizing medium and late stages of trophectoderm (TE) maturation. They correspond to the gene expression of 20 cells in each stage; representative on average of 82% of the total cells. Pseudo-perturbations referring to 10 (entry) genes expression were connected (PKN information) to 14 genes (output) expression. The 20 entry-output gene expression configurations allowed us to infer 2 families of BNs (composed of 8 and 15 logical gates) characterizing medium and late TE developmental stages.

2   Method

2.1   Pipeline Overview

Our pipeline is based on background notions stated in Appendix and its main steps, illustrated in supplementary material[1], are: (*i*) PKN reconstruction, (*ii*) experimental design construction, and (*iii*) BNs inference.

PKN Reconstruction is achieved by querying the Pathway Commons database, using pyBRAvo [6], with an initial gene list. Briefly, given a list of genes relevant to the case study, pyBRAvo explores recursively predecessors genes and outputs a signed-directed graph. The reconstructed PKN is then reduced to only include genes and their interactions measured in the scRNAseq data, as well as protein complexes associated with the genes to maintain their connectivity. The resulting PKN comprises nodes selected as input, intermediate, and readout.

---

[1] https://github.com/mathieubolteau/scRNA2BoNI/tree/master/ISBRA_2023_Supp.

Experimental Design Construction. This step constructs an experimental design from the reduced PKN and the scRNAseq data of the two studied cell classes (see an example in supplementary material). The experimental design is composed of: (i) pseudo-perturbations, which are binarized expression values for input and intermediate genes in chosen cells whose value is identical in both cell classes, and (ii) readout observations, which are normalized expression values for readout genes in the chosen cells of both cell classes. To capture the diversity of genes expression in scRNAseq data for each class, we implement a logic program to maximize the number of different pseudo-perturbations for k genes, given a set of input and intermediate genes (see Sect. 2.3). The resulting experimental design is based on the inputs, intermediates, and readouts of the PKN obtained in the previous step.

BN Inference. We infer BNs for each class using Caspo [13]. Given a PKN and an experimental design, Caspo learns a family of BNs compatible with the network's topology and the experimental design data. Caspo learns minimal (in size) BNs which minimize the error between their readouts predictions and experimental measures. In our framework, Caspo proposes specific BNs for each class. This is obtained thanks to the experimental design identified in the previous step, where a maximal number of entry-output associations is proposed with common entry gene values in both classes (pseudo-perturbations), and (maximally) different output gene values.

2.2 Experimental Data Preprocessing

We used single-cell data from [10], which measures the expression of $\sim 20,000$ genes across 1529 cells. Since we focused on genes in the PKN, our dataset comprised 125 genes (111 input and intermediates, and 14 readouts). We considered only cells at medium and late TE stage; therefore we had a total of 680 cells.

First, we discretize raw gene expression data of input and intermediate PKN nodes (see Sect. 2.1, PKN Reconstruction) by considering a gene expressed if at least 2 reads are identified in the raw data. Here, we denote by e_{ij} (resp. r_{ij}) is the binarized (resp. raw) expression of the gene j for the cell i. We have $e_{ij} = 0$ if $r_{ij} < 2$, and $e_{ij} = 1$ otherwise.

Second, we normalize the raw expression of genes related to PKN readouts (see Sect. 2.1). We denote by n_{ij} the normalized expression of the gene i for the cell j. We have $n_{ij} = (r_{ij} - min)/(max - min)$ where min (resp. max) is the minimum (resp. maximum) expression value of all readout genes across all cells.

2.3 Experimental Design Construction - Algorithm

This algorithm receives an integer k, as a parameter, limiting the number of genes to be selected. Its input data is the preprocessed scRNAseq matrix for input, intermediate, and readout PKN genes. The algorithm retrieves (i) a maximal number of pseudo-perturbations, which identify cells associations between two classes holding identical expression values for a set of k genes, and (ii) cell associations which maximize the readout difference across (redundant) cells associations. The details of this algorithm are presented below.

Maximizing the Number of Pseudo-perturbations. The input of this method is a binary matrix, E, where e_{ij} represents the presence (or activity level) of gene j for cell i (see Sect. 2.2). The output is a subset of genes and cells that adhere to various constraints, ensuring their pseudo-perturbations are balanced between the different classes (see supplementary material, experimental design example). Let us denote by C, the complete set of cells; and by G, the complete set of genes in our experimental data. Each cell is uniquely associated with one class (either A or B); $C = A \uplus B$. We use the binary matrix, E, to define the relation I^G, $I^G(c_i) = \{g_j \in G | e_{ij} = 1\}$. $I^G(c_i)$ thus represents the active genes, belonging to G, for cell c_i. If $G' \subset G$, then the restriction of I^G to G' is defined by $I^{G'}(c_i) = I^G(c_i) \cap G'$.

Problem Formulation. Given an association matrix E, associating a set G of genes to a set C of cells, where C is composed of cells belonging to 2 disjoint sets (classes) A and B; and given a parameter k limiting the number of selected genes, find a subset G' of genes and the largest subset C' ($C' = A' \uplus B' \subset C$, where $A' \subset A$ and $B' \subset B$) satisfying the three following constraints:

1. The size of G' is fixed to k (parameter). For large instances $k << |G|$.
2. $\forall c_1, c_2 \in A'$ (resp. B'), $c_1 \neq c_2$, we verify that $I^{G'}(c_1) \neq I^{G'}(c_2)$.
3. $\forall c_1 \in A'$ (resp. B'), $\exists c_2 \in B'$ (resp. A'), such that we verify $I^{G'}(c_1) = I^{G'}(c_2)$.

From this result, for each $c_i \in C'$ we define a binary vector b^i, such that for $j \in \{1, \cdots, k\}$, $b_j^i = 1$ (resp. $b_j^i = 0$) if gene $g_j \in I^{G'}(c_i)$ (resp. $\notin I^{G'}(c_i)$). b^i is called a pseudo-perturbation. Notice that since the sets G' and C' are not unique, there may exist several pseudo-perturbations vectors.

Constraints Justification. The imposed constraints are crucial in light of the entire framework, which handles Boolean network inference and single-cell data. *Constraint 1* reduces the search space, improves computational efficiency, and simplifies the subsequent step of learning Boolean networks. *Constraint 2* prevents redundancy in gene selection from different cells within the same class. This is essential due to the abundance of zero values and redundancy in single-cell data. *Constraint 3* promotes similarity in gene expression values between the two distinct classes. This alignment enables meaningful comparative analysis during the subsequent step of Boolean network inference. Despite the inherent evolutionary differences between cells belonging to different classes, selecting genes with similar expression values allows us to impose comparable entry conditions on the system, facilitating accurate modeling of the distinct regulatory mechanisms at play. Finally, selecting a larger number of pseudo-perturbations provides more information, enriching the Boolean network inference step and allowing for exploring various regulatory mechanisms.

Maximizing Readout Difference. Pseudo-perturbations identified by the previous algorithm relate cells in A' to those in B'. However, different cell relations may exist for the same pseudo-perturbation vector.

Problem Formulation. Given a set of pseudo-perturbation binary vectors, O, and given the matrix of preprocessed scRNAseq data of normalized readout values, find the sets of cells A'^* and B'^*, associated by all pseudo-perturbation vectors in O, that maximize the difference of readout vectors, $r^{A'*}$ (for readouts of cells in A'^*) and $r^{B'*}$ (for readouts of cells in B'^*).

Algorithm. For each vector b in the set of optimal pseudo-perturbations, relating cells c_1 (in A') and c_2 (in B'):

1. Compute a set of *redundant cells* for each class. This involves identifying cells in class A with an identical binarized vector b, denoted as set R_b^A, and likewise for class B denoted as R_b^B. Both sets, R_b^A and R_b^B, include cells c_1 and c_2 respectively.
2. Iterate across all pairs of cells in $R_b^A \times R_b^B$, and calculate the difference of readout genes values while keeping the maximal difference.

We retrieve an association of each optimal pseudo-perturbation to a vector of normalized readouts expression that maximizes the difference between the two classes. Additionally, we calculate the *representativity score* for the optimal pseudo-perturbations by considering the number of redundant cells. Let n^A be the number of cells in class A, and let O be the set of Boolean vectors in all optimal pseudo-perturbations for class A. The representativity score S^A for class A is defined as follows:

$$S^A = \frac{\sum_{b \in O} |R_b^A|}{n^A} \times 100. \tag{1}$$

2.4   Implementation and Software Availability

The complete framework was implemented in an open-source system scRNA2BoNI available at: https://github.com/mathieubolteau/scRNA2BoNI. scRNA2BoNI uses Answer Set Programming (ASP) [1] as logical modeling and constraint solving paradigm to *identify the maximal number of pseudo-perturbations* and Python for the *maximization of readout difference*. ASP is used to model problems from NP and provides state-of-the-art solvers that propose exact solutions for optimization problems and allow enumeration of all optimal or pseudo-optimal solutions. For our study, we used clasp [5]. On a computer cluster comprising 160 CPUs and 1.5 To of RAM, given an association matrix comprising expression of 111 genes for 680 cells, our pipeline allows us to generate 20 pseudo-perturbations in 65 h. This corresponds to a pseudo-optimal solution for this problem that is not unique. The ASP program of this algorithm is provided in the supplementary material. The complexity of our program can be analyzed considering two factors that create the search space: (*i*) the selection of k genes from a total set of G genes, and (*ii*) the choice of pairs of cells. That is, for each possible selection of k genes, an amount of c associations between cells in classes A and B (where the values of the k genes coincide) has to be tested to discard redundancies within the same class. The maximum value for c is $|A| \times |B|$; which represents associating all cells in both classes. clasp performs backjump and conflict-driven learning, optimizing the search space; thus, our estimate measures a worse case. The estimated complexity for the worst-case (see Eq. 2) implies that our algorithm is exponential on the number of considered genes and cells from our scRNAseq dataset.

$$\mathcal{O}(\binom{|G|}{k} \times 2^{|A| \times |B|}) \tag{2}$$

3   Results

Our data and results are available as supplementary material at: https://github.com/mathieubolteau/scRNA2BoNI/tree/master/ISBRA_2023_Supp.

3.1   Pseudo-perturbations Identification - Different Size Benchmarks

We tested our algorithm on 4 toy datasets (see specifications in Table 1, datasets $A - D$). We also applied our program on 2 entire datasets: phosphoproteomics data, measuring averaged cell population protein expression (dataset P) from [2] and scRNAseq data (dataset SC) from [9]. Our results are shown in Table 1. We can see that using dataset B we identified 5 optimal pseudo-perturbations with identical input and intermediate genes expression for both classes. These 5 different Boolean vectors of pseudo-perturbations represent the expression behavior of 83% (resp. 100%) of the cells in class *early TE* (resp. *medium TE*) for the $k = 5$ selected genes (see Eq. 1). On datasets $A - B$, we found an optimal solution, whereas on datasets $C - SC$, suboptimal ones. Our results enable us to advise potential users on expected computation times based on their dataset sizes. For datasets P and SC, we found up to 23 and 20 pseudo-perturbations, respectively. The representativity of selected patients in the phosphoproteomics data (21% and 45%) is vastly lower than the representativity of selected cells in the scRNAseq case study (75% and 89%), suggesting more redundancies in scRNAseq data. Our method is thus applicable for selecting optimal pseudo-perturbations from scRNAseq data.

3.2   Discrimination of the Medium and Late Trophectoderm Stages

PKN Reconstruction. We used 438 transcription factor (TF) genes involved in human embryonic development as input for pyBRAvo to build the PKN (see supplementary data for further details). The PKN is composed of 327 nodes and 475 edges, with only 28 of the 438 initial TFs found in Pathway Commons [12]. We then reduced the network to 191 nodes (84 input, 27 intermediate, 14 readout genes, and 66 complexes) and 285 edges, limited to genes measured in scRNAseq data and complexes linked to these genes (see supplementary material).

Table 1. Maximizing number of pseudo-perturbations applied to 6 case studies.

Dataset	Source	Classes (C1;C2)	m	Cells or patients (C1;C2)	k	Execution time[1]	Different Boolean vectors	Representative score S^2 (C1;C2)
A	artificial	C1 ; C2	10	10 (5;5)	3	0.105 s	3	50;60
B	subset of single-cell data	early TE ; medium TE	30	24 (12;12)	5	11.379 s	5	83;100
C	subset of single-cell data	early TE ; medium TE	100	50 (25;25)	10	5h*	11	76;80
D	subset of single-cell data	early TE ; medium TE	120	200 (100;100)	15	5h*	18	40;37
P	phosphoproteomics data from [2]	CR ; PR	79	191 (136;55)	10	96h*	23	21;45
SC	single-cell data	medium TE ; late TE	111	680 (348;332)	10	65h*	20	75;89

m refers to the number of input and intermediate genes or proteins. CR = Complete Remission ; PR = Primary Resistant (cf. to [2]). [1] Tests were performed on a computer cluster comprising 160 CPUs and 1.5 To of RAM. [2] see Eq. 1. * Execution time corresponds to the fixed timeout.

Experimental Design Construction. We generated pseudo-perturbations for the experimental design using the method described in Sect. 2.3, which employed the set of input and intermediate genes from the reduced PKN, comprising 111 genes. Our analysis focused on the expression of these genes across 680 cells, which were identified to be in medium and late TE developmental stages (see Table 1, dataset SC).

We tested different values of k, the number of selected genes, similar to those used in [13]. We observed the number of pseudo-perturbations generated after 30 h of calculation on a computer cluster and computed the representativity score for each k value. Based on our results, $k = 10$ was the best compromise between a high number of pseudo-perturbations and a high representativity score (see Fig. 1A). This value was also used in [2], supporting our decision.

Our method produced 20 pseudo-perturbation Boolean vectors, which paired medium and late TE cells to maximize the expression value difference of 14 readout genes. In Fig. 1C, we present the experimental design composed of 24 genes: 7 inputs genes (in green), 3 intermediate genes (in red), and 14 readouts (in blue). Each row represents a pseudo-perturbation (on the left, ordered from most to least representative) and its readout observations. Note that each vector is unique. We observe some readout genes with minimal variations (mean of expression difference between both stages less than 0.06), *e.g. DEC1* or *SOD1*, and some readout genes where a significant variation (mean of expression difference between both stages greater than 0.30) is observed, *e.g. CEBPB, CEBPD* or *GSR*. These last also appear in the learned BNs (see Fig. 1B).

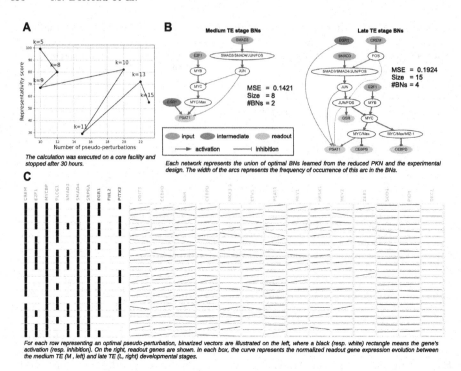

Fig. 1. Medium and late TE discrimination findings. **A.** Impact of k on the number of different pseudo-perturbations and their representativity in the dataset. **B.** Inferred BNs. **C.** Visualization of the experimental design. (Color figure online)

BN Inference. We used the generated experimental design combined with the reduced PKN to infer BNs specific to medium and late TE using the Caspo software. Caspo proposes BNs that match the PKN topology and have an optimal (minimal) mean square error (MSE) between the Boolean prediction of readout nodes (given the Boolean input states) and their experimental measurement. The Caspo used parameters are presented in the supplementary material.

Figure 1B illustrates the union of learned BNs for the medium and late TE developmental stages, respectively, which are compatible with the fixed fitness value. The size of the learned BNs is equal to 8 for medium TE and 15 for late TE. The optimal MSE for the learned BNs equals 0.1421 and 0.1924, respectively. The medium TE family has 2 BNs, while the late TE one has 4. The execution time for both classes is comparable. These two families of BNs exhibit distinct differences in their gene behaviors within cell types. Interestingly, the late TE BNs connect more input and readout genes than the medium TE BNs. Both classes of BNs share two input genes, $SMAD3$ and $E2F1$, as well as one intermediate gene, $EGR1$, while only one common readout, $PSAT1$, is present. Notably, most of the interactions (without considering their sign) of the medium TE BNs are included in the late TE BNs. While both medium and late TE BNs propose different regulatory mechanisms for PSAT1, the medium TE BNs suggest an activation path from SMAD3. In contrast, the late TE BNs propose two inhibition paths from the same input. Likewise, an inhibition path from *E2F1* to *PSAT1* is proposed in medium TE BNs, while an activation path between these genes is proposed for late TE BNs. This path is, however, subject to the presence or absence of SMAD3. Seemingly the *PSAT1* readout was measured differently in the same pseudo-perturbation configuration involving genes *E2F1* and *PSAT1*. Late TE BNs exhibit supplementary readout genes, namely *GSR*, *CEBPB*, and *CEBPD*, indicating that the readout measurements matched the late TE BNs prediction, given the selected pseudo-perturbation Boolean vectors. However, medium TE BNs could not predict the observed measurements with minimal error on these three genes. Consequently, late TE regulatory mechanisms appear more complex than medium TE ones.

4 Discussion and Conclusion

In this paper, we propose an original framework to compute families of Boolean networks compatible with scRNAseq data and prior regulatory knowledge. Our method generates Boolean networks comparing two different conditions. We applied the implemented framework to human embryo development to study the difference between cell behavior at a medium and late TE developmental stage. Despite the lack of in vitro perturbation data and the sparsity of single-cell datasets, our method yields meaningful results.

As significant results, we developed an algorithm to obtain pseudo-perturbations from scRNAseq data demonstrating scalability and efficiency through benchmarking with datasets of varying sizes. The worst-case search complexity for the real case study was of $\binom{111}{10} \times 2^{348 \times 332} = 3.26 \times 10^{34793}$, and our partial results were generated in 65h. We prove that our algorithm allows for more diverse pseudo-perturbation sets than the state-of-the-art method [2] (see supplementary material), which studied cell population-averaged measurements. We can simulate real perturbations by identifying pseudo-perturbations and proposing more precise (such as Boolean) computational models. Our method identified 20 pairs of cells with Boolean expressions coinciding with selected genes, representing of 75% and 89% of the complete set of cells in medium and late TE developmental stages, respectively.

Using diverse pseudo-perturbations sets, we generate families of Boolean networks to distinguish medium and late TE developmental stages in human embryonic development. The BNs propose Boolean functions derived from the Pathway Commons database to model gene regulation mechanisms. Late TE cells exhibit a more complex BN structure (size 15 vs. 8) than medium TE cells. These findings are consistent with the fact that late TE requires a gain of biological function to help the embryo implant in the endometrium. Differently, from methods that propose a single computational model of averaged cells, our method includes a subset of 20 cells for each stage and learns optimal families of BNs representing the diversity of expression mechanisms within this cell subset for each stage.

Acknowledgments. This work was partially supported by funds from the Agence Nationale de la Recherche [ANR-20-THIA-0011 to M.B., ANR-20-CE17-0007 to L.D. and M.B.]. We are most grateful to the Bioinformatics Core Facility of Nantes BiRD, member of Biogenouest, Institut Français de Bioinformatique (IFB) (ANR-11-INBS-0013) for the use of its resources and for its technical support. Some of the experimentations were performed using HPC resources from the GLiCID computing center.

Appendix

Boolean Network (BN). A Boolean network B, of dimension n is defined as $B = (N, F)$ where: $N = \{v_1, \ldots, v_n\}$ is a finite set of nodes (variables or genes) and $F = \{f_1, \ldots, f_n\}$ is a set of Boolean functions $f_i : \mathbb{B}^n \to \mathbb{B}$, with $\mathbb{B} = \{0, 1\}$, describing the evolution of variable v_i.

Influence Graph (IG). An IG is denoted by $G = (V, E, \sigma)$ with $V = \{v_1, \ldots, v_n\}$ the set of nodes, $E \subseteq V \times V$ the set of directed edges, and $\sigma \subseteq E \times \{+1, -1\}$ the signs of the edges.

In the context of gene regulation, $j \to i$ means that the change of j in time influences the level of i. Edges $j \to i$ are labeled with a sign, where $+1$ (resp. -1) indicates that j tends to increase (decrease) the level of i. The IG derived from regulatory knowledge bases, is called a *Prior-Knowledge Network* (PKN). The PKN serves as the initial base for generating multiple BNs that adhere to its topology. So that each node in the PKN corresponds to a gene and has an on/off state determined by the Boolean function defined by the BN. Different BNs can have the same IG, while a BN can only be assimilated to a single IG.

Within the PKN, we identify three types of genes. An *input gene*, which is a gene without any predecessor; an *intermediate gene*, with predecessor(s) and successor(s); and a *readout gene*, without any successor. Input and intermediate genes refer to the part of the PKN that can be stimulated (externally or internally), they can also be referred to as system *entries*. While readouts are the part of the system that can be observed, they can be referred to as the system *output*.

Pseudo-perturbations. Usually perturbation data is required to discover Boolean mechanisms within a system. This data comes in the form of *on/off* values of entries associated with output values. However, in the human embryonic development context, perturbing the system is not feasible for obvious reasons. Therefore, we introduce the notion of *pseudo-perturbations*, which refers to artificial perturbations derived from the (unperturbed) gene expression observations.

References

1. Baral, C.: Knowledge Representation, Reasoning, and Declarative Problem Solving. Cambridge University Press, New York (2003)

2. Chebouba, L., Miannay, B., Boughaci, D., Guziolowski, C.: Discriminate the response of acute myeloid leukemia patients to treatment by using proteomics data and answer set programming. BMC Bioinf. **19**(2), 15–26 (2018). https://doi.org/10.1186/s12859-018-2034-4

3. Chevalier, S., Noël, V., Calzone, L., Zinovyev, A., Paulevé, L., Paulevé: Synthesis and simulation of ensembles of Boolean networks for cell fate decision, pp. 193–209 (2020). https://doi.org/10.1007/978-3-030-60327-4_11

4. Dunn, S.J., Li, M.A., Carbognin, E., Smith, A., Martello, G.: A common molecular logic determines embryonic stem cell self-renewal and reprogramming. EMBO J. **38**(1), e100003 (2019). https://doi.org/10.15252/embj.2018100003

5. Gebser, M., Kaufmann, B., Schaub, T.: Conflict-driven answer set solving: from theory to practice. Artif. Intell. **187–188**, 52–89 (2012). https://doi.org/10.1016/j.artint.2012.04.001

6. Lefebvre, M., Gaignard, A., Folschette, M., Bourdon, J., Guziolowski, C.: Large-scale regulatory and signaling network assembly through linked open data. Database 2021 (2021). https://doi.org/10.1093/database/baaa113

7. Luecken, M.D., Theis, F.J.: Current best practices in single-cell RNA-seq analysis: a tutorial. Mol. Syst. Biol. **15**(6), e8746 (2019). https://doi.org/10.15252/msb.20188746

8. McInnes, L., Healy, J., Melville, J.: UMAP: Uniform Manifold Approximation and Projection for Dimension Reduction (2020). https://doi.org/10.48550/arXiv.1802.03426

9. Meistermann, D., et al.: Integrated pseudotime analysis of human pre-implantation embryo single-cell transcriptomes reveals the dynamics of lineage specification. Cell Stem Cell **28**(9), 1625-1640.e6 (2021). https://doi.org/10.1016/j.stem.2021.04.027

10. Petropoulos, S., et al.: Single-cell RNA-Seq reveals lineage and X chromosome dynamics in human preimplantation embryos. Cell **165**(4), 1012–1026 (2016). https://doi.org/10.1016/j.cell.2016.03.023

11. Qiu, X., et al.: Reversed graph embedding resolves complex single-cell trajectories. Nature Methods 2017 **14**(10), 979–982 (2017). https://doi.org/10.1038/nmeth.4402

12. Rodchenkov, I., et al.: Pathway commons 2019 update: integration, analysis and exploration of pathway data. Nucleic Acids Res. **48**(D1), D489–D497 (10 2019). https://doi.org/10.1093/nar/gkz946

13. Videla, S., Saez-Rodriguez, J., Guziolowski, C., Siegel, A.: caspo: a toolbox for automated reasoning on the response of logical signaling networks families. Bioinformatics **33**(6), 947–950 (2017). https://doi.org/10.1093/bioinformatics/btw738

Enhancing t-SNE Performance for Biological Sequencing Data Through Kernel Selection

Prakash Chourasia$^{(\boxtimes)}$, Taslim Murad, Sarwan Ali, and Murray Patterson

Georgia State University, Atlanta, GA, USA
{pchourasia1,tmurad2,sali85,mpatterson30}@gsu.edu

Abstract. The genetic code for many different proteins can be found in biological sequencing data, which offers vital insight into the genetic evolution of viruses. While machine learning approaches are becoming increasingly popular for many "Big Data" situations, they have made little progress in comprehending the nature of such data. One such area is the t-distributed Stochastic Neighbour Embedding (t-SNE), a general-purpose approach used to represent high dimensional data in low dimensional (LD) space while preserving similarity between data points. Traditionally, the Gaussian kernel is used with t-SNE. However, since the Gaussian kernel is not data-dependent, it only determines each local bandwidth based on one local point. This makes it computationally expensive, hence limited in scalability. Moreover, it can misrepresent some structures in the data. An alternative is to use the isolation kernel, which is a data-dependent method. However, it has a single parameter to tune in computing the kernel. Although the isolation kernel yields better performance in terms of scalability and preserving the similarity in LD space, it may still not perform optimally in some cases. This paper presents a perspective on improving the performance of t-SNE and argues that kernel selection could impact this performance. We use 9 different kernels to evaluate their impact on the performance of t-SNE, using SARS-CoV-2 "spike" protein sequences. With three different embedding methods, we show that the cosine similarity kernel gives the best results and enhances the performance of t-SNE.

Keywords: t-SNE · Dimensionality Reduction · Kernel Methods · Visualization · Embedding · SARS-CoV-2 · Spike Sequence Analyses

1 Introduction

In Machine Learning (ML), kernels are frequently used to solve challenges involving calculating object similarity between pairs of objects. The t-distributed Stochastic Neighbour Embedding (t-SNE) [15] is a popular solution researchers use to reduce this dimensionality. Its goal is to project high-dimensional datasets

---

P. Chourasia, T. Murad and S. Ali—Equal Contribution.

© The Author(s), under exclusive license to Springer Nature Singapore Pte Ltd. 2023
X. Guo et al. (Eds.): ISBRA 2023, LNBI 14248, pp. 442–452, 2023.
https://doi.org/10.1007/978-981-99-7074-2_35

into lower-dimensional spaces while maintaining data point similarities, as indicated by the Kullback-Liebler (KL) divergence. This paper reviews the use of different kernels and their impact on t-SNE for SARS-Cov-2[1] sequence data and its visualization in low dimension. We also assess these kernels in terms of classification and clustering. The vast global spread of COVID-19 spurred this, pushing viral sequence research into the "Big Data" sphere. This presents challenges since highly dimensional data cannot be used directly for ML solutions.

Since spike protein sequences cannot be used directly as input to machine learning (ML) models, we must first convert the sequences into a fixed-length numerical representation. For this purpose, using feature engineering-based methods is a popular choice, as proposed in many studies recently [1,2,10,12]. It has been shown that different embeddings can yield different results in terms of classification [3] and clustering [4–6,16] of SARS-CoV-2 spike sequences.

The t-SNE is a method to visualize high-dimensional data by mapping each point to a low-dimensional space (2 or 3 dimensions). In the literature, the Gaussian kernel is used by default for t-SNE-based visualization [25]. However, recent studies [8,25] show that the Gaussian kernel may not always be the best choice for t-SNE-based visualization, as it is computationally expensive and could perform worse than the isolation kernel (a data-dependent kernel). We argue in this paper that even using the isolation kernel [25] may not be the better option when dealing with biological sequences. When evaluating the t-SNE method both subjectively and objectively, we demonstrate that numerous kernels outperform the Gaussian and isolation kernel using embedding techniques(three feature engineering-based embeddings). In this paper, following are our contributions:

1. We show that the cosine similarity-based kernel is a better choice for t-SNE as compared to 8 other kernel methods, including Gaussian and isolation kernels, in the case of SARS-CoV-2 spike sequences.
2. We evaluate the performance of the t-SNE model on both objective and subjective criteria and report results for several kernel computation approaches.
3. We show that the cosine similarity kernel is better in terms of computational runtime and pairwise distance preservation in low dimensions as compared to the Gaussian and isolation kernel on SARS-CoV-2 sequences. Therefore, it could be a potential candidate for efficient t-SNE computation for the eventually larger sets of biological sequences [19].

The rest of the paper is organized as follows: Sect. 2 discusses the related work. Section 3 and Sect. 4 discuss the methods used to compute the kernel matrix and detail for computing the t-SNE using the kernel matrix. Section 5 contains details of the different embeddings we use. Section 6 contains experimentation and data statistics. In Sect. 7, we evaluate the impact of different kernels on t-SNE. Finally, we conclude the paper in Sect. 8.

---

[1] The SARS-Cov-2 virus is the cause of the global COVID-19 pandemic.

2   Related Work

Data visualization is an important task. Using t-SNE, originally introduced in [15], has made this task easy. Authors in [3,8] use t-SNE to visualize different variants in the coronavirus protein sequence data. It has also been found that clustering the COVID-19 protein sequences using k-means is also related to the patterns shown in the t-SNE plots [4,6,7,20].

Authors in [9] proposed Symmetric stochastic neighbor embedding (SNE) to get the 2D representation of the high dimensional data. An extension of t-SNE, called Heavy-tailed SNE, is proposed in [24], which considers different embedding similarity functions. Authors in [22] propose a method called t-Distributed Stochastic Triplet Embedding, based on similarity triplets to consider similar points and discard dissimilar points in the embeddings. Some efforts have been made previously to speed up the computation of t-SNE [21] using tree-based algorithms. Authors in [25] show that using the isolated kernel within the t-SNE could improve the visualization in 2D and its runtime compared to the Gaussian. A decentralized data stochastic neighbor embedding (dSNE) is proposed in [17], visualizing the decentralized data. The differentially private dSNE (DP-dSNE) version is proposed in [18]. Although contemporary t-SNE methods are effective on well-known datasets like MNIST, it is unclear if those methods will be just as effective when applied to biological protein sequences.

3   Kernel Matrix Computation

Table 1 describes different methods we use to compute the kernel matrix.

Table 1. Methods used to compute Kernel Matrix

Kernel	Formula
Cosine similarity	$k(x,y) = \frac{\|x\|\|y\| \times \cos(\theta)}{\|x\|\|y\|} = \frac{x \cdot y}{\|x\|\|y\|}$ (1)
Linear	$k(x,y) = x^T y + c$ (2)
Polynomial	$k(x,y) = \left(x^T y + r\right)^d$ (3)
Gaussian	$\mathcal{K}(x,y) = \exp\left(\frac{-\|x-y\|^2}{2\sigma^2}\right)$ (4)
Isolation [25]	$K_\psi(x,y \mid D) = \mathbb{E}_{H_\psi(D)}[\!|\!\mathcal{K}(x,y \in \theta[z] \mid \theta[z] \in H)]$ (5)
Laplacian	$k(x,y) = \exp\left(-\frac{1}{2\sigma^2}\|x-y\|_1\right)$ (6)
Sigmoid	$k(x,y) = \tanh\left(\gamma x^\top y + c_0\right)$ (7)
Chi-squared	$k(x,y) = \exp\left(-\gamma \sum_i \frac{(x_i - y_i)^2}{x_i + y_i}\right)$ (8)
Additive-chi-squared	$k(x,y) = \sum_i \frac{2x_i y_i}{x_i + y_i}$ (9)

4   Using Kernel Matrix for t-SNE

The t-SNE method takes an $n \times n$ kernel matrix as input, where n is the total number of sequences (embedding vectors), and produces a low dimensional representation d, where $d = \{1, 2, \cdots, n-1\}$. More formally, given the high dimensional (HD) data, the idea of t-SNE is to represent the data in low dimensions (LD) while preserving the pairwise distance between the embedding vectors i.e. to keep the distances between points in LD Y as close as to in HD X. The t-SNE approach works as follows:

Compute Conditional Probability or Pairwise Affinities. The first step in t-SNE is to calculate the Euclidean distances of each point from all other points in high dimensions. This can be done using different kernel functions. Later this distance between data points is converted into conditional probabilities, also known as pairwise affinities or similarity matrices.

High Dimensional Probability Computation. The conditional probability can be gathered to give joint distribution on pairs of points. This is gathered into a symmetric matrix, and returned joint probability can be written as given in Eq. 10 (as also given in [23]):

$$p_{j|i} = \frac{\exp\left(-\left\|x_i - x_j\right\|^2 / 2\sigma_i^2\right)}{\sum_{k \neq i} \exp\left(-\left\|x_i - x_k\right\|^2 / 2\sigma_i^2\right)} \tag{10}$$

Initial Solution Sampling. Initial solution Y is sampled with random initial values. These values are optimized to give the best lower dimensional representation of data points.

Compute Low Dimensional Joint Probability. Similar to conditional probability in the HD, we compute it in the LD. Finally, gather these to get the low dimensional joint probability, which can be written as follows (also mentioned in [18]):

$$Q_{ij} = \begin{cases} 0 & j = i \\ \frac{\left(1 + \|\mathbf{y}_i - \mathbf{y}_j\|^2\right)^{-1}}{\sum_{k \neq l}\left(1 + \|\mathbf{y}_k - \mathbf{y}_l\|^2\right)^{-1}} & j \neq i \end{cases} \tag{11}$$

Compute KL Divergence and Gradient. For these two distributions P in an HD and Q in an LD, to measure the distance between them, we use KL divergence. It is used to find the variation or distribution among the distances in the data points. KL divergence is computed as:

$$J = min \quad KL(P\|Q) = \sum_i \sum_j p_{ij} \log \frac{p_{ij}}{q_{ij}} \tag{12}$$

where J is the cost function, P and Q represents two different dimensions, and y_i, y_j are points in Q. Take the derivative and calculate gradient descent Eq. 13 (as mentioned in [23]) to get the minimum from J

$$\nabla = \frac{\delta J}{\delta y_i} = 4 \sum_j (p_{ij} - q_{ij})(y_i - y_j)\left(1 + \|y_i - y_i\|^2\right)^{-1} \tag{13}$$

and keep updating Y for minimum J Eq. 14. We apply t-distribution in this Stochastic Neighborhood Embedding (SNE). Applying t-distribution on Q low dimension gives us a longer tail to give better visualization.

$$y^{(t)} = y^{(t-1)} + \eta\nabla + \alpha(t)\left(y^{(t-1)} - y^{(t-2)}\right) \tag{14}$$

where $t \in T$ represents t-time iterations.

In summary, the kernel function plays an important role here to give distances between the points in the original data. The workflow of t-SNE computation using the kernel matrix as input is illustrated in Algorithm 1.

Algorithm 1. t-SNE Computation.

1: **Input:** (KM, dim)
   $KM = x[i][j], x_{ij} \in R^n$ ▷ KM => Kernel Matrix
   dim: number of dimension ▷ output dimension
2: **Output:** $Y = [y_1, y_2, y_3, \ldots, y_N], y_i \in R^{dim}$
3: **Function:** tSNE(KM, dim)
4: $Y = matrix(n, dim)$ ▷ randomly initialize output matrix
5: //Now initialize optimization parameters
6: $I = 1000$ ▷ Iterations
7: $\eta = 500$ ▷ Learning rate
8: $\alpha = 0.5$ ▷ momentum
9: $P = matrix(n \times n)$ ▷ probability matrix in HD
10: **for** $i \leftarrow 1$ to n **do**
11:    **for** $j \leftarrow 1$ to n **do**
12:       $P_{ij} = $ COMPUTEPROBABILITY($KM_{i,j}$) ▷ from Eq 10
13: $Q = matrix(n \times 2)$ ▷ probability matrix in LD
14: **for** $k \leftarrow 1$ to I **do** ▷ Iteration loop
15:    **for** $i \leftarrow 1$ to n **do**
16:       **for** $j \leftarrow 1$ to n **do**
17:          $Q_{ij} = $ COMPUTEPROBABILITY($Y_{i,j}$) ▷ Eq. 11
18:    $J = $ COMPUTEKLDIVERGENCE(P, Q) ▷ Compute KL Divergence from Eq 12
19:    $\nabla = $ COMPUTEGRADIENT(J) ▷ using Eq 13
20:    $Y = $ UPDATEOUTPUT(Y, α, η, ∇) ▷ using Eq 14
21:    **if** ($k == 250$) **then** $\alpha = 0.8$ ▷ momentum changed at iteration
22: **return** Y

---

5   Feature Embeddings Generation

This section describes the three embedding methods we use to convert the biological sequences into a fixed-length numerical representation.

1. One Hot Encoding (OHE) - To convert the amino acids into numerical representation, we use OHE [3,12].
2. Spike2Vec [2] - It generates a fixed-length numerical representations using the concept of k-mers (also called n-gram). It uses the sliding window concept to generate substrings (called mers) of length k (size of the window).
3. Minimizers - Using the $k$mer, the minimizer is computed as a substring (mer) of length m (where $m < k$) within that $k$mer. It is a lexicographical minimum in the forward and reverse order of the k-mer.

6   Experimental Setup

In this section, we discuss the dataset statistics followed by the goodness metrics used to evaluate the performance of t-SNE. All experiments are performed on Intel (R) Core i5 system with a 2.40 GHz processor and 32 GB memory.

6.1   Data Statistics

The dataset we use, we call the Spike7k dataset, consists of sequences of the SARS-CoV-2 virus and is taken from the well-known database GISAID [11]. It has 22 unique lineages as the label with the following distribution: B.1.1.7 (3369), B.1.617.2 (875), AY.4 (593), B.1.2 (333), B.1 (292), B.1.177 (243), P.1 (194), B.1.1 (163), B.1.429 (107), B.1.526 (104), AY.12 (101), B.1.160 (92), B.1.351 (81), B.1.427 (65), B.1.1.214 (64), B.1.1.519 (56), D.2 (55), B.1.221 (52), B.1.177.21 (47), B.1.258 (46), B.1.243 (36), R.1 (32).

6.2   Evaluating t-SNE

For objective evaluation of the t-SNE model, we use a method called k-ary neighborhood agreement (k-ANA) method [25]. The k-ANA method (for different k nearest neighbors) checks the neighborhood agreement (Q) between HD and LD and takes the intersection on the numbers of neighbors. More formally:

$$Q(k) = \sum_{i=1}^{n} \frac{1}{nk} |kNN(x_i) \cap kNN(x_i')| \tag{15}$$

where $kNN(x)$ is set of k nearest neighbours of x in high-dimensional and $kNN(x')$ is set of k nearest neighbours of x in corresponding low-dimensional.

We use a quality assessment tool that quantifies the neighborhood preservation and is denoted by $R(k)$, which uses Eq. 15 to evaluate on scalar metric whether neighbors are preserved [13] in low dimensions. More formally:

$$R(k) = \frac{(n-1)Q(k) - k}{n - 1 - k} \tag{16}$$

$R(k)$ represents the measurement for k-ary neighborhood agreement. Its value lies between 0 and 1, the higher score represents better preservation of the neighborhood in LD space. In our experiment, we computed $R(k)$ for $k \in 1, 2, 3, ..., 99$ then considered the area under the curve (AUC) formed by k and $R(k)$. Finally, to aggregate the performance for different k-ANN, we calculate the area under the $R(k)$ curve in the log plot (AUC_{RNX}) [14]. More formally:

$$AUC_{RNX} = \frac{\Sigma_k \frac{R(k)}{k}}{\Sigma_k \frac{1}{k}} \tag{17}$$

where AUC_{RNX} denotes the average quality weight for k nearest neighbors.

7   Subjective and Objective Evaluation of t-SNE

This section discusses the performance of t-SNE in subjective (using 2D scatter plots) and objective (using AUC_{RNX}) ways for different embeddings and kernels.

7.1   Subjective Evaluation

To visually evaluate the performance of t-SNE, we use different embedding methods and plot the 2D visual representation to analyze the overall structure of the data. Figure 1 shows the top 2 performing kernels in terms of AUC_{RNX} score for respective embedding. Similarly, Fig. 2 shows the worse 2 performing kernels.

OHE [12]. We analyzed the t-SNE plots for one-hot embedding for different kernel methods. For the Alpha variant (B.1.1.7), we can see that Gaussian, Isolation, Linear, and Cosine kernel can generate a clear grouping. However, for the other variants with a small representation in the dataset (e.g. B.1.617.2 and AY.4), we can see that the Cosine and linear kernels are better than the Gaussian and Isolation. This could mean that the Gaussian and Isolation tend to be biased toward the more representative class (alpha variant).

Spike2Vec [2]. The t-SNE plots for Spike2vec-based embeddings using different kernel methods are evaluated. It is similar to OHE, where the Gaussian and isolation kernels almost perfectly group the alpha (B.1.1.7) variant. However, all other variants are scattered around in the plot.

Minimizer. Similarly, in the t-SNE plots for minimizer embedding for different kernel methods for the Gaussian and isolation kernel, we can observe similar behavior for the alpha (B.1.1.7) variant.

We also show the 3D plots for t-SNE using the Cosine similarity kernel in Fig. 3 for Spike2Vec and Minimizers-based embedding. We can see for Spike2Vec, the Alpha variant shows clear grouping. Similarly, the delta and epsilon variant also contains a few small groups.

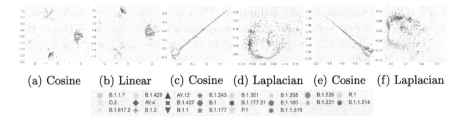

(a) Cosine   (b) Linear   (c) Cosine   (d) Laplacian   (e) Cosine   (f) Laplacian

Fig. 1. t-SNE plots for top-performing kernel methods for different embedding. This figure is best seen in color. (a), (b) is from OHE. (c), (d) are from Spike2Vec. and (e), (f) are from Minimizer encoding.

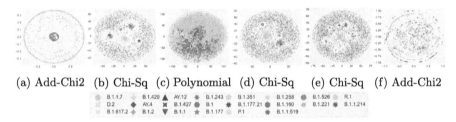

(a) Add-Chi2   (b) Chi-Sq   (c) Polynomial   (d) Chi-Sq   (e) Chi-Sq   (f) Add-Chi2

Fig. 2. t-SNE plots for worst-performing kernel methods for different embedding. This figure is best seen in color. (a), (b) is from OHE. (c), (d) are from Spike2Vec. and (e), (f) are from Minimizer encoding.

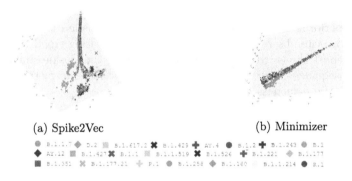

(a) Spike2Vec   (b) Minimizer

Fig. 3. t-SNE 3d plots using Cosine similarity kernel.

7.2   Objective Evaluation of t-SNE

The objective evaluation of tSNE is done using Eq. 17. The goodness of t-SNE using kernel computation runtime and AUC_{RNX} for different embedding approaches are reported in Table 2. An interesting insight, which we can observe, is that the Cosine similarity-based kernel outperforms all other kernel methods in terms of kernel computational runtime and AUC_{RNX} value for all embedding methods. This means that the Cosine similarity-based kernel could be scaled easily on a bigger dataset and its neighborhood agreement in high dimensional data vs. low dimensional data is highest. Therefore, we can conclude that for the biological protein sequences, using the cosine similarity kernel is better than the Gaussian or isolation kernel (authors in [25] argue that using the isolation kernel with t-SNE is better).

Table 2. AUC_{RNX} values for t-SNE using different kernel and encoding methods on **Spike7k datasets.** The kernel is sorted in descending order with the best at the top and the best values are shown in underlined and bold.

OHE			Spike2Vec			Minimizer		
Kernel	Kernel comp. time	AUC_{RNX}	Kernel	Kernel comp. time	AUC_{RNX}	Kernel	Kernel comp. time	AUC_{RNX}
Cosine similarity	**37.351**	**0.262**	Cosine similarity	**15.865**	**0.331**	Cosine similarity	**17.912**	**0.278**
Linear	51.827	0.260	Laplacian	721.834	0.260	Laplacian	983.26	0.242
Gaussian	94.784	0.199	Gaussian	54.192	0.235	Sigmoid	27.813	0.229
Sigmoid	57.298	0.190	Linear	30.978	0.197	Gaussian	65.096	0.206
Laplacian	2250.30	0.184	Isolation	24.162	0.189	Isolation	27.266	0.172
Polynomial	57.526	0.177	Sigmoid	30.037	0.168	Polynomial	28.666	0.166
Isolation	39.329	0.161	Additive-chi2	495.221	0.121	Linear	27.503	0.165
Chi-squared	1644.86	0.131	Chi-squared	495.264	0.104	Additive-chi2	576.92	0.125
Additive-chi2	1882.45	0.110	Polynomial	28.427	0.059	Chi-squared	921.63	0.095

7.3   Runtime Analysis

The runtime for computing different kernel matrices with increasing sequences is shown in Fig. 4a and its zoomed version in Fig. 4b. We can see the Cosine similarity outperforms all other kernel methods. The Gaussian takes the longest. Moreover, we can observe linear overall runtime increasing trend for most kernels. We report the t-SNE computation runtime for the Cosine similarity with increasing sequences in Fig. 4c. We can see a linear increase in the runtime as we increase the sequences.

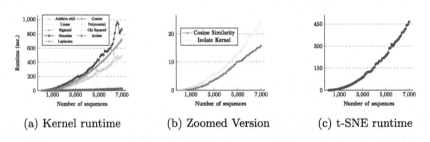

(a) Kernel runtime   (b) Zoomed Version   (c) t-SNE runtime

Fig. 4. (a) and (b) shows the kernel computation Runtime for an increasing number of sequences (using Spike2Vec-based embedding). (c) shows the t-SNE computation runtime with increasing sequences (using Spike2Vec-based embedding). This figure is best seen in color.

8   Conclusion

In this paper, we modify the original tsne algorithm and show its performance using different kernel methods. We demonstrate that the cosine similarity-based kernel performs best among 9 kernels for t-SNE-based visualization. We show that, rather than using Gaussian or isolation kernel (as argued in previous works), the cosine similarity kernel yields better computational runtimes (hence better scalability) and improves the performance of t-SNE. In the future, we will explore other biological data to evaluate the performance of the reported kernels. We also plan to use different embedding and kernel methods to explore the impact on the classification and clustering results for the biological sequences.

References

1. Ali, S., Bello, B., Chourasia, P., et al.: PWM2Vec: an efficient embedding approach for viral host specification from coronavirus spike sequences. MDPI Biol. **11**(3), 418 (2022)
2. Ali, S., Patterson, M.: Spike2vec: an efficient and scalable embedding approach for covid-19 spike sequences. In: International Conference on Big Data (Big Data), pp. 1533–1540 (2021)
3. Ali, S., Sahoo, B., Ullah, N., Zelikovskiy, A., Patterson, M., Khan, I.: A k-mer based approach for SARS-CoV-2 variant identification. In: Wei, Y., Li, M., Skums, P., Cai, Z. (eds.) ISBRA 2021. LNCS, vol. 13064, pp. 153–164. Springer, Cham (2021). https://doi.org/10.1007/978-3-030-91415-8_14
4. Ali, S., Tamkanat-E-Ali, et al.: Effective and scalable clustering of SARS-CoV-2 sequences. In: International Conference on Big Data Research (ICBDR), pp. 1–8 (2021)
5. Ali, S., Zhou, Y., Patterson, M.: Efficient analysis of covid-19 clinical data using machine learning models. arXiv preprint arXiv:2110.09606 (2021)
6. Chourasia, P., Ali, S., Ciccolella, S., Della Vedova, G., Patterson, M.: Clustering SARS-CoV-2 variants from raw high-throughput sequencing reads data. In: Bansal, M.S., et al. (eds.) ICCABS 2021. LNCS, vol. 13254, pp. 133–148. Springer, Cham (2022). https://doi.org/10.1007/978-3-031-17531-2_11
7. Chourasia, P., Ali, S., Ciccolella, S., Vedova, G.D., Patterson, M.: Reads2vec: efficient embedding of raw high-throughput sequencing reads data. J. Comput. Biol. **30**(4), 469–491 (2023)
8. Chourasia, P., Ali, S., Patterson, M.: Informative initialization and kernel selection improves t-SNE for biological sequences. arXiv preprint arXiv:2211.09263 (2022)
9. Cook, J., Sutskever, I., et al.: Visualizing similarity data with a mixture of maps. In: Artificial Intelligence and Statistics. PMLR (2007)
10. Corso, G., Ying, Z., et al.: Neural distance embeddings for biological sequences. In: Advances in Neural Information Processing Systems, vol. 34, pp. 18539–18551 (2021)
11. GISAID (2021). https://www.gisaid.org/. Accessed 29 Dec 2021
12. Kuzmin, K., Adeniyi, A.E., et al.: Machine learning methods accurately predict host specificity of coronaviruses based on spike sequences alone. Biochem. Biophys. Res. Commun. **533**(3), 553–558 (2020)

13. Lee, J.A., Peluffo-Ordóñez, D.H., Verleysen, M.: Multi-scale similarities in stochastic neighbour embedding: reducing dimensionality while preserving both local and global structure. Neurocomputing **169**, 246–261 (2015)

14. Lee, J.A., Renard, et al.: Type 1 and 2 mixtures of kullback-leibler divergences as cost functions in dimensionality reduction based on similarity preservation. Neurocomputing **112**, 92–108 (2013)

15. Van der Maaten, L., Hinton, G.: Visualizing data using t-SNE. J. Mach. Learn. Res. **9**(11) (2008)

16. Melnyk, A., et al.: From alpha to zeta: identifying variants and subtypes of SARS-CoV-2 via clustering. J. Comput. Biol. **28**(11), 1113–1129 (2021)

17. Saha, D.K., Calhoun, V.D., Panta, S.R., Plis, S.M.: See without looking: joint visualization of sensitive multi-site datasets. In: IJCAI, pp. 2672–2678 (2017)

18. Saha, D.K., et al.: Privacy-preserving quality control of neuroimaging datasets in federated environment. Hum. Brain Mapp. **43**(7), 2289–2310 (2022)

19. Stephens, Z.D., et al.: Big data: astronomical or genomical? PLoS Biol. **13**(7), e1002195 (2015)

20. Tayebi, Z., Ali, S., Patterson, M.: Robust representation and efficient feature selection allows for effective clustering of SARS-CoV-2 variants. Algorithms **14**(12), 348 (2021)

21. Van Der Maaten, L.: Accelerating t-SNE using tree-based algorithms. J. Mach. Learn. Res. **15**(1), 3221–3245 (2014)

22. Van Der Maaten, L., Weinberger, K.: Stochastic triplet embedding. In: IEEE International Workshop on Machine Learning for Signal Processing, pp. 1–6 (2012)

23. Xue, J., Chen, Y., et al.: Classification and identification of unknown network protocols based on CNN and t-SNE. In: Journal of Physics: Conference Series, vol. 1617, p. 012071 (2020)

24. Yang, Z., King, I., Xu, Z., Oja, E.: Heavy-tailed symmetric stochastic neighbor embedding. In: Advances in Neural Information Processing Systems, vol. 22 (2009)

25. Zhu, Y., Ting, K.M.: Improving the effectiveness and efficiency of stochastic neighbour embedding with isolation kernel. J. Artif. Intell. Res. **71**, 667–695 (2021)

Genetic Algorithm with Evolutionary Jumps

Hafsa Farooq[(✉)] [ID], Daniel Novikov [ID], Akshay Juyal,
and Alexander Zelikovsky[(✉)] [ID]

Department of Computer Science, Georgia State University, Atlanta, GA 30302, USA
hfarooq5@student.gsu.edu, alexz@gsu.edu

Abstract. It has recently been noticed that dense subgraphs of SARS-CoV-2 epistatic networks correspond to future unobserved variants of concern. This phenomenon can be interpreted as multiple correlated mutations occurring in a rapid succession, resulting in a new variant relatively distant from the current population. We refer to this phenomenon as an *evolutionary jump* and propose to use it for enhancing genetic algorithm. Evolutionary jumps were implemented using C-SNV algorithm which find cliques in the epistatic network. We have applied the genetic algorithm enhanced with evolutionary jumps (GA+EJ) to the 0–1 Knapsack Problem, and found that evolutionary jumps allow the genetic algorithm to escape local minima and find solutions closer to the optimum.

Keywords: Evolution · Knapsack Problem · Genetic Algorithm · Epistatic network

1 Introduction

The unprecedented density of SARS-CoV-2 sequencing data allows to follow the viral evolution much closer than in pre-pandemic time [4,8]. Epistatic networks of SARS-CoV-2 constructed on GISAID data, contain densely linked subgraphs of mutations which correspond to known variants of concern, and also allow us to predict and early detection of future variants [7]. The network has non-additive phenotypic effects and their vertices are single nucleotide polymorphism(SNPs) and its edges are correlated pair of mutations, where dense subgraphs have high density of connectivity among their vertices. It is remarkable that altered phenotype of variants of concern (VOC) do not appear gradually since one cannot observe intermediate variants containing substantial subsets of mutations defining the VOC. Such phenomenon was previously observed in Paleontology and referred as punctuated equilibrium. This phenomenon can be interpreted as if multiple correlated mutations occur in a rapid succession, resulting in appearing of a novel variant which is relatively distant from the closest representative of the current population. In this paper, we propose to apply this evolutionary mechanism (referred as evolutionary jumps) to enhance genetic algorithm (GA).

X. Guo et al. (Eds.): ISBRA 2023, LNBI 14248, pp. 453–463, 2023.
https://doi.org/10.1007/978-981-99-7074-2_36

Genetic algorithm mimics the natural evolution and select the fittest individuals which further reproduce using genetic operators. The drawback of genetic algorithm as well as other local optimization methods that they can stuck in local minima. We propose to rectify this drawback by applying evolutionary jumps when no significant improvement is achieved for several generations by GA. Instead of dense subgraphs in the epistatic network, our approach is to find cliques that's why we use C-SNV algorithm (C-SNV) [5]. It identifies cliques (maximal complete subgraphs) and use them to assemble viral variants present in the sequencing data. When GA stuck for a number of generations, we run C-SNV on all individual solutions constructed so far to identify new individuals with all mutations corresponding to identified cliques.

In order to evaluate the quality of the GA and compare with the proposed enhancements, we applied it to hard instances of 0–1 Knapsack Problem recently proposed in [3]. Since the simple GA solutions can be infeasible or extended, we first enhance GA with repairing and packing (GA-RP) and then further introduce evolutionary jumps (GA-EJ). Our experiments show that GA-RP significantly outperform the simple GA on all instances while GA-EJ further improves GA-RP for harder instances, i.e., for cases where finding optimal solution requires large runtime and GA-RP stuck significantly far away from the optimum.

Section 2 describes a genetic algorithm and proposed enhancements using evolutionary jumps. Section 3 applies GA to the 0–1 Knapsack Problem and gives details of enhancing GA with repairing and packing as well as evolutionary jumps. Section 4 describes hard problem instances and compares results achieved by GA, GA with repairing and packing (GA-RP), and GA with repairing and packing & evolutionary jumps (GA-EJ).

2   Genetic Algorithm with Evolutionary Jumps

2.1   Simple Genetic Algorithm

Genetic algorithm is a metaheuristic inspired by the process of natural evolution relying on biologically inspired operators such as mutation, crossover and selection [6]. GA is a heuristic search-based evolutionary algorithm developed by John Holland in 70's. Holland developed an electronic organism named chromosomes consisting of binary encoded strings [2] or unit entity known as gene. Those randomly generated binary encoded strings based chromosomes are also called individual solutions, and these potential solutions altogether are the initial population.

After the creation of initial population, evolution begins. The fitness function evaluation is performed for each individual of the population. The fitness score represents the ability of the individual to compete for mating and its quality in the solution. The individuals with higher fitness values are chosen for mating pool, called parents. After selecting the best fitted individuals from the population, selected individuals perform reproduction.

Crossover is a process of combining genetic material of parents by inheriting their traits in offspring. Crossover randomly chooses a point or locus in the individuals and exchange before and after sub-strings of individuals to create offspring. For example for single point crossover, consider the two individuals and crossover point at the 5th position of the individual, shown in Fig. 1(a).

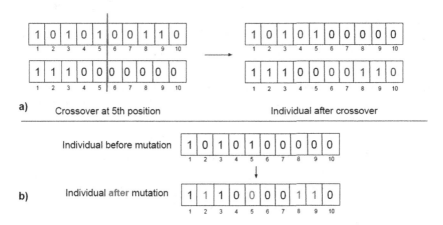

Fig. 1. a) Two individuals are performing crossover. Red line represents the point of crossover. b) Mutation is performed on 2nd, 5th, 8th and 9th gene of the individual. (Color figure online)

The offspring 1 has first five genes from individual 1 and next five genes from individual 2. Similarly, offspring 2 has left side of genes from individual 2 and right side of genes from individual 1.

After crossover, individuals undergo mutation. The mutation operator changes one or more genes randomly. It changes the gene value from 1 to 0 or vice versa, shown in Fig. 1(b). The type of crossover, mutation and its probability can be defined and depends on the problem under experimentation. Both the parents and the offspring now comprise the next generation of the genetic algorithm, where the process repeats. GA terminates after the fitness is not improved for predefined number of generations or number of generations exceeds a given number.

2.2   Punctuated Equilibrium and Epistatic Network of SARS-CoV-2

The genomic evolution of SARS-CoV-2 shows that the rate of mutations is not constant – the gradual relatively slow evolution is replaced with brief and fast bursts resulting in emerging of new viral variants with altered phenotypes including increased transmissibility. Such new variants (e.g., Alpha- and Omicron-variants) are referred as Variants of Concern (VOCs) or Variants of Interest (VOIs)). This phenomenon has been labeled as punctuated equilibrium. It has been shown recently that punctuated equilibrium events for SARS-CoV-2 which we refer to as *evolutionary jumps* can be predicted from its epistatic network [7].

Following [7], we define an epistatic network as a graph with vertices corresponding to mutated positions, e.g. single nucleotide polymorphisms (SNPs) or single amino-acid variations (SAVs). Two vertices are connected if the mutations in the the corresponding positions i and j are significantly more frequently observed in the same haplotype than they are expected if the mutations would be independent.

Formally, let 0 and 1 denote the reference and mutated alleles in positions i and j, respectively. Assuming that positions are biallelic, each possible haplotype h in positions i and j belong to the set $\{00, 0110, 11\}$. Let O_h (resp. $E(h)$) be the observed (resp. expected) number of haplotypes h in the sequencing data. It has been proved in [7] that if haplotype 11 is not viable (does not produce descendants), then

$$E_{00} * E_{11} \leq E_{01} * E_{10} \tag{1}$$

In the epistatic network we connect to vertices if the corresponding haplotype 11 is viable, i.e., when the $O_{00} * O_{11}$ is significantly larger than $O_{01} * O_{10}$.

Recently, it has been shown that evolutionary jumps in SARS-CoV-2 evolution correspond to dense subgraphs of the epistatic network [7]. Formally, the densest subgraph, i.e. the one with the maximum ratio of edges over vertices, frequently consists of mutations that differentiate an emerging viral variant from the reference. Therefore, rather than performing a random combination of mutations, evolutionary jumps include multiple mutually linked mutations.

2.3   Enhancement of GA with Evolutionary Jumps

The genetic algorithm uses selection pressure to push future generations closer to the optimum, with limited differences between consecutive generations. It can be observed that standard genetic algorithm is prone to getting stuck in local minima, because crossovers and mutations alone are not enough to escape them. Therefore, we propose to enhance the genetic algorithm with evolutionary jumps. Evolutionary jumps involve the appearance of new individual solutions in the population, which include genes or mutations that are observed to be correlated in previous generations.

Our procedure decides when to perform evolutionary jumps by monitoring the fitness of the best solution across generations. If the number of generations without fitness improvement exceeds a predefined threshold, then the result of an evolutionary jumps are added to the next generation. Instead of dense subgraphs in the epistatic network, our approach is to find cliques that's why we use C-SNV [5]. It identifies cliques (maximal complete subgraphs) and use them to assemble viral variants present in the sequencing data.

3   Application of Genetic Algorithm with Evolutionary Jumps to the 0–1 Knapsack Problem

3.1   The 0–1 Knapsack Problem

Given a set of items with weights and profits and a maximum capacity for the knapsack, the 0–1 Knapsack Problem asks for a subset of items that maximizes total profit without exceeding the knapsack capacity. A solution to the 0–1 Knapsack Problem is a vector with binary coordinates corresponding to items. The coordinate equals 1 when the corresponding item is selected, and 0 otherwise. We use the 0–1 Knapsack Problem as a benchmark to evaluate the performance of our proposed improvement to the genetic algorithm.

3.2   Implementation of Genetic Algorithm

As a base implementation of the genetic algorithm, we employ the PyGAD genetic algorithm Python library. [1] An initial population is created by randomly generating solutions that fill the knapsack up to capacity. For a fitness function, we use the sum of profits of the items included in the knapsack, unless the sum of their weights exceeds the knapsack capacity, in which case the solution receives the minimum fitness of −1.

In each generation, a tournament selection procedure identifies high-fitting solutions to be parents for the next generation. The chosen parents are grouped into pairs, and each pair is crossed-over and randomly mutated to produce a pair of offspring for the next generation, as shown in Fig. 1. This procedure repeats for the given number of generations.

3.3   Repairing and Packing

Throughout the execution of the standard genetic algorithm on the 0–1 Knapsack Problem, solutions frequently either exceed the knapsack capacity or are under-filled, meaning there are still items remaining which can fit in the solution without bringing it over capacity. Rather than discarding these solutions, we propose a procedure for repairing solutions that are over-filled, and for filling solutions that are under-filled. We call this procedure repairing and packing, illustrated in Fig. 2.

To repair and pack a given solution, the procedure begins by sorting the items of the problem instance such that all items which are included in the solution (1s) come first, and all non-included items (0s) come afterwards. Then, it randomly shuffles the included items and the non-included items separately. Ordering the array in this manner allows us to consider the items in a random order subject to constraint that all items included in the solution preceed those that are not included.

Our procedure repairs and packs solutions in a single pass through this sorted array of items. The procedure starts a new, empty solution, and begins iterating the sorted items array. While iterating through the included items region, it tries to add each item to the solution, and does so as long as the solution remains under capacity. If the item cannot fit (i.e., after its addition the total capacity exceeds the upper limit), we change it to 0, removing it from the knapsack. This ensures over-filled solutions will be brought back down to capacity. When we reach the non-included items region of the sorted array, we try to add those items to the solution as well, and do so as long as they fit. By applying the repairing and packing procedure, we can guarantee that solutions are not over-capacity, and that there are no items remaining which could still fit in the knapsack.

We apply this procedure to each solution throughout the execution of the genetic algorithm. The initial population, next generation offsprings, and evolutionary jumps solutions are all repaired-and-packed according to this procedure, ensuring each solution is both feasible and maximally filled.

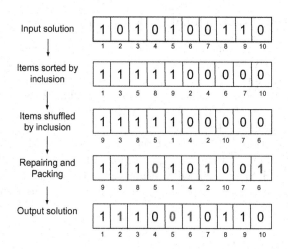

Fig. 2. Repairing and packing procedure on an example instance of 10 items

3.4   Evolutionary Jumps Implementation

Evolutionary Jumps involve the introduction of new individuals to the genetic algorithm population. These new solutions include items that are observed to be correlated in the past evolutionary history. Our procedure decides when to perform evolutionary jumps by monitoring the fitness of the best solution across generations. Each time we observe 10 consecutive generations without improvement to the best fitting solution, the evolutionary jump procedure is triggered.

To facilitate the evolutionary jumps, C-SNV is employed to find correlated pairs of mutations. C-SNV is a tool which finds characteristic haplotypes to describe a set of input sequences. Internally, the tool implements a procedure for identifying pairs of mutations which are correlated by high co-occurence in the input data.

Treating each solution vector as a sequence, we pass the entire evolutionary history, i.e., all solutions from all generations so far, as an input to C-SNV, encoding 0 as A and 1 as C, using a Fasta file format.

After finding correlated pairs of positions, C-SNV constructs an epistatic network over sequence sites with edges given by correlations, and finds cliques in this graph to reconstruct characteristic haplotypes for the evolutionary history. Each clique relates a set of items which have frequent pairwise co-occurrence. A haplotype is created from each clique, containing mutations for the sites that appeared in the clique. C-SNV typically returns 2 − 10 haplotypes. For each haplotype, we create a new knapsack solution to add to the population.

The procedure for creating knapsack solutions from C-SNV haplotypes begins with a solution vector containing 1 s where the haplotype had a 1. We observe 10 consecutive generations without improvement towards the best fitting solution, then evolutionary jump procedure triggers. Then, to each solution, we add the items included in the current best solution. In the result, the newly created solutions are guaranteed to take the items where the C-SNV haplotype contained a 1 for that position, and then additionally, they take the items included in the best solution observed so far, so long as those don't bring it over capacity. The newly created solutions each represent an evolutionary jump, and these solutions are added back to the genetic algorithm population, prior to starting the next generation. The new solutions are added by replacing the currently worst-fitting solutions with the new ones created by the jump procedure.

4 Results

4.1 Instances of the 0-1 Knapsack Problem

For validation of genetic algorithms on the 0–1 Knapsack Problem, we chose instances recently generated in [3]. It was proposed a new class of hard problem and shown that they are hard to solve to optimality. Many hard problem instances were not even possible to solve on a supercomputer using hundreds of CPU-hours. Out of 3240 instances we have selected the first ten instances which were solved to optimality.

The selected problem instances are listed in the Table 1 together with the nomenclature from [3]. Different letters in the names of the instances represents;

- n: Number of items of a problem instance
- c: Capacity of the knapsack
- g: Number of groups of items of a problem instance
- f: Approximate fraction of items in the last group
- ε: Noise parameter
- s: Upper bound for profits and weights of items in the last group

The runtime for a solver to reach an optimal solution is given in the column Runtime. Note that problem instances 1–5 are harder since they require significantly more runtime to reach optimality then problem instances 6–10.

Table 1. 0–1 Knapsack Problem Instances [3]

ID	Problem Instance	Optimal Fitness	Runtime(sec)
1	n_1000_c_10000000000_g_10_f_0.1_eps_0.0001_s_100	9999946233	2943
2	n_1000_c_10000000000_g_10_f_0.1_eps_0.0001_s_300	9999964987	6474
3	n_1000_c_10000000000_g_10_f_0.1_eps_0.01_s_100	9999229281	555
4	n_1000_c_10000000000_g_10_f_0.1_eps_0.01_s_200	9999239905	742
5	n_1000_c_10000000000_g_10_f_0.1_eps_0.01_s_300	9999251796	896
6	n_1000_c_10000000000_g_10_f_0.1_eps_0.1_s_100	9996100344	17
7	n_1000_c_10000000000_g_10_f_0.1_eps_0.1_s_200	9996105266	18
8	n_1000_c_10000000000_g_10_f_0.1_eps_0.1_s_300	9996111502	26
9	n_1000_c_10000000000_g_10_f_0.1_eps_0_s_100	9980488131	74
10	n_1000_c_10000000000_g_10_f_0.1_eps_0_s_200	9980507700	96

4.2   Parameter Tuning

We tuned the parameters of the genetic algorithm and the evolutionary jump procedure to optimize the output profit. In genetic algorithm, these parameters include the size of the population, number of generations, mutation probability, and so on. For evolutionary jumps, the parameters include the number of non-improving generations to wait before jumping, how sensitive C-SNV should be to determine links, and more. To find the optimal parameters for our problem instances, we applied multiple parameter configurations to each problem instance 10 times, observing the average performance under each configuration.

The finalized values of parameters are shown in the Table 2. Some parameters have several values, e.g., we report GA for 500, 1000, and 1500 generations.

Table 2. Parameters Table

GA Parameter	Value	C-SNV Parameter	Value
Population Size	1000	Jump Threshold	10
Num of Gen	500, 1000, 1500	Min.Gen.Wait	20
Num of Parents	500	Jump Type	Global Best
Crossover type	Single Point	C-SNV Timeout	120sec
Parent Type	Tournament	Threshold.Freq	0.01
K-Tournament	3(default),25	Threshold.Freq+	0.01
Mutation Type	Inversion	Memory	20 GB
Mutation Pr	0.02	Edge Limit	1000

4.3   Performance Comparison of GA, GA+RP, and GA+EJ

The results of our experiments are shown in Tables 3–5. "Min.Error" is the minimum difference between optimal fitness and the best fitness over 10 runs. "Avg.Error" is the average difference between optimal fitness and the best fitness over 10 runs. "Runtime" is shown in the minutes and its the average runtime over 10 runs.

We ran the GA and GA+RP on instances for K = 3, K = 25 and number of generations G = 500, 1000,1500 (see Tables 3–4) and GA+EJ just for K = 25 and number of generations G = 500 (see Table 5).

It is easy to see that GA+RP significantly outperform simple GA on all instances and all configurations. Also GA+RP with K = 25 outperforms GA+RP with K = 3 for all instances. Therefore we decided to run GA+EJ just for K = 25. For the harder instances 1,3,4,5 GA+EJ significantly outperform GA-RP even for G=1500 the novel eutionary enhancement of GA.

Table 3. Simple Genetic Algorithm Results

	Results	1	2	3	4	5	6	7	8	9	10
GA	Min.Error	4.7e5	1.4e6	9695	1.9e4	3.0e4	5914	9908	1.5e4	1.7e4	3.5e4
K = 3	Avg.Error	1.5e6	2.1e6	9798	1.9e4	3.0e4	4.3e6	4.6e6	4.3e6	1.8e4	3.6e4
G = 500	Runtime	28	29	39	37	33	58	62	63	28	27
GA	Min.Error	4.7e5	4.9e5	9620	1.9e4	2.9e4	5433	1.0e4	1.5e4	1.8e4	3.5e4
K = 3	Avg.Error	1.6e6	1.8e6	9721	1.9e4	3.0e4	2.7e6	5.4e6	4.3e6	1.8e4	3.6e4
G = 1K	Runtime	43	43	56	60	60	94	109	93	43	43
GA	Min.Error	4.7e5	4.9e5	9599	1.9e4	3.0e4	5953	1.0e4	1.4e4	1.7e4	3.5e4
K = 3	Avg.Error	1.1e6	6.9e5	9726	1.9e4	3.0e4	4.3e6	4.3e6	4.3e6	1.8e4	36174
G = 1.5K	Runtime	78	80	122	106	108	186	167	175	80	79
GA	Min.Error	4.7e5	4.8e5	9285	1.8e4	2.9e4	5498	3.9e6	1.3e4	1.7e4	3.4e4
K = 25	Avg.Error	4.7e5	4.9e5	9472	1.9e4	2.9e4	3.5e6	5.8e6	3.1e6	1.7e4	3.5e4
G = 500	Runtime	80	59	74	74	75	105	106	101	57	54
GA	Min.Error	4.7e5	4.8e5	9254	1.8e4	2.9e4	5286	9362	1.4e4	1.7e4	3.3e4
K = 25	Avg.Error	4.7e5	4.8e5	9414	1.9e4	2.9e4	3.1e6	3.5e6	1.9e6	1.7e4	3.5e4
G = 1K	Runtime	81	80	122	129	133	181	177	167	68	91
GA	Min.Error	4.7e5	4.8e5	9283	1.8e4	2.8e4	5396	9655	1.3e4	1.7e4	3.4e4
K = 25	Avg.Error	4.7e5	4.8e5	9428	1.8e4	2.9e4	3.9e6	3.1e6	3.1e6	1.7e4	35017
G = 1.5K	Runtime	159	159	194	200	214	286	263	275	133	126

Table 4. Results for Genetic Algorithm with Repairing and Packing

	Results	1	2	3	4	5	6	7	8	9	10
GA-RP	Min.Error	1102	3298	2662	4242	7433	71	167	235	4406	1.1e4
K = 3	Avg.Error	1157	3481	3170	5958	8577	120	275	400	6090	1.2e4
G = 500	Runtime	98	98	102	99	99	136	149	141	118	106
GA-RP	Min.Error	917	3181	2314	4063	6990	87	174	195	4556	8891
K = 3	Avg.Error	1084	3422	3026	5289	8531	115	255	371	5324	1.1e4
G = 1K	Runtime	203	197	189	188	201	244	241	243	217	218
GA-RP	Min.Error	915	3166	2353	4569	6777	67	163	184	3502	6445
K = 3	Avg.Error	1067	3395	2763	5294	8166	95	226	297	4778	9392
G = 1.5K	Runtime	281	282	294	292	291	397	390	400	307	334
GA-RP	Min.Error	874	2540	2742	4827	8388	14	36	48	306	573
K = 25	Avg.Error	1046	2901	3343	6160	9528	34	93	105	450	903
G = 500	Runtime	99	98	109	123	104	164	171	158	139	143
GA-RP	Min.Error	959	2335	2566	4586	5406	14	31	39	214	525
K = 25	Avg.Error	1037	2897	3120	5830	8867	26	55	75	361	956
G = 1K	Runtime	188	184	198	203	207	278	275	274	261	255
GA-RP	Min.Error	780	1451	2365	5010	7646	17	40	39	265	644
K = 25	Avg.Error	955	2642	3197	5670	9181	22	78	61	469	849
G = 1.5K	Runtime	283	299	316	318	341	472	473	462	408	407

Table 5. Results for Genetic Algorithm Evolutionary Jumps (GA+EJ)

	Results	1	2	3	4	5	6	7	8	9	10
GA-EJ	Min.Error	541	1950	1557	3972	6000	27	50	113	350	595
K = 25	Avg.Error	751	2425	2761	5329	8596	47	83	178	686	951
G = 500	Runtime	2234	1949	2081	1883	2171	3056	2829	3153	2441	2603

5   Conclusion

In this paper, we enhanced genetic algorithm with evolutionary jumps which simulated punctuated equilibrium phenomenon observed in SARS-CoV-2 sequencing data. We validated the enhanced genetic algorithm on hard instances of the 0–1 Knapsack Problem. Genetic algorithm for the Knapsack Problem was first improved by using repairing and packing method. We further enhance GA with evolutionary jumps that were implemented using CliqueSNV. Our experiments showed that evolutionary jumps significantly improve GA on very hard instance of the 0–1 Knapsack Problem.

References

1. Gad, A.F.: Pygad: An intuitive genetic algorithm python library (2021)
2. Holland, J.H.: Adaptation in natural and artificial systems. University of Michigan Press (1975)
3. Jooken, J., Leyman, P., De Causmaecker, P.: A new class of hard problem instances for the 0–1 knapsack problem. Eur. J. Oper. Res. **301**(3), 841–854 (2022)
4. Knyazev, S., et al.: Unlocking capacities of genomics for the COVID-19 response and future pandemics. Nat. Methods **19**(4), 374–380 (2022)
5. Knyazev, S., et al.: Cliquesnv: an efficient noise reduction technique for accurate assembly of viral variants from NGS data. bioRxiv 264242 (2020)
6. Mitchell, M.: An Introduction to Genetic Algorithms. MIT press, Cambridge (1998)
7. Mohebbi, F., Zelikovsky, A., Mangul, S., Chowell, G., Skums, P.: Community structure and temporal dynamics of SARS-CoV-2 epistatic network allows for early detection of emerging variants with altered phenotypes. bioRxiv, pp. 2023–04 (2023)
8. Novikov, D., Knyazev, S., Grinshpon, M., Icer, P., Skums, P., Zelikovsky, A.: Scalable reconstruction of SARS-CoV-2 phylogeny with recurrent mutations. J. Comput. Biol. **28**(11), 1130–1141 (2021)

Hetbisyn: Predicting Anticancer Synergistic Drug Combinations Featuring Bi-perspective Drug Embedding with Heterogeneous Data

Yulong Li⑩, Hongming Zhu⑩, Xiaowen Wang⑩, and Qin Liu$^{(\boxtimes)}$⑩

School of Software Engineering, Tongji University, Shanghai, China
1374803733@qq.com

Abstract. Synergistic drug combination is a promising solution to cancer treatment. Since the combinatorial space of drug combinations is too vast to be traversed through experiments, computational methods based on deep learning have shown huge potential in identifying novel synergistic drug combinations. Meanwhile, the feature construction of drugs has been viewed as a crucial task within drug synergy prediction. Recent studies shed light on the use of heterogeneous data, while most studies make independent use of relational data of drug-related biomedical interactions and structural data of drug molecule, thus ignoring the intrinsical association between the two perspectives. In this study, we propose a novel deep learning method termed HetBiSyn for drug combination synergy prediction. HetBiSyn innovatively models the drug-related interactions between biomedical entities and the structure of drug molecules into different heterogeneous graphs, and designs a self-supervised learning framework to obtain a unified drug embedding that simultaneously contains information from both perspectives. In details, two separate heterogeneous graph attention networks are adopted for the two types of graph, whose outputs are utilized to form a contrastive learning task for drug embedding that is enhanced by hard negative mining. We also obtain cell line features by exploiting gene expression profiles. Finally HetBiSyn uses a DNN with batch normalization to predict the synergy score of a combination of two drugs on a specific cell line. The experiment results show that our model outperforms other state-of-art DL and ML methods on the same synergy prediction task. The ablation study also demonstrates that our drug embeddings with bi-perspective information learned through the end-to-end process is significantly informative, which is eventually helpful to predict the synergy scores of drug combinations.

Keywords: synergistic drug combinations · Deep learning · Heterogeneous data · Graph attention network · Self-supervised learning

© The Author(s), under exclusive license to Springer Nature Singapore Pte Ltd. 2023
X. Guo et al. (Eds.): ISBRA 2023, LNBI 14248, pp. 464–475, 2023.
https://doi.org/10.1007/978-981-99-7074-2_37

1   Introduction

In the field of cancer treatment, drug combination therapy [1] holds significant importance. The interactions of pairwise combinations of drugs can be divided into synergistic, additive, and antagonistic by comparing the effect of the drug combination with the sum of effects of drugs applied separately [2]. Synergistic drug combinations can often reduce the development of drug resistance [3] and minimize the occurrence of drug-related side effects [4] during the treatment process. However, distinguishing synergistic drug combinations from non-synergistic ones is challenging as the combination space expands rapidly with the discovery of new drugs.

Early studies on synergistic drug combinations are mostly based on clinical experience, which is time-consuming and labor-intensive, and may lead to unnecessary or even harmful treatments on patients [5]. Even applying high-throughput screening technology (HTS) [6] that enables efficient testing of cell lines *in vitro* is impossible to screen through the complete combination space, let alone that the technology is expensive to build [7]. Researchers have therefore turned to computational methods to predict synergistic drug combinations.

Except for computational models that are only available on specific drugs or cell lines [5], recent studies have shed light on methods based on machine learning (ML) and deep learning (DL). The most common workflow consists of obtaining the features for cell lines and drugs and predicting the synergy score with a ML or DL model. Previous studies employed various ML models [8–12] to predict the synergy of anticancer drug combinations. In recent years, the availability of large-scale synergy datasets [13] has provided a valuable resource for employing DL methods in drug combination prediction. Commonly adopted DL models include Deep Neural Network (DNN) [2,5], Residual Neural Network [14], and other interpretable DL models. Additionally, special techniques such as Ensemble Learning [15,16], Transfer Learning [17], and Tensor factorization [18] have been adopted to predicting drug combination synergy.

On the other hand, drug features play an essential role in the synergy prediction task. A classical and universal method for drug representation is to directly use molecular fingerprints [19] or molecular descriptors [5], which refers to a predefined feature vector containing substructural and physicochemical properties. Some researchers [2,5,11,12] collected the interactions between drugs and other biomedical entities (e.g. drug target, pathway etc.), and simply obtained the feature vector by sampling a binary digit for each interaction. Those feature-engineering-based methods offer easy access to fairly informative representations, whereas they might be greatly affected by prior assumptions of biochemical domain knowledge. Methods based on representation learning are proposed to alleviate this problem [17,20–23].

When it comes to how data is utilized in constructing drug features, it should be emphasized that the chemical structure of a drug determines how it functions, while the interactions between drugs and other biomedical entities represent known patterns of drug action. Both aspects should be considered for comprehensive drug features. Concerning the way that molecular-level data and drug-related interaction are used in drug feature constructing, previous studies on synergy prediction either use only one type of data or simply perform concatenation after representations from both aspects are extracted respectively. As the two aspects are intrinsically associated but differs a lot in terms of data, such methods might not be able to fully exploit the latent information when the features are directly used for subsequent synergy prediction. There is still potential for improvement by devising a drug representation learning model to simultaneously obtain information from both structural and relational heterogeneous data.

To improve the synergy prediction of anti-cancer drug combination from the perspective of constructing compact drug representations that are more expressive and informative, we hereby propose a deep learning model **HetBiSyn** (Drug **Syn**ergy Prediction featuring **Bi**-perspective Drug Embedding with **Het**erogeneous Data). In this paper, drug embeddings that integrate information from both aspects are learned within a self-supervised training process. Specifically, HetBiSyn constructs a graph for interactions between drug-related biomedical entities and multiple molecular-level graphs for drugs. Both types of graphs are heterogeneous, and separate Heterogeneous Graph Attention Networks (HGAT) are applied to each type of graph to embed information from different perspective. A contrastive learning module is designed to learn a unified embedding based on the output of the two HGATs, and the hard sample mining strategy is adopted to enhance the model. Besides, HetBiSyn utilizes gene expression profiles to construct the cell line features. Lastly, a DNN with Batch Normalization mechanism is designed to predict the synergy score of drug combinations on cell lines. We compared HetBiSyn with other popular ML and DL methods on the synergy dataset contributed by O'Neil, and the result demonstrates that HetBiSyn can achieve more accurate drug synergy prediction.

2   Materials and Methods

2.1   Synergy Dataset

A high-throughput drug combination screening dataset was obtained from O'Neil's research. The dataset encompass 583 pairwise drug combinations involving 38 distinct drugs tested against 39 human cancer cell lines. Preuer et al. [5] computed a synergy score for each sample using Loewe Additivity values, and divided all samples into 5 disjoint folds with an equal count of drug combinations.

2.2   Cell Line Features

The cell line features are extracted mainly based on the gene expression data. The gene expression files are fetched from the ArrayExpress database [24] (accession number: E-MTAB-3610). We adopted the Factor Analysis for Robust Microarray Summarization method [25] to implement quantile normalization and summarization on the gene expression data. The method also provides calls on whether a gene is informative, by which effective genes are selected for the feature construction of cell lines. In all, 3739 genes are screened out and z-score normalization is performed to produce the feature vector.

2.3   Construction of Drug-Related Graphs

To directly exploit information from different perspectives, we design two types of drug-related graphs from which the embeddings of drugs are learned jointly. The bioinformatic graph provides identified patterns of drug action by integrating the interactions between drugs and other biomedical entities, while molecular-level graphs reveal the structural and chemical particulars inside a drug molecule.

Drug-Related Heterogeneous Bioinformatic Graph. Bioinformatic graphs, also called bioinformatic networks, are widely used in various drug-related problems, especially in extracting complex hidden features that implicate proven patterns of drug actions. Here we construct a heterogeneous graph $G_{Bio} = (V, E)$, in which each node v in the node set V belongs to a biomedical entity type t_v in a type set T, and each edge e in the edge set E belongs to a relation type $t_e \in S_t \times S_t$. As defined, there is at most 1 edge between 2 nodes, and all edges are set bidirectional in practice. Only the largest connected component of G_{bio} is retained so that information can be propagated through every single node. Also, we collect the biomedical entities and their relationships from Luo et al.'s work [26] and DrugBank (Version 3.0), and supplement the drug-target-interaction data with UniProtKB so that all 38 drugs in the synergy dataset are involved in G_{bio}.

Heterogeneous Molecular-Level Graph. To exploit drug properties from a microscopic perspective, we construct a graph G_{mol} for **each** drug in G_{bio} at molecular level. First, a molecular graph G_{mg} is generated for each drug by treating the atoms as nodes and the bonds between them as edges. All atom nodes are considered to be of the same type though they are initialized with different atomic features (e.g. chirality, formal charge, partial charge, etc.) [27], while edges vary in types according to the original bond types (e.g. single, double, aromatic, etc.). The edges in G_{mg} are bidirectional since chemical bonds are unbiased. We use the RDKit tool to convert the SMILES string of a drug into a molecule object for subsequent operations, and drugs that do not have a SMILES or cannot be converted are abandoned from G_{bio}. Inspired by Fang [28], we augment G_{mg} to leverage the associations between atoms that are not directly connected with bonds but share fundamental chemical attributes. By histogramizing the continuous attributes of atoms and converting them into discrete labels, totally 107 attributes of atoms and 17 relation types are devised. These attributes are then added to G_{mol} as another type of nodes, while their relations with the atoms in G_{mg} are modeled as different types of directed edges pointing to atoms.

2.4   HetBiSyn

In this paper, a novel deep learning method named HetBiSyn is proposed to predict synergy scores of drug combinations on cell lines. The overview of our method is shown in Fig. 1.

Heterogeneous Graph Attention Networks for Drugs. As shown in figure(B), the essential step for drug representation learning is to extract an intermediate embedding from both G_{bio} and G_{mol} for a drug, which would be continuously updated throughout the subsequent self-supervised learning process. Heterogeneous graph attention network (HGAT) [29] is a node representation learning model that can generate dense embedding while retaining information about network topology and meta-path importance with insights into heterogeneity. HetBiSyn set up two separate HGATs for G_{bio} and G_{mol} to exploit inter-entity and intra-molecular information, namely $HGAT_{macro}$ and $HGAT_{micro}$. The detailed derivation about how HGAT works within our study can be found in supplementary file (Sect. 1).

It is worth noting is that only the embedding for each atom or property node is obtained through training $HGAT_{micro}$, which cannot be directly referred to individual drugs. In order to present the drug embedding by micro-view, average pooling is conducted on the atom node vectors for each drug as whole-graph embedding.

Drug Representation Learning Based on Contrastive Learning. The subsequent step of drug embedding learning is to leverage the output of the two networks to train a unified embedding balancing both perspective. The key idea comes that the representations of the same drug generated from the two networks shall be as similar as possible, while drugs showing great distinction shall have differentiated embeddings. Under this assumption, we form a binary contrastive learning task aiming at estimating the performance of the outputs and optimizing the recurrent training process.

Let $Z(G, d_i)$ denote the output embedding of drug d_i from either HGAT. For each drug d_i in G_{bio}, $S(d_i, d_i) = [Z(HGAT_{macro}, d_i)||Z(HGAT_{micro}, d_i)]$ is defined as a positive sample that is labeled 1, while $S(d_i, d_j)$ given $i \neq j$ is defined as a negative sample that is labeled 0. The concatenated vector S is sampled twice in reverse order with respect to the two networks to generate both positive and negative samples. A simple deep neural network, denoted as DNN_{clf}, is set up for binary classification. We use binary cross-entropy loss with sum as reduction for the loss function. The loss function can be described as:

$$loss = \sum_{i=1}^{n}[y_i \cdot log(p_i) + (1 - y_i) \cdot log(1 - p_i)]; p_i = \frac{1}{1 + e^{-x_i}} \tag{1}$$

where x_i and y_i respectively represents the predicted label value and the true label value of a sample. It is worth noting that the drug embedding learning

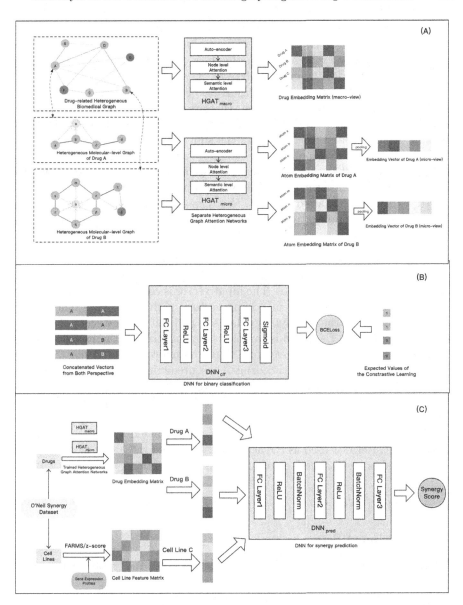

Fig. 1. Overview of HetBiSyn. (A) Two HGATs as shown integrate information from the drug-related heterogeneous biomedical graph G_{bio} and the heterogeneous molecular-level graphs G_{mol} respectively, yielding drug embedding at different perspectives. (B) Embedding of drugs from different perspectives are paired to form a contrastive learning task, where DNN_{clf} is set up for binary classification. (C) HetBiSyn predicts the synergy score of a drug combination on a specific cell line using the DNN with the batch normalization mechanism.

module is an end-to-end process, as the loss computed here is used to update DNN_{clf} as well as the two HGATs through back propagation. After multiple rounds of training, the representation of a drug can be inferred by averaging the output of its embedding from both HGATs.

Furthermore, the sampling strategy is also improved by hard negative mining. In other words, we try to find drugs that are alike and challenge the classifier to label their combination correctly. We collected a 167-dimension MACCS fingerprint, which is often used to assess the similarity between molecules, for all the drugs in G_{bio}, and calculated the Tanimoto similarity for each pair of drugs. A drug d_j having the most similarity with another drug d_i indicates this duo may compose a hard negative sample, while other randomly taken samples are viewed as common negative samples. We take positive samples, hard negative samples and common negative samples at a certain ratio for training DNN_{clf}.

Synergy Score Prediction. After the cell line features and drug embeddings are obtained, a regression model is designed for predicting cell-line-specific synergy scores of drug combinations. The input vector is constructed by sequentially concatenating the feature or embedding vectors of the two drugs and the cell line in a data point from the synergy dataset, and the synergy score of this trio serves as the output. Each trio is sampled twice in terms of the input vector by exchanging the order of drug vectors, since the network should not differentiate between permutations of two drugs. The prediction model is implemented by a feed forward DNN composed of three fully connected (FC) layers and two batch normalization layers in between, denoting as DNN_{pred}. The number of neurons in each FC layer is funnel-shaped as we have the most neurons in the first FC layer and one neuron representing the predicted synergy score in the last FC layer. ReLU is used as the activation function for the first two FC layers. DNN_{pred} takes mean square error loss as its loss function.

3   Result

3.1   Experiment Setup

We resort to the mean square error (MSE) and the root mean square error (RMSE) for the main evaluation metric. The Pearson correlation coefficient (PCC) between the predictions and the ground truth is also adopted. As the experiments are conducted under a 5-fold cross-validation approach, we present the mean and the standard deviation of each evaluation metric across the 5-fold dataset. Hyper-parameter settings are documented in the supplementary file (Sect. 2).

3.2   Performance Comparison with Other Models

To objectively appraise the performance of HetBiSyn, we compare HetBiSyn with some representative models on the same synergy dataset with a 5-fold cross validation. Four ML methods including Elastic Net [30], Support Vector Regression (SVR) [31], Random Forest [32] and XGBoost [33] are adopted by using the same drug and cell line features for input. We implemented these methods with sklearn and retained the default hyper-parameters. We also selected 4 DL methods for comparison, including DeepSynergy [5], MatchMaker [34], AuDNNSynergy [2] and DFFNDDS [35]. We adopted their feature generating process and validate them upon our dataset, as described in the supplementary file (Sect. 3).

Results of the experiment of the comparison among the methods above are shown in Table 1. The best and second best performance are shown in bold and italic respectively. HetBiSyn achieves the lowest MSE of 225.90 among all compared methods, which is 11.58% less than DeepSynergy, 6.31% less than the AuDnnSynergy which achieves the second lowest and 20.90% less than XGBoost. The PCC of HetBiSyn, which is the second best across all methods compared, also shows a strong correlation between the model's prediction and the ground truth. The result demonstrates the advantage of HetBiSyn on the synergy prediction task, and the possible reasons are: 1)Drug feature is more informative under the end-to-end self-supervised learning framework of HetBiSyn that integrates data from multiple aspects. 2) The DNN for final prediction may identify the nonlinear patterns in the synergy dataset better in comparison with other DL methods.

Table 1. Results of method comparison on the synergy score prediction task

Type	Method	MSE	RMSE	PCC
DL	HetBiSyn	**225.90±31.90**	**15.00±1.02**	*0.74±0.03*
	DeepSynergy	255.49	15.91±1.56	0.73±0.04
	MatchMaker	254.37±37.70	15.93±1.17	0.68±0.03
	AuDNNSynergy	*241.12±43.52*	*15.46±1.44*	**0.74±0.04**
	DFFNDDS	242.37±34.21	15.53±1.07	0.76±0.02
ML	Elastic Net	407.06±48.23	20.18±1.33	0.47±0.03
	SVR	338.57±53.39	18.40±1.48	0.58±0.03
	Random Forest	312.75±44.01	17.68±1.13	0.61±0.02
	XGBoost	285.60±44.40	17.16±1.31	0.68±0.02

3.3 Ablation Study

To further inspect how the use of heterogeneous data from different perspective contributes to the prediction result in our model, a series of variants of HetBiSyn are designed for comparison mainly by altering the drug feature construction process as:

- **HetBiSyn-Bio.** Only the data of drug-related biomedical interactions are utilized. Metapath2vec [36] is applied on the G_{bio} we constructed to extract drug features. We use the implementation provided by DGL [37].
- **HetBiSyn-MolHGT.** Only the data of molecular structure of drugs are utilized. We adopt a molecular representation learning framework MolHGT [38] to obtain features from heterogeneous graphic data, which treat atoms and bonds as different types of nodes and edges to extract representations.
- **HetBiSyn-MG.** Similar to the original HetBiSyn but the molecular graph G_{mg} mentioned in 2.3 is used instead. Atom properties are not considered so that the structural information completely emerges from the molecular graph.
- **HetBiSyn-Concat.** Drug features are constructed by concatenating the representation from HetBiSyn-Bio and HetBiSyn-MG.

All the variants are set to generate a drug feature of 128 dimensions except for HetBiSyn-Concat which is doubled by concatenating. The DNN for synergy prediction is same for each variant. The results are displayed in Table 2 .

Table 2. Results of the ablation study

Method	MSE	RMSE	PCC
HetBiSyn	**225.90±31.90**	**15.00±1.02**	**0.74±0.03**
HetBiSyn-Bio	246.75±48.02	15.71±1.52	0.68±0.02
HetBiSyn-MolHGT	248.24±41.91	15.76±1.42	0.67±0.02
HetBiSyn-MG	229.98±35.22	15.17±1.08	0.73±0.03
HetBiSyn-Concat	236.93±40.05	15.39±1.35	0.70±0.02

HetBiSyn-MG is designed to be a fair comparison to HetBiSyn-MolHGT, as they both depend on heterogeneous molecular graphs in terms of obtaining structural information. HetBiSyn-MG performs better because information from the macro-view perspective is also considered. It can be inferred that using data from either single perspective would not make better performance than integrating them even using simple concatenation. Furthermore, methods of fusing data from the micro-view and macro-view also affect the result of prediction. Though HetBiSyn-Concat outperforms other variants based on single perspective data, the original HetBiSyn shows an advantage to it even having less feature dimensions, which proves that the end-to-end self-supervised learning process of drug feature may better integrate the hidden information from both perspective.

4   Conclusion

In this paper, we propose a new DL based method for predicting anti-cancer synergistic drug combinations named HetBiSyn. HetBiSyn models the drug-related interactions between biomedical entities and the structure of drug molecules into different heterogeneous graphs, and designs a self-supervised learning framework to obtain a unified drug embedding that simultaneously contains information from both perspective. In details, two separate heterogeneous graph attention networks are adopted for the two types of graph, whose outputs are utilized to form a contrastive learning task for drug embedding that is enhanced by hard negative mining. We also obtain cell line features by exploiting gene expression profiles. Finally Hetbisyn uses a DNN with batch normalization to predict the synergy score of a combination of two drugs on a specific cell line. The experiment results show that our model outperforms other state-of-art DL and ML methods on the same synergy prediction task. Besides, the ablation study demonstrates that our drug embeddings with bi-perspective information learned through the end-to-end process is significantly informative and expressive, which is helpful to predict the synergy scores of drug combinations.

References

1. Mokhtari, R.B., Homayouni, T.S., Baluch, N., et al.: Combination therapy in combating cancer. Oncotarget **8**(23), 38022 (2017)
2. Zhang, T., Zhang, L., Payne, P.R.O., Li, F.: Synergistic drug combination prediction by integrating multiomics data in deep learning models. Methods Mol. Biol. **2194**, 223–238 (2021). PMID: 32926369. https://doi.org/10.1007/978-1-0716-0849-4_12
3. Liu, J., Gefen, O., Ronin, I., et al.: Effect of tolerance on the evolution of antibiotic resistance under drug combinations. Science **367**(6474), 200–4 (2020)
4. Kruijtzer, C., Beijnen, J., Rosing, H., et al.: Increased oral bioavailability of topotecan in combination with the breast cancer resistance protein and P-glycoprotein inhibitor GF120918. J. Clin. Oncol. **20**(13), 2943–2950 (2002)
5. Preuer, K., Lewis, R.P., Hochreiter, S., Bender, A., Bulusu, K.C., Klambauer, G.: DeepSynergy: predicting anti-cancer drug synergy with deep learning. Bioinformatics **34**(9), 1538–1546 (2018)
6. Lehár, J., Krueger, A.S., Avery, W., et al.: Synergistic drug combinations tend to improve therapeutically relevant selectivity. Nat. Biotechnol. **27**(7), 659–666 (2009)
7. Macarron, R., Banks, M.N., Bojanic, D., et al.: Impact of high-throughput screening in biomedical research. Nat. Rev. Drug Discov. **10**, 188–95 (2011)
8. Li, X., Yingjie, X., Cui, H., et al.: Prediction of synergistic anti-cancer drug combinations based on drug target network and drug induced gene expression profiles. Artif. Intell. Med. **83**, 35–43 (2017)
9. Li, H., Li, T., Quang, D., et al.: Network propagation predicts drug synergy in cancers. Cancer Res. **78**(18), 5446–57 (2018)
10. Low, Y.S., Daugherty, A.C., Schroeder, E.A., et al.: Synergistic drug combinations from electronic health records and gene expression. J. Am. Med. Inform. Assoc. **24**(3), 565–76 (2017)
11. Jeon, M., Kim, S., Park, S., et al.: In silico drug combination discovery for personalized cancer therapy. BMC Syst. Biol. **12**(2), 59–67 (2018)
12. Celebi, R., Bear Don't Walk, O., Movva, R., et al.: In-silico prediction of synergistic anti-cancer drug combinations using multi-omics data. Sci. Rep. **9**(1), 1–10 (2019)
13. O'Neil, J., Benita, Y., Feldman, I., et al.: An unbiased oncology compound screen to identify novel combination strategies. Mol. Cancer Ther. **15**(6), 1155–62 (2016)

14. Xia, F., Shukla, M., Brettin, T., et al.: Predicting tumor cell line response to drug pairs with deep learning. BMC Bioinf. **19**(18), 71–9 (2018)
15. Ding, P., Yin, R., Luo, J., et al.: Ensemble prediction of synergistic drug combinations incorporating biological, chemical, pharmacological, and network knowledge. IEEE J. Biomed. Health Inform. **23**(3), 1336–45 (2019)
16. Singh, H., Rana, P.S., Singh, U.: Prediction of drug synergy score using ensemble based differential evolution. IET Syst. Biol. **13**(1), 24–9 (2019)
17. Kim, Y., Zheng, S., Tang, J., et al.: Anticancer drug synergy prediction in understudied tissues using transfer learning. J. Am. Med. Inform. Assoc. **28**(1), 42–51 (2021)
18. Sun, Z., Huang, S., Jiang, P., et al.: DTF: deep tensor factorization for predicting anticancer drug synergy. Bioinformatics **36**(16), 4483–9 (2020)
19. Rogers, D., Hahn, M.: Extended-connectivity fingerprints. J. Chem. Inf. Model. **50**, 742–754 (2010)
20. Ekşioğlu, I., Tan, M.: Prediction of drug synergy by ensemble learning. arXiv:2001.01997 (2020)
21. Bai, Y., Gu, K., Sun, Y., Wang, W.: Bi-level graph neural networks for drug-drug interaction prediction. arXiv:2006.14002 (2020)
22. Winter, R., Montanari, F., Noé, F., Clevert, D.: Learning continuous and data-driven molecular descriptors by translating equivalent chemical representations. Chem. Sci. **10** (2019). https://doi.org/10.1039/C8SC04175J
23. Yang, J., Xu, Z., Wu, W., et al.: GraphSynergy: a network-inspired deep learning model for anticancer drug combination prediction. J. Am. Med. Inform. Assoc. **28**(11), 2336–2345 (2021)
24. Iorio, F., Knijnenburg, T.A., Vis, D.J., et al.: A landscape of pharmacogenomic interactions in cancer. Cell **166**(3), 740–54 (2016)
25. Hochreiter, S., Clevert, D.-A., Obermayer, K.: A new summarization method for Affymetrix probe level data. Bioinformatics **22**(8), 943–9 (2006)
26. Wan, F., Hong, L., Xiao, A., et al.: NeoDTI: neural integration of neighbor information from a heterogeneous network for discovering new drug-target interactions. Bioinformatics **35**(1), 104–111 (2019)
27. Chuang, K.V., Keiser, M.J.: Comment on 'predicting reaction performance in C-N cross-coupling using machine learning.' Science **362**(6416) (2018). American Association for the Advancement of Science (AAAS)
28. Fang, Y., et al.: Molecular contrastive learning with chemical element knowledge graph. In: Proceedings of the AAAI Conference on Artificial Intelligence. Molecular Contrastive Learning With Chemical Element Knowledge Graph. Proceedings of the AAAI Conference on Artificial Intelligence (2022)
29. Wang, X., Ji, H., Shi, C., et al.: Heterogeneous graph attention network. In: The World Wide Web Conference, WWW 2019, pp. 2022–32. Association for Computing Machinery, New York (2019)
30. Zou, H., Hastie, T.: Regularization and variable selection via the elastic net. J. R. Stat. Soc. Series B Stat. Methodol. **67**(2), 301–20 (2005)
31. Drucker, H., Burges, C.J., Kaufman, L., Smola, A., Vapnik, V.: Support vector regression machines. In: Advances in Neural Information Processing Systems, vol. 9, pp. 155–161 (1997)
32. Breiman, L.: Random forests. Mach. Learn. **45**(1), 5–32 (2001)
33. Chen, T., Guestrin, C.: XGBoost: a scalable tree boosting system. In: Proceedings of the 22nd ACM SIGKDD International Conference on Knowledge Discovery and Data Mining, pp. 785–794. Association for Computing Machinery (ACM), New York, NY, United States (2016)

34. Halil, K., Oznur, T., Ercument, C.: Matchmaker: a deep learning framework for drug synergy prediction. IEEE/ACM Trans. Comput. Biol. Bioinf., 1. https://doi. org/10.1109/TCBB.2021.3086702

35. Xu, M., Zhao, X., Wang, J., et al.: DFFNDDS: prediction of synergistic drug combinations with dual feature fusion networks. J. Cheminf. **15**. https://doi.org/ 10.1186/s13321-023-00690-3

36. Dong, Y., Chawla, N.V., Swami, A.: metapath2vec: scalable representation learning for heterogeneous networks. In: Proceedings of the 23rd ACM SIGKDD International Conference on Knowledge Discovery and Data Mining (2017)

37. Wang, M., Zheng, D., Ye, Z., et al.: Deep graph library: a graph-centric, highly-performant package for graph neural networks. arXiv preprint arXiv:1909.01315

38. Deng, D., Lei, Z., Hong, X., et al.: Describe molecules by a heterogeneous graph neural network with transformer-like attention for supervised property predictions. ACS Omega **7**(4), 3713–3721. https://doi.org/10.1021/acsomega.1c06389

Clique-Based Topological Characterization of Chromatin Interaction Hubs

Gatis Melkus[(✉)], Sandra Silina, Andrejs Sizovs, Peteris Rucevskis, Lelde Lace, Edgars Celms, and Juris Viksna

Institute of Mathematics and Computer Science, University of Latvia, Raina bulvaris 19, Riga, Latvia
gatis.melkus@lumii.lv

Abstract. Chromatin conformation capture technologies are a vital source of information about the spatial organization of chromatin in eukaryotic cells. Of these technologies, Hi-C and related methods have been widely used to obtain reasonably complete contact maps in many cell lines and tissues under a wide variety of conditions. This data allows for the creation of chromatin interaction graphs from which topological generalizations about the structure of chromatin may be drawn. Here we outline and utilize a clique-based approach to analyzing chromatin interaction graphs which allows for both detailed analysis of strongly interconnected regions of chromatin and the unraveling of complex relationships between genomic loci in these regions. We find that clique-rich regions are significantly enriched in distinct gene ontologies as well as regions of transcriptional activity compared to the entire set of links in the respective datasets, and that these cliques are also not entirely preserved in randomized Hi-C data. We conclude that cliques and the denser regions of connectivity in which they are common appear to indicate a consistent pattern of chromatin spatial organization that resembles transcription factories, and that cliques can be used to identify functional modules in Hi-C data.

Keywords: chromatin interaction graphs · network biology · ensemble Hi-C data · chromatin hubs · transcription factory

1 Introduction

The spatial organization of chromatin in the eukaryotic genome is an increasingly relevant topic of study in molecular biology. While seldom covered in classic models of gene regulation, the spatial proximity of cis-regulatory elements, the formation of heterochromatin and euchromatin as well as a vast number of other phenomena have profound influence on gene expression, which has led to the adoption of the term "nucleome" to cover the totality of the molecules

Supported by the Latvian Council of Science project lzp-2021/1-0236.

X. Guo et al. (Eds.): ISBRA 2023, LNBI 14248, pp. 476–486, 2023.
https://doi.org/10.1007/978-981-99-7074-2_38

and interactions involved [19]. These processes are mediated by biochemical, genetic and epigenetic factors that combine to create a dynamic regulatory landscape with several distinct layers of regulation such as chromatin compartments, topologically associating domains (TADs) and chromatin loops [6,8,23]. While research has illuminated some of the basic mechanisms behind these processes, a great deal of data having been gathered via methods such as Hi-C [13], a systematic understanding of the overall regulatory landscape remains challenging to establish.

A key approach in the analysis of spatiotemporal dynamics of chromatin is network analysis. Hi-C data lends itself to network analysis because contact maps from Hi-C experiments can be readily converted into adjacency matrices that are suitable as the basis of a network that can be further studied. Network analysis has produced several insights into chromatin organization, including the very concept of topologically associating domains [19]. TADs are one of many features that fall under the broad umbrella of a "chromatin hub", a nexus of interactions between different regions of chromatin (as seen in TADs) or between chromatin and other molecules (as seen in transcription factories and similar formations) [18]. An extensive array of methods currently exists for the purpose of network-based inference of various features in Hi-C data as well as for the prediction of interactions within a given dataset [19,24], and the extensive array of data available has even led to the development of meta-networks of Hi-C interactions meant to be used for the inference of nucleome-genome interactions (such as mapping eQTLs to chromatin interactions) [16]).

In recent years there has been interest in the concept of highly interconnected areas of chromatin as determinants of genomic function. There has been work centering on concepts such as cliques of topologically associating and lamina-associated domains [4,15] which are thought to coordinate genomic function at long ranges, as well as dense chromatin hubs as key determinants of cancer development [20], to name some examples. Here we utilize cliques to study chromatin hubs within healthy human tissues and cells in order to obtain a concrete idea of what, if any, biological features can be confidently mapped to their specific topologies.

2   Methods

2.1   Source Data

The datasets used to construct our chromatin interaction graphs include a promoter capture Hi-C dataset of 17 human blood cell types [7], a promoter capture Hi-C dataset of 27 human tissue types [9], additional Hi-C data for 20 tissues matching the human tissue type pcHi-C data from the 3DIV database [10]. All of the datasets were used in their processed contact list form, filtering for contacts of suitably high significance: a CHiCAGO score (a specific measure for assessing capture Hi-C, see [3]) greater than 5 for blood cell pcHi-C, an adjusted p-value greater than 0.7 for tissue pcHi-C, a -log p-value greater than 10 for tissue Hi-C. Where possible, these thresholds were selected according to guidelines provided by the authors and supplementary data. Where these were not available, a threshold providing suitable graph density for topological study was chosen instead. In the data sets used for our study, interchromosomal interactions had been pre-filtered out of the source data, and therefore our framework does not currently account for these.

2.2   Graph Generation and Analysis

We utilize filtered contact lists for distinct tissues (or cell types) to generate undirected graphs where genomic loci are rendered as nodes and the contacts between them are rendered as links. Each graph $G_t(V_t, E_t)$ is constructed for every chromosome and every tissue type $t \in T$. For each chromosome, graphs from different tissues are merged to form a comprehensive graph $G(V, E)$. Each link within E has a property – a list of tissue types for which the same chromatin interaction is present in their respective graphs G_t.

In other words, a link $e \in E$ is present if it is a part of at least one tissue-specific graph: $\exists t \in T (e \in E_t)$. Furthermore, every link from any tissue-specific graph is included in E, i.e., $\forall t \forall e \, (e \in E_t) \rightarrow e \in E$. A link e in the graph G is said to have tissue type t if it is present in the graph for tissue type t ($e \in E_t$).

We define the 'length' of a link as the distance (in base pairs) between the midpoints of the interacting chromosome segments, and the 'length distribution' as a function that maps a given length to the number of links of that length in the graph.

We also utilize the concept of a 3-vertex clique, or 'C3', which is a clique comprising three links that all share at least one common tissue. More formally, a triplet of nodes (A, B, D) forms a C3 if these nodes share a common tissue across their interactions $(A, B) \in E_t$, $(B, D) \in E_t$, and $(A, D) \in E_t$. We say that a C3 (A, B, D) includes a link e if e is one of its three links. Because every other n-vertex clique in the graph is encompassed by the full set of 3-vertex cliques, our work in this paper focuses on C3, detecting larger cliques indirectly as dense, countable aggregations of C3.

The 'tissue degree' of a C3, denoted as $DegC3(A, B, D)$, is defined as the number of tissues that share the same interactions that form the C3. If the tissue degree of a triplet of nodes is greater than 0 ($DegC3(A, B, D) > 0$), these nodes form a C3.

2.3   Randomization

To our knowledge, no method currently exists for randomizing Hi-C graphs that preserves both node degrees and link length distributions, which are key considerations in testing our clique distribution. Therefore, we develop our own method here to affirm the validity of our findings. Starting with a graph $G(V, E)$, where nodes represent chromosome segments and links correspond to chromatin interactions, we generate a randomized graph $G'(V, E')$. This graph maintains the following properties: the difference of E and E' is greater than or equal to a parameter q, the number of edges in E and E' are equal, the nodes V and their degrees remain unchanged, and the distribution of link lengths (chromatin interaction lengths) in both graphs remains as similar as feasible.

All links within the graph G are categorized into *link length groups*, each containing links of similar lengths (lengths correspond to distances of interactions on a chromosome). To keep the number of groups and number of links in each group balanced, we empirically chose the initial number of links within each *link length group* to be approximately $\log_2(|V|) \cdot \log_2(|E|) \cdot p$. Here, we chose a p of 0.25, having found that this produced link length groups of manageable size.

Our method for creating the randomized graph G' involves iteratively *swapping* link ends of the initial graph G (Fig. 1). For instance, given links (A, B) and (C, D) in G, we produce links (A, D) and (B, C), or (A, C) and (B, D), in G'. This guarantees the preservation of the original edge count and node degrees, while the selection of links for each swap is done in a manner that strives to maintain a comparable distribution of link lengths across groups as in the initial graph.

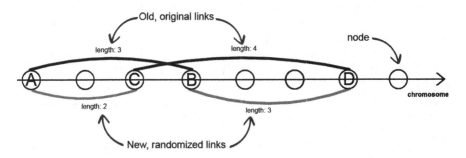

Fig. 1. Essential principle of the randomization algorithm outlined. The endpoints of links A-B and C-D are swapped to create new links A-C and B-D, preserving the node degrees.

For a prospective *swap* of two links, we compute a *score* indicating the improvement in the *link length distribution* in the randomized graph G' if the swap is performed. The swap is valid only if both new, randomized links are not in E and are not in E', and their lengths are not longer that the longest link in E. If $D(G)$: (*link length group l*) $\rightarrow$ (*number of links in group l in graph G*) is the link length distribution function in graph G, and if graph G'' would be obtained after the prospective swap of two links, then the score for the prospective swap is calculated $score = \sum_{l \in link\ length\ groups} |D(G'')(l) - D(G)(l)|$. A lower score implies a closer resemblance of G'' to the original link length distribution of G.

Before each swap we stochastically select links either (a) to increase the difference $|E \setminus E'|$ or (b) to improve the *link length distribution* in *link length groups*, using varying probabilities that depend on the state of G'. In case of (a), we randomly choose one link e: $e \in E \wedge e \in E'$. In case of (b), we randomly choose one link e: $e \in E' \wedge e \in l \wedge (D(G')(l) - D(G)(l) > 0)$, where l is the length group of link e. After that we randomly pool numerous other links from E', then calculate score as described above for swapping e with each of the pooled links. Finally, the chosen link e is *swapped* with a pooled link with one of the best scores.

The process is finished when either enough links are swapped ($|E \setminus E'| \geq q$) or no valid swap can be performed for a number of attempts that is comparable to $|E|$. Here we attempted to randomize all links in the data, but in practice this was not possible, as short links tend to be saturated - every possible link in close proximity already exists, and therefore no swap can be made within the link length group.

2.4   Supplementary Data and Validation

After obtaining a full set of cliques within our chosen datasets we then drew upon additional sources to analyze the topological features we had obtained.

Firstly, we performed Gene Ontology Enrichment Analysis (GOEA) using the GOATools library [11] and, in line with its specifications, analyzing our dataset using the basic acyclic version of the Gene Ontology database [2]. In this case, we searched for Gene Ontology terms that were overrepresented in a study gene list compared to a population gene list.

We utilized Ensembl gene lists [5] to assign genes to nodes in the graph G. Nodes located in chromosomal regions that contain particular genes are assigned that gene. Subsequently, we cliques $C3$, identified a subset of nodes that form these topological elements, and collated a set of genes present in at least one node of these cliques, then conducted the GOEA. For instance, we collected all $C3$ elements with a tissue degree of 2 or more, and took the genes present in at least one such $C3$ to form the study gene list (the input). This list was then compared to the population gene list (the background), which consisted of all genes found in at least one link in the graph.

Afterward, we also tested our cliques for enrichment in certain chromatin annotations. For this purpose we employed epigenomic data published by the BLUEPRINT consortium [1] (packaged with the original data) for our blood cell pcHi-C data and the ChromHMM core 15-state model from Roadmap Epigenomics [12]. For the ChromHMM annotations, we consolidated the different categories, which is to say variants of particular annotations as well as weak and strong varieties of annotations, into a final list of 9 annotations: transcription start sites (TSS), transcription (TX), enhancers (ENH), zinc finger sites (ZNF), heterochromatin (HET), bivalent transcription start sites (TSS BIV), bivalent enhancers (ENH BIV), Polycomb-mediated repression (REP) and quiescent chromatin (QUI). Much like with our GO analysis, we tested the prevalence of these annotations (whether the annotation is present in the chromatin regions covered) within links and cliques in the dataset, and compared these with the 10 tissues in our data (in both Hi-C and pcHi-C) that overlapped with epigenomic coverage: lung (LG), aorta (AO), spleen (SX), adrenal gland (AD), pancreas (PA), bladder (BL), ovary (OV), small intestine (SB), liver (LI), psoas muscle (PO). For the BLUEPRINT annotations, we did not consolidate any annotations and employed the full set for CTCF sites (CTCF), distal enhancers (Distal), DNase sites (DNase), proximal enhancers (Proximal), transcription factor binding sites (TFBS) and transcription start sites (TSS).

In all cases we obtained the normalized counts of each individual feature for the covered subset of our data for the total number of links and the links that are part of cliques, and then compared these via paired Wilcoxon test for each annotation. We then calculated ratios that showed the relative enrichment for each particular annotation in our cliques versus the links in each individual dataset overall.

3   Results and Discussion

3.1   Topological Properties of Hi-C Interaction Graphs

While the datasets used showed considerable variability in link count, topology and prevalence of cliques, there was some striking overlap observable in three of our datasets, primarily around chromosome 6 (Fig. 2). Here, the vast majority of cliques occurred in one particular region of the chromosome, matching up to the location of HLA and histone gene clusters. Similar overabundance of cliques in one location in the chromosome was observed on chromosome 19. These features matched up well in our data, especially in the case of the chromosome 6 peak.

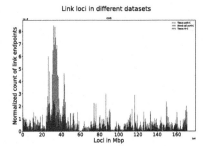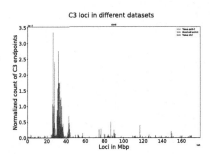

Fig. 2. Distributions of link and C3 cycle/clique endpoints across chromosome 6 for our blood pcHi-C, tissue pcHi-C and tissue Hi-C datasets, showing considerable overlap in clique-rich areas. Endpoints are nodes in the graph that are part of the feature in question, i.e. each clique has 3 endpoints while each link has 2.

The most obvious property of these cliques is that they occur chiefly in particularly dense areas of chromatin contacts, as can be seen in chromosome 6 where such contacts in one single connected component comprise the vast majority of interactions found throughout the entirety of the chromosome. The existence of a large region of interconnected chromatin specifically at chromosome 6 does not appear to be a technical artifact, as the presence of a transcriptionally coordinated and correspondingly spatially organized HLA genomic cluster is attested in both Hi-C [14] and older in-situ fluorescent hybridization [25] studies. This matches up with its appearance in several unrelated Hi-C datasets in our study, albeit with a minor offset.

3.2   Preservation of Topological Features in Randomized Graphs

Our randomization approach proved effective in preserving the overall structure of our primary datasets of interest, showing a nearly identical distribution of link lengths as well as link endpoints across the breadth of the chromosome regardless of the original distribution of link lengths in the dataset. At the same time, the randomization also produced a significant decrease in clique prevalence across the board, showing decreases of 50% or more (Fig. 3).

The decrease in cliques was much more pronounced in the tissue Hi-C dataset compared to the sparser pcHi-C datasets, though in all cases the vast majority of the clique depletion appeared to come from short-range interactions rather than long-range ones. This difference in performance makes sense given that the randomization algorithm does not generate new nodes, and so sparser and more discrete matrices generated by pcHi-C would preserve more of their original structure given the lower variety of nodes and links available for swapping.

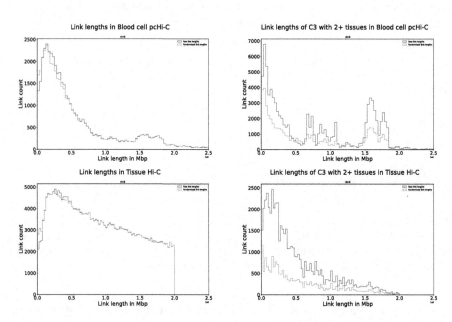

Fig. 3. Distributions of overall link lengths (left) and link lengths in C3 cliques found in 2 or more tissues (right) for blood cell pcHi-C (top) and tissue Hi-C (bottom). All data shown is for chromosome 6.

However, even in our maximally similarly distributed model it is clear that the clique structures observed, while partially explicable through a much larger abundance of links and high node degrees, are nevertheless overabundant in chromosomal hotspots. Moreover, the same principle also appears to hold outside of chromosomal hotspots, as clique abundance noticeably decreases across our three datasets (tissue Hi-C, tissue pcHi-C and blood cell pcHi-C) after randomization. As such, cliques could potentially serve as a method for identifying significantly dense parts of the graph that are less obvious than the HLA chromatin loop in chromosome 6, though the functional implications of this density still need to be established.

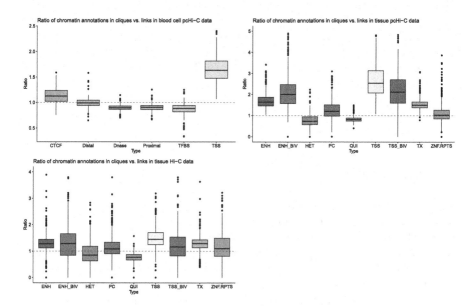

Fig. 4. Ratios of normalized counts of chromatin annotations within links/cliques across the three main datasets.

3.3 Cliques as Indicators of Functional Modules in Graphs

Our GO enrichment analysis revealed significant enrichment of certain ontologies in cliques compared to links, though not in even numbers across all three datasets. Of these, the chromosome 6 peak was by far the most consistent across datasets, and not all chromosomes appeared to have significant enrichment of any ontologies, especially in our tissue Hi-C data where enriched ontologies were particularly sparse. Of the enriched ontologies found, most were in large categories such as DNA binding, the RNA metabolic process, nucleosome components and cell adhesion. Notably, GOEA located enriched annotations for olfactory receptor activity in chromosome 14, another chromosomal hotspot identified by Liu et al. [14].

Our study of enrichment in annotations showed striking results (as seen in Fig. 4). While almost all annotations showed statistically significant deviation from the original datasets' prevalences in our cliques (generally with p-values lower than 0.001 per paired Wilcoxon test), examining the ratios clearly showed that cliques appeared to be consistently enriched for transcription start sites, with a median 1.5...3x enrichment. Heterochromatin and quiescent annotations were found to be rarer inside of cliques in general. Enhancers and active transcription, by comparison, were generally either modestly enriched (1...1.5x) or statistically indistinguishable from their prevalence in the dataset as a whole.

Judging from these results, the chromatin hubs our approach has identified most seem to resemble transcription factories, which are large aggregations of cooperatively transcribed genes characterized by aggregation of RNA polymerases [18,22] and, in more recent models, liquid-liquid phase separation [17,18]. They also bear a significant resemblance to "rich club" loci originally identified by Sandhu et al. [19,21], being both highly connected and, as our GOEA analysis showed, notably connected to core cellular processes. This is an encouraging result, as our cliques appear to consistently identify these features across datasets with no functional information, but further work is necessary to refine the approach for closer analysis.

4   Conclusions

In this study we utilized cliques in order to study various Hi-C datasets. We found that cliques are useful in identifying clusters of genes that match the general characteristics of transcription factories, both in obvious "chromosomal hotspots" and around the genome overall. These features can be identified with some consistency across several datasets, and occasionally show overlap in the case of particularly dense groupings, such as around the HLA cluster on chromosome 6. We foresee making use of cliques in future work to further elaborate on the structure and function of chromatin hubs, most likely through integrating our observations with other data such as ChIA-PET using RNA polymerase II to test the transcription factory hypothesis more directly, but also through adapting our methods to new Hi-C data.

5   Additional Resources

The resources used to generate the findings of this paper can be found at https:// github.com/IMCS-Bioinformatics/HiCCliqueGraphs, including scripts for generating our processed data from our sources and the full results of our GOEA and annotation studies.

References

1. Adams, D., Altucci, L., et al.: BLUEPRINT to decode the epigenetic signature written in blood. Nat. Biotechnol. **30**(3), 224–226 (2012). https://doi.org/10.1038/nbt.2153
2. Ashburner, M., Ball, C.A., et al.: Gene Ontology: tool for the unification of biology. Nat. Genet. **25**(1), 25–29 (2000). https://doi.org/10.1038/75556
3. Cairns, J., Freire-Pritchett, P., et al.: CHiCAGO: robust detection of DNA looping interactions in Capture Hi-C data. Genome Biol. **17**(1), 127 (2016). https://doi.org/10.1186/s13059-016-0992-2
4. Collas, P., Liyakat Ali, T.M., et al.: Finding friends in the crowd: three-dimensional cliques of topological genomic domains. Front. Genet. **10**, 602 (2019). https://doi.org/10.3389/fgene.2019.00602
5. Cunningham, F., Allen, J.E., et al.: Ensembl 2022. Nucleic Acids Res. **50**(D1), D988–D995 (2022). https://doi.org/10.1093/nar/gkab1049
6. Grubert, F., Srivas, R., et al.: Landscape of cohesin-mediated chromatin loops in the human genome. Nature **583**(7818), 737–743 (2020). https://doi.org/10.1038/s41586-020-2151-x
7. Javierre, B.M., Sewitz, S., et al.: Lineage-specific genome architecture links enhancers and non-coding disease variants to target gene promoters. Cell **167**(5), 1369–1384.e19 (2016). https://doi.org/10.1016/j.cell.2016.09.037
8. Jha, R.K., Levens, D., Kouzine, F.: Mechanical determinants of chromatin topology and gene expression. Nucleus **13**(1), 95–116 (2022). https://doi.org/10.1080/19491034.2022.2038868
9. Jung, I., Schmitt, A., et al.: A compendium of promoter-centered long-range chromatin interactions in the human genome. Nat. Genet. **51**(10), 1442–1449 (2019). https://doi.org/10.1038/s41588-019-0494-8
10. Kim, K., Jang, I., et al.: 3DIV update for 2021: a comprehensive resource of 3D genome and 3D cancer genome. Nucleic Acids Res. **49**(D1), D38–D46 (2021). https://doi.org/10.1093/nar/gkaa1078
11. Klopfenstein, D.V., Zhang, L., et al.: GOATOOLS: a python library for gene ontology analyses. Sci. Rep. **8**(1), 10872 (2018). https://doi.org/10.1038/s41598-018-28948-z
12. Kundaje, A., Meuleman, W., et al.: Integrative analysis of 111 reference human epigenomes. Nature **518**(7539), 317–330 (2015). https://doi.org/10.1038/nature14248
13. Lieberman-Aiden, E., Van Berkum, N.L., et al.: Comprehensive mapping of long-range interactions reveals folding principles of the human genome. Science **326**(5950), 289–293 (2009). https://doi.org/10.1126/science.1181369
14. Liu, L., Li, Q.Z., et al.: Revealing gene function and transcription relationship by reconstructing gene-level chromatin interaction. Comput. Struct. Biotechnol. J. **17**, 195–205 (2019). https://doi.org/10.1016/j.csbj.2019.01.011
15. Liyakat Ali, T.M., Brunet, A., et al.: TAD cliques predict key features of chromatin organization. BMC Genom. **22**(1), 499 (2021). https://doi.org/10.1186/s12864-021-07815-8
16. Lohia, R., Fox, N., Gillis, J.: A global high-density chromatin interaction network reveals functional long-range and trans-chromosomal relationships. Genome Biol. **23**(1), 238 (2022). https://doi.org/10.1186/s13059-022-02790-z
17. Maeshima, K., Tamura, S., et al.: Fluid-like chromatin: toward understanding the real chromatin organization present in the cell. Curr. Opin. Cell Biol. **64**, 77–89 (2020). https://doi.org/10.1016/j.ceb.2020.02.016

18. Mora, A., Huang, X., et al.: Chromatin hubs: a biological and computational outlook. Comput. Struct. Biotechnol. J. **20**, 3796–3813 (2022). https://doi.org/10.1016/j.csbj.2022.07.002

19. Pancaldi, V.: Network models of chromatin structure. Curr. Opin. Genet. Dev. **80**, 102051 (2023). https://doi.org/10.1016/j.gde.2023.102051

20. Sanalkumar, R., Dong, R., et al.: Highly connected 3D chromatin networks established by an oncogenic fusion protein shape tumor cell identity. Sci. Adv. **9**(13) (2023). https://doi.org/10.1126/sciadv.abo3789

21. Sandhu, K.S., Li, G., et al.: Large-scale functional organization of long-range chromatin interaction networks. Cell Rep. **2**(5), 1207–1219 (2012). https://doi.org/10.1016/j.celrep.2012.09.022

22. Sutherland, H., Bickmore, W.A.: Transcription factories: gene expression in unions? Nat. Rev. Genet. **10**(7), 457–466 (2009). https://doi.org/10.1038/nrg2592

23. Szabo, Q., Bantignies, F., Cavalli, G.: Principles of genome folding into topologically associating domains. Sci. Adv. **5**(4) (2019). https://doi.org/10.1126/sciadv.aaw1668

24. Tao, H., Li, H., et al.: Computational methods for the prediction of chromatin interaction and organization using sequence and epigenomic profiles. Brief. Bioinf. (2021). https://doi.org/10.1093/bib/bbaa405

25. Volpi, E., Chevret, E., et al.: Large-scale chromatin organization of the major histocompatibility complex and other regions of human chromosome 6 and its response to interferon in interphase nuclei. J. Cell Sci. **113**(9), 1565–1576 (2000). https://doi.org/10.1242/jcs.113.9.1565

Exploring Racial Disparities in Triple-Negative Breast Cancer: Insights from Feature Selection Algorithms

Bikram Sahoo[1($\boxtimes$)] ⓘ, Temitope Adeyeha[1] ⓘ, Zandra Pinnix[2] ⓘ, and Alex Zelikovsky[1($\boxtimes$)] ⓘ

[1] Department of Computer Science, Georgia State University, Atlanta, GA 30302, USA
{bsahoo1,tadeyeha1,alexz}@gsu.edu
[2] Department of Biology and Marine Biology, University of North Carolina at Wilmington, Wilmington, NC 28403, USA
pinnixz@uncw.edu

Abstract. Triple-negative breast cancer (TNBC) is a challenging subtype with pronounced racial disparities, more prevalent in African American (AA) women. We employed diverse feature selection algorithms, including filters, wrappers, and embedded methods, to identify significant genes contributing to these disparities. Notably, genes such as *LOC90784, LOC101060339, XRCC6P5,* and *TREML4* consistently emerged using correlation and information gain-based filter methods. Our two-stage embedded-based approach consistently highlighted *LOC90784, STON1-GTF2A1L,* and *TREML4* as crucial genes across high-performing machine learning algorithms. The unanimous selection of *LOC90784* by all three filter methods underscores its significance. These findings offer valuable insights into TNBC's racial disparities, aiding future research and treatments.

Keywords: Feature Selection · Triple-Negative Breast Cancer · Racial Disparity · Gene Expression

1 Introduction

Triple-negative breast cancer (TNBC) is an aggressive subtype, notably devoid of estrogen and progesterone receptors and HER2 amplification. It constitutes 15–20% of U.S. breast cancer diagnoses, marked by the poorest survival rates among subtypes [4,17]. Disconcertingly, TNBC exhibits pronounced racial disparities, with African American women facing bleaker outcomes than their European American counterparts. These disparities arise from a complex interplay of biological and socioeconomic factors [12,14,20]. At the molecular level, TNBC is exceptionally heterogeneous, categorized into subtypes like Basal-like 1, Basal-like 2, immunomodulatory, mesenchymal, mesenchymal stem-like, and luminal androgen receptor. Recently, M and MSL subtypes were excluded from this classification [19]. This inherent diversity challenges comprehension of the heightened

X. Guo et al. (Eds.): ISBRA 2023, LNBI 14248, pp. 487–497, 2023.
https://doi.org/10.1007/978-981-99-7074-2_39

mortality in African American women compared to European Americans. Two hypotheses emerge: distinct molecular characteristics in TNBC among racial groups or contributions from environmental and socioeconomic factors. Genomic investigations uncover molecular distinctions, illuminating this multifaceted issue [15,19].

Current research in TNBC identified a few genes signatures, mutations, and deregulated genes in tumor micro-environment (TME) to understand the racial disparity in AA and EA [1,14,18]. However, the identification of key biomarker for AA that is significantly deregulated compared to EA women in TNBC is still needed the current change to tackle this racial disparity. To address this issue, computational biologists can take advantage of numerous open-source RNA-seq gene expression data from sources like GEO [5], TCGA [6] by using robust machine learning (ML), and deep learning (DL) supervised classification algorithms [9,11,13]. Because recently computational biologist shifted their approaches to identify differential gene expression using traditional parametric and non-parametric statistical method based on logarithmic values of fold change (logFC) between different groups For instance, DESeq [3], DESeq2 [10], edgeR [16], and voom [8] because of high false positive and fasle negative rates in prediction of DEGs. ML methods classify genes based on expression. SVM and LR distinguish colon cancer and other cancers. Random Forest classifies genes in microarray data. An empirical study evaluated DTC, LR, NB, RFC, and SVC for gene expression classification. DL with transfer learning can classify novel data by learning complex relationships among training data features in one end-to-end system. CNNs, a type of DL, utilize convolutional approaches in internal layers, enabling computation, learning of non-linear relations, and feature extraction in both image and non-image data. CNNs, like DeepInsight, are powerful tools for classification, leveraging hierarchical filtering, weight sharing, and neighborhood information. They excel in tasks such as gene expression and text data feature extraction [7,21].

The aim of the current study is to employ a comprehensive methodology for feature selection to identify crucial genes. Initially, we established a baseline model to assess the data's predictive capabilities. Subsequently, we applied three feature selection algorithms: filters, wrappers, and embedded methods. Filter-based selection methods focus on features or genes chosen through correlation and information gain-based techniques. Wrapper-based selection differs, as it detects feature dependencies. Wrappers can be categorized into Recursive Feature Elimination (RFE) and Forward Selection. Due to computational costs, forward selection wasn't pursued here. Embedded methods combine RFE and Forward Selection, addressing high-dimensional data challenges. Unlike traditional methods, embedded selection doesn't require a predetermined feature count; it determines the optimal number during evaluation. By analyzing results from these methodologies, our study seeks to identify significant features, enhancing selection accuracy and robustness. This is crucial in addressing racial disparities in triple-negative breast cancer among African American (AA) and European American (EA) women.

2 Data

In this study, we analyzed gene expression data from breast cancer patients using publicly available RNA sequencing data. We processed the RNA-seq data, focusing on triple-negative breast cancer (TNBC) by selecting samples negative for ER, PR, and HER2 receptors. This filtering yielded 145 TNBC samples, and our analysis specifically examined racial disparities between European American (EA) and African American (AA) women. We reduced our sample size to 128 TNBC samples, comprising 87 EA and 41 AA women, for the final analysis. Data sources included [5, 6].

3 Methods

3.1 Baseline Model Generation and Data Pre-processing

3.1.1 Baseline Model Generation: The baseline models were generated using the LazyClassifier module from the Lazy Predict library [2]. This module integrates a range of methods including Ensemble, Gradient Boosting, Graph-based, Instance-based, Linear, Naive Bayes, Neural Network, Non-linear, Prototype-based, Support Vector Machine, and Tree-based models. The objective is to evaluate the performance of these algorithms by computing Accuracy, Balanced Accuracy, ROC AUC, and F1 Score using 5-fold cross-validation, as seen in Table 1.

3.1.2 Data Pre-processing

3.1.2.1 Dropping of Highly Correlated Columns: To enhance machine learning model performance by mitigating the impact of repetitive columns in the dataset, a combination of Pairwise column correlation and variance inflation factor techniques was applied, leading to the identification and removal of 912 highly repetitive columns, causing only a marginal 2% variance reduction in the baseline model.

3.1.2.2 Data Normalization: Crucial data normalization techniques, including Min-Max and standard scaling, were employed for numeric variable transformation, resulting in comparable performance improvements with a minor 2% variance from the baseline model; this, combined with non-repetitive columns, establishes a novel accuracy benchmark for future analyses.

3.2 Feature Selection

3.2.1 Filter. Filters, used in machine learning for feature selection, assess feature relevance via statistical metrics, eliminating features with low scores; they are efficient but reliant on specific metrics. In this study, we explored correlation-based and information gain-based filters.

3.2.1.1 Correlation - Based. We utilized the Pearson correlation coefficient (ρ) to identify the top genes, specifically limiting our selection to the top 10 genes (X_i), that demonstrate a significant correlation with the target variable (Y). In this study, the target variable represents the race of women with Triple-Negative Breast Cancer (TNBC) and is categorized as either African American or European American. The Pearson correlation coefficient (ρ) measures the linear relationship between two variables and is calculated using the following formula:

$$\rho = \frac{\sum (X_i - \mu_X)(Y - \mu_Y)}{\sqrt{\sum (X_i - \mu_X)^2} \sqrt{\sum (Y - \mu_Y)^2}} \tag{1}$$

where X_i and Y are variables, and μ_X and μ_Y are the mean values of X_i and Y, respectively. The coefficient ρ ranges from -1 to 1, with a value of 1 indicating a perfect positive linear correlation, -1 indicating a perfect negative linear correlation, and 0 indicating no linear correlation. By exclusively selecting features (X_i) with strong positive ($\rho > 0$) or negative ($\rho < 0$) correlations, we effectively reduced the dimensionality of the data to 10 genes. This reduction in dimensionality helps eliminate less relevant variables and focuses on those strongly associated with the target variable (Y), which, in this case, is the race of TNBC women. By retaining only the most correlated features, we aim to capture the most important information for cancer diagnosis.

3.2.1.2 Information Gain - Based. Information gain-based feature selection is a widely used filter-based technique in machine learning that relies on information theory to evaluate the relevance of each gene (X_i) in relation to the race of the TNBC women (Y). The information gain (IG) quantifies the reduction in entropy or uncertainty of the target variable resulting from the inclusion of a specific feature. Mathematically, entropy is a measure of the average amount of information or uncertainty in a random variable. The entropy of the target variable (Y), denoted as $H(Y)$, is calculated using the formula:

$$H(Y) = -\sum P(Y = y) \log P(Y = y) \tag{2}$$

where $P(Y = y)$ represents the probability of the target variable taking the value y. The conditional entropy of the target variable given a feature X_i, denoted as $H(Y|X_i)$, measures the remaining uncertainty in the target variable after considering the feature X_i. It is computed as:

$$H(Y|X_i) = -\sum P(X_i = x_i) \sum P(Y = y|X_i = x_i) \log P(Y = y|X_i = x_i) \tag{3}$$

where $P(X_i = x_i)$ represents the probability of feature X_i taking the value x_i, and $P(Y = y|X_i = x_i)$ represents the conditional probability of the target variable Y taking the value y given that feature X_i has the value x_i. In this study, we employed information gain-based feature selection to identify the top-10 features (X_i) that exhibit the highest information gain. By calculating the information gain for each feature using the formula $IG(X_i) = H(Y) - H(Y|X_i)$, we quantified their individual contributions in reducing the uncertainty or entropy associated with the target variable.

3.2.2 Wrappers. Wrappers, while superior in feature selection due to their model-based approach and ability to detect feature dependencies, can become computationally intensive, as seen in Recursive Feature Elimination and Forward Selection; we excluded the latter due to computational constraints.

3.2.2.1 Recursive Feature Elimination. In Recursive Feature Elimination (RFE), the optimal number of features is determined through an iterative elimination process. The goal of RFE is to select the most informative subset of features or genes by iteratively removing the feature with the lowest weight or rank, as determined by a machine learning algorithm. We iteratively eliminated the feature with the lowest weight or rank until we obtained a subset of 10 features or genes.

Mathematically, let $X = [X_1, X_2, ..., X_n]$ represent the set of input features or genes used to study the racial disparity in TNBC, and let Y denote the target variable, which in this case is the race of the TNBC women, either African American (AA) or European American. At each iteration, the machine learning algorithm assigns a weight or rank to each feature or gene, indicating its importance in the model. Let $w = [w_1, w_2, ..., w_n]$ represent the weights/ranks assigned to the features. In each iteration, the feature with the lowest weight or rank, $min(w)$, is eliminated from the feature set. The model's performance is then assessed using a performance metric, such as accuracy, error rate, or any other appropriate evaluation measure.

Iteratively eliminating features until reaching the set number, k (10 in this study), the Recursive Feature Elimination (RFE) algorithm concludes. The chosen subset of k features becomes the optimal set for analysis or model building. RFE enhances model performance, diminishes input dimensionality, and augments generalization. It aids in understanding racial disparities in TNBC.

3.2.2 Embedded. In this section, we present a novel approach that combines Recursive Feature Elimination (RFE) and Forward Selection to address the computational challenges arising from high-dimensional gene expression data. Unlike traditional filter and wrapper methods, our approach does not require a predetermined number of features or genes for the final model. This flexibility allows us to determine the optimal number of features during the evaluation process, enabling a deeper understanding of the racial disparity in TNBC (Triple-Negative Breast Cancer).

Our approach can be summarized in the following two steps:

Step 1 - Recursive Feature Elimination (RFE)
We employ a customized version of RFE to reduce the dimensionality of the data. This can be represented as:

$X^{(1)} = $ Original dataset

$X^{(k)} = $ Reduced dataset at iteration k, $[k = 1, 2,\ldots, K]$

$K = $ Total number of iterations

where each iteration involves fitting the dataset $X^{(k)}$ to a machine learning model, evaluating the model performance, extracting feature importance, and eliminating features with importance below the median.

Step 2 - Forward Selection

After dimensionality reduction, we apply Forward Selection on the reduced dataset $X^{(K)}$ to identify the most predictive features. This can be represented as:

$$X^{(K)} = \text{Reduced dataset after RFE}$$

Selected Features = Features selected through Forward Selection

The overall methodology employed in this study encompasses a systematic exploration of all conceivable feature combinations, accompanied by a comprehensive evaluation of their performance using cross-validation. This evaluation process incorporates a diverse range of machine learning algorithms, including Linear Models, Ensemble models, and other models. Logistic Regression is specifically utilized for Linear Models, while the results for Ensemble Models are directly presented, featuring Extra Trees, Random Forest Classifier, LGBM Classifier, and AdaBoost to eliminate redundancy. Furthermore, the analysis incorporates the SVM model as an additional component.

4   Results

4.1   Exploring Filter Methods for Identifying Crucial Genes

We employed the filter method to pinpoint genes distinguishing African American and European American TNBC women. This method evaluates feature relevance using statistical metrics and removes those with low scores. Two filter types, correlation-based and information gain-based, were considered. Combining both methods yielded a set of 10 genes and features. Notably, using this reduced set, machine learning algorithms demonstrated improved classification accuracy. For instance, the XGBClassifier achieved a 94% Balanced Accuracy, while the LGBMClassifier maintained a Balanced Accuracy of 90%, as seen in Fig. 1. Likewise, with the information gain-based filter, the top 10 features, including *LOC90784, LOC101060339, XRCC6P5, EIF4G2, LOC51240, FBXL5, SURF6, FDX1L, TREML4*, and *TMEM102*, led to a 96% Balanced Accuracy for both RandomForestClassifier and SGDClassifier. Notably, *LOC90784, LOC101060339, XRCC6P5*, and *TREML4* were common among the top 10 features in both filter methods.

Table 1. The classification results for African American (AA) and European American (EA) women with triple-negative breast cancer, considering all the features/genes, were obtained through various machine learning algorithms. The metric values are presented below, with the optimal values highlighted in bold for ease of interpretation.

Category	Algorithm	Accuracy	Balanced Accuracy	ROC AUC	F1 Score	Time Taken
Ensemble	AdaBoostClassifier	0.92	0.88	0.88	0.92	15.28
	ExtraTreesClassifier	0.85	0.80	0.80	0.84	1.68
	BaggingClassifier	0.87	0.88	0.88	0.88	5.07
	CalibratedClassifierCV	0.74	0.58	0.58	0.67	2.44
	RandomForestClassifier	0.79	0.71	0.71	0.78	1.83
Gradient Boosting	**LGBMClassifier**	**0.92**	**0.90**	**0.90**	**0.92**	**13.83**
	XGBClassifier	0.87	0.81	0.81	0.87	8.77
Graph-based	LabelSpreading	0.31	0.50	0.50	0.14	1.28
	LabelPropagation	0.31	0.50	0.50	0.14	1.30
Instance-based	KNeighborsClassifier	0.79	0.69	0.69	0.77	1.30
Linear	LinearSVC	0.69	0.69	0.69	0.70	1.55
	LinearDiscriminantAnalysis	0.79	0.74	0.74	0.79	2.46
	LogisticRegression	0.85	0.80	0.80	0.84	1.89
	RidgeClassifier	0.85	0.80	0.80	0.84	1.24
	RidgeClassifierCV	0.85	0.80	0.80	0.84	1.33
	PassiveAggressiveClassifier	0.67	0.67	0.67	0.68	1.47
	SGDClassifier	0.64	0.65	0.65	0.65	1.35
Naive Bayes	GaussianNB	0.62	0.49	0.49	0.58	1.43
	BernoulliNB	0.77	0.72	0.72	0.77	1.34
Neural Network	Perceptron	0.64	0.65	0.65	0.65	1.29
Non-linear	QuadraticDiscriminantAnalysis	0.44	0.45	0.45	0.45	1.62
Prototype-based	NearestCentroid	0.74	0.70	0.70	0.74	1.21
Support Vector Machine	NuSVC	0.69	0.52	0.52	0.61	1.81
	SVC	0.67	0.48	0.48	0.55	1.95
Tree-based	ExtraTreeClassifier	0.72	0.70	0.70	0.73	1.26
	DecisionTreeClassifier	0.79	0.78	0.78	0.80	1.78

4.2 Exploring Wrapper Method for Identifying Crucial Genes

We utilized the Recursive Feature Elimination (RFE) wrapper method for iterative feature selection. It identified an optimal set of 10 genes, including *BCL6, LOC101060339, DDX51, FOXD4L6, LOC101060247, KRT7, LOC90784, RNF35, LOC155153*, and *DKFZp434E1822*, critical for distinguishing African American and European American TNBC patients. Notably, LinearSVC Classifier achieved the highest balanced accuracy at 96%, as seen in Fig. 1. These 10 genes exhibited minimal overlap with filter and embedded methods, though *LOC101060339, DDX51, LOC90784*, and *LOC155153* were common selections.

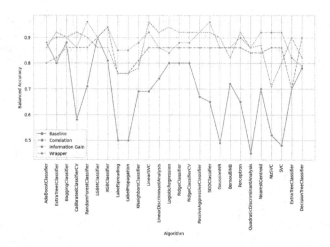

Fig. 1. The figure shows balanced accuracy in classifying African American (AA) and European American (EA) TNBC women using full gene set baseline and top 10 genes via correlation-based, Information Gain-based filters, and Recursive Feature Elimination.

4.3   Exploring Embedded Method for Identifying Crucial Genes

Incorporating Recursive Feature Elimination (RFE) in Stage 1 and Forward Selection in Stage 2, this approach effectively tackles high-dimensional data complexities.

Our findings reveal that Logistic Regression and SVC achieved the highest Balanced Accuracy at 99% each during Stage 2, employing embedded feature selection. Compared to the baseline model (Fig. 2), Logistic Regression selected nine features, including *CENPU, KBF2, LOC155153, LOC649506, LOC90784, STON1-GTF2A1L, TREML4, TUBB8,* and *ZNF702P*. Similarly, SVC, with the embedded approach, identified nine features: *DDX51, FDH, HEXIM1, LOC101060247, LOC90784, PKDREJ, STON1-GTF2A1L, TREML4,* and *ZFP64*. Notably, three features, *LOC90784, STON1-GTF2A1L,* and *TREML4,* were common to both algorithms and significantly contributed to African American and European American TNBC classification.

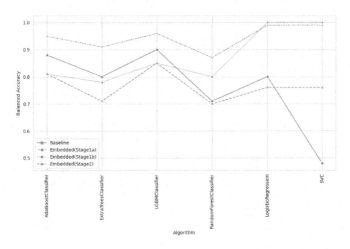

Fig. 2. The figure depicts balanced accuracy in classifying African American (AA) and European American (EA) TNBC women, achieved by various machine learning algorithms. Results stem from the full gene set baseline and top genes selected through an embedded method.

5   Conclusion

In this study, we meticulously analyzed RNAseq data from African American (AA) and European American (EA) women diagnosed with triple-negative breast cancer (TNBC). Our approach hinged on deploying cutting-edge machine learning techniques and feature selection methods, all with the goal of pinpointing the pivotal genes that differentiate these demographic groups. Employing both Filter and Wrapper methods, we extracted a succinct set of 10 vital features, consistently highlighting genes such as *LOC90784, LOC101060339, XRCC6P5,* and *TREML4*. While no singular algorithm consistently dominated, Random-ForestClassifier, SGDClassifier, and LinearSVC notably achieved a commendable 96% balanced accuracy. Further exploration with Recursive Feature Elimination unveiled minimal gene overlap compared to filter-based selection. A two-stage embedded selection approach illuminated Logistic Regression and SVC as top performers with 99% balanced accuracy, sharing *LOC90784* and *TREML4* as key contributors. These insights provide a crucial foundation for future investigations into the complex landscape of racial disparities in TNBC.

References

1. Genomic analysis of racial differences in triple negative breast cancer. Genomics **111**, 1529–1542 (2019). https://doi.org/10.1016/j.ygeno.2018.10.010
2. lazypredict (2022). https://pypi.org/project/lazypredict/
3. Anders, S., Huber, W.: Differential expression analysis for sequence count data. Genome Biol. **11** (2010). https://doi.org/10.1186/gb-2010-11-10-r106. https://genomebiology.biomedcentral.com/articles/10.1186/gb-2010-11-10-r106

4. Cho, B., et al.: Evaluation of racial/ethnic differences in treatment and mortality among women with triple-negative breast cancer. JAMA Oncol. **7**, 1016 (2021). https://doi.org/10.1001/jamaoncol.2021.1254

5. geo: Home - geo - NCBI (2019). https://www.ncbi.nlm.nih.gov/geo/

6. National Cancer Institute: The cancer genome atlas program (TCGA) - NCI (2022). https://www.cancer.gov/ccg/research/genome-sequencing/tcga

7. Kakati, T., Bhattacharyya, D.K., Kalita, J.K., Norden-Krichmar, T.M.: DEGnext: classification of differentially expressed genes from RNA-seq data using a convolutional neural network with transfer learning. BMC Bioinformatics **23** (2022). https://doi.org/10.1186/s12859-021-04527-4

8. Law, C.W., Chen, Y., Shi, W., Smyth, G.K.: voom: precision weights unlock linear model analysis tools for RNA-seq read counts. Genome Biol. **15**, R29 (2014). https://doi.org/10.1186/gb-2014-15-2-r29

9. Liu, S., et al.: Feature selection of gene expression data for cancer classification using double RBF-Kernels. BMC Bioinformatics **19** (2018). https://doi.org/10.1186/s12859-018-2400-2

10. Love, M.I., Huber, W., Anders, S.: Moderated estimation of fold change and dispersion for RNA-seq data with DESeq2. Genome Biol. **15** (2014). https://doi.org/10.1186/s13059-014-0550-8

11. Mahendran, N., Durai Raj Vincent, P.M., Srinivasan, K., Chang, C.Y.: Machine learning based computational gene selection models: a survey, performance evaluation, open issues, and future research directions. Frontiers Genet. **11** (2020). https://doi.org/10.3389/fgene.2020.603808

12. Makhani, S.S., Bouz, A., Stavros, S., Zucker, I., Tercek, A., Chung-Bridges, K.: Racial and ethnic inequality in survival outcomes of women with triple negative breast cancer. Cureus (2022). https://doi.org/10.7759/cureus.27120

13. Mori, Y., et al.: Deep learning-based gene selection in comprehensive gene analysis in pancreatic cancer. Sci. Rep. **11** (2021). https://doi.org/10.1038/s41598-021-95969-6

14. Newman, L.A., Kaljee, L.M.: Health disparities and triple-negative breast cancer in African American women. JAMA Surg. **152**, 485 (2017). https://doi.org/10.1001/jamasurg.2017.0005

15. Prakash, O., Hossain, F., Danos, D., Lassak, A., Scribner, R., Miele, L.: Racial disparities in triple negative breast cancer: a review of the role of biologic and non-biologic factors. Frontiers Public Health **8** (2020). https://doi.org/10.3389/fpubh.2020.576964

16. Robinson, M.D., McCarthy, D.J., Smyth, G.K.: edgeR: a bioconductor package for differential expression analysis of digital gene expression data. Bioinformatics **26**, 139–140 (2009). https://doi.org/10.1093/bioinformatics/btp616

17. Sahoo, B., Pinnix, Z., Sims, S., Zelikovsky, A.: Identifying biomarkers using support vector machine to understand the racial disparity in triple-negative breast cancer. J. Comput. Biol. (2023). https://doi.org/10.1089/cmb.2022.0422

18. Sahoo, B., Sims, S., Zelikovsky, A.: An SVM based approach to study the racial disparity in triple-negative breast cancer. In: Bansal, M.S., et al. (eds.) ICCABS 2021. LNB, vol. 13254, pp. 163–175. Springer, Cham (2022). https://doi.org/10.1007/978-3-031-17531-2.13

19. Siddharth, S., Sharma, D.: Racial disparity and triple-negative breast cancer in African-American women: a multifaceted affair between obesity, biology, and socioeconomic determinants. Cancers **10**, 514 (2018). https://doi.org/10.3390/cancers10120514

20. Siegel, S.D., Brooks, M.M., Lynch, S.M., Sims-Mourtada, J., Schug, Z.T., Curriero, F.C.: Racial disparities in triple negative breast cancer: toward a causal architecture approach. Breast Cancer Res. **24** (2022). https://doi.org/10.1186/s13058-022-01533-z

21. Tadist, K., Najah, S., Nikolov, N.S., Mrabti, F., Zahi, A.: Feature selection methods and genomic big data: a systematic review. J. Big Data **6**(1), 1–24 (2019). https://doi.org/10.1186/s40537-019-0241-0

Deep Learning Reveals Biological Basis of Racial Disparities in Quadruple-Negative Breast Cancer

Bikram Sahoo[1]([✉])(iD), Zandra Pinnix[2](iD), and Alex Zelikovsky[1]([✉])(iD)

[1] Department of Computer Science, Georgia State University,
Atlanta, GA 30302, USA
{bsahoo1,alexz}@gsu.edu
[2] Department of Biology and Marine Biology, University of North Carolina
at Wilmington, Wilmington, NC 28403, USA
pinnixz@uncw.edu

Abstract. Triple-negative breast cancer (TNBC) lacks crucial receptors. More aggressive is quadruple-negative (QNBC), which lacks androgen receptors. Racial disparities emerge, with African Americans facing worse QNBC outcomes. Our study deploys deep neural networks to identify QNBC ancestral biomarkers. Achieving 0.85 accuracy and 0.928 AUC, the model displays robust learning, optimized through hyperparameter tuning. Top genes are chosen via ANOVA rankings and hypothesis testing, highlighting *ABCD1* as significant post-correction. Effect sizes suggest important shifts in other genes. This approach enhances QNBC understanding, particularly racial aspects, potentially guiding targeted treatments.

Keywords: Quadruple-negative breast cancer (QNBC) · Androgen receptor (AR) · Racial Disparity · Neural Networks · Binary Classification · Feature Engineering

1 Introduction

Breast cancer's intricacies challenge oncology, notably triple-negative breast cancer (TNBC), lacking ER, PR receptors and HER2 amplification. TNBC constitutes 15–20% of US cases, with the lowest survival rates [3,21,22]. Quadruple-negative breast cancer (QNBC), an even tougher subtype, lacks TNBC's receptors and androgen receptor (AR). QNBC's aggressiveness garners research attention [7,8,19]. Studies reveal AR expression disparities in TNBC across ancestral backgrounds, favoring AR-positive TNBC in European Americans, while African Americans tend towards AR-negative expression [5,11]. Aggressive AR-negative TNBC (QNBC) is prominent in African American women, manifesting distinct molecular subtypes like BL1, BL2, and IM. These findings drive investigation into AR expression, QNBC's aggression, and racial ancestry, exploring

QNBC's clinical characteristics in African American and European American women [2,4,10]. Contemporary investigations into quadruple-negative breast cancer (QNBC) have unveiled certain biomarkers with the aim of comprehending the racial disparities existing within African American (AA) and European American (EA) populations [4,5,18]. Nevertheless, the quest for a pivotal biomarker within AA women, showing pronounced deregulation compared to their EA counterparts in the context of QNBC, remains a pressing need to address this racial imbalance. To confront this challenge, computational biologists can harness the wealth of open-source RNA-seq gene expression data, available from repositories like the Gene Expression Omnibus (GEO) [6] and The Cancer Genome Atlas (TCGA) [9]. This can be achieved by leveraging robust machine learning (ML) and deep learning (DL) supervised classification algorithms [14,16,17], which are increasingly replacing traditional parametric and non-parametric statistical methods based on logarithmic values of fold change (logFC) for differential gene expression analysis due to their high rates of false positive and false negative predictions [1,13,15,20]. ML approaches enable gene classification based on their expression patterns, while support vector machines (SVM) and logistic regression (LR) discern between colon cancer and other cancer types. In addition, random forest (RF) algorithms effectively classify genes in microarray data. Notably, deep learning (DL) with transfer learning has emerged as a potent tool for classifying novel data by learning intricate relationships within training data features in an integrated system. Convolutional neural networks (CNNs), a subtype of DL, capitalize on convolutional techniques within internal layers, enabling efficient computation, nonlinear relationship learning, and feature extraction across both image and non-image data. CNNs, exemplified by tools like DeepInsight, are adept at classification tasks, utilizing hierarchical filtering, weight sharing, and neighborhood information to excel in feature extraction from gene expression and text data [12,23].

This study aims to identify pivotal biomarkers for unraveling racial disparities between European American (EA) and African American (AA) women with QNBC, employing advanced deep neural network technology. Curating RNA-seq data involves strategies like oversampling to tackle class imbalance. Meticulous hyperparameter tuning optimizes the neural network, while Lasso regularization enhances predictive power by highlighting essential features. ANOVA rankings guide the selection of top features, validated through statistical tests. Metrics systematically evaluate model performance, addressing imbalanced data and refining configurations. Statistical tests ascertain feature significance in distinguishing EA and AA classes. This methodology targets data imbalance, model refinement, and performance evaluation, with potential to address racial disparities in quadruple-negative breast cancer among AA and EA women.

2 Data

We analyzed gene expression data from breast cancer patients, utilizing open-source RNA sequencing with log2 median-centering. Processed RNA-seq data was sourced from [6,9]. Breast cancer subtypes are distinguished by hormone-receptor status, assessed via immunohistochemistry. Our focus was on a subtype, quadruple-negative breast cancer (QNBC), a subset of triple-negative breast cancer (TNBC). We filtered TNBC samples, selecting those negative for ER, PR, HER2 receptors, or HER2 equivocal by FISH. We also identified AR-negative samples using median AR gene expression. This yielded 64 samples (39 EA, 25 AA) from 128 TNBC samples, revealing differences between European American (EA) and African American (AA) women. The dataset comprises 20,531 genes or features.

3 Methods

3.1 Data Preprocessing

The raw dataset contains features in string format and target labels encoded as 'EA' and 'AA'. In the data preprocessing stage, the string features are converted to numeric float values using a feature conversion function, denoted as F_{conv}. The function F_{conv} maps each unique string value to a unique numeric float value, thus ensuring compatibility with machine learning algorithms that require numerical inputs.

To encode the target labels as integers, the LabelEncoder class is utilized. This class assigns the label 'EA' to the integer 0 and the label 'AA' to the integer 1. Mathematically, the encoding process can be represented as follows:

$$\text{Encoded Target Label} = \begin{cases} 0 & \text{if target label is 'EA'} \\ 1 & \text{if target label is 'AA'} \end{cases} \quad (1)$$

After preprocessing the raw data, stratification is employed to split the dataset into training and testing sets. Stratification ensures that the class distribution in both sets is representative of the original dataset. Specifically, the data is split into 80% training and 20% testing sets while preserving the relative proportion of 'EA' and 'AA' labels. This is expressed mathematically as:

Training Set,

$$D_{\text{train}} = \{(\mathbf{x}_i, y_i) \mid \mathbf{x}_i \text{ is a feature vector and } y_i \text{ is the encoded target label}\} \quad (2)$$

where $i = 1, 2, ..., \lfloor 0.8 \times N \rfloor$
Testing Set,

$$D_{\text{test}} = \{(\mathbf{x}_j, y_j) \mid \mathbf{x}_j \text{ is a feature vector and } y_j \text{ is the encoded target label}\}, \quad (3)$$

where $j = \lfloor 0.8 \times N \rfloor + 1, \lfloor 0.8 \times N \rfloor + 2, ..., N$ (N is the total number of samples in the dataset.)

To address the imbalanced distribution of the two classes, the ADASYN (Adaptive Synthetic) oversampling technique is applied to the training set. ADASYN generates synthetic examples for the minority class (label 'AA') by interpolating between existing samples. The synthetic examples are added to the training set, effectively increasing the number of 'AA' samples. This helps prevent the model from biasing towards the majority class, enabling it to learn robust patterns from both classes. Mathematically, the ADASYN oversampling process can be expressed as:

$$D_{\text{train\_oversampled}} = \text{ADASYN}(D_{\text{train}}) \tag{4}$$

where $D_{\text{train\_oversampled}}$ is the training set after applying ADASYN.

Overall, the encoding, stratification, and ADASYN oversampling steps facilitate effective preprocessing of the raw data into formats suitable for modeling and evaluation in an imbalanced binary classification problem. The validation split, separating the data into training and testing sets, protects against information leakage and enables a rigorous assessment of the model's generalization performance on unseen data.

3.2   Model Architecture and Training

A deep neural network classifier is built using the Keras API with Tensorflow as the backend. The model architecture consists of densely-connected or fully-connected layers, mathematically represented as:

$$\text{Layer}_i = \text{ReLU}(\mathbf{W}_i \cdot \text{Layer}_{i-1} + \mathbf{b}_i) \tag{5}$$

where Layer_i is the output of layer i, $\mathbf{W}_i$ is the weight matrix for layer i, $\mathbf{b}_i$ is the bias vector for layer i, and $\text{ReLU}(\cdot)$ is the Rectified Linear Unit activation function, defined as $\text{ReLU}(x) = \max(0, x)$.

To prevent overfitting, dropout regularization is applied after each hidden layer. Dropout randomly sets a fraction of the activations to zero during training. Mathematically, dropout can be expressed as:

$$\text{Layer}'_i = \text{Dropout}(\text{Layer}_i, \text{dropout\_rate}) \tag{6}$$

where Layer'_i is the output of layer i after dropout, and dropout_rate is the fraction of activations to drop, typically in the range of 0.2 to 0.7.

The output layer contains a single node with a sigmoid activation function to generate probabilities for the binary classification task. The sigmoid activation maps the output to the range $[0, 1]$, representing the probability that the input belongs to class 1 (positive class).

To train the neural network, the ADAM optimization algorithm is employed with binary cross-entropy loss. ADAM is an adaptive learning rate optimization algorithm that efficiently updates the model weights during training. Binary cross-entropy loss measures the dissimilarity between predicted probabilities and true labels for binary classification tasks.

Optimal model architecture and hyperparameters are determined through grid search k-fold stratified cross-validation. Hidden layers, dropout rate, and learning rate are tuned for the highest area under the ROC curve (AUC) on validation data. AUC reflects model's discrimination ability between classes. Dropout and AUC-based selection curb overfitting, enabling effective pattern learning with a smaller dataset. The approach aims for robust binary classification and strong predictive performance.

3.3   Feature Selection and Analysis

To identify the most predictive features from the high-dimensional dataset, a combination of filtering and statistical testing methods are employed. First, highly correlated features are removed by analyzing the correlation matrix and dropping variables with correlations exceeding 0.9. Mathematically, the correlation coefficient between two features X_i and X_j is given by:

$$\text{Corr}(X_i, X_j) = \frac{\text{Cov}(X_i, X_j)}{\sigma(X_i)\sigma(X_j)} \tag{7}$$

where $\text{Cov}(X_i, X_j)$ is the covariance between X_i and X_j, and $\sigma(X_i)$ and $\sigma(X_j)$ are the standard deviations of X_i and X_j respectively.

This process eliminates redundant features, as highly correlated features often convey similar information.

Next, the training data is oversampled with SMOTEENN (SMOTE + Edited Nearest Neighbors) to balance classes for the next stage. SMOTE generates synthetic examples for the minority class by interpolating between existing samples, while Edited Nearest Neighbors removes noisy samples from the dataset. The combination of SMOTE and Edited Nearest Neighbors effectively increases the number of minority class instances, enhancing the model's ability to learn from both classes.

Lasso regularization provides an initial filtering of features by removing those with zero coefficients, identifying a subset of important variables. Mathematically, the Lasso regularization term is defined as:

$$\text{Lasso}_\lambda(\beta) = \sum_{i=1}^{N}(y_i - \mathbf{X}_i \cdot \beta)^2 + \lambda \sum_{j=1}^{p} |\beta_j| \tag{8}$$

where N is the number of samples, p is the number of features, y_i is the target variable, $\mathbf{X}_i$ is the feature vector for sample i, β is the vector of regression coefficients, and λ is the regularization parameter. By setting some coefficients to zero, Lasso performs feature selection.

ANOVA F-values from SelectKBest are then used to rank features and select the top 50. ANOVA assesses the variance between class means and within class variances, providing a measure of feature significance. SelectKBest selects the top k features based on the highest F-values.

The predictive value of the selected features is statistically evaluated using several tests. Cohen's d effect size quantifies the degree of difference between the class means for each feature. Mathematically, Cohen's d is defined as:

$$\text{Cohen's d} = \frac{\text{Mean}_1 - \text{Mean}_2}{\text{Pooled Standard Deviation}} \tag{9}$$

where Mean_1 and Mean_2 are the means of the feature values for the positive and negative classes, and the pooled standard deviation accounts for variability within both classes.

Mann-Whitney U test gauges significant feature distribution differences in binary classes. Non-normally distributed data suits this rank-sum test. False discovery rate correction prevents inflated significance due to multiple testing. The approach yields informative feature ranking and quantifies their discriminative power.

3.4   Model Evaluation

The performance of the final model is comprehensively evaluated on the held-out test set to assess its real-world generalization ability. Classification metrics, including accuracy, precision, recall, F1 score, and AUC, are reported to provide a multi-faceted view of model performance across different thresholds.

The accuracy of the model is defined as the ratio of correctly classified samples to the total number of samples in the test set:

$$\text{Accuracy} = \frac{\text{Number of Correctly Classified Samples}}{\text{Total Number of Samples}} \tag{10}$$

Precision represents the ability of the model to correctly identify positive class samples among all samples predicted as positive:

$$\text{Precision} = \frac{\text{True Positives}}{\text{True Positives} + \text{False Positives}} \tag{11}$$

Recall, also known as sensitivity or true positive rate, quantifies the model's ability to identify positive class samples among all actual positive samples:

$$\text{Recall} = \frac{\text{True Positives}}{\text{True Positives} + \text{False Negatives}} \tag{12}$$

The F1 score is the harmonic mean of precision and recall, providing a balanced assessment of the model's performance on the positive class:

$$\text{F1 Score} = 2 \times \frac{\text{Precision} \times \text{Recall}}{\text{Precision} + \text{Recall}} \tag{13}$$

AUC (Area Under the ROC Curve) quantifies class discrimination across thresholds, computed by plotting True Positive Rate (Recall) against False Positive Rate (1 - Specificity). AUC ranges from 0 to 1, with higher values indicating better classification. Model architecture and optimal hyperparameters, including hidden layers, dropout rate, and learning rate, are revealed through cross-validation. Rigorous evaluation on untouched test data confirms reported performance's validity for new data. This practice adheres to industry benchmarks, ensuring robust model evaluation and informed data-driven decisions.

4   Results

4.1   Performance Evaluation of the Deep Learning Model

The optimal architecture of the deep learning model demonstrates remarkable performance in classifying quadruple-negative breast cancer (QNBC) cases among European American (EA) and African American (AA) populations. High consistency in precision, recall, and F1-scores is observed, with EA achieving 0.83 and AA achieving 0.86 (Table 1). These results yield an impressive overall accuracy of 0.85. The model's balanced classification ability is supported by macro and weighted F1-scores of 0.85. Furthermore, with an AUC of 0.928 on the test set, the model effectively distinguishes between the two classes. Characterized by two dense layers and a dropout layer, the architecture captures intricate data patterns, showcasing potential for accurate prediction. This well-designed model holds promise for advancing quadruple-negative breast cancer classification.

Table 1. Performance Metrics for Class Labels EA and AA

Class Label	Precision	Recall	F1-score
EA	0.83	0.83	0.83
AA	0.86	0.86	0.86

4.2   Identification of Key Molecular Features (Genes)

Using the SelectKBest feature selection technique, a subset of the top 50 genes is unveiled (Fig. 1), derived from ANOVA F-values. F-scores gauge gene expression mean differences, higher scores indicating better classification discrimination. Notably, *ACADL* scores 22, and the 50th gene *LOC100130426* scores 1.6. F-scores over 2–3 signify informative genes. This gene list, with scores, highlights key molecular attributes influencing classification. Exploring biological functions and pathways of top genes could reveal insights into differentiating European American (EA) and African American (AA) QNBC women. Scores quantify gene predictive power, assessing significance within the analytical framework.

4.3   Statistical Validation of Selected Features (Genes)

The top 50 genes were subjected to statistical scrutiny to assess their discriminatory capacity between European American (EA) and African American (AA) QNBC women. Employing the non-parametric Mann-Whitney U test,

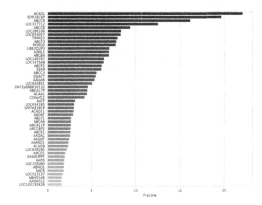

Fig. 1. Top 50 selected features with their corresponding F-scores, determined through a feature selection process for European American (EA) and African American (AA) QNBC women. Higher F-scores indicate greater influence on the predictive model.

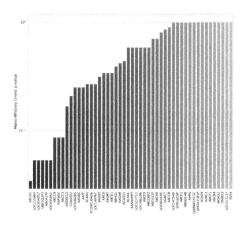

Fig. 2. This figure displays U-test p-values for top features selected via deep learning-based choice, comparing EA and AA QNBC women. It plots genomic attributes (x-axis) against p-values (y-axis) with a log scale for enhanced visualization.

we examined gene expression distribution disparities between groups. With an alpha of 0.05, solely the *ABCD1* gene exhibited significant class differentiation post-FDR correction (FDR-adjusted p = 0.03), shown in Fig. 2. Cohen's d effect size analysis unveiled predominantly small to medium effects ($|d| < 0.5$), while *LOC654057*, *UBE2Q2P3*, *LOC645851*, and *ABCD1* displayed larger magnitudes (> 0.7) in Fig. 3. Notably, while *ABCD1* surpassed the significance threshold after multiple comparisons, effect sizes suggest biologically significant changes in select genes, warranting further exploration of their functional roles. Statistical testing objectively evaluates predictive genes arising from machine learning analyses.

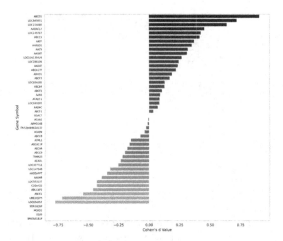

Fig. 3. This figure displays Cohen's d effect sizes, indicating feature importance for EA and AA differentiation. Gene symbols (y-axis) denote attributes by descending values (x-axis). Clear y-labels, gridlines, and compact design enhance insights into class distinction.

5 Conclusion

This study showcases the synergy of integrated machine learning and statistical testing, revealing molecular factors behind racial disparities in quadruple-negative breast cancer. The deep neural network achieves impressive accuracy, balancing precision, recall, and AUC exceeding 0.9. This signifies robust learning in distinguishing African American and European American QNBC tumors. Hyperparameter optimization yields a concise yet potent model architecture. Rigorous statistical analyses validate top genes from ANOVA selection. While *ABCD1* alone survives multiple hypothesis correction, other genes show meaningful shifts. The study exemplifies combining deep learning's predictive strength with hypothesis testing's objectivity, deriving insightful knowledge from complex biomedical data.

Reliable biomarkers result from rigorous evaluation, class imbalance handling, and standard feature engineering. Expanding datasets, time points, and network analyses could enhance insights. Integrating advanced analytics, statistical rigor, and clinical relevance advances QNBC understanding, potentially aiding prognosis, diagnostics, and treatments considering European American and African American QNBC differences.

References

1. Anders, S., Huber, W.: Differential expression analysis for sequence count data. Genome Biol. **11** (2010). https://doi.org/10.1186/gb-2010-11-10-r106. https://genomebiology.biomedcentral.com/articles/10.1186/gb-2010-11-10-r106
2. Angajala, A., et al.: Quadruple negative breast cancers (QNBC) demonstrate subtype consistency among primary and recurrent or metastatic breast cancer. Transl. Oncol. **12**, 493–501 (2019). https://doi.org/10.1016/j.tranon.2018.11.008
3. Cho, B., et al.: Evaluation of racial/ethnic differences in treatment and mortality among women with triple-negative breast cancer. JAMA Oncol. **7**, 1016 (2021). https://doi.org/10.1001/jamaoncol.2021.1254
4. Davis, M., et al.: AR negative triple negative or "quadruple negative" breast cancers in African American women have an enriched basal and immune signature. PloS One **13**, e0196909 (2018)
5. Gasparini, P., et al.: Androgen receptor status is a prognostic marker in non-basal triple negative breast cancers and determines novel therapeutic options. PLoS ONE **9**, e88525 (2014). https://doi.org/10.1371/journal.pone.0088525
6. geo: Home - geo - NCBI (2019). https://www.ncbi.nlm.nih.gov/geo/
7. Hon, J.: Breast cancer molecular subtypes: from TNBC to QNBC (2016)
8. Huang, M., Wu, J., Ling, R., Li, N.: Quadruple negative breast cancer. Breast Cancer **27**(4), 527–533 (2020). https://doi.org/10.1007/s12282-020-01047-6
9. National Cancer Institute: The cancer genome atlas program (TCGA) - NCI (2022). https://www.cancer.gov/ccg/research/genome-sequencing/tcga
10. Jinna, N., et al.: Racial disparity in quadruple negative breast cancer: aggressive biology and potential therapeutic targeting and prevention. Cancers **14**, 4484 (2022). https://doi.org/10.3390/cancers14184484. https://www.mdpi.com/2072-6694/14/18/4484
11. Jovanović, B., et al.: A randomized phase II neoadjuvant study of cisplatin, paclitaxel with or without everolimus in patients with stage II/III triple-negative breast cancer (TNBC): responses and long-term outcome correlated with increased frequency of DNA damage response gene mutations, TNBC subtype, AR status, and KI67. Clin. Cancer Res. Official J. Am. Assoc. Cancer Res. **23**, 4035–4045 (2017). https://doi.org/10.1158/1078-0432.CCR-16-3055. https://pubmed.ncbi.nlm.nih.gov/28270498/
12. Kakati, T., Bhattacharyya, D.K., Kalita, J.K., Norden-Krichmar, T.M.: DEGnext: classification of differentially expressed genes from RNA-seq data using a convolutional neural network with transfer learning. BMC Bioinformatics **23** (2022). https://doi.org/10.1186/s12859-021-04527-4
13. Law, C.W., Chen, Y., Shi, W., Smyth, G.K.: voom: precision weights unlock linear model analysis tools for RNA-seq read counts. Genome Biol. **15**, R29 (2014). https://doi.org/10.1186/gb-2014-15-2-r29
14. Liu, S., et al.: Feature selection of gene expression data for cancer classification using double RBF-kernels. BMC Bioinformatics **19** (2018). https://doi.org/10.1186/s12859-018-2400-2
15. Love, M.I., Huber, W., Anders, S.: Moderated estimation of fold change and dispersion for RNA-seq data with DESeq2. Genome Biol. **15** (2014). https://doi.org/10.1186/s13059-014-0550-8
16. Mahendran, N., Durai Raj Vincent, P.M., Srinivasan, K., Chang, C.Y.: Machine learning based computational gene selection models: a survey, performance evaluation, open issues, and future research directions. Frontiers Genet. **11** (2020). https://doi.org/10.3389/fgene.2020.603808

17. Mori, Y., et al.: Deep learning-based gene selection in comprehensive gene analysis in pancreatic cancer. Sci. Rep. **11** (2021). https://doi.org/10.1038/s41598-021-95969-6

18. Muhammad, A., et al.: Potential epigenetic modifications implicated in triple- to quadruple-negative breast cancer transition: a review. Epigenomics **14**, 711–726 (2022). https://doi.org/10.2217/epi-2022-0033

19. Newman, L.A., et al.: Hereditary susceptibility for triple negative breast cancer associated with western sub-Saharan African ancestry. Ann. Surg. **270**, 484–492 (2019). https://doi.org/10.1097/sla.0000000000003459

20. Robinson, M.D., McCarthy, D.J., Smyth, G.K.: edger: a bioconductor package for differential expression analysis of digital gene expression data. Bioinformatics **26**, 139–140 (2009). https://doi.org/10.1093/bioinformatics/btp616

21. Sahoo, B., Pinnix, Z., Sims, S., Zelikovsky, A.: Identifying biomarkers using support vector machine to understand the racial disparity in triple-negative breast cancer. J. Comput. Biol. (2023). https://doi.org/10.1089/cmb.2022.0422

22. Sahoo, B., Sims, S., Zelikovsky, A.: An SVM based approach to study the racial disparity in triple-negative breast cancer. In: Bansal, M.S., et al. (eds.) ICCABS 2021. LNB, vol. 13254, pp. 163–175. Springer, Cham (2022). https://doi.org/10.1007/978-3-031-17531-2.13

23. Tadist, K., Najah, S., Nikolov, N.S., Mrabti, F., Zahi, A.: Feature selection methods and genomic big data: a systematic review. J. Big Data **6**(1), 1–24 (2019). https://doi.org/10.1186/s40537-019-0241-0

CSA-MEM: Enhancing Circular DNA Multiple Alignment Through Text Indexing Algorithms

André Salgado[1,2]($\boxtimes$) iD, Francisco Fernandes[1]($\boxtimes$) iD,
and Ana Teresa Freitas[1,2]($\boxtimes$) iD

[1] Instituto de Engenharia de Sistemas e Computadores: Investigação e
Desenvolvimento (INESC -ID), R. Alves Redol 9, 1000-029 Lisbon, Portugal
andre.salgado@inesc-id.pt, francisco.fernandes@ist.utl.pt,
ana.freitas@tecnico.ulisboa.pt
[2] Instituto Superior Técnico (IST/UL), Av. Rovisco Pais 1, 1049-001 Lisbon,
Portugal

Abstract. In the realm of Bioinformatics, the comparison of DNA
sequences is essential for tasks such as phylogenetic identification, comparative genomics, and genome reconstruction. Methods for estimating sequence similarity have been successfully applied in this field. The
application of these methods to circular genomic structures, common in
nature, poses additional computational hurdles. In the advancing field of
metagenomics, innovative circular DNA alignment algorithms are vital
for accurately understanding circular genome complexities. Aligning circular DNA, more intricate than linear sequences, demands heightened
algorithms due to circularity, escalating computation requirements and
runtime. This paper proposes CSA-MEM, an efficient text indexing algorithm to identify the most informative region to rotate and cut circular
genomes, thus improving alignment accuracy. The algorithm uses a circular variation of the FM-Index and identifies the longest chain of non-
repeated maximal subsequences common to a set of circular genomes,
enabling the most adequate rotation and linearisation for multiple alignment. The effectiveness of the approach was validated in five sets of mitochondrial, viral and bacterial DNA. The results show that CSA-MEM
significantly improves the efficiency of multiple sequence alignment, consistently achieving top scores compared to other state-of-the-art methods. This tool enables more realistic phylogenetic comparisons between
species, facilitates large metagenomic data processing, and opens up new
possibilities in comparative genomics.

Keywords: Circular DNA · Multiple Alignment · Text Indexing

1 Introduction

Recent advances in metagenomics have propelled the field to new frontiers, allowing researchers to explore microbial communities with unprecedented depth and

X. Guo et al. (Eds.): ISBRA 2023, LNBI 14248, pp. 509–517, 2023.
https://doi.org/10.1007/978-981-99-7074-2_41

breadth [13]. However, as datasets become larger and more intricate, the challenge of addressing circular DNA alignment, efficient data processing, accurate taxonomic classification, and integration of multiomics data is of paramount importance [11]. Overcoming these challenges will not only enhance our understanding of microbial ecosystems, but also pave the way for innovative applications in fields such as biotechnology, environmental science, and personalised medicine [6].

With the expansion in scope and scale of the metagenomics field comes the urgent need to tackle the challenges posed by the analysis of large metagenomic datasets. One of the main challenges is the efficient alignment of circular DNA within these complex datasets. Circular genomes are common in many microorganisms, and accurately aligning them is crucial for understanding their structure and function. Circular DNA molecules, known as plasmids, play an important role in conferring adaptive advantages, such as antibiotic resistance and virulence, to bacteria [5]. Circular mitochondrial DNA plays an essential role in the survival and energy production of eukaryotic cells [18] and has long been used for phylogenetic analyses [17]. Smaller structures known as extrachromosomal circular DNA are considered a hallmark of genomic flexibility in eukaryotes [21]. Furthermore, the circular nature of DNA in some viruses has a major impact on their replication and infection strategies [20]. Understanding these complex mechanisms is crucial for the development of antiviral therapies and vaccines [23].

Existing DNA alignment algorithms, often designed for linear genomes, such as ClustalW [19], may struggle to handle the unique characteristics of circular DNA, leading to misinterpretations and inaccuracies in the analysis [22]. Furthermore, as metagenomic datasets increase in size and complexity, issues related to data management and analysis, processing speed, and computational resources become increasingly pressing. Efficient algorithms are needed to address challenges related to assembly quality, binning, and functional annotation, which are vital for extracting meaningful biological information from the sheer volume of metagenomic data [16].

The special importance of circular DNA presents a unique challenge in comparing its sequences. Because circular sequences can start from any point, it adds complexity. This distinctiveness feature makes traditional linear-centric multiple alignment algorithms inadequate because they lack the adaptability to effectively cope with the inherent circular structure. The outcome is a potential loss of critical genetic information during the alignment process [9].

Beyond identifying the optimal rotation for each sequence in a multiple circular DNA alignment, it is also necessary to address the challenge of handling large volumes of data. In this context, it is necessary to develop new algorithms that provide better solutions in terms of space and time efficiency.

Efforts to reconcile the circular-to-linear disparity have given rise to remarkable methods. Cyclope [15] is a software designed to enhance the alignment of multiple circular sequences. However, the cubic runtime of the pairwise alignment step becomes a limiting factor in practical scenarios. CSA [9] is an algorithm based on a circular version of a generalised suffix tree. The algorithm identifies the largest chain of non-repeated longest subsequences common to a set of circular DNA sequences to determine their optimal rotations. Although very efficient, it is limited to 32 input sequences and relies on an outdated suffix tree data structure.

Other types of methods include BEAR [2], which extends existing algorithms for circular and fixed-length approximate string matching [3]. It calculates edit distances and rotations between all pairs of sequences and then uses agglomerative hierarchical clustering to determine the most suitable rotations. A similar and more recent approach, MARS [1], is an heuristic method that computes all pairwise cyclic edit distances using a distance measurement algorithm based on q-grams [11]. It then performs classic progressive alignment of sequence profile pairs using a guide tree to refine the rotations. However, this progressive nature and the dependency on dynamic programming algorithms may render these last methods slower and less efficient when dealing with longer sequences or larger datasets.

1.1 Contribution

To effectively tackle the challenges associated with circular multiple sequence alignment on large datasets, we introduce CSA-MEM, in which we propose: (1) the adaptation of advanced data structures such as the FM-Index [10], namely a circular modification based on the implementation used in slaMEM [8] to achieve a computationally efficient exact solution for circular sequence matching, and (2) an effective identification of the longest chain of non-repeated maximal exact matches (MEMs) common to a set of circular DNA sequences, in an approach similar to the CSA tool [9]. This way, the CSA-MEM algorithm allows for a seamless rotation and linearisation process for multiple circular alignment purposes.

2 Methods

2.1 Basic Notions

We generally consider that a text is appended with a terminator sentinel symbol '$' which is lexicographically smaller than all other characters in its alphabet Σ. For a string T of length n, the Suffix Array (SA) [14] of T is an array of integers $SA[1, n]$ where each element $SA[i]$ corresponds to the starting position in T of the i-th lexicographically smallest suffix of the string. $SA[i]$ points to the position in string T where the suffix with lexicographic rank i begins.

The Longest Common Prefix array LCP is an integer array of length n which stores information about the length of the longest common prefixes (lcp) between consecutive suffix pairs in SA [12]. It is defined as $LCP[i] = lcp(T[SA[i - 1], n], T[SA[i], n])$ for $i \neq 1$ and 0 otherwise.

The Burrows-Wheeler Transform (BWT) [4] is a data structure that consists of a reversible transformation that rearranges the original characters of T into a new string more suitable to text processing and data compression methods. In the conceptual matrix of all the lexicographically sorted rotations of a string, the BWT matches its last column L. This corresponds to the character immediately preceding each suffix starting at position $SA[i]$ in the string T and is formally defined as $L[i] = T[SA[i] - 1]$ when $SA[i] \neq 1$ and $L[i] = \$$ otherwise.

The FM-Index [10] is an indexing data structure built on top of the BWT which can be used to search and process large volumes of text efficiently. In addition to the BWT and SA arrays, the FM-index also uses a summation array which stores the *rank* of each character in the alphabet, meaning the number of occurrences of character c in the BWT up to position i, and represented by the function $rank_c(L, i)$. The inverse operation, $select_c(L, j)$, returns the position i in the BWT corresponding to the j-th occurrence of character c. To *locate* and *count* the occurrences of a specific pattern P, the FM-index employs a backward search strategy and maintains two pointers, identifying the start and end index positions of runs of consecutive suffixes starting with the current matched string, which are updated by iteratively applying the *LF-mapping* procedure [10].

Maximal Exact Matches (MEMs) are substrings which simultaneously belong to both a reference text T and a query text R, and that cannot be extended in either direction without producing a mismatch, i.e. $R[i, j] = Q[i', j']$ and $R[i-1] \neq Q[i'-1] \wedge R[j+1] \neq Q[j'+1]$. This type of substrings is often used in genomic comparison as they provide common blocks between the sequences that can be used as anchors to detect similar regions in the alignments. Such MEMs can be retrieved by matching query Q over the FM-Index of text R using an efficient algorithm, such as slaMEM [8].

2.2   Circular FM-Index

In the pursuit of identifying the optimal rotation for the circular DNA sequence alignment, CSA-MEM strategically leverages MEMs across multiple sequences, akin to the procedural essence of the slaMEM algorithm [8]. To seamlessly accommodate circularity, substantial changes were made to the underlying data structures within the algorithmic framework. Both SA and BWT were subject to meticulous modifications, enabling the uninterrupted searching of sequences even upon encountering the predefined terminal points of the reference sequence. This pivotal enhancement empowers the method to sustain its search for substrings beyond sequence boundaries, deeply enriching its circular DNA analysis capabilities.

The circular nature of the DNA sequence was considered by ignoring the terminator character present in standard SA. The modified SA retains the original order of all non-rotated substrings, thus enabling the construction of a circular BWT, an essential step for the efficient MEM detection.

The circular BWT also omits the terminator character from the transform, avoiding complications arising from circularity, but maintaining unaffected the FM-Index algorithm retaining its efficiency.

An example is shown in Fig. 1 displaying the regular linear BWT for the text $TGCCTTTG\$$ and its circular version without the terminator symbol. In this example, the BWT is $GGCTTT\$TC$ and the circular BWT is $GCTTGTTC$. An illustrative example of sequence matching can also be found in the project's repository.

SA	F							L	LCP
9	$	T	G	C	C	T	T	T	G	0
3	C	C	T	T	T	G	$	T	G	0
4	C	T	T	T	G	$	T	G	C	1
8	G	$	T	G	C	C	T	T	T	0
2	G	C	C	T	T	T	G	$	T	1
7	T	G	$	T	G	C	C	T	T	0
1	T	G	C	C	T	T	T	G	$	2
6	T	T	G	$	T	G	C	C	T	1
5	T	T	T	G	$	T	G	C	C	2

SA	F						L	LCP
3	C	C	T	T	T	G	T	G	0
4	C	T	T	T	G	T	G	C	1
2	G	C	C	T	T	T	G	T	0
8	G	T	G	C	C	T	T	T	1
1	T	G	C	C	T	T	T	G	0
7	T	G	T	G	C	C	T	T	2
6	T	T	G	T	G	C	C	T	1
5	T	T	T	G	T	G	C	C	2

Fig. 1. Linear and circular versions of rotations, SA, LCP and BWT for the example string '*TGCCTTTG$*'.

2.3 Most Significant Common Subsequence Chain of MEMs

As explained in the CSA paper [9], the key method to determine optimal rotations for aligning a set of sequences involves identifying the longest chain of shared subsequences present in all the data. This is crucial, as it marks the most informative region universally present across all sequences.

Since MEMs are common subsequences of different sequences, to unearth the most significant chain for circular alignment within these sequences, a pivotal strategy lies in extracting the longest contiguous chain of MEMs that is shared across all of them. Initiating this process involves iterative removal of the smallest coincident MEMs from each sequence individually, since a subset of the sequence characters can be present in any number of parts of another sequence. This strategic pruning serves to eliminate less impactful substring matches, thereby directing focus towards more substantial segments that substantially contribute to a more adequate rotation. It should be noted that the resultant MEMs do not exhibit any overlapping characteristics. This inherent attribute simplifies the subsequent analysis stage, which ensures a distinct demarcation between the common MEMs within the sequences. For a more comprehensive understanding, the pseudo-code is available in the project's repository at GitHub.

3 Results and Discussion

In order to evaluate our approach, we conducted tests on five distinct datasets. The first three sets contained mitochondrial DNA (mtDNA) for 16 primates, 12 mammals, and a more diverse set of 19 sequences. This third set is composed of a mixture between the 16 primates and the addition of 3 more distant evolutionary species (Drosophila melanogaster, Gallus gallus, and Crocodylus niloticus). The fourth set contains viral DNA in the form of circular genomes of 9 viruses recently discovered in a small metagenomics study [7]. The fifth set was used to evaluate the tool's efficiency on extensive datasets, in this case consisting on 35 sequences of Escherichia coli (E. coli). The first two datasets were previously used to benchmark CSA, BEAR and MARS in their respective works. The third dataset was also used in both CSA and BEAR. Some statistics of each of these datasets are available in Table 1.

Table 1. General properties of each one of the datasets used in the benchmarks.

Dataset	Number of sequences	Average size (bp)	Total size (KB)
Mammals	12	16,777	204
Primates	16	16,581	269
Diverse	19	16,759	323
Viruses	9	3,130	29
E. coli	35	5,703,382	190,370

The quality of the rotations was assessed using linear multiple sequence alignment, by feeding the rotated sequences into ClustalW to obtain the alignment score. An example of the multiple sequence alignment produced by ClustalW on one of the datasets, before and after rotation with CSA-MEM is presented in Fig. 2. The gaps are represented in blue, and alignment conservation increases from green to red. The sequence start and end positions meet at the top centre of their circular representations. The gaps and unmatched portions at both loose ends of the unrotated sequences are clearly visible in the first image, after which they fade away and fit together to produce a much more meaningful alignment after the sequences are rotated, in the second image.

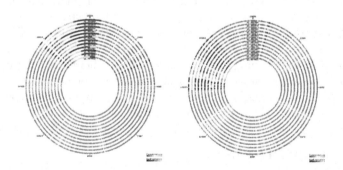

Fig. 2. Circular representation of the multiple sequence alignment output of one of the datasets before (left) and after rotating (right) the sequences with CSA-MEM.

The size of the consensus alignment profile was captured, since a more compact consensus usually corresponds to fewer gaps and a better alignment. Time and memory requirements were also measured by capturing the wall-clock running time and the peak resident-set memory size, using Linux shell scripts. All benchmark tests were conducted on a dedicated server system equipped with an Intel (R) Xeon(R) Silver 4214R CPU @ 2.40 GHz with 256 GB of RAM running the Linux Ubuntu 20.04.4 LTS x86_64 distribution, ensuring a stable and consistent platform for all experiments.

3.1   Benchmarks

CSA-MEM was evaluated against CSA, Cyclope, and MARS on the five mentioned datasets using the described metrics to assess their respective performance. Table 2 presents the results for the first four datasets. It was not possible to test BEAR algorithm due to compilation errors.

Table 2. Comparative benchmarks between the four analysed circular sequence alignment software tools on four distinct datasets considering different metrics.

		Alignment score	Consensus length (bp)	Running time (s)	Memory used (MB)
Primates	CSA-MEM	**10474101**	17446	1	**30**
	CSA	10471600	**17444**	1	91
	Cyclope	10472551	17447	2820	4670
	MARS	10468302	17458	215	1183
Mammals	CSA-MEM	**5067971**	18608	<1	**30**
	CSA	5049488	**18202**	<1	75
	Cyclope	5020605	18268	1620	4797
	MARS	5033469	18460	168	1340
Diverse	CSA-MEM	13002485	19784	7	**30**
	CSA	12942609	**19751**	<1	103
	Cyclope	**13132115**	19889	4260	5419
	MARS	13031632	20172	342	1562
Viruses	CSA-MEM	259034	4005	10	**30**
	CSA	248401	3980	**<0.5**	33
	Cyclope	**276949**	**3963**	34	340
	MARS	257741	4051	6	96

On the smaller datasets, CSA-MEM achieves the highest alignment scoring results in the first and second datasets, while being the second best in the remaining ones. Although Cyclope shows good scoring results, the computational costs are prohibitive for it to be used with current large data volumes. For instance, in the Diverse dataset the requirements of Cyclope in terms of both time and space are more than 1000x those of CSA-MEM. In terms of alignment consensus size, CSA consistently produces the most compact alignments in most tests. This means that the chaining algorithm used in CSA preserves more common blocks than the one used in CSA-MEM and the other tools.

CSA and CSA-MEM are the fastest algorithms due to their reliance on text indexing data structures and their strategy based on common substrings. In some cases, CSA achieves better running times than CSA-MEM, possibly due to the more elaborate and time-consuming construction step of the FM-Index in CSA-MEM compared to the construction of the suffix tree variation in CSA. Both

MARS and Cyclope are less efficient due to the use of time-consuming dynamic programming and progressive pairwise alignment algorithms. CSA-MEM memory consumption is always around 30MB, indicating that this should be the overhead for maintaining its data structures for the relatively small datasets considered.

In the context of larger datasets, such as the E. coli dataset, CSA-MEM demonstrates a great advantage in terms of both processing time and memory usage. CSA-MEM was capable of generating the optimal rotation in under 2 minutes, consuming less than 70 MB of memory. In sharp contrast, the alternative methods did not produce any results even after extended hours of processing. As a consequence of these outcomes, we have omitted their presentation in the table.

4   Conclusion

Understanding the genetic relationships and evolutionary history encoded in circular DNA molecules has a profound impact on human health, environmental studies, and understanding the complexity and diversity of life. The CSA-MEM tool demonstrates that the use of efficient indexing data structures and string matching algorithms for circular sequences yields superior benchmark scores with minimal computational demands. This novel approach not only enhances rotation strategies, but also optimises space and time complexities, enabling a more thorough analysis of circular DNA sequences. The future work will include building a database with various datasets for circular genomes that can be used to characterise the boundaries of the algorithms tested in comparative genomics studies. The software source code, scripts and datasets used in this work are available for download at: https://github.com/andre99salgado/CSA-MEM.

Acknowledgement. The authors acknowledge the support of Fundação para a Ciência e a Tecnologia, projects PRELUNA (Grant PTDC/CCIINF/4703/2021) and UIDB/50021/2020.

References

1. Ayad, L.A., Pissis, S.P.: MARS: improving multiple circular sequence alignment using refined sequences. BMC Genomics 18(1), 1–10 (2017)
2. Barton, C., Iliopoulos, C.S., Kundu, R., Pissis, S.P., Retha, A., Vayani, F.: Accurate and efficient methods to improve multiple circular sequence alignment. In: Bampis, E. (ed.) SEA 2015. LNCS, vol. 9125, pp. 247–258. Springer, Cham (2015). https://doi.org/10.1007/978-3-319-20086-6_19
3. Barton, C., Iliopoulos, C.S., Pissis, S.P.: Fast algorithms for approximate circular string matching. Algorithms Mol. Biol. 9, 1–10 (2014)
4. Burrows, M.: A block-sorting lossless data compression algorithm. SRS Res. Rep. 124 (1994)
5. Carattoli, A.: Plasmids and the spread of resistance. Int. J. Med. Microbiol. 303(6), 298–304 (2013)

6. Dulanto, C.A., Dekker, J.P.: From the pipeline to the bedside: advances and challenges in clinical metagenomics. J. Infect. Dis. **221**(Supplement 3), S331–S340 (2019)

7. Fehér, E., Mihalov-Kovács, E., Kaszab, E., Malik, Y.S., Marton, S., Bányai, K.: Genomic diversity of CRESS DNA viruses in the eukaryotic Virome of swine feces. Microorganisms **9**(7), 1426 (2021)

8. Fernandes, F., Freitas, A.T.: slaMEM: efficient retrieval of maximal exact matches using a sampled LCP array. Bioinformatics **30**(4), 464–471 (2014)

9. Fernandes, F., Pereira, L., Freitas, A.T.: CSA: an efficient algorithm to improve circular DNA multiple alignment. BMC Bioinformatics **10**(1), 1–13 (2009)

10. Ferragina, P., Manzini, G.: Opportunistic data structures with applications. In: Proceedings 41st Annual Symposium on Foundations of Computer Science, pp. 390–398. IEEE (2000)

11. Grossi, R., Iliopoulos, C.S., Mercas, R., et al.: Circular sequence comparison: algorithms and applications. Algorithms Mol. Biol. **11**(12) (2016)

12. Gusfield, D.: An "increment-by-one" approach to suffix arrays and trees. Report. CSE-90-39, Computer Science Division, University of California, Davis (1990)

13. Laudadio, I., Fulc, V., Stronati, L., Carissimi, C.: Next-generation metagenomics: methodological challenges and opportunities. OMICS **23**(7), 327–333 (2019)

14. Manber, U., Myers, G.: Suffix arrays: a new method for on-line string searches. SIAM J. Comput. **22**(5), 935–948 (1993)

15. Mosig, A., Hofacker, I.L., Stadler, P.F.: Comparative analysis of cyclic sequences: viroids and other small circular RNAs. In: Lecture Notes in Informatics. Proceedings German Conference on Bioinformatics (2006)

16. Pan, S., Zhao, X.M., Coelho, L.P.: SemiBin2: self-supervised contrastive learning leads to better MAGs for short- and long-read sequencing. Bioinformatics **39**(Supplement 1), i21–i29 (2023)

17. Pereira, L., et al.: The diversity present in 5140 human mitochondrial genomes. Am. J. Hum. Genetics **84**(5), 628–640 (2009)

18. Pohjoismäki, J.L.O., Goffart, S.: Of circles, forks and humanity: topological organisation and replication of mammalian mitochondrial DNA. BioEssays **33**(4), 290–299 (2011)

19. Thompson, J.D., Gibson, T.J., Higgins, D.G.: Multiple sequence alignment using ClustalW and ClustalX. Curr. Protoc. Bioinformatics **1**, 2–3 (2003)

20. Tisza, M.J., et al.: Discovery of several thousand highly diverse circular DNA viruses. Elife **9** (2020)

21. Yang, L., et al.: Extrachromosomal circular DNA: biogenesis, structure, functions and diseases. Signal Transduct. Target. Ther. **7**(1), 342 (2022)

22. Zhang, Y., Zhang, Q., Zhou, J., Zou, Q.: A survey on the algorithm and development of multiple sequence alignment. Briefings Bioinformatics **23**(3) (2022)

23. Zhao, L., Rosario, K., Breitbart, M., Duffy, S.: Chapter three - eukaryotic circular rep-encoding single-stranded DNA (cress DNA) viruses: ubiquitous viruses with small genomes and a diverse host range. In: Advances in Virus Research, vol. 103, pp. 71–133 (2019)

A Convolutional Denoising Autoencoder for Protein Scaffold Filling

Jordan Sturtz[1], Richard Annan[1], Binhai Zhu[2], Xiaowen Liu[3],
and Letu Qingge[1(✉)]

[1] Department of Computer Science, North Carolina A&T State University,
Greensboro, NC, USA
{jasturtz,rkannan}@aggies.ncat.edu, lqingge@ncat.edu
[2] Gianforte School of Computing, Montana State University,
Bozeman, MT, USA
bhz@montana.edu
[3] John W. Deming Department of Medicine, Tulane University,
New Orleans, LA, USA
xwliu@tulane.edu

Abstract. De novo protein sequencing is a valuable task in proteomics,
yet it is not a fully solved problem. Many state-of-the-art approaches
use top-down and bottom-up tandem mass spectrometry (MS/MS) to
sequence proteins. However, these approaches often produce protein scaffolds, which are incomplete protein sequences with gaps to fill between
contiguous regions. In this paper, we propose a novel convolutional
denoising autoencoder (CDA) model to perform the task of filling gaps
in protein scaffolds to complete the final step of protein sequencing. We
demonstrate our results both on a real dataset and eleven randomly generated datasets based on the MabCampath antibody. Our results show
that the proposed CDA outperforms recently published hybrid convolutional neural network and long short-term memory (CNN-LSTM) based
sequence model. We achieve 100% gap filling accuracy and 95.32% full
sequence accuracy on the MabCampth protein scaffold.

Keywords: De Novo Protein Sequencing · Convolutional Layer ·
Denoising Autoencoder · Protein Scaffold Filling

1 Introduction

Protein sequencing plays an important role in many aspects of proteomics,
including identification of structure and functions of proteins, new protein
biomarkers, construction of phylogenetic tree to find evolutionary relationship
and new drug design. De novo protein sequencing refers to the process of determining the primary structure of proteins directly without inferring the full
sequence by merely matching against an existing protein database. Complete
de novo protein sequencing remains a challenging problem in bioinformatics.

This work is supported by the National Science Foundation of the United States under
Award 2307571, 2307572 and 2307573.

Every protein can be defined by its unique sequence of amino acids, which is called its primary structure. Proteins are comprised of 20 different amino acids. We use the term "peptide" to refer to small multi-amino acid sub-units of proteins. The goal of peptide or protein sequencing is to determine the complete unique sequence of amino acids in a peptide or protein. In general, peptide or protein sequencing from mass spectrometry can refer to either de novo sequencing or database searching. With database searching, once a mass spectrum is generated, it is compared to databases of known peptides or proteins to retrieve the sequence with the closest matching mass spectrum. Often, these databases will include only proteins or peptides generated from genomic data [10]. Many proteins of interest are not included in such databases, especially those that are not directly inscribed in genomes such as monoclonal antibodies. Even if a protein sequence is known, it is often still desirable to perform de novo sequencing to discover novel proteoforms [11]. For instance, proteoforms may be created by post-translational modifications, which occur when amino acids of proteins undergo a process of proteolytic cleavage which alters the amino acid in the primary structure by adding a modifying group [8,9]. De novo protein sequencing has been used for many purposes, including full sequencing of proteins, to sequence endogenous peptides [12,13], to characterize mutations in antibodies [14], and to perform proteomic analysis of novel organisms not found in protein databases.

We organize our paper as follows. In Sect. 2, we discuss the problem statement and gap challenges that motivate our research, deficiencies in existing approaches. In Sect. 3, we introduce the methodology that we will use to develop a new convolutional denoising autoencoder (CDA) model as a solution. In Sect. 4, we discuss in detail our proposed CDA model, including data preprocessing, model architecture and hyperparameters tuning steps. In Sect. 5, we show our experimental prediction results both on the original real MabCampth scaffold data and simulation data. Finally, we conclude our paper and discuss the future directions.

2 Preliminaries

The Protein Scaffold Filling (PSF) Problem: Given a complete target protein sequence S and the scaffold T, fill the missing amino acids in the scaffold T such that $Score(S, T)$ is maximized, where function $Score$ is the total number of one-to-one matches of amino acids between S and T.

The protein scaffold filling problem has been shown to be polynomial solvable in $O(n^{26})$ time [4]. In [4], the authors proposed several practical algorithms based on greedy algorithm, dynamic programming and local search. These algorithms rely on high quality homologous reference proteins. As reported in [4], these algorithms run in a reasonable amount of time when gaps are small. Thus, our goal is to investigate deep learning approaches to the same problem to improve our accuracy, especially when gaps are large or the homologous reference proteins are dissimilar to proteins scaffolds produced in a lab.

Most recently in 2022, the authors [7] developed several deep learning models based on CNN and LSTM models for the PSF problem and achieved high accuracy when filling the gaps in the MabCampath scaffold dataset. The basic idea behind this approach is to iteratively predict each amino acid in sequence by deploying a model that can predict the next amino acid given the preceding K amino acids. From left to right, when a gap is encountered in the protein scaffold, the model predicts the next amino acid as a replacement for that gap. This process is repeated until all gaps are filled. The authors trained a forward model and reverse model so they can predict gaps at the end of any protein scaffold. For training data, the authors query for homologous sequences to their scaffold protein, then generate all kmers of each training instance. Each kmer represents a single training instance input, and the amino acid after the kmer in the sequence is the training output. So, for example, from the sequence DIQMSPIL..., the following input-output pairs would be generated: (DIQMS, P), (IQMSP, I), (QMSPI, L). The authors trained various CNN-LSTM hybrid models to compare their accuracy.

Though their reported accuracy is higher than that reported in [4], this approach suffers from a few flaws. First, since the model is a kmer sequence-based approach, any errors in inference are likely to propagate, leading to subsequent incorrect inferences. See Fig. 1 for an illustration. If this issue is indeed a significant problem for the sequence-based approach, it suggests that such approaches will tend to do worse when the gaps to fill between contigs of a protein scaffold are particularly large.

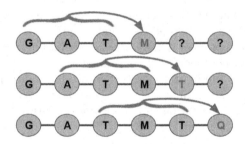

Fig. 1. CNN-LSTM model illustration [7]. Since during inference the model predicts only the next amino acid, if it makes a poor prediction, it will feed that poor prediction into the next inference step, causing future inferences to be unreliable

In this paper, our goal is to develop a deep learning model that can accurately predict the missing amino acids in gaps of the scaffold while improving on the approach described in [7] by also correcting incorrect amino acids in the existing scaffold.

3   Methodology

The approach we use is a convolutional denoising autoencoder (CDA) trained on homologous sequences of our given scaffold. The motivation behind an autoencoder in general is that it imputes all the missing amino acids at once, which is different from the iterative sequence-based approach described in [7]. Not only can the CDA predict gaps in the scaffold but it can also correct any incorrect amino acids in the scaffold. In contrast, the LSTM models designed in [7] can only predict the missing amino acids in the gaps of the scaffold.

Autoencoders. Autoencoders are neural networks that learn how to reconstruct its input through the composition of an encoder and a decoder [1]. Typically, the idea is to encode the original input into a lower dimensional space and then decode the compressed representation into the original input.

Denoising Autoencoders. Simple autoencoders suffer from the problem that the autoencoder may simply learn an identity function, which produces trivially useless results [2]. A common solution to this problem is to intentionally corrupt the original input in some way by adding some kind of "noise" to the data. The goal of the autoencoder, then, is to learn how to denoise the corrupted input, which produces a more robust representation that avoids trivial solutions [3]. A model trained on corrupted inputs can learn an internal representation that can correct those defects.

Convolutional Layers. Convolutional layers in a neural network are useful whenever the input contains hidden features created by the relationships among neighboring components of the input. In this way, convolutional layers can be viewed as automatic feature extractors. Since the dataset consists of sequences of amino acids, it is a reasonable hypothesis that there are meaningful features to extract among neighboring values of each sequence.

Pooling and Upsampling. Pooling is in general a useful technique to reduce model complexity to speed up training. In our case, pooling is how the model achieves the compression characteristic of autoencoders. The model convolves the original input to extract features, then compresses those feature maps with pooling into a reduced dimensional space. The decoder portion of the autoencoder performs inverse convolutions and upsampling to produce the final sequence length of the training data.

4   The Proposed Convolutional Denoising Autoencoder Model

4.1   Data Collection

The protein scaffold we use to evaluate our proposed model is the light chain of alemtuzumab (MabCampath). In [5], the authors generated the MabCampath scaffold data by combining top-down and bottom-up tandem mass spectrometry. This scaffold includes five contigs and six contiguous gaps of missing amino acids. The main steps of generating the MabCampth scaffold consists of converting raw spectra to a prefix residue mass (PRM) spectra, spectral selection and merging, improving the top-down spectrum using bottom-up spectra, spectra mapping, gap filling by extension and gap filling by mass matching. More technical details about generating the MabCampath protein scaffold can be found in [5]. The scaffold information can be seen in Fig. 2, in which the red colored dash line represents gaps in the scaffold and the other red characters are non-gap errors in the scaffold. We feed the scaffold into NCBI's Protein Blast Server [6] to retrieve 1000 homologous sequences as our training data.

```
Target Sequence
DIQMTQSPSSLSASVGDRVTITCKASQNIDKYLNWYQQKPGKAPKLLIYNTNNL
QTGVPSRFSGSGSGTDFTFTISSLQPEDIATYYCLQHISRPRTFGQGTKVEIKR
TVAAPSVFIFPPSDEQLKSGTASVVCLLNNFYPREAKVQWKVDNALQSGNSQES
VTEQDSKDSTYSLSSTLTLSKADYEKHKVYACEVTHQGLSSPVTKSFNRGEC
Protein Scaffold
---MTQSPSSISASVGDRVTITCK---NIDKYINWYQQKPGKAPKIIIYNTNNI
QTGVPSRF---G----FTFTI-----------YCIQHISRPRTFGQGTKVEIKR
SIAAPSVFIFPPSDEQIKSGTASVVCIINNFYPREAQPRRKVDNAIQSGNSQES
VTEQDSKDSTYSISSTITISKADYEKHKVYACEVTHQGISSPVTKSFN----
```

Fig. 2. Dashes are missing amino acids, i.e., gap errors. The other red-colored characters are non-gap errors in the given Protein Scaffold. Target Sequence is a ground truth sequence that we will predict. (Color figure online)

As our model depends on padding shorter protein sequences with empty amino acids, we also prune the collected training data by limiting the lengths of acceptable training sequences to those where the length is between 95% to 105% of the length of the target. In this way, we reduce the required amount of padding in our training data to allow for varied sequence lengths while also minimizing biases that may occur due to the model learning the noise of the extra padding. To get a sense for the quality of training data for each test scaffold, we choose the homologous sequences with the range of 205–224 lengths, the range of 98%-100% query coverage, and the range of 44%-89% percent identical similarity among sequences in each training dataset. The query coverage refers to the percentage of the queried sequence that is covered by the returned sequence, whereas the percent similarity refers to the percent of one-to-one matches in the sequence alignments.

4.2   Data Preprocessing

One-Hot Encoding. In general, there are two ways to represent categorical data. The first method is label-encoding, in which each category is assigned a numerical value. The second method is one-hot encoding, in which each category is represented by a binary vector where the position of the 1 in the binary vector represents the category of the datum.

One-hot encoding is often a preferred method for categorical data and it is the type of encoding we choose here. Thus, our network must learn a representation where the full input dataset is a tensor of shape (samples, sequence_length, classes).

Noisification. To add noise to our input data, we add a new class label to represent emptiness. Thus, in data preprocessing, a percent P of the amino acids are replaced by the empty class represented by blank.

Padding. Not all sequences in the training data will have the same lengths. To feed these sequences into a neural network, it is therefore necessary to employ a strategy to either pad or truncate training sequences to get a fixed length. We opt to pad each training sequence with empty amino acids until the lengths reach the maximum length sequence in the training data. Let S be the maximum length of the sequence in the training data. It is important that pooling layers in our model cause a reduction in the size of the feature maps such that upsampling in the decoding phase produces the same shape as our target inputs. For instance, suppose S is 211 and the neural network has two pooling layers. In this case, the encoder will produce a length of 52: $\lfloor \lfloor 211/2 \rfloor /2 \rfloor = 52$. But if a shape of 52 is then upsampled in the decoder, it produces an output length of 208: $52*2*2 = 208$. We want the output of the neural network to have a length of 211 to match the length of the input. To solve this technical problem, we increment S until $S \mod L = 0$, where L is the product of the shapes of the pooling layers.

The Model Architecture. The final model architecture is illustrated in Fig. 3. There are two convolutional layers in the encoder, each followed by max pooling and dropout layers. Likewise, there are two inverse convolutional layers in the decoder followed by upsampling and dropout layers. We split our dataset into training and validation of 85% and 15% respectively. The more details about the model architecture can be found in Fig. 3. Noise and padding are added to the model input, then it is one-hot encoded before running through the encoder, which ultimately compresses the input into a reduced dimensional space. The decoder portion of the neural network reconstructs the input using upsampling. Dropout is added to reduce overfitting. The model hyperparameters are listed in Table 2. Our developed code can be found from https://github.com/astonish24/QinggeLab_ISBRA23_paper.

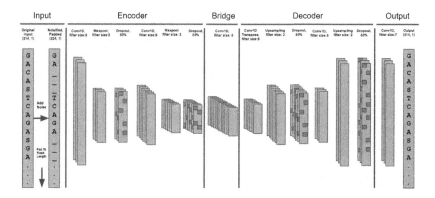

Fig. 3. The proposed convolutional denoising autoencoder (CDA) model architecture.

4.3 Simulation Data

A protein scaffold produced using MS/MS will contain errors both in its contigs (what we will call *non-gap* errors) as well as gaps that need to be filled between contigs to complete the entire protein (what we will call *gap* errors). The number of amino acids to fill between contigs as well as the number of total errors will vary from protein scaffold to protein scaffold. For this reason, and because we are interested in comparing our results to the up-to-date sequence-based approach described in [7], we generate random protein scaffolds from our target protein sequence. Note that, for validation purpose, we know our target sequence that we are constructing.

We generate eleven new protein scaffolds using combinations of three values for errors percentage (20%, 30%, and 40%) and four values (4, 6, 8, 10) for the number of contiguous gaps size in each scaffold. To maintain realistic artificially generated protein scaffolds, we split the percent error into a ratio of 60/40 for gap and non-gap errors respectively, which roughly corresponds to the ratio present in the protein scaffold produced by [5].

Once the eleven protein scaffolds are generated, we collect training data by querying the National Center for Biotechnology Information (NCBI) Protein BLAST server to retrieve the top 1000 most similar reference proteins [6]. Table 1 shows each training dataset and the range of percent identical similarity in the returned reference proteins. The protein scaffolds with smaller values for reference similarity are likely to have worse results, since the training data will be based on less similar reference proteins.

Table 1. Generated Protein Scaffolds and Training Similarity

ID	# Contiguous Gaps	% Incorrect	Reference Similarity
1	6	20%	80.4% - 68.2%
2	8	20%	87.8% - 70.9%
3	10	20%	92.5% - 73.9%
4	4	30%	75.7% - 64.0%
5	6	30%	71.4% - 61.2%
6	8	30%	80.2% - 62.3%
7	10	30%	71.2% - 57.7%
8	4	40%	88.1% - 65.1%
9	6	40%	68.2% - 60.1%
10	8	40%	68.2% - 60.6%
11	10	40%	67.4% - 53.5%

5   Results and Comparison

We compare the performance of our proposed model with the recently developed hybrid CNN-LSTM [7] in terms of gap filling accuracy and full sequence accuracy. The gap filling accuracy is computed by dividing the number of correct predictions on missing gaps by the number of missing gaps in the scaffold, where we use the target sequence as a ground truth sequence. The full sequence accuracy is the percentage of one-to-one matches between the full prediction and the target protein. Note that the CNN-LSTM model only predict the missing amino acids in the gaps. While our proposed denoinsing autoencoder model not only predict the missing amino acids in the gaps but also it has an ability to correct the amino acids in the scaffolds which is obtained from bottom-up and top-down methods. Also, in the bottom-up and top-down methods, it cannot distinguish the same weight amino acids I and L. However, our proposed model is able to correctly identify both I and L in the predicted sequence.

Table 2. The CDA Hyperparameters

learning_rate	3.061E-4
dropout_percent	0.50
bridge_filters	160
conv_filters1	46
conv_filters2	90
conv_filter_size1	5
conv_filter_size2	9
bridge_filter_size	5
final_filter_size	7
kmer_size	15
noise_percent	40%

5.1   Results on the MabCampath Scaffold

We run both our proposed CDA and the CNN-LSTM based model [7] discussed in Sect. 2 on the original MabCampath scaffold. Both models did not appear to display any overfitting. Figure 5 shows training and validation accuracy for both models, and Fig. 6 shows training and validation losses for both models.

We also display the predictions for both the CDA and the CNN-LSTM models on the original scaffold protein in Fig. 4. In this figure, the green colored amino acids are correctly predicted amino acids and the red colored amino acids are incorrectly predicted amino acids from both CDA and CNN-LSTM models. From our proposed model, we also achieve 100% gap filling accuracy as the CNN-LSTM model produced in [7]. While for the full sequence accuracy, our model obtain 95.32% accuracy compared with the target sequence which outperforms the CNN-LSTM model's 89.7% accuracy [7].

The non-gap accuracy, which is the percentage of correct predictions on non-gap region in the protein scaffold with respect to the target sequence. The non-gap accuracy will always be 0% for the sequence-based approach, since the sequence-based approach cannot in principle attempt to correct non-gap errors. On the other hand, since the CDA imputes the full protein sequence, which is taken as the prediction for all amino acids, the autoencoder may at times incorrectly change amino acids that should not have been altered. It is for this reason that we display the full sequence accuracy.

MabCampath Protein Scaffold	
CDA Full Acc: 95.327%	DIQMTQSPSSLSASVGDRVTITCRASQDIDNYLNWYQQKPGKAPKLLIYDASNL QTGVPSRFSGSGSGTDFTFTISSLQPEDIATYYCLQHYNYPYTFGQGTKVEIKR TVAAPSVFIFPPSDEQLKSGTASVVCLLNNFYPREAKVQWKVDNALQSGNSQES VTEQDSKDSTYSLSSTLTLSKADYEKHKVYACEVTHQGLSSPVTKSFNRGEC
CNN-LSTM Full Acc: 89.7%	DIQMTQSPSSISASVGDRVTITCKASQNIDKYINWYQQKPGKAPKIIIYNTNNI QTGVPSRFSGSGSGTDFTFTIGSLQPEDFATYYCIQHISRPRTFGQGTKVEIKR SIAAPSVFIFPPSDEQIKSGTASVVCIINNFYPREAQPRRKVDNAIQSGNSQES VTEQDSKDSTYSISSTITISKADYEKHKVYACEVTHQGISSPVTKSFNRGEC

Fig. 4. MabCampath Protein Scaffold Predictions from CDA and CNN-LSTM Models

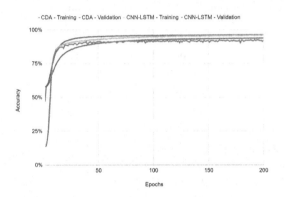

Fig. 5. CDA and CNN-LSTM Training and Validation Accuracies

5.2   Results on Simulation Datasets

To further demonstrate the performance of our proposed CDA and CNN-LSTM [7] model, we test both models on the generated scaffolds as described in Sect. 4.3. The CDA outperforms the sequence-based CNN-LSTM approach on 10 out of the 11 datasets in terms of full sequence accuracy. The chart in Fig. 7 compares the full-sequence accuracies. Our proposed CDA model has a better prediction accuracy for full sequence comparison with the target sequence. The main reason is that CDA is able to predict the missing amino acids in the gaps, also it can fix the errors in the non-gaps regions of the constructed scaffold. While CNN-LSTM model does not have such capability. It only focus on predicting the missing amino acids in the gaps of the scaffold. The CNN-LSTM model approach cannot in principle correct non-gap errors, so the non-gap accuracy is always 0%. The CDA model, on the other hand, suffers from the deficiency that since it outputs a full sequence to be used for its full prediction, it may inadvertently change amino acids that should not be changed. In fact, on the one generated scaffolds (scaffold #9), the CNN-LSTM model achieves higher full sequence accuracy. The reason the lower full sequence accuracy of CDA is merely that the CDA changes too many amino acids that should have remained the same.

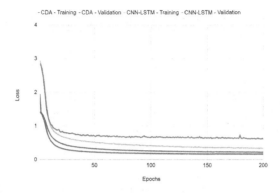

Fig. 6. CDA and CNN-LSTM Training and Validation Loss

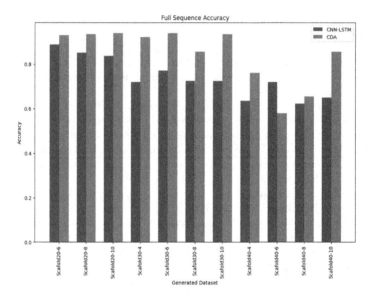

Fig. 7. A Comparison Result Between CDA and CNN-LSTM on Simulation Datasets

6   Conclusion

De novo protein sequencing from mass spectrometry data is still a hard problem in proteomics. Current state-of-the-art approaches are still unable to completely sequence proteins accurately. In this paper, we show that we can apply deep learning methods to aid in a final step in de novo protein sequencing, namely filling gaps in the protein scaffold. Moreover, we have shown that our CDA model is able to perform this task more accurately than the sequence-based approach [7], which also outperforms the existing combinatorial algorithms based on dynamic programming, local search and greedy methods described in [4]. The advantage of this approach is that it is far simpler once the model is built to perform the inference needed to fill the gaps. This simplicity avoids the potential deficiency we identified with the sequence-based approach that predicts one amino acid after another. We conclude that if the constructed scaffold with higher accuracy and smaller gaps, the deep learning based approaches can produce more higher accuracy on protein sequencing predictions. For the future work, we will test our model on the more real protein scaffold dataset and explore other machine learning models for the protein sequencing problem.

Acknowledgement. This work is supported by the National Science Foundation of the United States under Award 2307571, 2307572 and 2307573. We thank anonymous reviewers for their insightful comments and inputs.

References

1. Bengio, Y.: Deep learning of representations for unsupervised and transfer learning. In: Proceedings Of ICML Workshop On Unsupervised And Transfer Learning, pp. 17–36 (2012)
2. Vincent, P., Larochelle, H., Bengio, Y., Manzagol, P.: Extracting and composing robust features with denoising autoencoders. In: Proceedings Of The 25th International Conference On Machine Learning, pp. 1096–1103 (2008)
3. Vincent Pascalvincent, P., Laroceh, L., Autoencoders, H.: Learning useful representations in a deep network with a local denoising criterion pierre-antoine manzagol. J. Mach. Learn Res. **11**, pp. 3371–3408 (2010)
4. Qingge, L., Liu, X., Zhong, F., Zhu, B.: Filling a protein scaffold with a reference. IEEE Trans. Nanobiosci. **16**, 123–130 (2017)
5. Liu, X., et al.: De novo protein sequencing by combining top-down and bottom-up tandem mass spectra. J. Proteome Res. **13**, 3241–3248 (2014)
6. National Center for Biotechnology Information Blast. https://blast.ncbi.nlm.nih.gov/Blast.cgi?PAGE=Proteins
7. Sturtz, J., Zhu, B., Liu, X., Fu, X., Yuan, X., Qingge, L.: Deep learning approaches to the protein scaffold filling problem. In: 2022 IEEE 34th International Conference On Tools With Artificial Intelligence (ICTAI), pp. 1055–1061 (2022)
8. Ramazi, S., Allahverdi, A., Zahiri, J.: Evaluation of post-translational modifications in histone proteins: a review on histone modification defects in developmental and neurological disorders. J. Biosci. **45**(1), 1–29 (2020). https://doi.org/10.1007/s12038-020-00099-2
9. Ramazi, S., Zahiri, J.: Post-translational modifications in proteins: resources, tools and prediction methods. Database (2021)
10. Eng, J.K., McCormack, A.L., Yates, J.R.: An approach to correlate tandem mass spectral data of peptides with amino acid sequences in a protein database. J. Am. Soc. Mass Spectrom. **5**(11), 976–989 (1994). https://doi.org/10.1016/1044-0305(94)80016-2
11. Smith, L., Kelleher, N.: Proteoform: a single term describing protein complexity. Nat. Methods **10**, 186–187 (2013)

12. Alhaider, A., et al.: Through the eye of an electrospray needle: mass spectrometric identification of the major peptides and proteins in the milk of the one-humped camel (Camelus dromedarius). J. Mass Spectrom. **48**, 779–794 (2013)

13. Viala, V., et al.: Pseudechis guttatus venom proteome: insights into evolution and toxin clustering. J. Proteomics **110**, 32–44 (2014)

14. Costa, D., et al.: Sequencing and quantifying IgG fragments and antigen-binding regions by mass spectrometry. J. Proteome Res. **9**, 2937–2945 (2010)

Simulating Tumor Evolution
from scDNA-Seq as an Accumulation
of both SNVs and CNAs

Zahra Tayebi, Akshay Juyal, Alexander Zelikovsky, and Murray Patterson[✉]

Georgia State University, Atlanta, GA, USA
{ztayebi1,ajuyal1,alexz,mpatterson30}@gsu.edu

Abstract. Ever since single-cell sequencing (scDNA-seq) was coined 'method of the year' in 2013, it has provided many insights into the evolution of tumors, viewed as a branching process of accumulating cancerous mutations that initiated with a single driver mutation — a model of clonal evolution which has been theorized almost half a century ago (Nowell, 1976). With this, is seen an explosion of methods for inferring the histories of such evolution, often in the form of a phylogenetic tree, from single-cell sequencing data. While the first methods modeled such evolution as an accumulation of point mutations (SNVs), copy number aberrations (CNAs, i.e., duplications or deletions of large genomic regions) are an important factor to consider. As a result, later methods began to bolster cancer phylogeny inference with bulk sequencing data, to account for CNAs. Despite the dozens of such inference methods available, there still does not exist much in the form of a unified benchmark for all such methods.

This paper moves to initiate such a benchmark, which can be built upon, by proposing a simulator which models both SNVs and CNAs jointly in generating an evolutionary scenario which can be interpreted as a scDNA-seq/matched bulk sample pair. The simulator models the accumulations of SNVs, and the duplication or deletion of chromosomal segments. We test this simulation on three methods: (a) a method which accounts for SNVs only, and under the infinite sites assumption (ISA), (b) a second more general method which models only SNVs, but allows for relaxations to the ISA, and (c) a third most general method which accounts for both SNVs and CNAs (and violations to the ISA). Results are consistent with the generality of these methods. This work is a step in the direction of developing a de-facto benchmark for cancer phylogeny inference methods.

Keywords: single-cell sequencing · tumor phylogeny ·
single-nucleotide variants (SNVs) · copy number aberrations (CNAs) ·
benchmarking

1 Introduction

Cancer is a complex disease driven by genomic alterations that result in uncontrolled cell growth and spread [27,34]. These alterations accumulate

X. Guo et al. (Eds.): ISBRA 2023, LNBI 14248, pp. 530–540, 2023.
https://doi.org/10.1007/978-981-99-7074-2_43

non-randomly, leading to diverse cancer cell populations evolving over time [2,29], see Fig. 1. Understanding cancer's evolutionary dynamics is vital for effective therapies targeting specific genomic changes [13,26]. Genomic alterations, including copy number aberrations (CNAs) and single nucleotide variations (SNVs), play critical roles in cancer progression [1,3,20,22]. CNAs involve changes in chromosome copy numbers, causing imbalances that disrupt genes and promote cancer [35]. SNVs, altering individual DNA sequence nucleotides, impact gene function and cellular pathways [5,15].

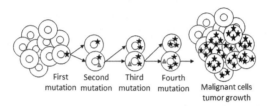

First
mutation

Second
mutation

Third
mutation

Fourth
mutation

Malignant cells
tumor growth

Fig. 1. The evolution of cancer progression

Cancer progression follows distinct patterns in CNAs and SNVs accumulation [18,36]. Genomic changes occur stepwise, building on each other, leading to diverse cancer cell populations [36]. This diversity enables cancer cells to adapt, survive, and resist therapies [30]. Researchers employ computational algorithms and statistical models to reconstruct cancer's phylogenetic tree and analyze CNAs and SNVs from biopsy-derived tumor samples [11,19]. The goal is to comprehend tumor evolution, relationships among cell populations, and drug resistance emergence [4]. The cancer phylogenetic tree construction considers CNAs and SNVs from parent to child nodes, modeling dynamic genome changes during cancer evolution [33,38]. Understanding CNAs and SNVs interplay is crucial for targeted therapies [31]. Identifying key genetic changes driving tumor growth and metastasis enables precise, effective treatments [14].

Despite various methods [6,8,16,24,33,38] to reconstruct tumor phylogenies from single-cell and bulk sequencing, a unified benchmark is lacking. This benchmark is crucial for tool comparison and assessment. To address this gap, a simulator incorporating CNAs and SNVs was developed. Well-known phylogenetic tree reconstruction methods were evaluated using simulated data. Results indicate the simulator captures realistic cancer progression dynamics involving CNAs and SNVs. The contributions include introducing a phylogenetic tree benchmark through simulation, evaluating established tools on this benchmark, and showing consistent results aligning with expectations.

This paper is structured as follows: Sect. 2 outlines related work, Sect. 3 presents the proposed approach, Sect. 4 describes the experimental setup, Sect. 5 presents and discusses results, and Sect. 6 concludes and suggests future work.

2   Related Work

Research studies have focused on understanding the complex interplay between copy number aberrations (CNAs), single nucleotide variations (SNVs), and cancer progression [3,9,23]. The aim is to unravel the mechanisms driving cancer development and evolution. To explore cancer dynamics, computational algorithms and statistical models have been developed for reconstructing cancer phylogenetic trees [19,24,32], considering CNAs and SNVs' roles in shaping tumor genomes [33,38]. Evaluating these methods often necessitates known ground truth. Synthetic data with known ground truth has emerged as a popular approach for assessing method performance, providing an effective alternative for accuracy evaluation. SCSsim, based on MALBAC, simulates single cells introducing SNVs, indels, and CNAs [37]. However, it lacks phylogenetic tree simulation, limiting its use for studying cancer evolution. SCSIM simulates both single cells and bulk data, yet lacks phylogenetic tree representation [12], hindering the portrayal of cell relationships during cancer progression. SimSCSnTree generates synthetic single-cell DNA sequencing data, incorporating CNVs and SNVs in cancer cells [25]. However, the tool's models lack detailed explanations and its adoption seems limited.

3   Proposed Approach

Generating a cancer progression tree of Copy Number Aberrations (CNAs) and Single Nucleotide Variations (SNVs) is a computational model used to represent the evolutionary history of a tumor. It models how cancer cells acquire genomic alterations over time, leading to disease progression and heterogeneity. Here, we design a simulator which generates such trees, which can be used to benchmark tumor phylogeny reconstruction methods.

The simulator utilizes a random process to generate a random tree that represents the evolutionary history of a tumor, considering the Copy Number Alterations (CNAs) and Single Nucleotide Variations (SNVs). To begin, we initialize a list of Copy Number Profiles (CNPs), which consists of two sublists (for each allele of the human diploid chromosome), see Fig. 2 (a). These sublists are populated with randomly chosen elements representing SNVs on the particular alleles. This initialization process is performed in a separate function.

Next, we employ Algorithm 1 to generate a random tree, Fig. 2 (b). This algorithm takes a parameter n, indicating the desired number of nodes in the tree. It starts by creating a root node and then iteratively adds new nodes with random parent nodes until the specified number of nodes is reached. The algorithm extracts the edges from the resulting tree structure and returns them as a list of tuples, where each tuple represents an edge between a parent and its child node. Then, we initialized a dictionary that represents the cancer progression tree with information about the parent-child relationships and the CNPs (each containing some SNVs) associated with each node, see Algorithm 2.

Algorithm 2 lines 10 and after outline the subsequent steps, which involve iterating through the children of nodes(parent) specified by the node(key) in the info dictionary. For each child node, an empty list ϕ is initialized. Iterations through the list values of the parent node's lists are made. A random operation (addition, deletion, or no-change) is then applied, these operations introduce variations in the copy number. Depending on the option, a corresponding function (addition, deletion, or nochange) is called to modify the copy number, which is then appended to the modified ϕ and in turn updated to the child node's lists using the values from the updated ϕ. This operation is done recursively on each child node to perform the same process down the hierarchy. After the process is completed the info dict is updated with the generated copy numbers for each node.

For example, in Fig. 2 (c), the first bin b_1 undergoes the deletion option, resulting in the removal of one list and its corresponding mutations. On the other hand, the second and third bins, b_2 and b_3, experience the addition option, where new point mutations are introduced, and a new list is appended to model a CNA.

Figure 3 provides an example of a final tree that reflects the updated CNPs resulting from the process of CNAs and the accumulation of SNVs.

Algorithm 1. RandomTreeGenerator

Input: number of nodes n
Output: tree, edges
1: tree ← Node(0)   ▷ creating the root node of the tree with a value of 0
2: nodes ← [tree]   ▷ initializing the nodes list with the root node
3: **for** i ← 0 to n **do**
4:     parent ← random.choice(nodes)   ▷ parent randomly selected from list
5:     node ← Node(i, parent=parent)   ▷ node created and attached to parent
6:     nodes.append(node)   ▷ newly created node added to list
7: **end for**
8: edges ← tree.edges   ▷ extracting the edges from the given tree
9: **return** tree, edges

---

4   Experimental Setup

The experimental pipeline was built using Snakemake [28]. The tool and the pipeline are available online for reproducibility.[1]

4.1   Simulating SCS Data

With our tool, we simulated a total of 50 trees, each with $n = 20$ nodes, where each tree was generated considering $m = 2$ bins. At the initial stage, the number of copies was set to two, while the number of SNVs in each new step was set to one. This results in an average number of total mutations of 80 (20 nodes × 2

---

[1] https://github.com/murraypatterson/scDNA-seq-sim.

Algorithm 2. TreeInfoGenerator

    Input: edges
    Output: info   ▷ the information dictionary of tree
  1: info ← {}
  2: **for** i **in** len(edges) **do**
  3:     info[i] ← {'parent':'','children':[],'lists':[]}   ▷ updating dictionary by keys
  4: **end for**
  5: **for** e **in** edges **do**
  6:     info[e[0]]['children'].append(e[1])   ▷ assigning second item of edge list as child
  7:     info[e[1]]['parent']=e[0]   ▷ assigning first item of edge list as parent
  8: **end for**
  9: **function** EVALUATECNPs($node = 0$)
10:     **if** length of info[node]["children"] > 0 **then**
11:         **for** *childNode* **in** info['node']['children'] **do**
12:            ϕ ← []   ▷ Initialize CNPs lists
13:            **for** list **in** info[childNode]['lists'] **do**
14:               option ← random.choice   ▷ randomly choosing an option
15:               **if** option is addition **then**
16:                   New_lists ← ADDITION(list)
17:               **else if** option is deletion **then**
18:                   New_lists ← DELETION(list)
19:               **else if** option is no change **then**
20:                   New_lists ← NOCHANGE(list)
21:               **end if**
22:               ϕ.append(New_lists)   ▷ appending the updated CNPs list
23:            **end for**
24:            info[childNode]["lists"]= ϕ
25:            EVALUATECNPs($node = childNode$)
26:         **end for**
27:     **end if**
28: **end function**
29: **return** info   ▷ returning info dictionary of the tree with updated CNPs

---

bins × 2 alleles with one SNV added for each allele, and given that an addition is as likely as a deletion), which was the maximum number we could consider for each phylogeny inference tool to complete in a reasonable amount of time (of 24 h on a single input).

We then attached 100 cells randomly to the 20 copy number profiles (CNPs) of each tree, each cell being "sampled" from the clone represented by the CNP — similarly, 100 was the maximum number of cells we could consider to ensure timely phylogeny inference. To simulate a real-case scenario, we then added noise to these 100 cells according to typical false negative ($\alpha = 0.1$), false positive ($\beta = 10^{-4}$) and dropout ($\mu = 0.2$) rates [6,16,24]. Note that a false negative constitutes flipping a non-zero entry to zero, a false positive (which is rare) flipping a zero entry to one (the most common multiplicity), and a dropout setting the entry to unknown (?). The process resulted in essentially a mutational (SNV) profile — a matrix of 100 cells by 80 mutations (on average) with multiplicities.

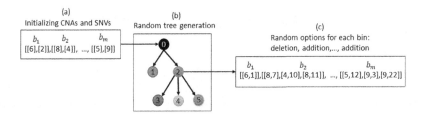

Fig. 2. The generation of CNAs and SNVs from a parent node to a child node. (a) each chromosomal "bin" b_i has two alleles, each with a point mutation. (b) a random tree topology is generated. (c) with respect to its parent 0, in this child 2, the bin b_1 undergoes a deletion of the second copy [2] (and accumulation of SNV 1 in copy [6]), bin b_2 undergoes an addition: duplication of copy [8] (and accumulations of SNVs 7, 10 and 11 in all existing and newly created copies), and bin b_m undergoes a duplication of copy [9] along with added SNVs in each copy).

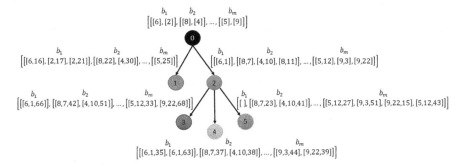

Fig. 3. Cancer progression tree visualization with CNAs and SNVs

Since the tree represents the underlying ground truth process which generated this observed mutational profile, the goal is then to see how well different cancer phylogeny inference methods infer such a tree.

4.2 Cancer Phylogeny Inference Methods

To evaluate our proposed system, we employed three methods that utilize some combination of CNAs and SNVs for phylogeny tree reconstruction. These methods were chosen, because they follow a trajectory of increasing generality in terms of the data considered and/or models used. The details of each method are as follows:

SCITE. [17] is a computational algorithm that deduces a tumor's evolutionary history from single-cell mutation profiles using a flexible Markov chain Monte Carlo (MCMC) approach. It considers the noise and incompleteness of these profiles due to experimental factors and sequencing errors. Although SCITE constructs a phylogeny based solely on mutational profiles, without multiplicity information, it addresses noise-related issues, partially accounting for the effects of copy number aberrations (CNAs). It's particularly valuable for studying tumor heterogeneity due to its noise-handling capabilities and accurate reconstructions.

SASC. [7] (Simulated Annealing Single-Cell inference) is a computational method designed to infer cancer progression from Single-Cell Sequencing (SCS) data while considering mutation losses using the Dollo-k model. The goal of SASC is to reconstruct the evolutionary trajectory of cancer cells and identify the sequence of mutational events that occur during tumor development. SASC makes use of only the mutational profiles (without multiplicities), however it attempts to build a Dollo-k phylogeny [10] (and also allowing for some noise). Since a Dollo-k phylogeny allows for some backmutations (k backmutations per site, to be precises), this more explicitly accounts for the artifacts [21] of CNAs.

PhISCS. [24] (Phylogeny Inference using Subclonal Copy number and Single-cell sequencing data) is a computational method that reconstructs tumor phylogenies by integrating single-cell and bulk sequencing data. It combines single-cell data for SNVs and bulk data for CNAs to capture tumor heterogeneity and subclonality. PhISCS is the most comprehensive among methods like SCITE and SASC, utilizing mutational profiles (without multiplicities) along with CNAs information from matched bulk sequencing. Its model accommodates mutational loss in the phylogenetic tree if supported by CNAs in the bulk sample.

4.3   Running the Experiments

Since the simulation process resulted essentially in mutational (SNV) profiles with multiplicities, we needed to transform this into the appropriate input for each tool. To produce the mutational profile (without multiplicities) which is taken as input by all methods, we simply cast all multiplicities to 1 in our simulated mutational profiles (with multiplicities), *i.e.*, by setting all entries > 1 to 1, retaining absence (0) and dropout (?) entries. To produce the matched bulk sample for PhISCS, following the format in the example on its GitHub webpage[2], we set column `MutantCount` to the multiplicity of the mutation in the simulated profile, and `ReferenceCount` to 5 times the maximum multiplicity of any mutation, to ensure a large enough reference readcount (see Snakemake pipeline for details).

Then, for SCITE, we ran it with parameters $\alpha = 0.1$, $\beta = 10^{-4}$ with a MCMC chain length `-l <INT>` of 900000 following its GitHub webpage[3]. For SASC, we ran it with parameters $\alpha = 0.1$, $\beta = 10^{-4}$, $k = 1$ (allowing for a Dollo-1 phylogeny), and $d = 5$, the expected number of total mutational losses in the tree, according to [6].

---

[2] https://github.com/sfu-compbio/PhISCS.
[3] https://github.com/cbg-ethz/SCITE.

4.4   Evaluating the Tree Inference

Another reason for choosing these methods is that they are easy to use, and output directly a tree in Graphviz format[4]. This allowed comparison to the simulated ground truth tree from whence the input data were generated. To assess the accuracy, $(TP + TN)/(TP + TN + FP + FN)$, we used the Ancestor-descendant accuracy and the different lineages accuracies, which compute the four rates as follows:

Ancestor-descendant accuracy. For each pair of mutations in an ancestor-descendant relationship in the ground truth tree, it is a true positive (TP) if it is also conserved in the interred tree, and a false negative (FN) otherwise. Conversely, if a pair of mutations are *not* in an ancestor-descendant relationship in the ground truth tree, but are in such a relationship in the inferred tree, it is a false positive (FP), otherwise it is a true negative (TN).

Different lineages accuracy. This measure is analogous to the ancestor-descendant accuracy, except for that it concerns pairs of mutations which are in different lineages. Note that a pair of mutations is either in an ancestor-descendant relationship or a different lineages relationship.

5   Results and Discussion

In this section, we first report some statistical properties of the data simulated by our tool, and then the results of evaluating SCITE, SACS and PhISCS on this simulation.

Properties of the Simulated Data. We first mention a few statistical properties of the data that were generated with our simulator. A mutational profile with multiplicities had 100 cells (by design) and an average ($\pm$ std.dev.) of 79.74 (± 23.91) mutations, and had mutational multiplicities ranging between 2 and 5. The average number of mutational losses (mutations which are present in some copy number profile in the parent, but absent in the child) was 21.60 (± 5.90) on average ($\pm$ std.dev.), which is rather high. Finally, k (in the Dollo-k model, that is, the maximum number of losses of a particular mutation on independent branches of the tree) was 3 on average, which is also rather high. Note that this can be tuned by adjusting the probability that a deletion operation occurs (Sec. 3). Finally, while the simulator models losses, it does not model recurrences (the apparition of the same mutation on independent branches of the tree), however, this could be easily added as well.

Phylogeny Inference Methods. Of the 50 inputs, 5 of them could not complete with PhISCS in the 24-hour time limit imposed, none of which had outlying statistical properties, of those reported above. We report all results on the remaining 45 inputs. Figure 4 depicts the distribution of the ancestor-descendant and different lineage accuracies, respectively, of the methods on the 45 trees in

---

[4] https://graphviz.org/.

the form of box plots. Not surprisingly, the most general method, PhISCS, which uses information from both SNVs and CNVs is able to perform as well as, or better than, the other two methods. It should be noted, however, that PhISCS incurred much higher runtimes than the other two methods as well. Finally, while it is not surprising that SASC performed among the best on the different lineages accuracy, it is surprising that it performed less well on the ancestor-descendant accuracy measure. This could be due to the elevated number of losses (and a k of 3 on average). Because of this, we also ran SASC with parameter d unset (which is unbounded by default), and the trees it inferred had an average number ($\pm$ std.dev.) of 22.18 ($\pm$7.82) losses, which is quite close to the true number of losses in ground truth. The ancestor-descendant and different lineages accuracies did not change much in this case, however.

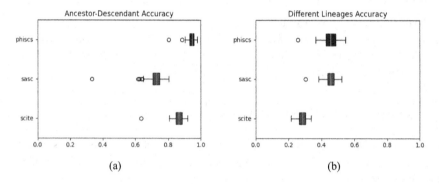

(a)   (b)

Fig. 4. The distribution of Ancestor-Descendant (resp. Different lineages) accuracy of each tree inference method on the datasets in the form of a box plot.

6   Conclusion

In this study, we present a novel simulator for cancer phylogenetic trees. Our approach aims to accurately simulate the progression of cancer by incorporating patterns of CNAs and SNVs. We evaluate the effectiveness of our proposed method by utilizing the simulated trees as input for three well-known phylogenetic tree reconstruction methods. This allows us to assess the performance and reliability of our simulator in capturing the complex dynamics of cancer progression. Future work can focus on the integration of other genomic alterations such as structural variants, epigenetic changes, and gene fusions which play significant roles in cancer progression. A more extensive experimental study will also be carried out. We believe this is a start in the right direction of benchmarking cancer phylogeny inference from scDNA-seq/bulk sequencing data.

References

1. Alexandrov, L.B., et al.: Signatures of mutational processes in human cancer. Nature **500**(7463), 415–421 (2013)
2. Anderson, K., Lutz, C., et al.: Genetic variegation of clonal architecture and propagating cells in leukaemia. Nature **469**(7330), 356–361 (2011)
3. Beroukhim, R., Mermel, C.H., Porter, D., et al.: The landscape of somatic copynumber alteration across human cancers. Nature **463**(7283), 899–905 (2010)
4. Bollen, Y., Stelloo, E., van Leenen, P., et al.: Reconstructing single-cell karyotype alterations in colorectal cancer identifies punctuated and gradual diversification patterns. Nat. Genet. **53**(8), 1187–1195 (2021)
5. Chan, K.A., et al.: Cancer genome scanning in plasma: detection of tumorassociated copy number aberrations, single-nucleotide variants, and tumoral heterogeneity by massively parallel sequencing. Clin. Chem. **59**(1), 211–224 (2013)
6. Ciccolella, S., et al.: Inferring cancer progression from Single-Cell sequencing while allowing mutation losses. Bioinformatics **37**(3), 326–333 (2020). https://doi.org/10.1093/bioinformatics/btaa722
7. Ciccolella, S., et al.: Inferring cancer progression from single-cell sequencing while allowing mutation losses. Bioinformatics **37**(3), 326–333 (2021)
8. El-Kebir, M.: SPhyR: tumor phylogeny estimation from single-cell sequencing data under loss and error. Bioinformatics **34**(17), i671–i679 (2018). https://doi.org/10.1093/bioinformatics/bty589
9. El-Kebir, M., Satas, G., Raphael, B.J.: Inferring parsimonious migration histories for metastatic cancers. Nat. Genet. **50**(5), 718–726 (2018)
10. Farris, J.S.: Phylogenetic analysis under Dollo's law. Syst. Biol. **26**(1), 77–88 (1977). https://doi.org/10.1093/sysbio/26.1.77
11. Fischer, A., Vázquez-García, I., Illingworth, C.J., Mustonen, V.: High-definition reconstruction of clonal composition in cancer. Cell Rep. **7**(5), 1740–1752 (2014)
12. Giguere, C., Dubey, H.V., Sarsani, V.K., Saddiki, H., He, S., Flaherty, P.: SCSIM: jointly simulating correlated single-cell and bulk next-generation DNA sequencing data. BMC Bioinform. **21**(1), 1–10 (2020)
13. Greaves, M., Maley, C.C.: Clonal evolution in cancer. Nature **481**(7381), 306–313 (2012)
14. Gupta, R.G., et al.: Intratumor heterogeneity: novel approaches for resolving genomic architecture and clonal evolution. Mol. Cancer Res. **15**(9), 1127–1137 (2017)
15. Huang, W., Skanderup, A.J., Lee, C.G.: Advances in genomic hepatocellular carcinoma research. Gigascience **7**(12), giy135 (2018)
16. Jahn, K., Kuipers, J., Beerenwinkel, N.: Tree inference for single-cell data. Genome Biol. **17**, 86 (2016)
17. Jahn, K., Kuipers, J., Beerenwinkel, N.: Tree inference for single-cell data. Genome Biol. **17**(1), 1–17 (2016)
18. Jolly, C., Van Loo, P.: Timing somatic events in the evolution of cancer. Genome Biol. **19**(1), 1–9 (2018)
19. Kang, S., et al.: SIEVE: joint inference of single-nucleotide variants and cell phylogeny from single-cell DNA sequencing data. Genome Biol. **23**(1), 248 (2022)
20. Khalil, A.I.S., Khyriem, C., Chattopadhyay, A., Sanyal, A.: Hierarchical discovery of large-scale and focal copy number alterations in low-coverage cancer genomes. BMC Bioinform. **21**, 1–22 (2020)

21. Kuipers, J., et al.: Single-cell sequencing data reveal widespread recurrence and loss of mutational hits in the life histories of tumors. Genome Res. **27**, 1885–1894 (2017)

22. Lawrence, M.S., et al.: Discovery and saturation analysis of cancer genes across 21 tumour types. Nature **505**(7484), 495–501 (2014)

23. Malikic, S., Jahn, K., Kuipers, J., Sahinalp, S.C., Beerenwinkel, N.: Integrative inference of subclonal tumour evolution from single-cell and bulk sequencing data. Nat. Commun. **10**(1), 2750 (2019)

24. Malikic, S., et al.: PhiSCS: a combinatorial approach for subperfect tumor phylogeny reconstruction via integrative use of single-cell and bulk sequencing data. Genome Res. **29**(11), 1860–1877 (2019). https://doi.org/10.1101/gr.234435.118, http://genome.cshlp.org/content/29/11/1860.abstract

25. Mallory, X.F., Nakhleh, L.: SimSCSnTree: a simulator of single-cell DNA sequencing data. Bioinformatics **38**(10), 2912–2914 (2022)

26. McGranahan, N., Swanton, C.: Biological and therapeutic impact of intratumor heterogeneity in cancer evolution. Cancer Cell **27**(1), 15–26 (2015)

27. Michor, F., Iwasa, Y., Nowak, M.A.: Dynamics of cancer progression. Nat. Rev. Cancer **4**(3), 197–205 (2004)

28. Mölder, F., et al.: Sustainable data analysis with snakemake. F1000 Res. 10 (2021). pMCID: PMC8114187

29. Nowell, P.C.: The clonal evolution of tumor cell populations: acquired genetic lability permits stepwise selection of variant sublines and underlies tumor progression. Science **194**(4260), 23–28 (1976)

30. Proietto, M., et al.: Tumor heterogeneity: preclinical models, emerging technologies, and future applications. Front. Oncol. **13**, 1164535 (2023)

31. Ren, X., Kang, B., Zhang, Z.: Understanding tumor ecosystems by single-cell sequencing: promises and limitations. Genome Biol. **19**(1), 1–14 (2018)

32. Sashittal, P., Zhang, H., Iacobuzio-Donahue, C.A., Raphael, B.: ConDoR: tumor phylogeny inference with a copy-number constrained mutation loss model. bioRxiv, version 1 (2023)

33. Satas, G., et al.: Scarlet: single-cell tumor phylogeny inference with copy-number constrained mutation losses. Cell Syst. **10**(4), 323–332 (2020)

34. Sperelakis, N.: Cell physiology sourcebook: a molecular approach. Elsevier (2001)

35. Tan, E.S., et al.: Copy number alterations as novel biomarkers and therapeutic targets in colorectal cancer. Cancers **14**(9), 2223 (2022)

36. Vergara, I.A., et al.: Evolution of late-stage metastatic melanoma is dominated by aneuploidy and whole genome doubling. Nat. Commun. **12**(1), 1434 (2021)

37. Yu, Z., Du, F., Sun, X., Li, A.: SCSsim: an integrated tool for simulating single-cell genome sequencing data. Bioinformatics **36**(4), 1281–1282 (2020)

38. Zaccaria, S., Raphael, B.J.: Characterizing allele-and haplotype-specific copy numbers in single cells with chisel. Nat. Biotechnol. **39**(2), 207–214 (2021)

CHLPCA: Correntropy-Based Hypergraph Regularized Sparse PCA for Single-Cell Type Identification

Tai-Ge Wang, Xiang-Zhen Kong, Sheng-Jun Li, and Juan Wang$^{(\boxtimes)}$

School of Computer Science, Qufu Normal University, Rizhao 276826, China
wangjuansdu@163.com

Abstract. Over the past decade, high-throughput sequencing technologies have driven a dramatic increase in single-cell RNA sequencing (scRNA-seq) data. The study of scRNA-seq data has widened the scope and depth of researchers' understanding of cellular heterogeneity. A prerequisite for studying heterogeneous cell populations is accurate cell type identification. However, the highly noisy and high-dimensional nature of scRNA-seq data poses a challenge to existing methods to further improve the success rate of cell type identification. Principal component analysis (PCA) is an important data analysis technique that is widely used to identify cell subpopulations. On the basis of PCA, we propose correntropy-based hypergraph regularized sparse PCA (CHLPCA) for accurate cell type identification. In addition to using correntropy to reduce the effect of noise, CHLPCA also considers higher-order relationships between samples by constructing the hypergraph, which compensates for the lack of local structure capture ability of PCA. Furthermore, we introduce the $L_{2,1/5}$-norm into the model to enhance the interpretability of principal components (PCs), which further improves the model performance. CHLPCA has superior clustering accuracy and outperforms the best comparative method by 5.13% and 8.00% for ACC and NMI metrics, respectively. The results of clustering visualization experiments also confirm that CHLPCA can better perform the cell type recognition task.

Keywords: Cell Type Identification · Correntropy · Principal Component Analysis · Hypergraph Regularization · Sparsity Constraint

1 Introduction

In recent years, single-cell RNA sequencing (scRNA-seq) technology has provided a new perspective for studying biological questions at the single-cell level, substantially advancing our understanding of biological systems. The primary utility of scRNA-seq technology is to detect cell heterogeneity. The prior requirement for studying heterogeneous cell populations is accurately identifying the cell type [1]. Cell type identification is an unsupervised clustering problem, and numerous clustering methods have been proposed for analyzing scRNA-seq data [2–4]. However, existing clustering methods lack the means to cope with the high-dimensional and noisy nature of scRNA-seq data, so developing new algorithms is necessary and challenging.

© The Author(s), under exclusive license to Springer Nature Singapore Pte Ltd. 2023
X. Guo et al. (Eds.): ISBRA 2023, LNBI 14248, pp. 541–551, 2023.
https://doi.org/10.1007/978-981-99-7074-2_44

Principal component analysis (PCA) [5] is a commonly used technique for dimensionality reduction; it maps high-dimensional data into a low-dimensional subspace while preserving the most important information from the original data. PCA has a wide range of applications in various fields. In bioinformatics, to discover as many new cell types as possible, Lall et al. proposed location-sensitive PCA (LSPCA) [6]. Pierson et al. presented ZIFA based on probabilistic PCA (PPCA) for analyzing the impact of dropout events in scRNA-seq data [7]. However, these PCA-based methods cannot further explore the local information hidden in nonlinear manifolds, which limits the model performance. Furthermore, PCA-based methods use the Frobenius norm as the default error function. The sensitivity of the Frobenius norm to noise and outliers in the data limits the performance of the model. In contrast, correntropy [8] considers the joint distribution of the data, which allows correntropy-based PCA methods to capture the underlying structure even in the case of noise and outliers [9].

The sparsity of the model is also a vital factor to be considered. The principal components (PCs) of PCA contain considerable redundant information, which makes PCs difficult to interpret. The sample features in sparse PCs are more distinct and are more likely to be clustered together, which provides a potential performance enhancement to the model. Therefore, obtaining sparse PCs is also crucial for PCA-based methods.

In this paper, we consider integrating correntropy, hypergraph regularization, and the $L_{2,1/5}$-norm into PCA and propose CHLPCA. Correntropy makes our model more robust and more effective in dealing with noise and outliers in the data. Hypergraph regularization enables our model to consider higher-order geometric information of the data and avoid the loss of valuable information when mining the local information of the data, thus producing a more accurate sample matrix and achieving accurate clustering. By imposing the sparsity constraint on the sample matrix of the model, we remove a large amount of redundant information from the PCs and improve the interpretability of the PCs. The sparse PCs provide potential performance improvements to our model. The experimental results show that the overall performance of CHLPCA has significant advantages over other PCA-based methods and various advanced single-cell clustering methods.

2 Related work

2.1 Principal Component Analysis

PCA is a matrix decomposition technique, and it expects to maximize the data variance during the mapping process to make the distribution of data points in the projection dimension as spread out as possible, thus retaining the most important information of the original data. Consider a data matrix $\mathbf{X} = (x_1, \ldots, x_n) \in \mathbf{R}^{m \times n}$, where m denotes the number of features and n represents the number of samples. The Frobenius norm of the matrix $\mathbf{X}$ is denoted as $\|\mathbf{X}\|_F$. The objective of PCA is to determine the product of matrices $\mathbf{Q}^T = (q_1, \ldots, q_k) \in \mathbf{R}^{k \times n}$ and $\mathbf{U} = (u_1, \ldots, u_k) \in \mathbf{R}^{m \times k}$ to make $\mathbf{X} \approx \mathbf{U}\mathbf{Q}^T$. The optimization problem that denotes the conventional PCA can be stated as follows:

$$\arg\min_{\mathbf{U}, \mathbf{Q}} \left\{ \left\| \mathbf{X} - \mathbf{U}\mathbf{Q}^T \right\|_F^2 \right\} s.t. \mathbf{Q}^T \mathbf{Q} = \mathbf{I}. \tag{1}$$

2.2 Hypergraph Regularization

As an improvement to the graph, the hypergraph is more flexible. Hypergraph allows its edges to connect multiple vertices, which makes it possible to represent higher-order data relationships. Moreover, the hypergraph can capture more information about node relationships and preserves complex data structures and relationships.

The data matrix $\mathbf{X}$ can be described as a hypergraph $\mathbf{G} = (\mathbf{V}, \mathbf{E}, \mathbf{W})$, where $\mathbf{V}$ is the set of vertices in the hypergraph, and $\mathbf{E}$ is the set of hyperedges. $\mathbf{W}$ is the weight set of the hyperedge, and the weight of each hyperedge $\mathbf{W}_i$ can be stated as follows:

$$\mathbf{W}_i = \sum_{v_j \in e_i} \exp\left(-\varsigma^{-2}\|v_i - v_j\|_2^2\right), \tag{2}$$

where ς represents the average distance among all vertices. Next, the correlation matrix $\mathbf{H}$ of the hypergraph $\mathbf{G}$ can be defined as

$$\mathbf{H}(v, e) = \begin{cases} 1, & \text{if } v \in e, \\ 0, & \text{otherwise.} \end{cases} \tag{3}$$

Then, we can define the degree of a vertex $d(v)$ as follows:

$$d(v) = \sum_{e \in E} w(e)\mathbf{H}(v, e), \tag{4}$$

where $w(e)$ is the corresponding weight of hyperedge e.

Moreover, $\delta(e) = \sum_v \mathbf{H}(v, e)$, $\delta(e)e$ denotes the degree of one hyperedge e, which is determined by the quantity of vertices contained in e. $\mathbf{D}_v$ is the vertex degree matrix of which values are related to $d(v)$. $\mathbf{D}_e$ is the diagonal matrix of which values are related to $\delta(e)$. Ultimately, we obtain the hypergraph Laplacian matrix containing higher-order relationships between the samples in the data matrix, which can be constructed as $\mathbf{L}_H = \mathbf{D}_v - \mathbf{H}\mathbf{W}(\mathbf{D}_e)^{-1}\mathbf{H}^T$. The hypergraph regularization is written as $tr(\mathbf{Q}^T\mathbf{L}_H\mathbf{Q})$, and $tr(\bullet)$ is the trace function of the data matrix. This form of regularization allows learned PCs by the model to more realistically reflect the intrinsic structure of the data, thus allowing the model to identify cell types more accurately.

2.3 Correntropy

Correntropy is a metric for assessing the nonlinearity and local similarity of random variables. Correntropy was initially applied to information theoretic learning (ITL) analysis and was gradually widely used in bioinformatics [10] due to its excellent noise reduction capability. The correntropy of x and y can be described by $\mathbf{C}(x, y) = \mathbf{E}\big[k(x, y)\big]$, where $k(\cdot, \cdot)$ is the kernel function satisfying Mercer theory and $\mathbf{E}[\cdot]$ is the mathematical expectation. In this paper, we employ the Gaussian kernel as the kernel function of the correntropy, which is formulated as

$$k_\sigma(x, y) = g(x - y) = \exp\left(-\frac{(x - y)^2}{2\sigma^2}\right), \tag{5}$$

where σ is the parameter of kernel bandwidth, and $\sigma > 0$. In practical applications, it is generally difficult to know the joint distribution function of the random variables x and y. Therefore, we calculate the correntropy of the sample in the following way:

$$\mathbf{C}_\sigma(x, y) = \frac{1}{n} \sum_{i=1}^{n} g(x_i - y_i). \tag{6}$$

In this way, the maximum correntropy criterion (MCC) can be obtained by maximizing the correntropy in (6). Different from the global measure—Frobenius norm, the correntropy is local and more robust while coping with data including outliers and noise. Thus, PCA methods incorporating MCC typically have better performance.

2.4 $L_{2,1/5}$-norm Constraint

Given a data matrix $\mathbf{X} = (x_1, \ldots, x_n) \in \mathbf{R}^{m \times n}$, the $L_{r,p}$-norm of $\mathbf{X}$ can be defined as $\|\mathbf{X}\|_{r,p} = \left(\sum_{s}^{m} \|x_l\|_r^p \right)^{\frac{1}{p}}$, where x_l represents the l-th row vector of $\mathbf{X}$. The L_r-norm of vector x is denoted as $\|\mathbf{X}\|_r$. When $r = 2$ and $p = 1$, $\|\cdot\|_{r,p}$ is converted to the $L_{2,1}$-norm. In our previous work [11], we used the $L_{2,p}$-norm to obtain sparse PCs and fixed the value of p to 1/5 by experiments. Compared to the $L_{2,1}$-norm, the $L_{2,1/5}$-norm promotes model sparsity more aggressively due to its smaller p-value.

3 Methods

3.1 The Objective Function of CHLPCA

To overcome the noisy and high-dimensional nature of scRNA-seq data, we propose a novel approach, CHLPCA, to identify cell types from scRNA-seq data accurately and efficiently. We replace the Frobenius norm with correntropy in the error function of CHLPCA to enhance the robustness of the model. To enable the proposed method is able to explore higher-order geometric relationships within the data, we introduce hypergraph regularization in the model. Moreover, we apply the $L_{2,1/5}$-norm constraint to the sample matrix $\mathbf{Q}$ to alleviate the problem of dense PCs in PCA, thus reducing the negative impact of redundant information on PCs and making it easier for cells with similar characteristics to cluster together.

Figure 1 illustrates the research framework of CHLPCA. The objective function of CHLPCA is as follows:

$$\arg\min_{\mathbf{U},\mathbf{Q}} \sum_{i=1}^{m} g\left(\sqrt{\sum_{j=1}^{n} \left(x_{ij} - (uq^T)_{ij} \right)^2} \right) + \alpha Tr\left(\mathbf{Q}^T \mathbf{L}_H \mathbf{Q} \right) + \beta \|\mathbf{Q}\|_2^{1/5}. \tag{7}$$

As shown in (7), α and β are utilized to balance the weights of the hypergraph regularization and the $L_{2,1/5}$-norm constraint, respectively. $\mathbf{L}_H$ is the hypergraph Laplacian matrix. $\|\mathbf{Q}\|_2^{1/5}$ is the sparsity constraint imposed on the sample matrix $\mathbf{Q}$.

The equation in (7) is not only nonquadratic, but also nonconvex. Therefore, it is quite challenging to optimize the solution directly. We apply the half-quadratic (HQ)

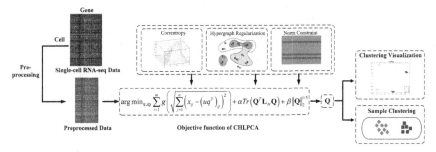

Fig. 1. The research framework of CHLPCA.

technique to solve the optimization problem [12]. By using the properties of convex conjugate functions, the convex conjugate function $\phi(\cdot)$ of $g(x)$ is denoted as

$$g(x) = \max_{e}\left(\frac{e}{\sigma^2}\|x\|^2 - \phi(e)\right), \tag{8}$$

for a given x, the maximum value can be obtained when $e = -g(x)$.

Then, we incorporate the concept presented in (8) into the objective function of CHLPCA (7), and we can obtain the following equation:

$$\arg\min_{U,Q}\sum_{i=1}^{m} g\left(-\frac{e_i}{\sigma^2}\left(\sum_{j=1}^{n}\left(x_{ij} - (uq^T)_{ij}\right)^2\right) - \phi(e_i)\right) + \alpha Tr\left(Q^T L_H Q\right) + \beta\|Q\|_2^{1/5}, \tag{9}$$

where $e = [e_1, \ldots, e_n]^T$ stands for the auxiliary vectors of the HQ technique.

Then, by fixing U and Q, we maximize the augmented objective function with respect to e and obtain

$$e_i = -g\left(\sqrt{\sum_{j=1}^{n}\left(x_{ij} - (uq^T)_{ij}\right)^2}\right) = -\exp\left(-\frac{1}{2\sigma^2}\left(\sum_{j=1}^{n}\left(x_{ij} - (uq^T)_{ij}\right)^2\right)\right), \tag{10}$$

where σ is the parameter of kernel bandwidth, commonly determined by experience. Then, we refer to the literature [10] and gain

$$\sigma = \sqrt{\frac{1}{m}\sum_{i=1}^{m}\sum_{j=1}^{n}\left(x_{ij} - (uq^T)_{ij}\right)^2}. \tag{11}$$

For the fixed σ, we reformulate the augmented objective function in (10) as

$$\arg\min_{U,Q}\sum_{i=1}^{m}\left(-\frac{e_i}{\sigma^2}\left(\sum_{j=1}^{n}\left(x_{ij} - (uq^T)_{ij}\right)^2\right)\right) + \alpha Tr\left(Q^T L_H Q\right) + \beta\|Q\|_2^{1/5}. \tag{12}$$

3.2 Optimization of CHLPCA

By transforming the optimization problem presented in (13) into a weighted PCA problem using the HQ technique, we provide a closed-form solution for the problem that ensures the stability of the iterations. The alternating direction multiplier (ADM) method [13] is used to update the matrices $\mathbf{U}$ and $\mathbf{Q}$. Only one variable is updated at each iteration in solving the objective function, while the other variables are fixed.

The optimization model (13) can be converted as follows:

$$\arg\min_{\mathbf{U},\mathbf{V}} Tr\left(\mathbf{X} - \mathbf{U}\mathbf{Q}^T\right)\mathbf{D}\left(\mathbf{X} - \mathbf{U}\mathbf{Q}^T\right)^T + \alpha Tr\left(\mathbf{Q}^T\mathbf{L}_H\mathbf{Q}\right) + \beta Tr\left(\mathbf{Q}^T\mathbf{S}\mathbf{Q}\right), \quad (13)$$

where $\mathbf{S} = diag\left\{1/10\|\mathbf{Q}_{1\cdot}\|_{1/5}^{2-1/5}, \ldots, 1/10\|\mathbf{Q}_{n\cdot}\|_{1/5}^{2-1/5}\right\}$, $\mathbf{S}$ represents the matrix of the $L_{2,1/5}$-norm. $\mathbf{D}$ is a diagonal matrix and its entries are

$$d_{ii} = -\frac{e_i}{\sigma^2} = \sigma^{-2}\exp\left(-\frac{1}{2\sigma^2}\left(\sum_{j=1}^{n}\left(x_{ij} - \left(uq^T\right)_{ij}\right)^2\right)\right). \quad (14)$$

We summarize the process of optimization iterations in Algorithm 1.

Algorithm 1. The alternating updating algorithm of CHLPCA.

Input: Data matrix $\mathbf{X}$, Hypergraph Laplacian matrix $\mathbf{L}_H$, parameters α, β, k, p

Output: matrix $\mathbf{U}, \mathbf{Q}$

Repeat

1: Compute $\mathbf{D}$ by $d_{ii} = -\dfrac{e_i}{\sigma^2} = \sigma^{-2}\exp\left(-\dfrac{1}{2\sigma^2}\left(\sum_{j=1}^{n}\left(x_{ij} - \left(uq^T\right)_{ij}\right)^2\right)\right)$.

2: Compute $\mathbf{U}$ by $\mathbf{U} = \mathbf{X}\mathbf{Q}$.

3: Compute $\mathbf{S}$ by $\mathbf{S} = diag\left\{1/10\|\mathbf{Q}_{1\cdot}\|_{1/5}^{2-1/5}, \ldots, 1/10\|\mathbf{Q}_{n\cdot}\|_{1/5}^{2-1/5}\right\}$.

4: Compute $\mathbf{A}$ by $\mathbf{A} = -\mathbf{X}^T\mathbf{D}\mathbf{X} + \alpha\mathbf{L}_H + \beta\mathbf{S}$.

5: Compute $\mathbf{Q}$ by $\mathbf{Q} = \left(q_1, \ldots, q_k\right)$.

until convergence

---

4 Results and Discussion

To verify the effectiveness of CHLPCA, we compare CHLPCA with six popular clustering algorithms on seven scRNA-seq datasets. We choose accuracy (ACC) [14] and normalized mutual information (NMI) [15] as metrics to evaluate the performance of these methods. Higher ACC and NMI scores indicate that the method performs better and identifies different cell types more accurately.

Table 1. Details of the seven scRNA-seq datasets.

Datasets	Species	Genes	Cells	Cell Types
Zheng	Homo sapiens	4776	500	3
Pollen	Homo sapiens	14805	249	11
Grover	Mus musculus	14739	135	2
MECS	Mus musculus	8989	182	3
Engel	Homo sapiens	23337	203	4
Buettner	Mus musculus	8989	182	3
Deng	Mus musculus	12548	135	7

4.1 Datasets

We conduct experiments on seven scRNA-seq datasets [16–21]. These scRNA-seq datasets are made up of cells with known labels, and their details are shown in Table 1.

4.2 Parameter Setting

The parameters α and β are responsible for balancing the contribution of hypergraph regularization and the $\mathbf{L}_{2,1/5}$-norm constraint. We apply the grid search approach to find the optimal combination of α and β on all datasets. We vary α and β in the range of 10^{-5} to 10^5 to ensure a full search of the parameter space. Moreover, for PCA, the determination of the optimal dimension k is crucial for cluster analysis. k represents the number of PCs that can capture the highest amount of variance in the data. This paper uses the gap statistic method [22] to select k. The combination of parameters is displayed in Table 2.

Table 2. Optimal parameter setting.

Datasets	α	β	k
Zheng	$10^{2.14}$	$10^{3.48}$	3
Pollen	$10^{1.00}$	$10^{3.77}$	11
Grover	$10^{3.00}$	$10^{3.00}$	2
MECS	$10^{1.62}$	$10^{1.30}$	3
Engel	$10^{1.95}$	$10^{0.00}$	4
Buettner	$10^{0.00}$	$10^{0.00}$	3
Deng	$10^{2.90}$	$10^{4.56}$	7

4.3   Clustering Results

To fully demonstrate the effectiveness of our method, we compare our method to gLPCA [23], MPSSC [2], NMFLRR [24], PgLPCA [25], SinNLRR [3], and SIMLR [4] on seven scRNA-seq datasets. Moreover, we include the proposed model without hypergraph regularization (CHLPCA$_{[1]}$) in our experiments to study the effect of the hypergraph structure on the performance of the model. The results are shown in Fig. 2.

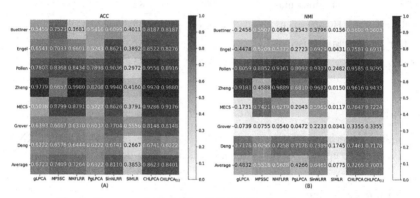

Fig. 2.   Heatmap of (A) ACC and (B) NMI values for all methods.

Compared with other methods, CHLPCA achieves the most satisfactory clustering performance on most datasets and has the highest average clustering accuracy. SinNLRR performs best in the comparison methods. Compared with SinNLRR, the average ACC and NMI of CHLPCA are 5.13% and 8.00% higher than those of SinNLRR, respectively. The possible reason for this is that CHLPCA has greater noise immunity and can mine the high-order geometric manifolds of the data. Furthermore, the overall performance of CHLPCA is superior to that of CHLPCA$_{[1]}$ due to the incorporation of hypergraph regularization. This observation suggests that the introduction of hypergraph regularization enables the model to mine complex higher-order data relationships more efficiently, thus significantly enhancing the model's performance. In conclusion, compared with other clustering methods, the performance advantages of our method are obvious, and it can better accomplish the cell type recognition task.

4.4   Clustering Visualization

The clustering visualization can reflect the information-mining ability of the model. We apply T-distribution stochastic neighborhood embedding (t-SNE) [26] to the sample matrix to observe the clustering performance of the model more visually. To fully display the clustering effects of methods built on different techniques, we compare CHLPCA with SinNLRR, MPSSC, gLPCA, and the original data. The visualization results are shown in Fig. 3. "Perplexity" is a critical parameter of t-SNE to control the weight distribution of neighboring points. We set it to the default value of 30 [27].

As shown in Fig. 3, CHLPCA effectively clusters the three types of cells in the MECS dataset into nonoverlapping cell clusters, and only a few data points are incorrectly clustered. In contrast, the results of the other methods are unsatisfactory. SinNLRR only successfully separates two types of cells, and the third is covered. On the Engel dataset, CHLPCA has much less overlap than the other methods. Notably, SinNLRR achieves a slightly higher ACC value than CHLPCA on the Engel dataset. Nevertheless, in the clustering visualization results, SinNLRR shows only two types of cells, and the other two are covered. Overall, CHLPCA is able to classify cell types more correctly and can better perform the task of cell type identification.

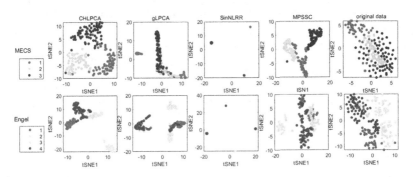

Fig. 3. Clustering visualization results.

5 Conclusion

We present a novel PCA approach in this paper, called CHLPCA. The presented approach reduces the negative impact of noise and outliers on the model performance by incorporating the correntropy. Moreover, CHLPCA uses hypergraph regularization and the $L_{2,1/5}$-norm to consider the higher-order geometric information of the data and obtain sparse PCs, which effectively enhances the clustering performance of the algorithm. Experimental results based on scRNA-seq data indicate that CHLPCA performs better than existing cell type identification methods.

However, our method still has room for improvement. In future studies, we will continue to apply various techniques to improve the performance of the method.

Acknowledgment. This work is supported by the National Natural Science Foundation of China (Grant Nos. 62172253).

References

1. Raman, P., et al.: A comparison of survival analysis methods for cancer gene expression RNA-sequencing data. Cancer Genet. **235**, 1–12 (2019)
2. Park, S., Zhao, H.: Spectral clustering based on learning similarity matrix. Bioinformatics **34**, 2069–2076 (2018)

3. Zheng, R., Li, M., Liang, Z., Wu, F.-X., Pan, Y., Wang, J.: SinNLRR: a robust subspace clustering method for cell type detection by non-negative and low-rank representation. Bioinformatics **35**, 3642–3650 (2019)

4. Wang, B., Zhu, J., Pierson, E., Ramazzotti, D., Batzoglou, S.: Visualization and analysis of single-cell RNA-seq data by kernel-based similarity learning. Nat. Methods **14**, 414–416 (2017)

5. Abdi, H., Williams, L.J.: Principal component analysis. Wiley Interdiscip. Rev.: Comput. Stat. **2**, 433–459 (2010)

6. Lall, S., Sinha, D., Bandyopadhyay, S., Sengupta, D.: Structure-aware principal component analysis for single-cell RNA-seq data. J. Comput. Biol. **25**, 1365–1373 (2018)

7. Pierson, E., Yau, C.: ZIFA: dimensionality reduction for zero-inflated single-cell gene expression analysis. Genome Biol. **16**, 1–10 (2015)

8. Liu, W., Pokharel, P.P., Principe, J.C.: Correntropy: properties and applications in non-Gaussian signal processing. IEEE Trans. Sig. Process. **55**, 5286–5298 (2007)

9. He, R., Hu, B.-G., Zheng, W.-S., Kong, X.-W.: Robust principal component analysis based on maximum correntropy criterion. IEEE Trans. Image Process. **20**, 1485–1494 (2011)

10. Yu, N., Wu, M.-J., Liu, J.-X., Zheng, C.-H., Xu, Y.: Correntropy-based hypergraph regularized NMF for clustering and feature selection on multi-cancer integrated data. IEEE Trans. Cybern. **51**, 3952–3963 (2020)

11. Wang, T.-G., Shang, J.-L., Liu, J.-X., Li, F., Yuan, S., Wang, J.: Joint L2,p-norm and random walk graph constrained PCA for single-cell RNA-seq data. Comput. Methods Biomech. Biomed. Eng. 1–14 (2023)

12. Nikolova, M., Chan, R.H.: The equivalence of half-quadratic minimization and the gradient linearization iteration. IEEE Trans. Image Process. **16**, 1623–1627 (2007)

13. Boyd, S., Parikh, N., Chu, E., Peleato, B., Eckstein, J.: Distributed optimization and statistical learning via the alternating direction method of multipliers. Found. Trends® Mach. Learn. **3**, 1–122 (2011)

14. Cai, D., He, X., Han, J.: Document clustering using locality preserving indexing. IEEE Trans. Knowl. Data Eng. **17**, 1624–1637 (2005)

15. McDaid, A.F., Greene, D., Hurley, N.: Normalized mutual information to evaluate overlapping community finding algorithms. arXiv preprint arXiv:1110.2515 (2011)

16. Zheng, G.X., et al.: Massively parallel digital transcriptional profiling of single cells. Nat. Commun. **8**, 14049 (2017)

17. Pollen, A.A., et al.: Low-coverage single-cell mRNA sequencing reveals cellular heterogeneity and activated signaling pathways in developing cerebral cortex. Nat. Biotechnol. **32**, 1053–1058 (2014)

18. Grover, A., et al.: Single-cell RNA sequencing reveals molecular and functional platelet bias of aged haematopoietic stem cells. Nat. Commun. **7**, 11075 (2016)

19. Buettner, F., et al.: Computational analysis of cell-to-cell heterogeneity in single-cell RNA-sequencing data reveals hidden subpopulations of cells. Nat. Biotechnol. **33**, 155–160 (2015)

20. Engel, I., et al.: Innate-like functions of natural killer T cell subsets result from highly divergent gene programs. Nat. Immunol. **17**, 728–739 (2016)

21. Deng, Q., Ramsköld, D., Reinius, B., Sandberg, R.: Single-cell RNA-seq reveals dynamic, random monoallelic gene expression in mammalian cells. Science **343**, 193–196 (2014)

22. Tibshirani, R., Walther, G., Hastie, T.: Estimating the number of clusters in a data set via the gap statistic. J. Roy. Stat. Soc.: Ser. B (Stat. Methodol.) **63**, 411–423 (2001)

23. Jiang, B., Ding, C., Luo, B., Tang, J.: Graph-Laplacian PCA: closed-form solution and robustness. In: Proceedings of the IEEE Conference on Computer Vision and Pattern Recognition, pp. 3492–3498. (2011)

24. Zhang, W., Xue, X., Zheng, X., Fan, Z.: NMFLRR: clustering scRNA-seq data by integrating nonnegative matrix factorization with low rank representation. IEEE J. Biomed. Health Inform. **26**, 1394–1405 (2021)
25. Feng, C.-M., Gao, Y.-L., Liu, J.-X., Zheng, C.-H., Yu, J.: PCA based on graph Laplacian regularization and P-norm for gene selection and clustering. IEEE Trans. Nanobiosci. **16**, 257–265 (2017)
26. Van der Maaten, L., Hinton, G.: Visualizing data using t-SNE. J. Mach. Learn. Res. **9** (2008)
27. Van Der Maaten, L.: Fast optimization for t-SNE. In: Neural Information Processing Systems (NIPS) 2010 Workshop on Challenges in Data Visualization. Citeseer (2010)

Author Index

Printed in the United States
by Baker & Taylor Publisher Services